Luminescence from Biological and Synthetic Macromolecules
Eighth Katzir Conference

ANNALS OF THE NEW YORK ACADEMY OF SCIENCES

Volume 366

LUMINESCENCE FROM BIOLOGICAL AND SYNTHETIC MACROMOLECULES
Eighth Katzir Conference

Edited by H. Morawetz and I.Z. Steinberg

The New York Academy of Sciences
New York, New York
1981

Library of Congress Cataloging in Publication Data

Aharon Katzir-Katchalsky Conference (8th : 1980 : New York, N.Y.)
Luminescence from biological and synthetic macromolecules.
(Annals of the New York Academy of Sciences ; v. 366)
1. Chemiluminescence--Congresses. 2. Bioluminescence--Congresses. 3. Macromolecules--Congresses. I. Morawetz, Herbert. II. Steinberg, I. Z., 1931– . III. Title. IV. Series. Q11.N5 vol. 366 [QD505.8] 500s [547.7] 81-9616 AACR2

WP
Printed in the United States of America
ISBN 0-89766-123-0 (cloth)
ISBN 0-89766-124-9 (paper)

ANNALS OF THE NEW YORK ACADEMY OF SCIENCES

VOLUME 366

June 8, 1981

LUMINESCENCE FROM BIOLOGICAL AND SYNTHETIC MACROMOLECULES*

EIGHTH KATZIR CONFERENCE

Editors and Conference Chairmen
H. Morawetz and I.Z. Steinberg

CONTENTS

Part I. Luminescence Studies of Synthetic Polymer Solutions

Luminescence Methods in the Study of Chain Statistics and Dynamics. *By* G. WEILL ... 1

Studies of Singlet Energy Transfer in Polymers by Time-Resolved Fluorescence Spectroscopy. *By* D.A. HOLDEN, W.A. RENDALL, AND J.E. GUILLET ... 11

Studies of Intramolecular Excimer Formation in Synthetic Polymers. *By* IAN SOUTAR ... 24

Fluorescence and Microbrownian Motion in Flexible Macromolecules. *By* CARLA CUNIBERTI AND ANGELO PERICO ... 35

Study of Conformation and Intramolecular Motility of Polypeptides in Solution by a Novel Fluorescence Method. *By* E. KATCHALSKI-KATZIR, E. HAAS, AND I.Z. STEINBERG ... 44

Triplet Probe Study on the Rates of Diffusion-Controlled Reactions in Solution. *By* ITARU MITA ... 62

The Dynamics of Polymer Cyclization 3. Excluded Volume Effects in the End-to-End Cyclization of Polystyrene Probed by Intramolecular Pyrene Excimer Formation. *By* A.E.C. REDPATH AND MITCHELL A. WINNIK ... 75

Excimer Formation in Poly- and Dichromorphic Molecules in Solution. *By* F.C. DE SCHRYVER, K. DEMEYER, M. VAN DER AUWERAER, AND E. QUANTEN ... 93

Multicomponent Fluorescence Decay in Vinyl Aromatic Polymers and Copolymers. *By* A.J. ROBERTS, D.V. O'CONNOR, AND D. PHILLIPS ... 109

Part II. Luminescence from Solutions of Biological Macromolecules

Light Scattering by Chromophores at Their Absorption Bands. *By* IZCHAK Z. STEINBERG AND JACOB ANGLISTER ... 125

* This series of papers is the result of a conference entitled Luminescence from Macromolecules—Biological and Synthetic, Eighth Katzir Conference, held on June 16–18, 1980 by The New York Academy of Sciences and co-sponsored by the Aharon Katzir-Katchalsky Center at the Weizmann Institute of Science, the Israel National Academy of Sciences and Humanities, and the Polymer Research Institute of the Polytechnic Institute of New York.

Fluorophores in Polar Media: Spectral Effects of the Langevin Distribution of Electrostatic Interactions. *By* ROBERT B. MACGREGOR, JR. AND GREGORIO WEBER 140

Orientational Effects in Intra- and Intermolecular Long Range Excitation Energy Transfer. *By* J. EISINGER, W.E. BLUMBERG, AND R.E. DALE 155

Rotational Diffusion of Biological Macromolecules by Time-Resolved Delayed Luminescence (Phosphorescence, Fluorescence) Anisotropy. *By* THOMAS M. JOVIN, MARTY BARTHOLDI, WINCHIL L.C. VAZ, AND ROBERT H. AUSTIN 176

Nanosecond Fluorometry and the Luminescence of Biological Macromolecules. *By* LUDWIG BRAND, J.B. ALEXANDER ROSS, AND WILLIAM R. LAWS 197

Insights into the Mode of Antibody Action from Intrinsic and Extrinsic Fluorescent Probes. *By* ISRAEL PECHT 208

The Mechanism of Negative Cooperativity in Rabbit Muscle Glyceraldehyde-3-Phosphate Dehydrogenase. *By* YOAV I. HENIS AND ALEXANDER LEVITZKI 217

A New Component in Protein Fluorescence. *By* J.W. LONGWORTH 237

Part III. Luminescence Studies of Cells and Cell Components

Fluorescence Studies on Chromatin Structure. *By* CHARLES R. CANTOR, ANN E. DIETERICH, AND CHRISTOPHER P. PRIOR 246

Spectroscopic Properties of Rhodamine B-labeled Thyroid Hormone. *By* HAROLD EDELHOCH, SHEVE-YANN CHENG, AND GAETANO IRACE 253

Rotational Diffusion of Rhodopsin in Membranes as Determined by Steady-State Polarized Photochromism. *By* C. PASTERNAK, D. CORDA, H. STIEVE, AND M. SHINITZKY 265

Dynamics of Hormone Receptors on Cell Membrane. *By* JOSEPH SCHLESSINGER 274

Conformational Dynamics of Tropomyosin in Solution: Evidence for Two Conformational States. *By* SHERWIN S. LEHRER, PHILIP GRACEFFA, AND DAVID BETTERIDGE 285

Luminescence Measurements of Macromolecular Mobility. *By* W.W. WEBB 300

Bioluminescence of Bacterial Luciferase. *By* J. WOODLAND HASTINGS AND SHIAO-CHUN TU 315

Part IV. Luminescence from Synthetic Polymers in Bulk

Solid-State Polymers as Studied by Fluorescence Polarization: Mobility and Orientation Measurements. *By* LUCIEN MONNERIE AND JEAN-PIERRE JARRY 328

Photo-Physical Processes in Organized Systems. *By* C. DAVID AND D. BAEYENS-VOLANT 341

Lanthanide Ion Fluorescence as a Probe of Ionomer Structure. *By* E. BANKS, Y. UEBA, AND Y. OKAMOTO 356

Energy Transfer in Films of Polymers with Aromatic Side-Groups. *By* WALTER KLÖPFFER 373

Excimer Fluorescence as a Molecular Probe of Polymer-Polymer Interactions in the Amorphous Solid State. *By* C.W. FRANK AND M.A. GASHGARI ... 387

Studies of Polymer Compatability by Nonradiative Energy Transfer. *By* H. MORAWETZ ... 404

Index of Contributors ... 413

Financial assistance was received from:

- ABBOTT LABORATORIES
- ALLIED CHEMICAL CORPORATION
- BECTON, DICKINSON & COMPANY
- DOW CHEMICAL COMPANY
- E.I. DU PONT DE NEMOURS & COMPANY
- EASTMAN KODAK COMPANY
- FIRESTONE
- GENERAL ELECTRIC COMPANY
- HOFFMANN-LA ROCHE, INC.
- INTERNATIONAL BUSINESS MACHINES CORPORATION
- E. LEITZ, INC.
- MILES LABORATORIES, INC.—AMES DIVISION
- MILES LABORATORIES, INC.—DOME LABORATORIES DIVISION
- MONSANTO COMPANY
- NATIONAL SCIENCE FOUNDATION, POLYMERS PROGRAM
- ORTHO DIAGNOSTICS, INC.
- STAUFFER CHEMICAL COMPANY, EASTERN RESEARCH CENTER
- USV PHARMACEUTICAL CORPORATION

PART I. LUMINESCENCE STUDIES OF SYNTHETIC POLYMER SOLUTIONS

LUMINESCENCE METHODS IN THE STUDY OF CHAIN STATISTICS AND DYNAMICS

G. Weill

Centre de Recherches sur les Macromolécules
et Université L. Pasteur
Strasbourg, France

Despite some occasional difficulties related to their bulkiness and additional degrees of freedom, *extrinsic* luminescent probes constitute an efficient tool in the study of chain statistics and dynamics. The simultaneous use of quantum yield, lifetime and polarization measurements opens many roads to the imagination. Some of those which have been explored by the author are presented below.

FLUORESCENCE QUENCHING AS A TOOL IN THE STUDY OF THE DISTRIBUTION OF LOCAL CONCENTRATIONS IN POLYMER SOLUTIONS AND AS A MODEL FOR SIDE CHAIN REACTIVITY[1]

Let us imagine a dilute solution of chain molecules as a suspension of swollen coils, occupying a fraction of volume ψ. If this molecule carries a weight fraction α of quenching side groups, the quantum yield of fluorescence ϕ of a microscopic fluorescent probe evenly distributed "inside" and "outside" of the coil may be written as

$$\phi = \phi_0 \left[(1 - \psi) + \psi \frac{1}{1 + K\dfrac{\alpha c}{\psi}} \right] \quad (1)$$

where c is the overall concentration in polymer and K a Stern Vollmer constant.

Since ψ is proportional to c, a classical plot of ϕ_0/ϕ (or I_0/I where I is the fluorescence intensity) versus c is expected to show an upward curvature from which $\psi(c)$ and K could be extracted. Similarly, the fluorescence decay should present two components corresponding to the unquenched outside probes and to the quenched inside probes. At higher concentrations, where the coils overlap, the concentration tends to be uniform and the Stern-Vollmer plot should become linear

$$\frac{\phi_0}{\phi} = 1 + K\alpha c \quad (2)$$

with a value of K reduced with respect to the monomeric quencher because of the reduced rate of collision with a polymer side group arising from both steric and diffusional restrictions.

As an illustration, the Stern-Vollmer plot for the quenching of diphenyl-anthracene (DPA) by dimethylparatoluidine (DMPT) and by a homo- and a copolymer of dimethylaminostyrene is given in FIGURE 1.

0077-8923/81/0366-0001 $01.75/0 © 1981, NYAS

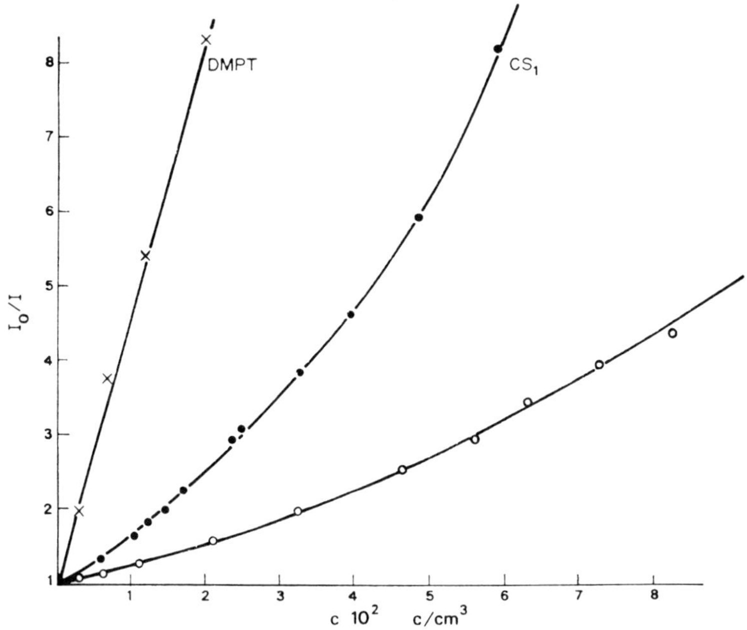

FIGURE 1. Stern-Vollmer plot for the quenching of diphenylanthracene by dimethylparatoluidine and a homo- and co-polymer (cs) of dimethylaminostyrene.

A more quantitative interpretation of the data and a test for the distribution of local concentrations require us to introduce a distribution of concentration inside the swollen coil. Using a Gaussian density of segments around the center of mass, Equation 1 becomes

$$\phi = \phi_0 \frac{\displaystyle\int_0^R \frac{4\pi r^2\,dr}{1 + K\alpha \frac{M}{N}\left(\frac{3}{2\pi\rho^2}\right)^{3/2} \exp \frac{3r^2}{2\rho^2}}}{\displaystyle\int_0^R 4\pi r^2\,dr} \tag{3}$$

where R is the radius of the spherical volume offered to each coil of molecular weight M

$$R = (3M/4\pi N)^{1/3} \tag{4}$$

and ρ the radius of gyration which can be calculated from the Flory-Fox relation for the intrinsic viscosity $[\eta]$

Taking $\frac{r}{\rho}$ as a variable

$$\phi = \phi_0 \frac{3\rho^2}{R^3} \int_0^{R/\rho} \frac{x^2\,dx}{1 + KA \exp - \frac{3x^2}{2}} \tag{5}$$

The experimental values of ϕ_0/ϕ can be redrawn as a function of $R/\rho = 1.46/c[\eta]$ together with a set of theoretical curves corresponding to different values of KA. The results are given in FIGURE 2 for a series of polymers and copolymers of dimethylaminostyrene. The agreement at high R/ρ is very good. The value of K calculated using $A = 2\alpha/[\eta]$ is found constant throughout a series of polymers or copolymers of different M and $[\eta]$. It is reduced by a factor of 5 (for copolymers) or 10 (for homopolymers) with respect to the value for DMPT. This gives an idea of the effect of steric and diffusional hindrance to side group reactivity.

The departure from the theoretical curves at $R/\rho = 1.3$ is a common feature to all systems. It is attributed to the transition from a dilute to a semidilute solution at the overlap concentration c^*. In scaling theories, c^* is often taken at $R/\rho = 1$. It is evident that it should depend slightly upon the measured physical quantity and this method is expected to be very sensitive to overlap.

The distribution of concentration can also be probed by fluorescence-decay experiments. If $I(t)$ is the flash emission curve, the decay of the intensity of fluorescence writes

$$I(t) = \int_0^t F(t) \int_0^R \exp - \left[\frac{t-T}{\tau_0} \left(1 + K \frac{2M}{N} \left(\frac{3}{2\pi\rho^2} \right) \right)^{3/2} \cdot \exp - \frac{3r^2}{2\rho^2} \right] \times 4\pi r^2 \, dr \, dT \qquad \textbf{(6)}$$

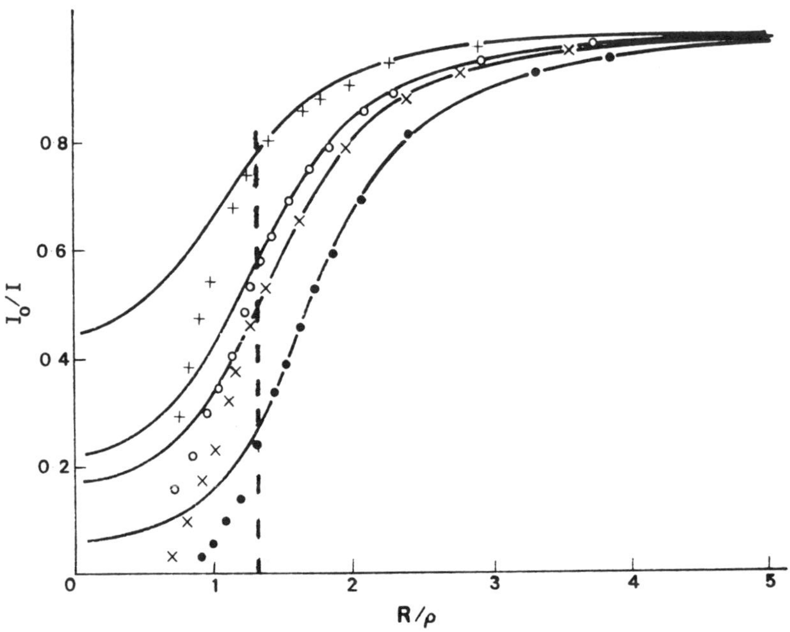

FIGURE 2. Ratio of the fluorescence intensities I_0 and I in the absence and presence of quencher plotted as a function of the ratio R/ρ (see text). Full curves are theoretical for a Gaussian distribution of concentrations in the individual polymer coils. +, ○, ×, ● correspond to different homo- and co-polymers of dimethylaminostyrene.

Using the preceding values of K, the calculated curve does not depart very much from a purely exponential decay as do the experimental curves. The calculated and experimental values of the time constant τ are in fairly good agreement.

FLUORESCENCE QUENCHING AS A TOOL IN THE STUDY OF CHAIN DYNAMICS[2]

Local motions in polymer chains can be studied by a number of different methods such as dielectric constant, *NMR*, spin labeling or fluorescence polarization. They all depend on one orientational functional $P_1(t) = < \cos \theta(t) >$ or $P_2(t) = 1/2 < 3 \cos^2 \theta(t) - 1 >$ where $\theta(t)$ is the angle of one vector linked to the chain with its initial position. This autocorrelation function is expected to be highly non-exponential, not principally because of some rotational freedom of a side group around the link to the main chain, but because of the one-dimensional character of the diffusional motion, which predicts in particular a $t^{-1/2}$ asymptotic behavior. Direct measurement of the decay of the two polarized components of fluorescence $I_{\parallel}(t)$ and $I_{\perp}(t)$ give access to $P_2(t)$ through the anisotropy factor:

$$r(t) = \frac{I(t) - I(t)}{I(t) + 2I(t)} = \frac{3\cos^2\beta - 1}{5} P_2(t) \tag{7}$$

where β is the angle between the absorption and emission transition moments $\vec{\mu}_a$ and $\vec{\mu}_e$.

Sub-nanosecond measurements become, however, very difficult because of the necessary deconvolution of $I(t)$ from $F(t)$, which covers typically a few nanoseconds for a conventional deuterium lamp. The use of picosecond flashes from Q-switched lasers or the very reproducible flash from synchrotron radiation with a Gaussian time dependance is a possible way to directly explore the short time behavior, but access to these instruments may be difficult. It occurred to us some years ago that, since the value of r observed under steady state illumination

$$\bar{r}(\tau) = r_0 \int_0^{\infty} \frac{3 \cos^2\theta(t) - 1}{2} \exp - \frac{t}{\tau}\, d\frac{t}{\tau} \tag{8}$$

is the Laplace transform of $r(t)$ for the value τ of the lifetime, one could modify τ by use of a suitable dynamic quencher and span this Laplace transform down to the very low values of τ permitted by the sensitivity of modern spectro-fluorimeters, as a much simpler way to test possible shapes of $P_2(t)$.

The result of such a measurement of $\bar{r}$ as a function of τ/τ_0 (where τ_0 is the lifetime in the absence of quencher) is shown on FIGURE 3 for DPA quenched by DMPT. For an exponential autocorrelation function, $P_2(t) = \exp - t/\tau_c$.

$$r^{-1}(\tau) = r_0^{-1}\left(1 + \frac{\tau}{\tau_c}\right) \tag{9}$$

In the plot of FIGURE 3, τ_c has been corrected for the small change of viscosity resulting from the addition of the quencher.

Application of the method to a polystyrene labeled with DPA in such a way that $\vec{\mu}_a$ and $\vec{\mu}_e$ are directed along the chain backbone leads to the plot of FIGURE 4. It is seen that the long time behavior can be represented as a function of $\tau^{-1/2}$ at large τ

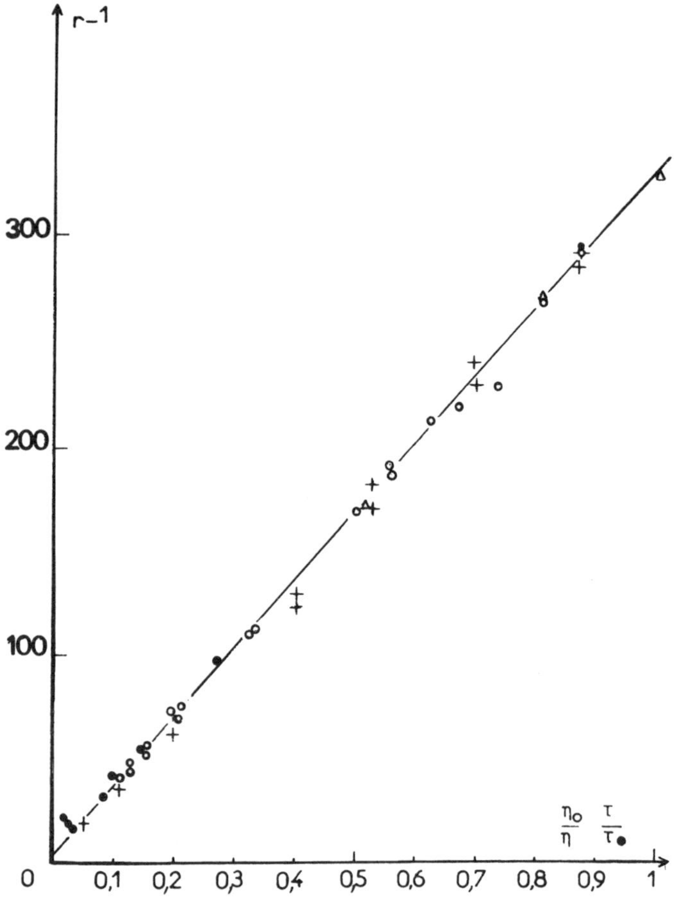

FIGURE 3. Plot of $\bar{r}^{-1}$ versus τ/τ_0, corrected for the changes in viscosity for diphenylanthracene quenched by CCl_4 (○), DMPT (●), and O_2 (Δ). * represent measurements as a function of viscosity at constant lifetime of fluorescence.

in good agreement with an expression of $P_2(t)$ for a one-dimensional diffusion

$$P_2(t) = \exp \frac{t}{\tau_c} \operatorname{erfc} \left(\frac{t}{\tau_c}\right)^{1/2} \tag{10}$$

leading to

$$\bar{r}(\tau)^{-1} = r_0^{-1}\left(1 + \left(\frac{\tau}{t_c}\right)^{1/2}\right) \tag{11}$$

From a comparison of the direct measurement of $r(t)$ and of the measurement of $\bar{r}(\tau)$ Valeur[3] has fitted similar data to an improved correlation function

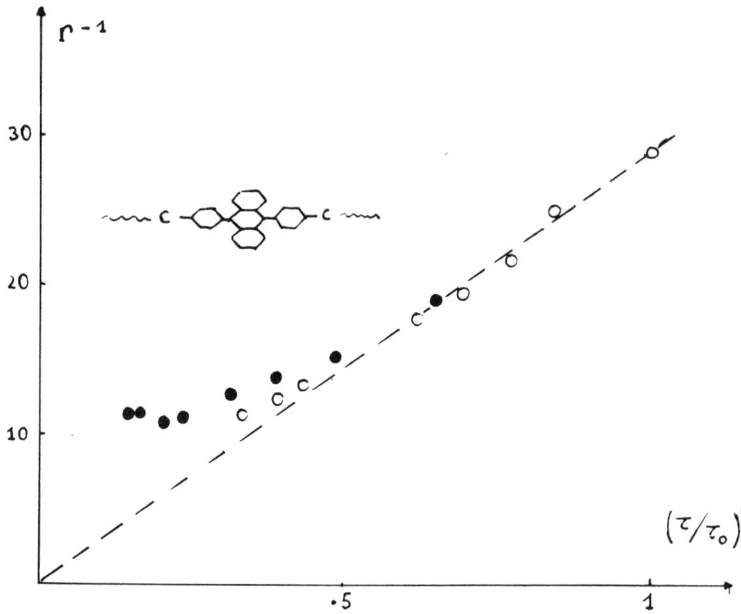

FIGURE 4. Plot of $\bar{r}^{-1}$ versus $(\tau/\tau_0)^{1/2}$ for a polystyrene labeled by diphenylanthracene in its backbone.

$$P_2(t) = \exp - \frac{t}{\theta} \exp \frac{t}{\tau_c} \operatorname{erfc} \left(\frac{t}{\tau_c}\right)^{1/2} \quad \textbf{(12)}$$

$$\bar{r}(\tau)^{-1} = r_0^{-1} \left(1 + \frac{\tau}{\theta} + \left[\frac{\tau}{\tau_c}\left(1 + \frac{\tau}{\theta}\right)\right]^{1/2}\right) \quad \textbf{(13)}$$

Fluorescence Polarization in an Electric Field: A New Electro-optic Effect for the Study of Rigid Molecules[4]

It seems particularly appropriate in this eighth Katzir Conference to mention a method that we have developed in the hope of contributing to one of the problems Aharon Katschalsky has been much interested in: the deformability of the layer of counterions around a rigid polyelectrolyte at high fields.[5]

Electro-optic methods can be used either to get information on the rotational dynamics of macromolecules, using the transient field free decay, or to get information on the orientation mechanism using the electric field dependence of the steady-state amplitude of the effect. The obtaining of precise orientation factors, and therefore of electric polarizability anisotropies $\Delta\alpha$ or of permanent dipole moments μ, implies the separation of electric and optical factors and accounts for the possible effects of polydispersity. As an example of the difficulties encountered, the curvature of the field dependence of the electric dichroism of DNA fragments has often been considered evidence of the existence of a permanent dipole moment (despite the antiparallel double helical structure) and of an appreciable base tilting (despite their perpendicular arrangement in the crystalline structure).

The effect of the orientation on the polarization of fluorescence is the basis of a new electro-optic method which, in combination with birefringence and dichroism, offers an interesting way to eliminate some of the aforementioned difficulties. The reasons are:

—it leads not only to the second but also to the fourth moment of the orientation function;

—the orientation of $\vec{\mu}_a$ (and $\vec{\mu}_e$) with respect to the molecular axis enters in a different way in the optical factor; and

—four polarized components of fluorescence can be recorded and used to check the self-consistency of the electric and optical factors.

For the case of DNA fragments obtained by sonication and labeled with intercalated Acridine orange (a dye with which the parallel increase of ϕ and τ upon intercalation manifests the existence of a single binding site) and assuming a perpendicular arrangement with respect to the DNA axis, we have shown that the general expression of the polarized components of fluorescence as a function of the orientation reduces to

$$V_v \propto \frac{1}{8}\langle\sin^4\theta\rangle(1 + 2\cos^2\gamma)$$

$$H_v \propto V_H \propto \frac{1}{4}\langle\sin^2\theta\rangle - \frac{1}{16}\langle\sin^4\theta\rangle(1 + 2\cos^2\gamma) \quad \textbf{(14)}$$

$$H \propto \frac{24}{64}[1 - \langle\sin^2\theta\rangle + \langle\sin^4\theta\rangle] + \frac{\cos^2\gamma}{64}[16 - 16\langle\sin^2\theta\rangle - 2\langle\sin^4\theta\rangle]$$

where γ is the time averaged angle between $\vec{\mu}_a$ and $\vec{\mu}_e$ deduced from the value of the fluorescence anisotropy in the absence of orientation

$$\bar{r} = \frac{3\cos^2\gamma - 1}{2} \quad \textbf{(15)}$$

and θ the angle of the molecular axis with the electric field E.

From the relative changes of V_v, H_v and H_H as a function of the orientation, we can get the field dependence of the two orientation factors

$$\phi = \frac{3\langle\cos^2\theta(t)\rangle - 1}{2}$$

$$\Delta = 1 - \frac{15}{8}\langle\sin^4\theta(t)\rangle \quad \textbf{(16)}$$

Their low field behavior is independent of the orientation mechanism

$$\phi = \frac{1}{15}AE^2$$

$$A = \left(\frac{\Delta\alpha}{RT} + \frac{\mu^2}{R^2T^2}\right) \quad \textbf{(17)}$$

$$\Delta = \frac{2}{21}AE^2$$

The ratio of Δ to ϕ is well respected in the low field experimental results of FIGURE 5. The rather abnormal curvature of the field dependence of ϕ and Δ at higher fields is not due to a permanent dipole moment orientation mechanism or to a polydis-

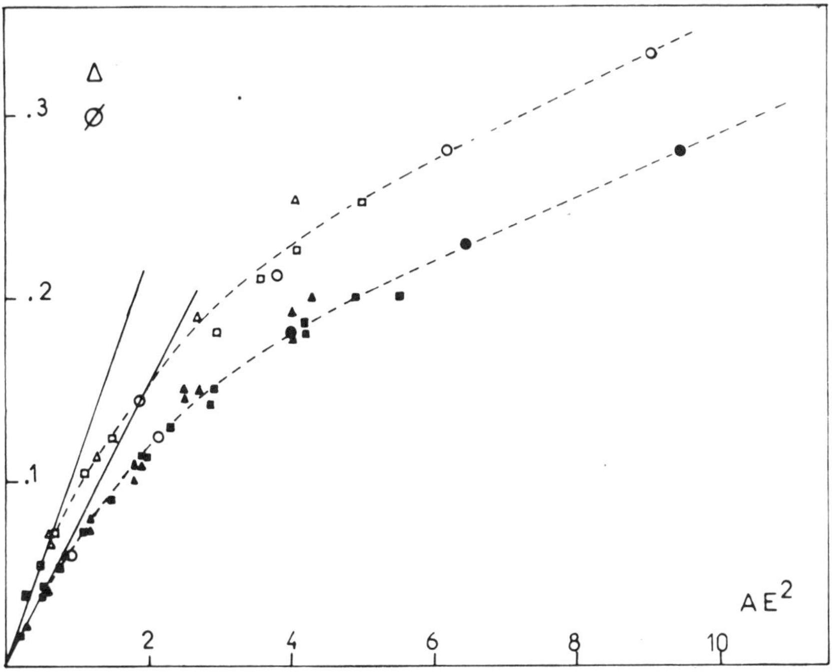

FIGURE 5. Field dependence of the two orientation factors ϕ and Δ normalized to their initial Kerr behavior for a series of DNA fragments obtained with 2′, 4′ and 6′ sonication.

persity effect (it is independent on the sonication time when normalized to the low field behavior), but to a *saturation of the induced dipole moment at a limiting value* μ_∞. It reflects the effect of counterion-counterion repulsions. Since it takes place at fields of ~5kV/cm where the orientation is far from complete, one can calculate the field dependence of ϕ and Δ corresponding to this new orientation mechanism. It differs from a true permanent dipole moment, since the electrostatic energy is not changed by a reversal of π in the field. ϕ and Δ are then predicted to be linear functions of E at intermediate fields

$$\phi = \frac{1}{8}\mu_\infty E/kT$$
$$\Delta = \frac{3}{16}\mu_\infty E/kT \qquad \textbf{(18)}$$

This ratio of ϕ to Δ corresponds fairly well to that found in the experiments as shown in FIGURE 6. Extrapolations at infinite orientation are then easily performed in full consistency with the original assumption of perpendicular arrangement. Reinterpretation of older data according to the new extrapolation scheme confirms that the limiting UV dichroism agrees well with a perpendicular arrangement of the base pairs.

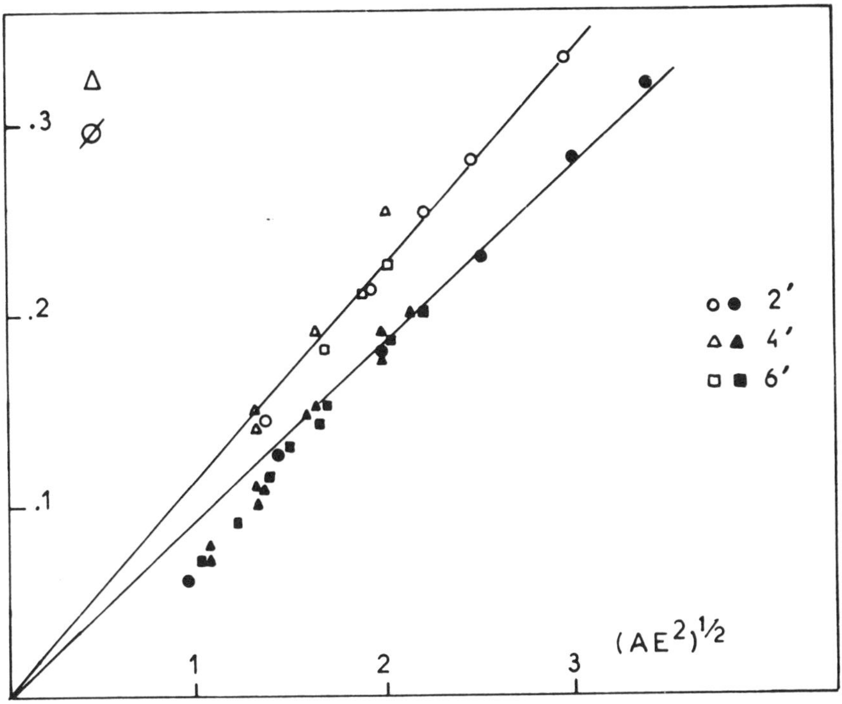

FIGURE 6. The transition from an E^2 to an E dependence for ϕ and Δ which reveals the saturation of the induced dipole moment.

PHOSPHORESCENCE OF POLYMERS IN SOLUTION AND THE PROBLEM OF ENERGY TRANSFER

Triplet-state energy transfer is well demonstrated in polymer solids. It is implicitly but generally accepted that energy transfer should preferentially proceed along the chain backbone, the long triplet lifetime overcoming the reduced transfer probability when compared to the singlet state. The demonstration of triplet energy transfer in single chains in solution, where the triplet lifetime is much reduced, relies generally on the comparison of the efficiency of triplet quenchers for the isolated side group or in the homopolymer. Comparison of the quenching efficiency of external or copolymerized quenching groups can also be used if proper account is taken of the local concentration inside the swollen coil. The determination of the quenching efficiency is generally based on the measurement of

—the quantum yield of a photochemical reaction (chain scission or production of photoproducts);

—the lifetime of the triplet state which is populated by an intense laser flash and whose decay is monitored by the $T - T^*$ absorption (laser flash photolysis).

From a comparison of the quenching of the photochemical reaction of the triplet state of benzophenone (BP) and polyvinylbenzophenone (PVB) with isopropanol in benzene,[6] we have concluded that there is no evidence for energy transfer in this polymer provided there is no drastic reduction of the triplet lifetime of the benzophenone side group in the polymer. Recently published[7] photophysical data on a similar reaction of BP and PVB with tetrahydrofuran lead to the conclusion that there is only a small decrease of this lifetime (2.27 μsec versus 3.57 μsec in pure benzene) but from the lifetime dependences in the presence of *THF* there seems to be an efficient energy transfer. Preliminary laser flash photolysis experiments performed by Dr. Fouassier in Mulhouse reveal that the triplet-state decay in PVB is highly non-exponential and deviates from first-order kinetics, a fact which is not taken in account in the former work which only reports and interprets the long-time (>1 μsec) decay. Such non-exponential decays have also been reported by Schnabel[8] and are interpreted as resulting from triplet-triplet annihilation due to collisions between distant parts of the molecules. This not only puts severe limitations on the use of intense flashes, but also reveals that the collisions between distant segments should also be considered an efficient process for energy transfer. This is not so surprising considering the rates of collision of the ends of short chains as reported in this conference. Low-intensity steady-state phosphorescence measurements should be—when possible—easier to interpret. This is the case for BP and PVB in carefully deoxygenated solutions. Preliminary experiments by Dr. Knoesel in Strasbourg (carefully subtracting the Raman scattering contribution for benzene solutions) lead to the following phosphorescence quantum yields

CCl_4	$\phi_{BP} = 5.7\ 10^{-3}$	ϕ_{PVB}: insoluble
C_6H_6	$\phi_{BP} = 1.4\ 10^{-3}$	ϕ_{PVB}: $5.5\ 10^{-3}$
$CHCl_3$	$\phi_{BP} = 4.9\ 10^{-4}$	ϕ_{PVB}: $8.4\ 10^{-4}$

The result obtained in CCl_4 is in agreement with other data and with the ratios of phosphorescence quantum yield in solution and in the glassy state. The result in C_6H_6 confirms that the lifetime of the triplet state is even bigger in PVB than in BP. The result in $CHCl_3$ reveals that the hydrogen abstraction reaction which causes the decrease in quantum yield of BP is hindered in PVB, another manifestation of the lack of efficient triplet transfer *at the time scale of the microsecond.*

This is in contrast with polyvinylphenylketone where all photochemical[9] and photophysical[10] data converge to demonstrate an efficient energy transfer at the time scale of 1 microsecond. Considering the above remarks, both the local conformations and the dynamic of the chain should be considered in the interpretation of this difference.

References

1. Duportail G., D. Froehlich & G. Weill. 1971. Eur. Polym. J. **7:** 977–987.
2. Bentz J. P., J. P. Beyl, G. Beinert & G. Weill. 1975. Eur. Polym. J. **11:** 711–718.
3. Valeur B. & L. Monnerie. 1976. J. Polym. Sci. **14:** 29–37.
4. Sokerov S. & G. Weill. 1979. Biophys. Chem. **10:** 161–171.
5. Neumann E. & A. Katchalski. 1972. Proc. Nat. Acad. Sci. USA **69:** 993–997.
6. Sanchez G., R. Knoesel & G. Weill. 1978. Eur. Polym. J. **14:** 485–488.
7. Kamachi M., Y. Kikuta & S. I. Nozakura. 1979. Polym. J. **11:** 273–277.
8. Schnabel W. 1979. Makromol. Chem. **180:** 1487–1495.
9. Hrdlovič P., I. Lukac & I. Paska. 1973. J. Polym. Sci. **C40:** 57.
10. Lougnot D. J., J. P. Fouassier & J. Faure. 1975. J. Chim. Phys. **72:** 125.

STUDIES OF SINGLET ENERGY TRANSFER IN POLYMERS BY TIME-RESOLVED FLUORESCENCE SPECTROSCOPY

D. A. Holden, W. A. Rendall, and J. E. Guillet*

Department of Chemistry
University of Toronto
Toronto, Canada M5S 1A1

INTRODUCTION

As biochemists' understanding of the process of photosynthesis increases, the question arises whether this understanding can be demonstrated by the development of a synthetic model system incorporating both the energy-gathering and light-converting functions of plant chloroplasts.[1,2] The discovery of energy transfer processes in man-made polymers which appear to mimic similar effects in chloroplast grana has produced much interest in this field.[3-6]

We have investigated singlet electronic energy transfer between naphthalene and anthracene chromophores bound to the same polymer chain.[5-7] Such labeled macromolecules constitute simple models of the energy-gathering part of the natural photosynthetic apparatus. The goal of this research was to reproduce the function and efficiency, but not the detailed structure, of the much more complex pigment layers of living plants.

Polymers of 1-naphthylmethyl methacrylate (NMMA) containing terminal anthracene traps or copolymerized 9-anthrylmethyl methacrylate (AMMA) were synthesized.

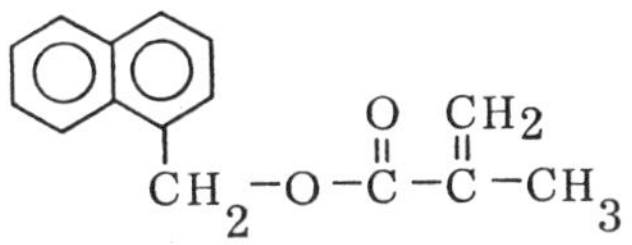

NMMA

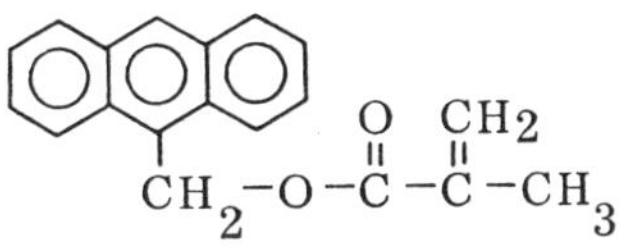

AMMA

In addition, a copolymer of acrylic acid (AA) with NMMA containing terminal anthracene units was prepared. The effects of solvent and molecular mobility on the efficiency of energy transfer between the two chromophores were evaluated quantitatively. The advantage of using a fluorescent acceptor such as anthracene is that useful information concerning the mechanism of energy transfer is contained in the fluorescence decay of the trap. Through time- and wavelength-resolved fluorescence studies, it was possible to study the time scale and the mechanism of energy transfer.

* Please address all correspondence to: Professor James Guillet, Department of Chemistry, University of Toronto, Toronto, Canada M5S 1A1.

0077-8923/81/0366-0011 $01.75/0 © 1981, NYAS

EXPERIMENTAL

Polymers

1-Naphthylmethyl methacrylate was prepared by the Schotten-Baumann condensation of 1-naphthalenemethanol with methacryloyl chloride and was purified by column chromatography on alumina. 9-Anthrylmethyl methacrylate was obtained in a similar manner from 9-anthracene-methanol and was recrystallized repeatedly from cyclohexane. Polymers were prepared by free radical polymerization in degassed benzene or dioxane, and were purified by multiple precipitation from benzene into methanol. Anthracene-terminated poly(NMMA) and poly(AA-NMMA) were obtained by polymerization in the presence of 5 mol % 9-bromomethyl anthracene. Full details of the preparation and characterization of polymer samples are published elsewhere.[6,7] The compositions and molecular weights of the copolymer samples are listed in TABLE 1.

Steady-State Fluorescence Spectra

Fluorescence spectra of samples in deoxygenated purified solvents were recorded using a Hitachi-Perkin Elmer MPF-2A spectrofluorimeter and are shown corrected for the wavelength dependence of the detector response.[7] Quantum yields were determined using 1-naphthol and 9-methylanthracene as standards.[8] Efficiencies of energy transfer from naphthalene to anthracene were determined from copolymer fluorescence spectra as described elsewhere[7] by reconstructing the naphthalene excimer band which lies partially obscured by anthracene fluorescence. FIGURE 1 shows a typical copolymer spectrum, illustrating the partially buried excimer emission band. The relative intensities I_N and I_A of naphthalene and anthracene fluorescence were measured, and the quantum yield χ of energy transfer was then calculated according to Equation 1:

$$\phi_N I_A/\phi_A I_N = \chi/(1 - \chi) \tag{1}$$

TABLE 1
PROPERTIES OF THE ANTHRACENE-CONTAINING POLYMER SAMPLES

Sample	Monomer 1	Monomer 2	Structure*	Mole % Anthracene		$\bar{M}_n$† ($\times 10^{-5}$)
				Monomer Mixture	Copolymer	
1	NMMA	9-bromomethyl anthracene	E	5	0.50	0.41
2	NMMA	AMMA	R	0.093	0.074	1.03
3	NMMA	AMMA	R	0.26	0.174	1.09
4	NMMA	AMMA	R	0.50	0.348	1.10
5	NMMA	AMMA	R	0.89	0.665	1.11
6	NMMA	AMMA	R	1.44	1.42	‡
7	NMMA	AMMA	R	2.27	2.06	‡

* E = anthracene end-group; R = random copolymer.
† By membrane osmometry in toluene.
‡ Polymer crosslinks during analysis.

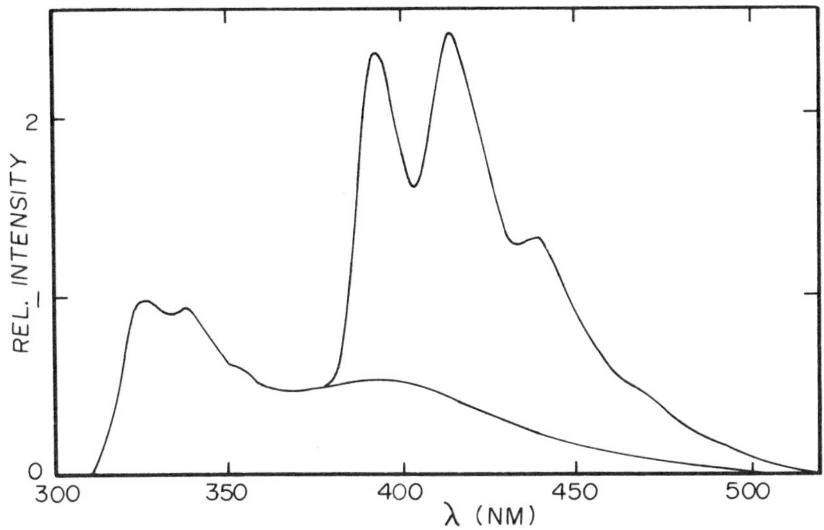

FIGURE 1. Fluorescence spectrum of a copolymer of NMMA with 0.348 mole % AMMA in deoxygenated THF solution at 25°C on excitation of the naphthalene chromophore at 280 nm. The spectrum shows the reconstruction of the partially obscured naphthalene excimer band.

In Equation 1, ϕ_N and ϕ_A are the fluorescence efficiencies of NMMA homopolymer and of the bound anthracene chromophore on direct excitation.

Fluorescence Decays and Time-resolved Fluorescence Spectra

Fluorescence decays and time-resolved fluorescence spectra were recorded by single photon counting.[5,6,9,10] For time-resolved spectroscopy the excitation source was a Thyratron-gated flash lamp operating at 25 kHz, 5.2 kV, 3.0 mm arc gap and 0.5 atmosphere of research grade N_2. The estimated time resolution under these excitation conditions is 2 ns. A Spex Minimate monochromator driven by a Rapid-Syn-28 PS stepping motor swept through the spectral region of interest in 0.14 nm increments with a 0.1 second dwell time at each step. At least 50 scans were superimposed in the multichannel analyzer memory in order to build up a sufficiently detailed spectrum. The estimated spectral bandwidth was 8 nm. The positions of the upper and lower time limits of each time-resolved spectrum relative to the maximum of the excitation pulse were determined by recording the decay curve in the normal pulse-height analysis mode of the multichannel analyzer. Naphthalene and anthracene chromophores were excited at 280 nm and 366 nm, respectively, through narrow bandpass interference filters.

RESULTS AND DISCUSSION

Steady-State Measurements

FIGURE 1 shows the fluorescence spectrum of a copolymer of NMMA containing 0.348 mol% AMMA in THF solution. Structured anthracene emission in the 390-450 nm region dominates the copolymer spectrum, even though the low anthracene content of the copolymer means that the extent of direct excitation of anthracene is negligible. Although the NMMA homopolymer fluorescence quantum yield is considerably smaller than that of copolymerized AMMA, differences in fluorescence efficiencies cannot account for the strong anthracene fluorescence. FIGURE 1 shows that singlet electronic energy transfer occurs from naphthalene to anthracene.

TABLE 2 lists values of the quantum efficiency χ of energy transfer from naphthalene to anthracene in NMMA-AMMA copolymers of different composition. High efficiences of energy transfer can be achieved at low mole fractions of bound acceptor. The value of χ of 84% at 2.06% AMMA content may be compared with estimated efficiencies of from 75%[11] to 96%[12] for transfer from the antenna pigments to the ca 1% reaction centres of natural photosystems. It is possible to create a macromolecule which absorbs light and transfers it to a bound trap with an efficiency approaching that observed in living systems. A polymer antenna molecule has advantages over energy-gathering systems based on micelles or vesicles[13] because the donor-acceptor distance is restricted in the former by the macromolecular chain. As a result, energy transfer remains efficient even when the bulk chromophore concentrations are 10^{-6} M or lower.

TABLE 3 illustrates the effect of solvent on the efficiency of energy transfer in an anthracene-terminated PNMMA sample. On going from good solvents such as THF or dioxane to poor solvents such as dioxane-methanol or poly(methyl methacrylate), a pronounced increase in χ is observed. The increase results from a decrease in the

TABLE 2

VARIATION OF THE EFFICIENCY χ OF ENERGY TRANSFER FROM NAPHTHALENE TO ANTHRACENE IN COPOLYMERS OF NMMA WITH AMMA.*

Sample†	Mole % AMMA	χ (%)
2	0.074	3.7
3	0.174	9.9
4	0.348	18
5	0.665	30
6	1.42	69
7	2.06	84

* Solutions in THF at 25 °C

† See TABLE 1 for numbering of samples.

TABLE 3

EFFECT OF SOLVENT ON EFFICIENCY χ OF ENERGY TRANSFER FROM NAPHTHALENE TO ANTHRACENE IN AN ANTHRACENE-TERMINATED POLY(NMMA) SAMPLE* AT 25°C

Solvent	χ (%)
THF	10
Dioxane	11
Dioxane/methanol (60/40 vv)	21
Dioxane/methanol (40/60 vv)	47
Dioxane/methanol (20/80 vv)	66
Poly(methyl methacrylate)	43

* Sample 1, TABLE 1.

average donor-acceptor distance produced by contraction of individual polymer coils in poor solvents. It is likely that energy transfer is sensitive to coil dimensions in much the same way as the extent of excimer formation.[14–16]

Copolymers of NMMA with AMMA are soluble in 2-methyl THF, which forms a glass at 77°K. FIGURE 2 shows that the fluorescence spectrum of NMMA homopolymer in 2-methyl THF at 77°K contains no component of emission from the naphthalene excimer. Nevertheless, en^rgy transfer still occurs in copolymers, as is evident from FIGURE 3. TABLE 4 lists values of χ for PNMMA-AMMA samples at 77°K. (These results were obtained with narrow excitation slit widths so that interference fromtriplet-triplet annihilation and delayed fluorescence was too small to detect.) The data of TABLE 4 lead to two significant conclusions. Energy transfer to anthracene from the naphthalene excimer is not the major mechanism of transfer, because the values of χ are almost as high under conditions where no excimers are present. Secondly, molecular motion does not play a predominant role in the mechanism of transfer. Energy transfer is nearly as efficient in a rigid glass as it is in fluid solution. It is evident that it is the electronic excitation which "diffuses," and that segmental diffusion is a comparatively small effect.

Energy Transfer in a Water-Soluble Copolymer

TABLE 5 shows the variation in χ with solvent composition for a copolymer of acrylic acid with 6.9 mol % NMMA terminated with one anthracene acceptor per

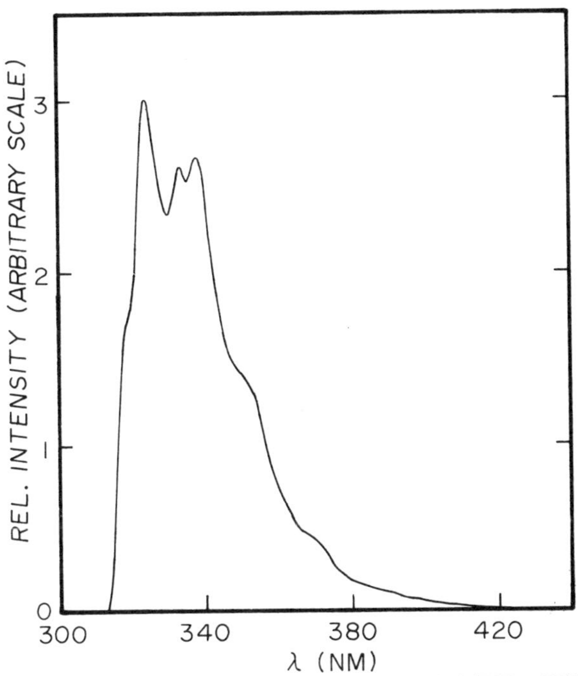

FIGURE 2. Fluorescence spectrum of poly(NMMA) in 2-methyl THF at 77°K on excitation at 280 nm.

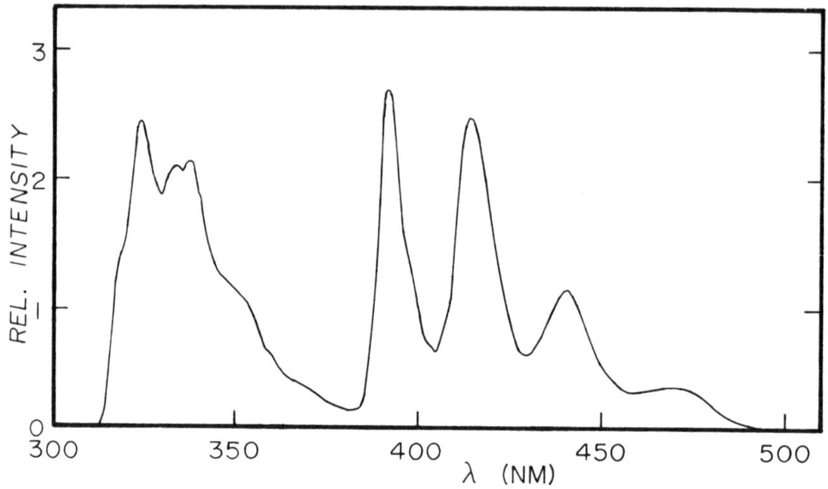

FIGURE 3. Fluorescence spectrum of a copolymer of NMMA with 0.665 mole % AMMA in 2-methyl THF at 77°K, on excitation of the naphthalene chromophore at 280 nm.

TABLE 4
VARIATION OF THE EFFICIENCY χ OF ENERGY TRANSFER FROM NAPHTHALENE TO ANTHRACENE WITH ANTHRACENE CONTENT IN COPOLYMERS OF NMMA WITH AMMA*

Sample†	Mole % AMMA	χ (%)
3	0.174	8
4	0.348	15
5	0.665	22
6	1.42	61
7	2.06	70

* Solutions in 2-Methyl THF at 77°K
† See TABLE 1 for numbering of samples.

TABLE 5
EFFECT OF SOLVENT ON EFFICIENCY χ OF ENERGY TRANSFER FROM NAPHTHALENE TO ANTHRACENE IN AN ANTHRACENE-TERMINATED COPOLYMER OF ACRYLIC ACID WITH 6.9 MOLE % NMMA

Solvent	χ (%)
Dioxane	12
Aqueous NaOH (pH = 12.0)	70

chain. A pronounced increase in χ is observed in going from dioxane to 0.02 M aqueous NaOH. Furthermore, substantial naphthalene excimer emission is observed in water, although it is almost absent in dioxane. These results suggest that aggregation of the hydrophobic NMMA units occurs in aqueous solution. Such a process could involve micellization (aggregation of NMMA units within individual coils), or "multimerization"[17] (aggregation of NMMA units from several chains with the hydrophobic groups functioning as transient ties between macromolecules). The absence of a change in χ with polymer concentration suggests that intra-chain "hydrophobic bonding" is the more pronounced effect. Efficient energy transfer in aqueous solution is a significant achievement in the development of a synthetic chloroplast, because it is expected that this solvent can function as an inexpensive reactant in an artificial photosynthetic system.

Transient Fluorescence Measurements

FIGURE 4 illustrates the time-resolved fluorescence spectrum of anthracene-terminated poly(NMMA) in toluene at 25°C, on excitation of the naphthalene chromophore at 292 nm. The contribution to the total spectrum from anthracene emission increases for the first 20 ns from the maximum of the exciting pulse, then decreases very slowly over the next 200 ns.

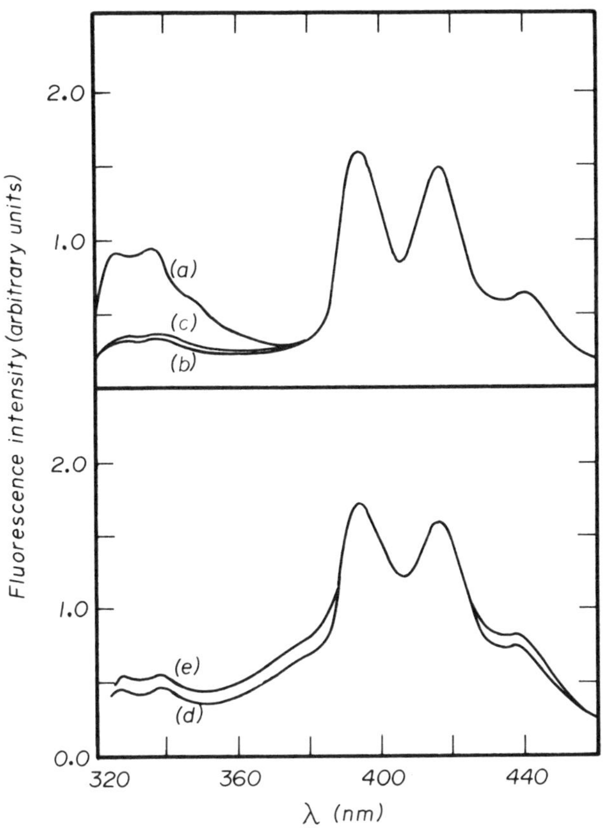

FIGURE 4. Time-resolved fluorescence spectra of anthracene-terminated PNMMA in toluene at 25.0°C on excitation at 292 nm. Lower and upper time limits in ns from maximum of the exciting pulse are (a) 0.0–2.2, (b) 11.5–13.8, (c) 21.0–24.0, (d) 44.3–48.6, and (e) 179.2–228.5. The curves are normalized to the same intensity at 390 nm.

In time-resolved spectroscopy, changes with time in the relative intensities of the emitting species may arise from energy transfer from one species to another, or simply from the different decay rates of the emitting species. If I_i is the relative intensity of species i at a representative wavelength in a time-resolved spectrum recorded between times t_1 and t_2, then I_i is a function of a decay law $I_i(t)$ convoluted with the lamp profile $G(t)$, as shown in Equation 2:

$$I_i = c_i \int_{t_1}^{t_2} [G * I_i](t)\,dt \qquad \textbf{(2)}$$

Equation 2 is a mathematical representation of a time-resolved spectrum.

If energy transfer occurs by static quenching, i.e., instantaneous one-step transfer from naphthalene to anthracene, then the energy transfer function follows the excitation pulse profile. We have performed calculations to test whether static quenching could predict the evolution of the time-resolved spectrum observed experimentally. The relative intensities of the three emitting species—naphthalene monomer, naphthalene excimer and anthracene—in time-resolved spectra at short time were used to evaluate the various c_i in Equation 2. Using the decay laws of NMMA homopolymer and of anthracene on direct excitation, time-resolved "spectra" were simulated for the same time-intervals as in FIGURE 4. The simulated spectra do not correspond at all with those shown in FIGURE 4, and instead predict that the contribution from anthracene emission should decrease rapidly from its initial high value and be negligibly small by 200 ns. We conclude that static quenching is not an appropriate description of the energy transfer process.

Fluorescence decay measurements in conjunction with time-resolved spectroscopy provide quantitative information about the time-scale of energy transfer. FIGURE 5

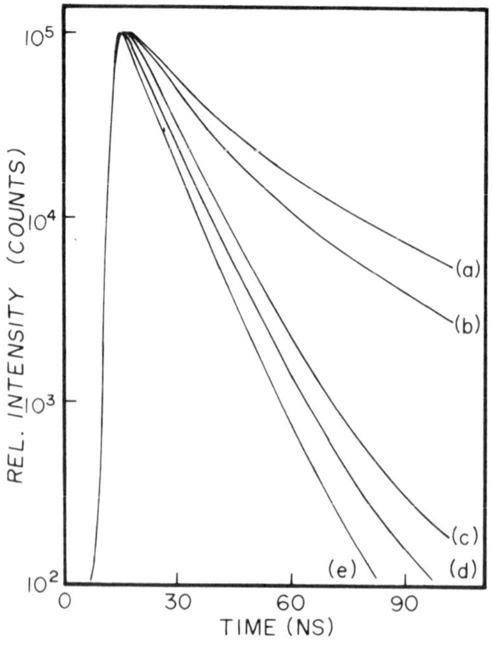

FIGURE 5. Fluorescence decays at 420 nm on 280 nm excitation of PNMMA-AMMA samples in nitrogenated THF at 25°C. The mole % AMMA in the copolymer samples is (a) 0.348, (b) 0.665, (c) 1.42, (d) 2.06. Curve (e) is the decay of the anthracene emission on direct excitation at 366 nm.

illustrates the decay of the 420 nm emission from PNMMA-AMMA copolymers in THF at 25°C. The initial decay rate is slower on energy transfer from naphthalene than it is on direct excitation of anthracene. The difference $\Delta\tau$ between the slope of the initial portion of the decay and the fluorescence lifetime of AMMA on 366 nm excitation decreases from 6.4 to 1.4 ns as the anthracene content increases from 0.35% to 2.06%. Thus energy transfer occurs on a nanosecond time scale in these copolymers.

At room temperature, at least part of the 420 nm emission arises from the naphthalene excimer, so that analysis of the decays of FIGURE 5 is prevented. Similar decay measurements were performed on 2-methyl THF solutions at 77°K. The analysis of the acceptor decays is made possible by the absence of superimposed excimer emission at this temperature. FIGURE 6 shows the variation of the acceptor decays with AMMA content. The decay of the AMMA fluorescence on energy transfer from naphthalene is non-exponential and much slower than that observed on direct excitation.

Energy Migration in PNMMA

Numerous authors have shown partial or complete depolarization of the emission from naphthalene-containing polymers, and have proposed that energy migration occurs in these systems.[18–21] Our evidence for energy migration between the naphthalene chromophores of poly(NMMA) is also obtained from fluorescence depolarization measurements.[22] On excitation with 300 nm polarized light, the excimer emission from films of NMMA homopolymer is completely depolarized (P = −0.002 ± 0.01). These films are completely transparent and show no strain-induced birefringence

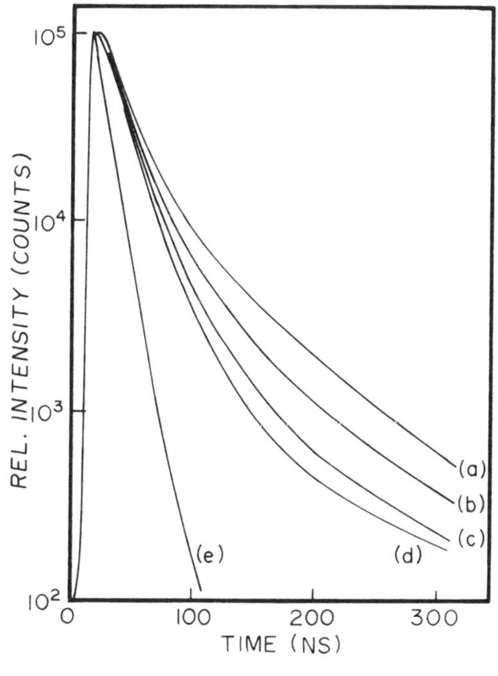

FIGURE 6. Decay of the fluorescence of the anthracene chromophore at 425 nm on 280 nm excitation of PNMMA-AMMA copolymers in 2-methyl THF at 77°K. The mole % AMMA in the copolymers is (a) 0.174, (b) 0.665, (c) 1.42, (d) 2.06. Curve (e) is the decay of the anthracene emission on direct excitation at 366 nm.

under crossed polarizers. No delayed luminescence was observed at room temperature. MacCallum[23] reports that the excimer fluorescence in poly(1-vinyl naphthalene) retains substantial polarization, so the depolarization observed in poly(NMMA) does not arise because the intrinsic depolarization of the naphthalene excimer is close to zero. The measured polarization (300 nm excitation) of the 340 nm emission of an isolated naphthalene molecule in a polystyrene film was 0.1, in agreement with literature values,[18] a result indicating that the chromophore is stationary on the timescale of fluorescence. Loss of memory by the excimer of the excitation polarization arises because the excimer is formed by energy migration between naphthalene monomer units to a site in or close to the excimer configuration. However, fluorescence depolarization measurements suffer from the disadvantage of being unable to differentiate between 10 and 10^5 migration steps since both lead to essentially complete depolarization.

Mechanism of Energy Transfer

It has been proposed by Guillet[24] that energy transfer from naphthalene to anthracene in these antenna polymers can be explained on the basis of (1) singlet energy migration between naphthalene chromophores lying outside the range of Förster transfer to the anthracene trap, followed by (2) direct single step Förster transfer to anthracene once the excitation reaches a naphthalene chromophore within the Förster range. This is illustrated schematically in FIGURE 7. The important problem is to determine the relative contributions of these two processes to the observed energy tranfer efficiencies.

In order to elucidate the problem, we have first calculated the expected decay curves if only single step Förster transfer occurs (process 2). For the case of Förster transfer from a long-lived donor to a short-lived acceptor, and assuming a random distribution of stationary chromophores, the acceptor fluorescence response function $I_A(t)$ is:[25,26]

$$I_A(t) = 2k_{AF}[D^*]_0 m'\{\exp(-k_A t)D(m') + \exp[-k_D t - 2c(k_D t)^{1/2}]D[(a't)^{1/2} - m']\} \quad (3)$$

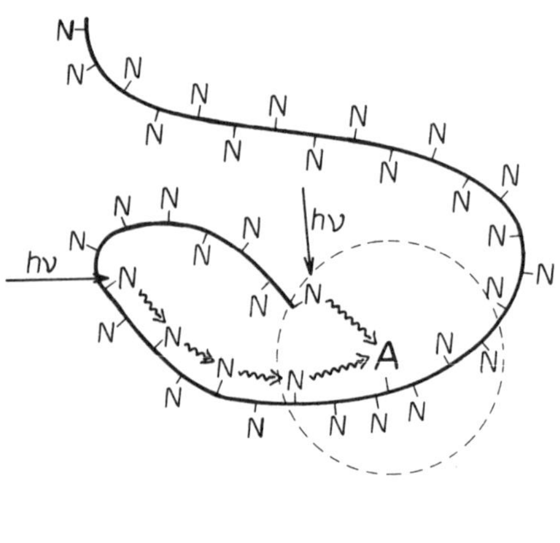

FIGURE 7. A conceptual illustration of energy transfer in an antenna polymer.

where

$$a' = k_A - k_D \tag{4}$$

$$m' = c[k_D/(k_A - k_D)]^{1/2} \tag{5}$$

$$c = [A]/c_0 \tag{6}$$

and $D(x)$ is the Dawson integral of x, defined by

$$D(x) = \exp(-x^2)\int_0^x \exp(u^2)du \tag{7}$$

In Equations 3 to 6, k_D and k_A are reciprocals of the donor lifetime in the absence of acceptor, and of the acceptor lifetime on direct excitation. These fluorescence lifetimes are 86 and 9.3 ns, respectively, for PNMMA-AMMA in 2-methyl THF at 77K. k_{AF} is the rate constant for fluorescence, $[D^*]_0$ is the initial concentration of photoexcited donor, $[A]$ is the acceptor concentration and c_0 is the Förster concentration[27] describing the range of the dipole-dipole interaction between the two chromophores. Appropriate values of the reduced concentration parameter c were calculated from the values of χ measured from the steady-state copolymer fluorescence spectra using Equation 8:

$$\chi = \pi^{1/2}c \exp c^2(1 - \text{erf}c) \tag{8}$$

where erf$c$ is the error function of c. The calculated acceptor decay curves are shown convoluted with an experimental excitation pulse profile in FIGURE 8. These calcu-

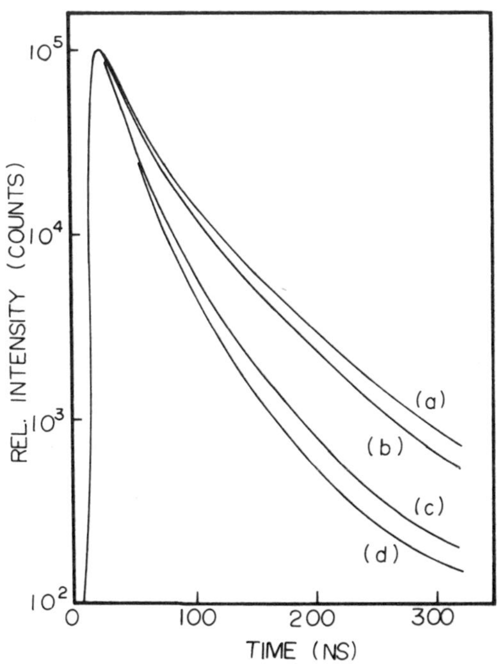

FIGURE 8. Calculated anthracene fluorescence decay curves at 425 nm on 280 nm excitation of PNMMA-AMMA samples in 2-methyl THF at 77°K, assuming one-step Förster transfer from naphthalene to anthracene. The curves are convoluted with an experimental lamp profile and are shown on the same time scale as in FIGURE 6. Values of the reduced acceptor concentration $[A]/c_0$, defined in Equation 6 are: (a) 0.0502, (b) 0.146, (c) 0.618, (d) 0.818.

lated curves closely resemble the experimental fluorescence decay curves illustrated in FIGURE 6. This treatment assumes that the chromophores attached to polymers are randomly distributed in space, which is not correct. We hope to refine the calculation to take in the effects of realistic chromophore distributions. The fact that acceptor fluorescence decay curves calculated assuming one-step Förster transfer between naphthalene and anthracene are similar to the experimental decay curves suggests that the distance over which energy migration occurs in PNMMA at 77K is not large relative to the 24 Å radius[28] of Förster transfer between naphthalene and anthracene. It is proposed that energy transfer occurs in these copolymers by one-step dipole-dipole transfer from naphthalene to anthracene, and that the effective range of this transfer is increased by a moderate number of discrete hops of the electronic excitation between naphthalene chromophores.

CONCLUSIONS

Efficient energy-gathering systems can be constructed by incorporating donors and acceptors into the same macromolecule. The effectiveness of these "molecular antennae" approaches that of the light harvesting pigment layers of living plants. Time-resolved fluorescence spectroscopy shows that energy transfer occurs on a nanosecond time scale. These dynamic fluorescence measurements suggest that the mechanism of energy transfer involves one-step Förster transfer from naphthalene to anthracene. Fluorescence depolarization measurements further suggest that the efficiency of this one-step transfer process may be enhanced by a moderate number of energy transfers between naphthalene chromophores.

ACKNOWLEDGMENTS

We are grateful to the Natural Sciences and Engineering Research Council of Canada for financial support of this research and for fellowship support (to D. A. Holden and W. A. Rendall).

REFERENCES

1. PORTER, G. & M. D. ARCHER. 1976. Interdiscip. Sci. Rev. **1:** 119–143.
2. PORTER, G. 1978. Proc. R. Soc. London, Ser. A **362:** 281–303.
3. KLÖPFFER, W. 1978. Spectrosc. Lett. **11:** 863–876.
4. MACCALLUM, J. R. 1978. Chem. Soc. (London), Ann. Reports A: 99–116.
5. HOYLE, C. E. & J. E. GUILLET. 1978. J. Polym. Sci., Polym. Lett. Ed. **16:** 185–191.
6. ASPLER, J. S., C. E. HOYLE & J. E. GUILLET. 1978. Macromolecules **11:** 925–929.
7. HOLDEN, D. A. & J. E. GUILLET. 1980. Macromolecules **13:** 289–295.
8. BERLMAN, I. B. 1971. *In* Handbook of Fluorescence Spectra of Aromatic Molecules. 2nd edit. Academic Press. New York.
9. LEWIS, C., W. R. WARE, L. T. DOEMENY & T. L. NEMZEK. 1973. Rev. Sci. Instrum. **44:** 107–114.
10. HOYLE, C. E., T. L. NEMZEK, A. MAR & J. E. GUILLET. 1978. Macromolecules **11:** 429–431.
11. TOMITA, G. & E. RABINOVITCH. 1962. Biophys. J. **2:** 483–499.
12. PORTER, G., C. J. TREDWELL, G. F. W. SEARLE & J. BARBER. 1978. Biochim. Biophys. Acta **501:** 232–245.
13. NOMURA, T., J. R. ESCABI-PEREZ, J. SUNAMOTO & J. H. FENDLER. 1980. J. Am. Chem. Soc. **102:** 1484–1488.

14. SOMERSALL, A. C. & J. E. GUILLET. 1973. Macromolecules **6:** 218–223.
15. ASPLER, J. S. & J. E. GUILLET. 1979. Macromolecules **12:** 1082–1088.
16. ABUIN, E. A., E. A. LISSI, L. GARGALLO & D. RADIC. 1979. Eur. Polym. J. **15:** 373–377.
17. SELB, J. & Y. GALLOT. 1980. Makromol. Chem. **181:** 809–822.
18. DAVID, C., D. BAEYENS-VOLANT & G. GEUSKENS. 1976. Eur. Polym. J. **12:** 71–76.
19. DAVID, C., N. PUTMAN DE LAVAREILLE & G. GEUSKENS. 1977. Eur. Polym. J. **13:** 15–16.
20. REID, R. F. & I. SOUTAR. 1977. J. Polym. Sci., Polym. Lett. Ed. **15:** 153–160.
21. REID, R. F. & I. SOUTAR. 1978. J. Polym. Sci., Polym. Phys. Ed. **16:** 231–244.
22. HOLDEN, D. A. & J. E. GUILLET. Unpublished results.
23. MACCALLUM, J. R. Private communication.
24. GUILLET, J. E. 1979. Polym. Prepr., Am. Chem. Soc., Div. Polym. Chem. **20**(1): 395–397.
25. HAUSER, M., U. K. A. KLEIN & U. GÖSELE. 1976. Z. Phys. Chem. (NF) **101:** 255–266.
26. HALLAM, A. & J. B. BIRKS. 1978. J. Phys. B **11:** 3273–3288.
27. FÖRSTER, T. 1949. Z. Naturforsch. A **4:** 321–327.
28. BERLMAN, I. B. 1973. *In* Energy Transfer Parameters of Aromatic Compounds. Academic Press. New York.

STUDIES OF INTRAMOLECULAR EXCIMER FORMATION IN SYNTHETIC POLYMERS

Ian Soutar

Chemistry Department
Heriot-Watt University
Riccarton, Currie, Edinburgh EH14 4AS, Scotland

INTRODUCTION

Macromolecular systems offer interesting possibilities for the study of photophysical phenomena such as the formation of excimers. Not only are there opportunities to obtain information regarding the nature of intramolecular interactions leading to excimer formation in polymers, the "energy harvesting" afforded by energy migration to potential excimer sites combined with the generation of high local concentrations of chromophores allows the study of excimer formation in species in which excimer is difficult to observe in low molar mass species. An example of the latter effect is the observation of excimer emission from polymers incorporating oxadiazole species[1] which had previously been supposed to be incapable of excimer formation as a consequence of mutual repulsion between excited-state and ground-state species.[2]

An intriguing problem in polymer photophysics is the extent to which excimer formation occurs as a consequence of interactions between chromophores located within close proximity on the polymer chain as opposed to those between chromophores which are brought into the correct alignment for complex formation through the coiling of the macromolecular chain. In the current paper, a variety of studies are described in which information is sought upon the nature of the mechanism of excimer formation in polymers derived from vinylaromatic monomers.

The work is subdivided into studies of the photophysical behavior of random copolymers of vinylaromatics, block copolymers of styrene and head-to-head polystyrene under steady-state excitation conditions and of transient excitation studies of copolymers of 1-vinylnaphthalene. The ability of copolymeric moieties to provide information upon intramolecular interactions in excimer formation has been demonstrated by several workers.[3-9]

RESULTS AND DISCUSSION

Random Copolymers Incorporating 1-Vinylnaphthalene or Styrene Derivatives as Chromophore (Steady-State Excitation)

Intramolecular excimer formation has been studied in copolymers of various vinylaromatic monomers in combination with a variety of "spectroscopically inactive" comonomers. The intramolecular concentration dependence of excimer emission intensity relative to that of the unassociated excited-state chromophore in macromolecular systems confers a greater degree of similarity between the intermolecular phenomenon than that involving intramolecular interactions in low molar

0077-8923/81/0366-0024$01.75/0 © 1981, NYAS

mass species. Consequently, it has been argued that the kinetic scheme of Birks[10] may be modified to model intramolecular excimer formation in copolymers.[7-9]

Under photostationary-state conditions, Birks'[10] scheme reduces to the following expression for the intensity ratio of dimer, I_D, to monomer, I_M, at constant temperature.

$$\frac{I_D}{I_M} = k\, k_{DM}[M] \quad (1)$$

where k is a composite rate coefficient, k_{DM}, is the rate coefficient for excimer formation and $[M]$ is the concentration of fluorescent species. Modification of Birks'[10] mechanism to encompass intramolecular excimer formation in polymers requires generation of a function descriptive of the concentration of potential excimer-forming sites and allowance for the extent to which such sites may be populated by intramolecular energy migration. For copolymers of 1-vinylnapthalene, 2-vinylnaphthalene or styrene and methyl methacrylate[7] and of 1-vinylnaphthalene and methyl acrylate,[8] it has been shown that adequate description I_D/I_M is afforded by the expression

$$\frac{I_D}{I_M} = k'\ \bar{1}_a f_{aa} \quad (2)$$

where f_{aa}, the fraction of aromatic pairs within the copolymer, is supposed to characterize the concentration of potential excimer sites and $\bar{1}_a$ is a term descriptive of energy migration as estimated by the extent of depolarization conferred upon the emission as a result of the energy transfer.[7,8] The validity of such a model implies that excimer formation in these copolymers results from the interaction of nearest-neighbor chromophores as previously proposed by David *et al.*[3] for the styrene-methyl methacrylate copolymer system.

In contrast, it has been suggested that in polymers with a greater degree of restriction upon interactions between chromophores in close proximity in the polymer chain, such as poly(1-naphthyl methacrylate), excimer formation will be dominated by interactions between species distant on the macromolecule which are brought into juxtaposition through coiling of the polymer.[11,12] In this connection, the dependence of excimer emission upon the thermodynamic nature of the solvent has been cited as corroborative evidence of such "long-range" effects, particularly in view of the reported solvent independence of excimer emission in polymers such as poly(1-vinylnaphthalene).[13] However, as Aspler and Guillet[12] have indicated, the reported solvent effect upon polystyrene[14] implies that adoption of the criterion of the existence of solvent effects as an indicator of the interactions involved in excimer formation is more complex than originally supposed. Since the compactness of the polymer coil will affect the efficiency of energy migration and, to a lesser extent, the concentration of aromatic species in conformations suitable for excimer formation, solvent effects are to be expected even in polymers in which excimer formation results from nearest-neighbor interactions.

FIGURE 1 shows the dependence of I_D/I_M upon the thermodynamic nature of the solvent expressed in terms of the solubility parameter δ for poly(1-vinylnaphthalene) and a copolymer of 1-vinylnapthalene and methyl methacrylate containing 47 mole percent vinylnaphthalene.[15] It is evident that the extent to which excimer emission is observed is a function of the nature of the solvent in polymers containing vinylnaphthalene derivatives, although the effect is less dramatic than that observed, for

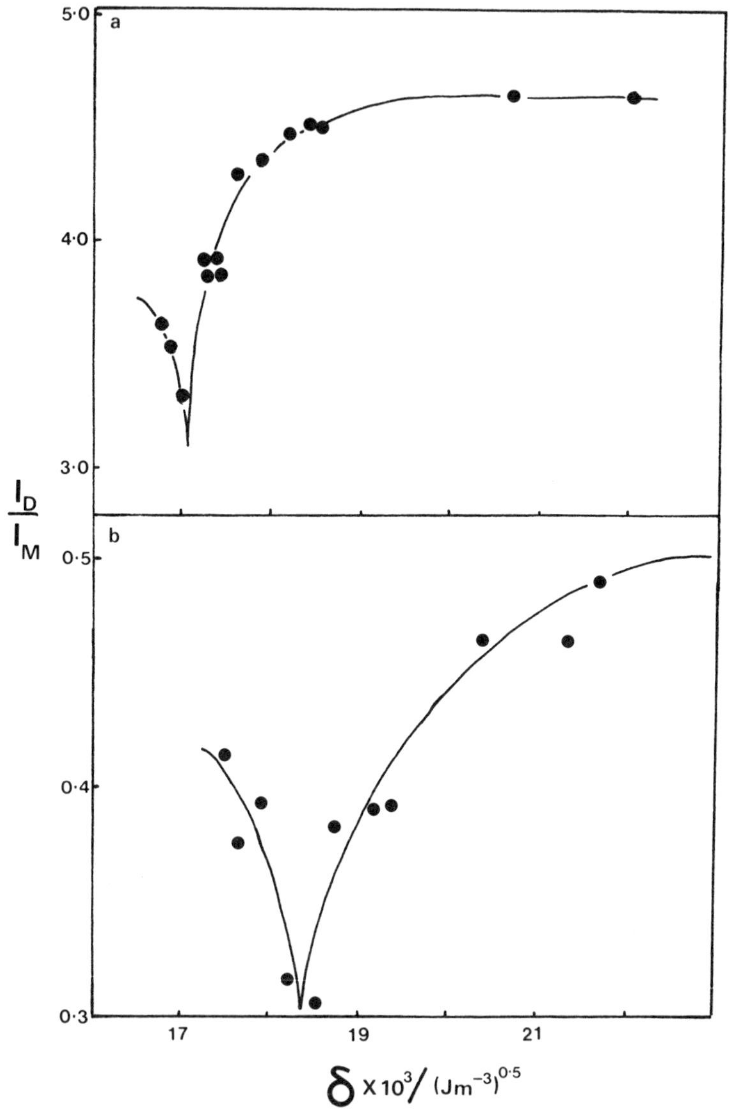

FIGURE 1. I_D/I_M as a function of solubility parameter δ for the mixed solvent systems toluene/methanol and toluene/cyclohexane. (a) poly(l-vinylnaphthalene) (b) 1-vinylnaphthalene/methyl methacrylate copolymer (0.47 mole fraction in vinylnaphthalene).

example, in poly(naphthyl methacrylate)s. A minimum in I_D/I_M is observed as the solubility parameter of the solvent matches that of the polymer (or that of the segment governing excimer formation).[15] Variations in I_D/I_M with solubility parameter have been used to investigate polymer compatibility in solid solutions.[16]

If contributions from long-range interactions are significant in the determination of the extent of excimer formation in solvents of poor thermodynamic compatibility with the polymer, it is to be expected that the function $\bar{I}_a \cdot f_{aa}$ will be incapable of description of I_D/I_M. The ability of Equation 2 to describe adequately the photophysical behavior of the 1-vinylnaphthalene/methyl methacrylate system regardless of solvent is demonstrated in FIGURE 2. Furthermore, the model has been validated in description of I_D/I_M over a substantial range of temperatures[15] and for the 1-vinylnaphthalene/methyl acrylate copolymer series[8] in which the enhanced flexibility of the polymer chains might have been expected to produce enhancement of long-range effects if such are important in these systems.

Styrene Block Copolymers and Head-to-Head Polystyrene

The study of intramolecular excimer formation in low molar mass systems[17] has shown that the $n = 3$ generalization of Hirayama[18] is valid for weakly interacting chromophores such as phenyl residues. Consequently, it is to be expected that, if

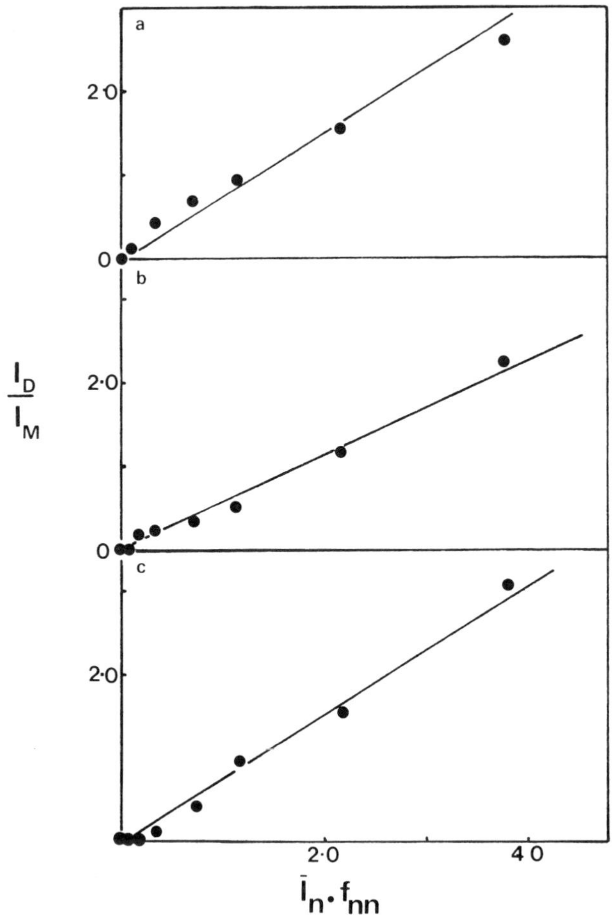

FIGURE 2. I_D/I_M as a function of $\bar{l}_n \cdot f_{nn}$ for the copolymer system l-vinylnapthalene/methyl methacrylate with solvent (a) toluene:methanol 4:1 by wt. (b) toluene (c) dichloromethane at 298° *K*.

long-range interactions are unimportant in polystyrenes, head-to-head polystyrene will not show excimer emission. Emission spectra of head-to-head polystyrene in mixed solvent systems contain no contribution to the spectral profile from excimeric species up to the point of polymer precipitation.[19] That chain coiling produces enhanced energy migration in the polymer is evident from the development in poor solvents of emission from low concentrations of 2,3-diphenyl-2-butene energy traps present in the polymer from the synthesis.[20] Similar effects are evident from studies of the thermal dependence of fluorescence of the head-to-head polystyrene in fluid solutions over a range of temperatures.[19]

In an attempt to gain further information upon the influence of non-nearest-neighbor interactions in polystyrenes we have prepared, by "living polymer" synthetic techniques, a series of polystyrenes; styrene-butadiene (SB) and styrene-butadiene-styrene-butadiene (SBSB) block copolymers of varying styrene (and butadiene) block lengths.

FIGURE 3 shows I_D/I_M as a function of the number of bonds between styrene residues. A smooth curve is obtained for all polymers provided the styrene sequences in the SBSB block copolymers are regarded as isolated blocks. The observed molar mass dependence of the excimer emission is in good agreement with that previously reported for polystyrenes of varying degree of polymerization.[21]

The independence of I_D/I_M upon the existence of a second styrene sequence in the SBSB block copolymers substantiates the conclusion, from studies on random copolymers described in the first section, that non-nearest neighbor interactions are

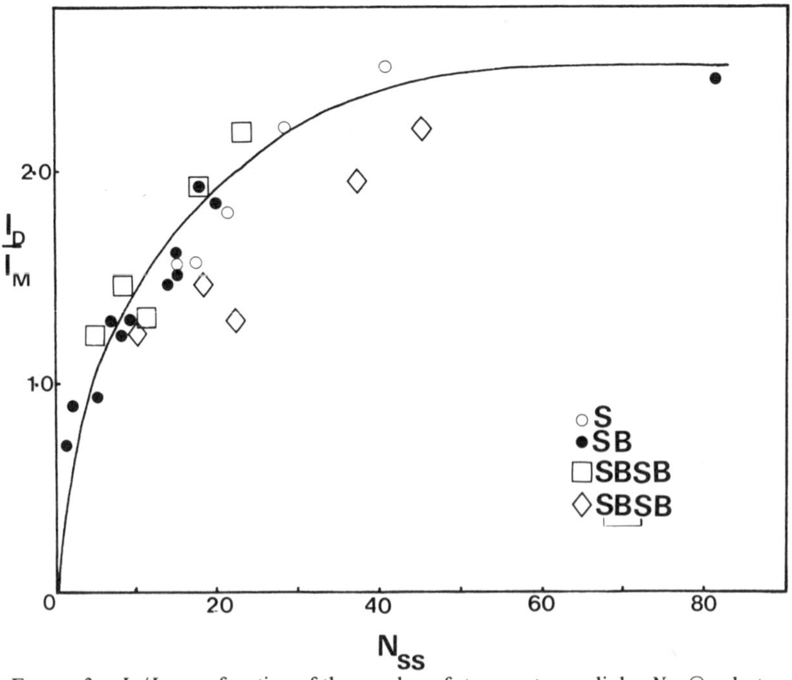

FIGURE 3. I_D/I_M as a function of the number of styrene-styrene links, N_{ss}. ○ polystyrenes. ● styrene-butadiene (SB) block copolymers. □ styrene-butadiene (SBSB) block copolymers treating styrene sequences as independent blocks. ◇ styrene-butadiene (SBSB) block copolymers counting all styrene species.

of minor importance in excimer formation in these vinylaromatic polymers. Corroborative evidence for such a conclusion is afforded by the absence of excimer emission in the spectra of alternating copolymers containing monomers which are respectively spectroscopically active and inactive in the frequency range of interest.[4,5]

Time-resolved Studies of Polymer Photophysics

The application of non-stationary-state excitation techniques to the study of copolymer systems is of considerable potential in the investigation of polymer photophysics. In particular, it is possible in principle to derive the individual rate coefficients for the decay processes in the photophysical reaction scheme for excimer formation through employment of extrapolation procedures analogous to those used in low molar mass systems.[10] Implementation of such a method is dependent upon, among other things, two criteria:

1. Functions descriptive of the intramolecular concentration excimer dependence in macromolecular systems must be determined to allow extrapolation to the appropriate concentration limits.
2. Either the basic kinetic scheme of Birks[10] must be shown to be applicable to polymer photophysical phenomena or must be suitably modified to accommodate deviations in kinetic behavior such that appropriate extrapolation plots may be constructed.

Condition 1 is satisfied if it is assumed that the concentration dependence of the excimer formation mechanism is described adequately by functions applicable under conditions of steady-state behavior. However, the second criterion appears to be subject to a number of restrictions. As part of a collaborative study with Professor Phillips' group at the University of Southampton, it has been shown that the basic photophysical reaction scheme[10] is inapplicable in a variety of macromolecular systems including polymers containing acenaphthylene,[22] 1-vinylnaphthalene,[23] naphthyl methacrylates[24] and 1-vinylpyrene.[24] In the systems studied to date, the decay curves in the regions of monomer and excimer emission, which are described according to the conventional decay scheme as the sum and difference respectively of two exponential functions, can be described adequately only by invoking triple exponential functions to fit the experimental data.[23,24]

According to the conventional decay scheme, individual rate parameters may be derived from the coefficientsλ_1 and λ_2, characteristic of the dual exponential decay, according to the procedures outlined in TABLE 1.

The existence of several cross correlations allows a check of the consistency of the method employed and, in the case of polymer systems, of the validity of the function employed to characterize the intramolecular excimer formation concentration dependence.

In the 1-vinylnaphthalene/methyl methacrylate copolymer system, the transient decay, analyzed at 330 nm (in the region of monomer emission) may be described according to Equation 3[23]

$$i_m(t) = A_1 \exp(-\lambda_1 t) + A_2 \exp(-\lambda_2 t) + A_3 \exp(-\lambda_3 t) \quad (3)$$

We have attached the significance of interacting monomer-excimer parameters to λ_1 and λ_2 through consideration of their dependence upon chromophore concentration.[23] λ_3 is assigned to the presence of a third species. Since time-resolved spectroscopy does not reveal the presence of more than two emitting species, generally

TABLE 1
PROCEDURES FOR DERIVATION OF RATE CONSTANTS

Method	Procedure	Function Derived	Derived Parameter
(a)	Measurement of unquenched monomer lifetime	k_M	k_M
(b)	Extrapolation of λ_1 to $[M] = 0$	k_M	k_M
(c)	$\frac{(\lambda_1 + \lambda_2)}{\partial[M]}$	k_{DM}	k_{DM}
(d)	$\frac{\partial\lambda_2}{\partial[M]} \rightarrow k_{DM}$ as $[M] \rightarrow 00$	k_{DM}	k_{DM}
(e)	$(\lambda_1 + \lambda_2) \rightarrow k_M + k_{MD} + k_D$ as $[M] \rightarrow 0$	$k_M + k_{MD} + k_D$	$k_{MD} + k_D$ {by combination with (a) or (b)}
(f)	$(\lambda_1\lambda_2) \rightarrow k_M (k_{MD} + k_d)$ as $[M] \rightarrow 0$	$k_M (k_{MD} + k_D)$	$k_{MD} + k_D$ {by combination with (a) or (b)}
(g)	$\frac{\partial(\lambda_1\lambda_2)}{\partial[M]} = k_{DM}k_D$	$k_{DM}k_D$	k_D {by combination with (c) or (d)}
(h)	$\lambda_1 \rightarrow k_D$ as $[M] \rightarrow 00$	k_D	k_D
(i)	(e) or (f) plus (g) or (h)	k_{MD}	k_{MD}

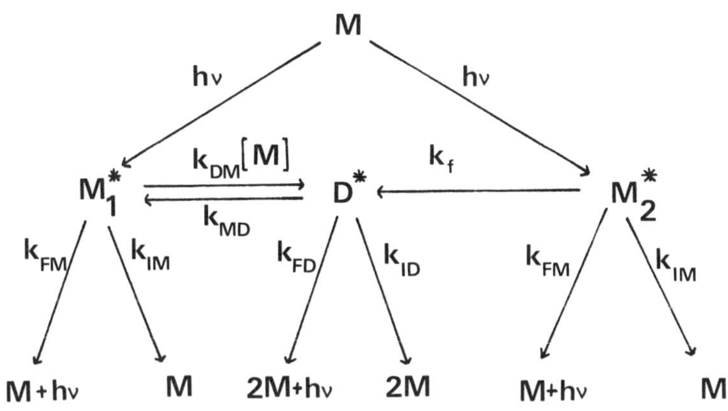

FIGURE 4. Photophysical reaction scheme.

assumed to be monomer and excimer in previous publications (e.g., references 7,8,25,26), and since the relative contribution of the species to the total fluorescence increases as the aromatic content of the polymer decreases, it is proposed that the third species is an unassociated chromophore which can not populate excimer by the "normal" mechanism of excitation transfer.[23]

One plausible explanation of the observed decay behavior is that the second monomer species, M_2*, is an isolated naphthyl species which can form excimers through long-range interactions through coiling of the polymer chain. The relatively minor contribution of such interactions to the overall fluorescence emission would account for their apparent unimportance in description of steady-state fluorescence measurements. The resultant kinetic scheme is shown in FIGURE 4.[23] According to

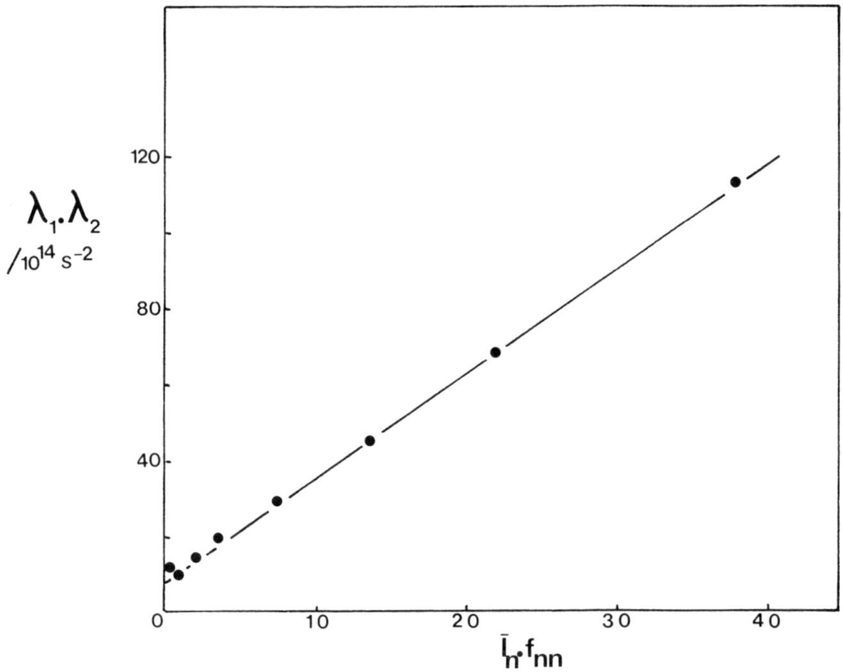

FIGURE 5. $\lambda_1 \cdot \lambda_2$ as a function of $\bar{l}_n \cdot f_{nn}$ for 1-vinylnapthalene/methyl methacrylate copolymers (tetrahydrofuran solvent, $298^0 K$).

TABLE 2
KINETIC DATA

Method	Kinetic Parameter	Value/10^7 s^{-1}
(b)	k_M	1.3
(c)	αk_{DM}	12.1
(d)	αk_{DM}	12.0
(e)	$k_{MD} + k_D$	6.5
(f)	$k_{MD} + k_D$	6.2
(g)	k_D	2.2
(h)	k_D	2.2
(i)	k_{MD}	4.2

this kinetic scheme, excimer and monomer decays will be described by triple exponential decays. Furthermore, it will be possible to describe the interactions between M_1* and D* by a modified Birks'[10] scheme. FIGURE 5 shows a typical fit to the empirical data in the form of a plot of $\lambda_1 \cdot \lambda_2$ as function of $f_{nn} \cdot \bar{l}_n$, the function descriptive of excimer site concentration in steady-state experiments.[7]

The kinetic parameters derived by the procedures outlined in TABLE 1 are listed in TABLE 2.

The assumptions involved in the application of the kinetic scheme and in the choice of concentration term are supported by the good agreement between kinetic coefficients derived by alternative extrapolations.

Since the function $\bar{l}_n \cdot f_{nn}$ is assumed to be proportional to the intramolecular concentration of excimer sites, the parameter k_{DM} depicting the rate coefficient of excimer formation is not capable of direct derivation from the transient decay data. However, the parameter αk_{DM} which is derived is characteristic of the process of excimer formation since α is merely a proportionality constant. In other words, the dependence upon intramolecular concentration has been removed to an extent which is not possible by the "single point" determinations which have been published to date for a variety of polymer systems.[12,26–28]

Experimental

Materials

Random copolymers incorporating 1-vinylnaphthalene or styrene were prepared by bulk and solution polymerization.[7,8] Reactivity ratios were determined according to the method of Kelen and Tüdos.[29]

Block copolymers of styrene and butadiene and polystyrenes of varying degrees of polymerization were prepared by living polymer techniques using butyl lithium as initiator in *THF* at $298^0 K$.

Head-to-head polystyrene was prepared as described in the literature.[20]

Techniques

Steady-state fluorescence measurements were performed on a Perkin Elmer *MPF* 3*L* spectrofluorimeter.

Fluorescence decay curves and time-resolved emission spectra were recorded on a high resolution fluorimeter employing an externally frequency-doubled, cavity-dumped dye laser (Coherent model 590) pumped using a 4 *W* argon-ion laser (Spectra Physics model 166) using a conventional single photon counting detection system. The operation of the fluorimeter has been described.[30,31]

Conclusions

Steady-state fluorescence measurements on a series of random copolymers of vinylnaphthalene and methyl methacrylate or methyl acrylate suggest that long-range interactions are unimportant in determination of the concentration of excimer species in such systems. Studies of the photophysical behavior of styrene-butadiene block copolymers of the type SBSB support this conclusion when related to photophysical data obtained using polystyrenes and SB block copolymers of varying styrene sequence lengths. The lack of excimer emission evinced by head-to-head polystyrene over a range of solvent and temperature conditions is consistent with the supposition that, for weakly interacting chromophores, long-range intramolecular interactions exhibit a limited degree of influence upon the process of excimer formation as studied under stationary-state conditions.

The use of pulsed laser sources for excitation has enabled accurate observation of the fluorescence decays of 1-vinyl-naphthalene/methyl methacrylate copolymers over a substantial range of intramolecular chromophore concentration. The conventional decay scheme applicable to intermolecular excimer formation in low molar mass systems has been shown to be inappropriate in description of the photophysical behavior of vinylnaphthalene copolymers. It is possible that such deviations from "expected" decay characteristics are due to the influence of long-range interactions in polymers. Further information on the plausibility of such a mechanism relative to alternative explanations will be gained from future studies on the influence of polymer chain flexibility, solvent and temperature upon the ease of excimer formation in a variety of copolymer systems.

SUMMARY

Investigations are described which relate to the influence upon the process of intramolecular excimer formation in vinylaromatic polymers of interactions between chromophores that are not located in close proximity along a polymer chain. The photophysical behavior of random copolymers of 1-vinylnaphthalene and methyl methacrylate, block copolymers of styrene and butadiene and head-to-head polystyrene would indicate that long-range interactions are not significant in the determination of the steady-state fluorescence characteristics of such systems.

Conversely, transient decay studies of 1-vinylnapthalene/methyl methacrylate copolymers suggest that long-range interactions may be invoked as one plausible explanation of the observed dependence of the decay profiles upon the intramolecular concentration of aromatic species.

The application of non-stationary-state techniques to the study of a series of copolymers in which the intramolecular concentration of chromophore is varied over a substantial range allows the determination of individual rate parameters in the photophysical scheme descriptive of excimer formation. While it is not possible to obtain an absolute value for the rate coefficient, k_{DM}, for excimer formation using the method, it is possible to remove the dependence upon local chromophore concentration to a degree not attainable in the "single point" determinations employed to date in polymer systems.

ACKNOWLEDGMENTS

The results presented in this paper are consequent upon a number of collaborative investigations. The author gratefully acknowledges the involvement of the following colleagues:

- Dr. R.F. Reid, ICI Corporate Lab., The Heath, Runcorn, Cheshire, U.K.
- Dr W.E. Lindsell and Mr. F.C. Robertson, Chemistry Department, Heriot-Watt University, Riccarton, Currie, Edinburgh, EH14 4AS. U.K.
- Professor D. Phillips* and Mr. A.J. Roberts,* The University, Southampton, S09 5NH. U.K.

The author wishes to thank the Science Research Council for financial support on the projects described in this paper, and the Royal Society and Carnegie Trust for support in respect to this presentation

REFERENCES

1. ANDERSON, R. A., I. MCINALLY & I. SOUTAR. To be published.
2. LAMI, H. & G. LAUSTRIAT, 1968. J. Chem. Phys. **48:** 1832.

* Present address: The Royal Institution, 21, Albemarle Street, London, W1X 4BS, UK.

3. David, C., M. Lempereur & G. Geuskens. 1973. Eur. Polym. J. **9:** 1315.
4. Fox, R. B., T. R. Price, R. F. Cozzens & W. H. Echols. 1974. Macromolecules **7:** 937.
5. Yokoyama, M., T. Tamamura, M. Atsumi, M. Yoshimura, Y. Shirota & H. Mikawa. 1975. Macromolecules **8:** 101.
6. Wang, Y.-C. & H. Morawetz. 1975. Makromol. Chem. Suppl. **1:** 283.
7. Reid, R. F. & I. Soutar. 1978. J. Polym. Sci. Polym. Phys. Ed. **16:** 231.
8. Anderson, R. A., R. F. Reid & I. Soutar. 1979. Eur. Polym. J. **15:** 925.
9. Reid, R. F. & I. Soutar. 1980. J. Polym. Sci. Polym. Phys. Ed. **18:** 457.
10. Birks, J. B. 1970. *In* Photophysics of Aromatic Molecules. John Wiley & Sons. New York.
11. Somersall, A. C. & J. E. Guillet. 1973. Macromolecules **6:** 218.
12. Aspler, J. S. & J. E. Guillet. 1979. Macromolecules **12:** 1082.
13. Fox, R. B., T. R. Price, R. F. Cozzens & J. R. McDonald. 1972. J. Chem. Phys. **57:** 534.
14. Nishihara, R. & M. Kaneko. 1969. Makromol. Chem. **124:** 84.
15. Reid, R. F. 1978. Studies in Macromolecular Photophysics. Ph.D. Thesis. Heriot-Watt University.
16. Frank, C. W. & M. A. Gashgari. 1979. Macromolecules **12:** 163.
17. Klöpffer, W. 1973. *In* Organic Molecular Photophysics. J. B. Birks, Ed. Vol. **1:** 357. John Wiley & Sons, New York.
18. Hirayama, F. 1965. J. Chem. Phys. **42:** 3163.
19. Lindsell, W. E., F. C. Robertson & I. Soutar. Eur. Polym. J. In press.
20. Inoue, H., M. Helbig & O. Vogl. 1977. Macromolecules **10:** 1331.
21. Ishii, T., T. Handa & S. Matsunaga. 1978. Macromolecules **11:** 40.
22. Phillips, D., A. J. Roberts & I. Soutar. 1980. J. Polym. Sci. Polym. Lett. Ed. **18:** 123.
23. Phillips, D., A. J. Roberts & I. Soutar. 1980. J. Polym. Sci. Polym. Phys. Ed. In press.
24. Phillips, D., A. J. Roberts & I. Soutar. To be published.
25. Ghiggino, K. P., R. D. Wright & D. Phillips. 1978. J. Polym. Sci. Polym. Phys. Ed. **16:** 1499.
26. Ito, S., M. Yamamoto & Y. Nishijima. 1976. Rep. Prog. Polym. Phys. Japan **19:** 421.
27. Ishii, T., T. Handa & S. Matsunaga. 1977. Makromol. Chem. **178:** 2351.
28. Ishii, T., T. Handa & S. Matsunaga. 1978. Macromolecules **11:** 40.
29. Kelen, T. & F. Tüdos. 1975. J. Macromol. Sci. **A9:** 1.
30. Ghiggino, K. P., A. J. Roberts & D. Phillips. 1980. J. Phys. E. Sci. Instrum. **13:** 446.
31. Roberts, A. J., D. V. O'Connor & D. Phillips. 1981. Ann. N. Y. Acad. Sci. **366:**.

FLUORESCENCE AND MICROBROWNIAN MOTION IN FLEXIBLE MACROMOLECULES

Carla Cuniberti and Angelo Perico

Centro Studi Chimico Fisici di Macromolecole Sintetiche e Naturali
Istituto di Chimica Industriale dell'Università di Genova
Genova, Italia

Since the pioneering work by Vala *et al.*[1] on polystyrene in 1965, intramolecular excimer formation in macromolecules has received increasing attention. The reaction has been almost exclusively studied in polymers and copolymers containing aromatic units in which the dominant contribution to the excimer formation arises from nearest-neighbor or from next-to-nearest-neighbor interactions.[2,3] These studies provided, in many cases, information on the rate of hindered rotation around the backbone bonds, thus furnishing a technique that can be used as an alternative to the variety of methods more traditionally utilized for investigating high-frequency motion in polymers.

By placing long-lived chromophores on far-distant units of a flexible macromolecule, the intramolecular excimer formation that takes place under diffusion-controlled conditions as a consequence of the conformational changes of the chain can be studied to gain information about macromolecular dynamical processes on the long-range scale. Pyrenyl groups placed at the two ends of poly(ethylene oxide) polymers allowed us to study the end-to-end cyclization dynamics at different chain lengths.[4] The reaction rate in this case falls rapidly to zero with increasing molecular weight because the first-order cyclization constant, k_t, behaves approximately as k_t = constant $\times M^{-1.5}$, as expected on the basis of theoretical considerations.[5] Measurable reaction effects were observed for molecular weights up to around 2×10^4, i.e., in this system, excimer formation can compete with the spontaneous fluorescence decay when the pyrenyl groups are at separation distances up to about 1300 bonds. If several pyrenyl labels are introduced on a long chain at a mean separation within this value, an enhanced reaction rate can be expected with respect to the corresponding end-to-end case, and information on the internal polymer motions occurring on the frequency scale typical of the chain separation between the interacting chromophores can be obtained. Thus, by changing the mean interchromophoric distance, a large spectrum of relaxation frequencies could be studied.

At low excitation intensity, the reaction system can be represented by one excited label surrounded by unexcited ones. In this case, simple considerations can depict the main features that may be expected from the first order rate constant k of intramolecular excimer formation under diffusion control. Assuming concurrent site-independent reactions, k is given by

$$k = \sum_i k_i p_i \qquad (1)$$

where p_i is the probability of finding one unexcited group at the position i along the

0077-8923/81/0366-0035$01.75/0 © 1981, NYAS

chain. Each k_i describes the reaction between two groups placed at the mean square distance $\langle h_i^2 \rangle$ and, assuming equal mutual diffusion constants for internal and end groups, it might be roughly approximated by the cyclization rate constant, k_t, of a chain with the same end-to-end distance. A theoretical evaluation of k_t was given by us in a recent study based on the theory of diffusion-controlled reactions developed by Wilemski and Fixman for bead-spring chains in the non-draining and free-draining limits.[5] Exactly in the asymptotic limit ($M \rightarrow \infty$) and with a good approximation for chains of finite length, it is k_t = constant $\times$ $(6D/\langle h^2 \rangle)$, where D is the polymer translational diffusion constant. Under non-draining conditions, $k_t = C(KT/\eta_0)\,\langle h^2 \rangle^{-3/2}$ and k_i can then be given by:

$$k_i = C\frac{KT}{\eta_0}(\langle h_i^2 \rangle)^{-3/2} \tag{2}$$

For an infinitely long polymer, characterized by an effective monomer length l_{eff} and a given number fraction of substituted monomeric units, w, the rate constant k can be written

$$k = C\frac{KT}{\eta_0}2\sum_{i=1}^{\infty} p_i(\langle h_i^2 \rangle)^{-3/2} \tag{3}$$

In order to obtain k from Equation 3, the probability p_i has to be specified. In the case of evenly spaced labels, p_i is unity only at distances that are multiples of $b = l_{eff}/w^{1/2}$ and zero elsewhere, and k becomes

$$k = C\frac{KT}{\eta_0}\frac{w^{3/2}}{l_{eff}^3}2\sum_{i=1}^{\infty} i^{-3\nu} \tag{4}$$

In the case of randomly distributed labels, p_i is equal to w, and k is

$$k = C\frac{KT}{\eta_0}\frac{w}{l_{eff}^3}2\sum_{i=1}^{\infty} i^{-3\nu} \tag{5}$$

Here the exponent ν takes into account excluded volume effects on $\langle h_i^2 \rangle^{1/2}$ and is 0.5 under theta conditions and around 0.6 in good solvents.[6] The sums in these two cases are the Riemann Zeta functions respectively ζ (1.5) and ζ (1.8). The environment affects the reaction by means of the ratio T/η_0, typical of a completely diffusive process, and by means of the excluded volume parameter ν. Thus, not only an increase of the viscosity of the solvent but also an increase of its thermodynamic power slows down the reaction rate. In fact, under the limiting conditions of Equations 4 and 5, a reduction factor $\zeta(1.8)/\zeta(1.5) = 0.72$ is expected in very good solvents with respect to theta solvents. It may be noted that Equations 4 and 5 give for k under theta conditions, 5.224 ($= 2\,\zeta(1.5)$) or $5.224/w^{1/2}$ ($= 2\,\zeta(1.5)/w^{1/2}$) times the end-to-end rate constant $k_t = C(KT/\eta_0)/b^3$, relative to a chain with a mean square end-to-end distance $b^2 = l_{eff}^2/w$. Thus, by enhancing noticeably the reaction rate, the method of intramolecular diffusion controlled excimer formation between internal labels would enable us to study the dynamical behavior of chain segments longer than in the case of a pair of terminal groups.

On the basis of the simplified model just discussed, the expected behavior of k with respect to its dependence on molecular weight and polymer structural parameters and of the viscosity and thermodynamic power of the solvent can be inferred. In fact, at finite molecular weight, the infinite upper limit of the sums of Equations 4 and 5 is replaced by a number m, proportional to the polymer molecular weight. The rate constant k then increases asymptotically with increasing molecular weight; for instance, under theta conditions, the behavior is that of the function $\zeta_m = \sum_{i=1}^{m} i^{-3/2}$ which increases asymptotically with m from 1 to 2.6. At sufficiently high molecular weight, the diffusive rate constant depends on the degree of substitution, w, on the distribution of the labels, on the chain parameter l_{eff} and on environmental properties but it is molecular weight independent.

Here we present some experimental results of a study of intramolecular excimer formation in poly(vinyl acetate) containing 1% of pyrenyl substituents (PVAc-PBA), whose chemical structure is

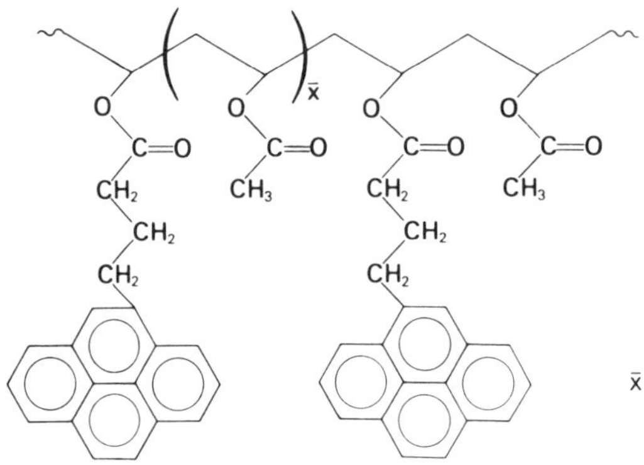

$\bar{x} = 100$

where the reaction process can be represented by the usual scheme:

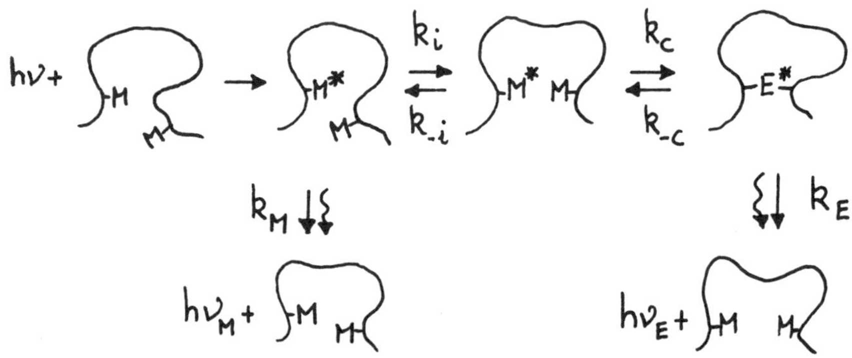

Under diffusion control and fast excimer decay ($k_{-i} << k_c$ and $k_e > k_{-c}$), the diffusive rate constant k is readily obtainable from photostationary fluorescence emission, either by measuring the monomer quantum yield quenching ratio

$$\frac{\Phi_M^0}{\Phi_M} = 1 + \tau_M^0 k \tag{6}$$

or by measuring the excimer to monomer quantum yields ratio

$$\frac{\Phi_E}{\Phi_M} = \frac{\Phi_E^0}{\Phi_M^0} \tau_M^0 k \tag{7}$$

where $\tau_M^0 = k_M^{-1}$ is the monomer intrinsic lifetime and Φ_M^0, Φ_E^0, respectively, the monomer and excimer quantum efficiencies.

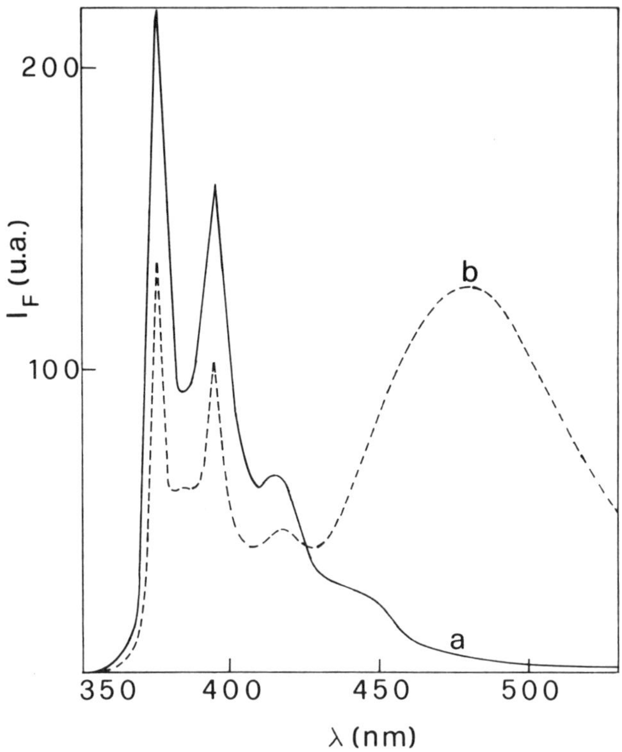

FIGURE 1. Emission spectra of PVAc-PBA (curve b) and of the model PBA-ester (curve a) in methanol at 20°C ([PBA] ≃ 10^{-6} mol l^{-1}). λ_{EX} = 345 nm.

The fluorescence emission spectrum of dilute PVAc-PBA solutions in degassed methanol (MeOH), shown in FIGURE 1, exhibits the structureless emission band centered at 480 nm typical of pyrene excimer fluorescence. The same figure shows also the spectrum of a low molecular weight ester of 4-(1-pyrenyl) butyric acid[4] in MeOH at low concentration (<10^{-5} *M*), where only monomer fluorescence appears.

The ratio F_E/F_M of the excimer to the monomer emission peaks measured respectively at 480 and 396 nm and corrected for the monomer emission at 480 nm, was found to represent correctly Φ_E/Φ_M. The dependence of excimer formation on the polymer molecular weight, on the solvent viscosity and on the solvent thermodynamic power was investigated by simply measuring the ratio F_E/F_M.

Polymer fractions with constant w (= 0.01) and molecular weight (M) ranging from 4.5 × 10^3 to 6.3 × 10^5 were obtained from the initial sample by gel permeation chromatography in tetrahydrofuran. The excimer to monomer intensity ratio was found to increase asymptotically with M, as expected, already reaching a constant value at $M \simeq 10^5$.[7] A fraction of PVAc-PBA with M = 1.04 × 10^5 was then employed for studying solvent effects.

The viscosity of the solvent was varied by mixing methanol (MeOH) with ethylene glycol (EG) or ethyl acetate (EA) with glycerol triacetate (GTA). The ranges of viscosity covered at 20°C were respectively 0.59–19.9 cP and 0.45–15.5 cP. The

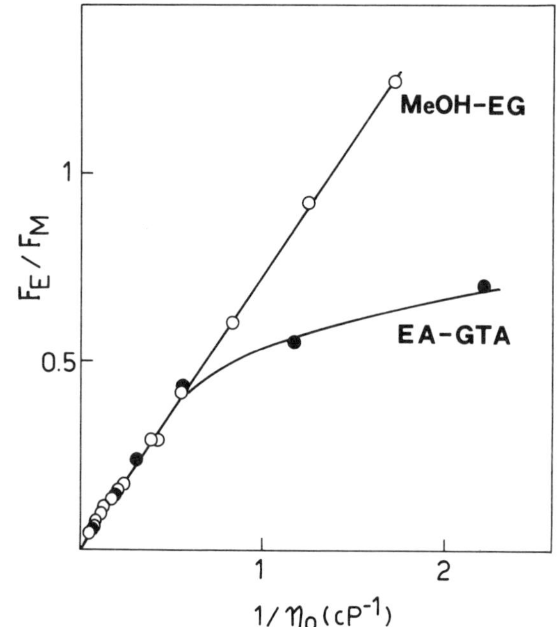

FIGURE 2. Viscosity dependence of the excimer to monomer intensity ratio of PVAc-PBA in MeOH-EG (○) and EA-GTA (●).

similar chemical nature of the components in each mixture should guarantee the absence of solvent effects on the quanity $(F_E^0/F_M^0) \times \tau_M^0$. The emission spectra of PVAc-PBA in EG and in GTA show excimer fluorescence only slightly higher than for the model compound. A quantitative estimation of the solvent viscosity effect can be observed in FIGURE 2, where the ratio F_E/F_M is plotted against the reciprocal viscosity $1/\eta_0$.

While in the system MeOH-EG, k is proportional to $1/\eta_0$ in the whole range of viscosity explored, a large deviation from linearity appears in EA-GTA for $\eta_0 < 1.5$ cP. For $\eta_0 > 1.5$ cP, all the data fall on the same line, thus showing that the differences between the results in the two kinds of mixtures in the low viscosity range cannot be due to substantial differences of the quantity $(F_E^0/F_M^0)\tau_M^0$. On the contrary, excluded volume effects on k could be the origin of the deviation, since at 20°C methanol is almost a theta solvent while ethyl acetate is a good solvent for PVAc.[8] The effect of the thermodynamic power of the solvent has been explored in more detail carrying out measurements in various pure solvents and solvent mixtures. The intrinsic viscosity of the polymer, $[\eta]$, has been taken to represent the coil expansion. FIGURE 3 shows that the product $(F_E/F_M)\,\eta_0\,[\eta]$ is approximately constant in the whole range of excluded volume explored.

Specific solvent effects on the spectroscopic behavior of the pyrenyl moieties cannot be completely ruled out, but we have seen that in many of the systems studied they gave negligible effects, certainly within the scatter of the data in FIGURE 3. On this basis, we can infer that, at constant solvent viscosity, k decreases with increasing excluded volume as $[\eta]^{-1}$, or as the cube of the reciprocal of the coil expansion

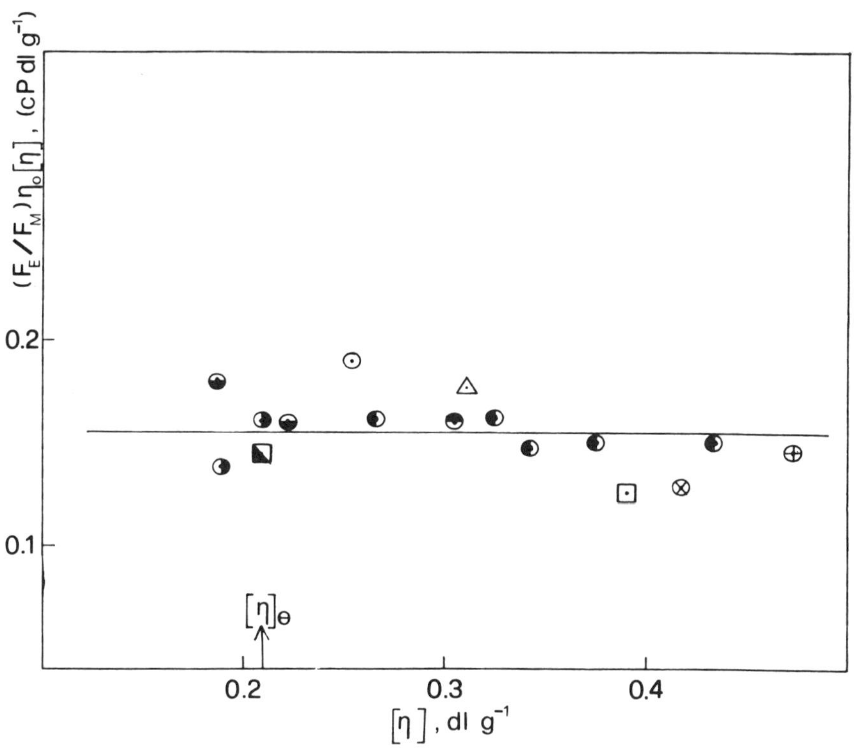

FIGURE 3. Plot of $F_E/F_M)\eta_0[\eta]$ vs. $[\eta]$ in MeOH (⊙), MeOH-EtOH (◑), MeOH-EG (◒), MeOH-THF (◐), MeOH-DIOX (◓), THF (⊗), DIOX (△), EA (⊡), EA-GTA (◪), DMF (⊕).

coefficient $\alpha\eta$, in qualitative agreement with the behavior of k expected on the basis of Equations 4 and 5. It may be noted that the maximum reduction factor obtained for k is around 0.3–0.4, a value much smaller than the predicted 0.72. A more sophisticated model for k must be adopted for the complete interpretation of the experimental data.

A consequence of the noticeable excluded volume effect is that care must be taken when interpreting temperature effects on the reaction in terms of "activation energy", as is customary. To illustrate this point, ln F_E/F_M is plotted against reciprocal temperature in FIGURE 4.

Two bad solvents, MeOH and EG, and a very good solvent, dimethylformamide (DMF), were employed. In DMF, where an increase of temperature does not produce further expansion of the coil, the slope of the low temperature side of the bell-shaped curve characteristic of the formation of the excimers gives an "activation energy" coincident with the activation energy of the viscous flow of the solvent. By contrast, in MeOH, which is a theta solvent for PVAc at 6°C, the increase of F_E/F_M with temperature is much lower than predicted by the simple reduction of the viscosity of the medium, thus showing that excluded volume plays an important role in this case. In the highly viscous solvent EG, where excimer formation is diffusion controlled in the whole range of temperature examined (up to 120°C) and polymer precipitation

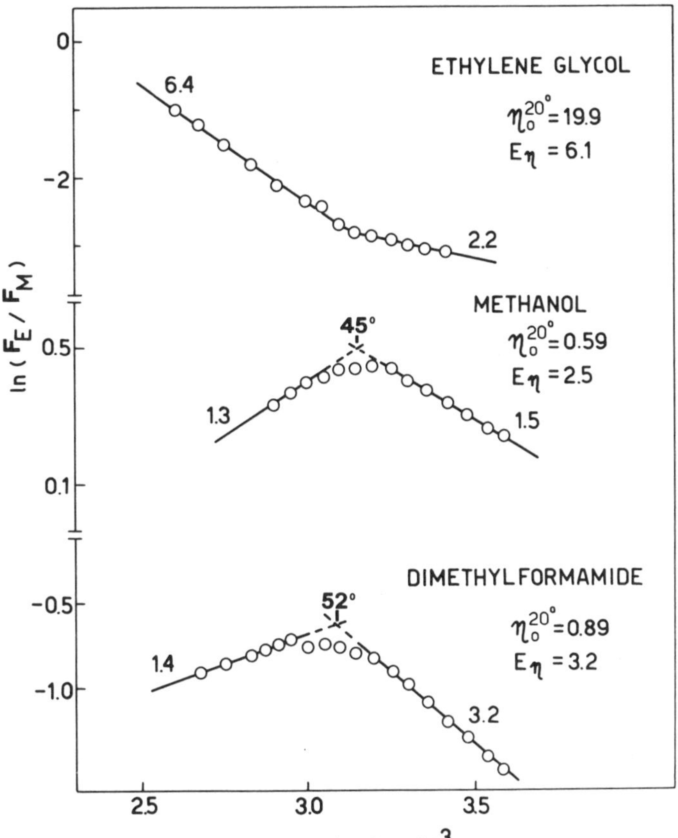

FIGURE 4. Temperature dependence of the excimer to monomer intensity ratio of PVAc-PBA in two bad solvents (MeOH and EG) and in a very good solvent (DMF).

has been observed below 17°C, the opposing effects of polymer expansion and solvent viscosity changes with temperature produce a non-linear dependence of $\ln(F_E/F_M)$ on $1/T$. The slope of the lower temperature side gives an "activation energy" of about 2 Kcal/mol, much lower than the 6 Kcal/mol that characterizes the change of solvent viscosity with temperature, thus showing that here the reaction behavior is dominated by the change in the thermodynamic power of the solvent. At the highest temperatures, where presumably further increase of polymer dimensions are of minor relevance, the slope reaches the value typical of the predominant role of the viscosity of the solvent.

The high-temperature side of the bell-shaped curve represents the region where the dissociation constant of the monomer-excimer equilibrium is rate-limiting and its slope gives the binding energy of the excimer. The values for the binding energy obtained in MeOH and DMF seem too small for typical pyrene excimers[9] and would indicate distorted excimer configuration. This conclusion is in contrast with the indication withdrawn from the temperatures of the peak, usually taken as a qualitative estimation of the excimer stability, that fall at somewhat higher values than for free pyrene molecules in solvents of comparable viscosity. Presumably, the high-temperature data are confined to a range where complete equilibrium is not yet reached.

The magnitude of the first-order rate constant k was evaluated in MeOH at 20°C, a low-viscosity, almost ideal solvent for PVAc-PBA, by measuring the monomer quenching ratio Φ_M^0/Φ_M. The monomer quantum efficiency Φ_M^0 was obtained by using as a model compound a low molecular weight ester of 4-(1-pyrenyl)butyric acid[4], and the value of the monomer lifetime τ_M^0—necessary to calculate k from Equation 6—was taken as 193 nsec.[10] This leads to $k = 1.37 \times 10^7$ sec^{-1}. This value can be compared with those calculated from Equations 4 and 5 with $\nu = 0.5$. The constant C was evaluated by us as 0.38 in the non-draining limit and 0.11 for partial-draining chains with the flexibility of polystyrene.[5] In the case of the regular distribution one obtains for k the values 4.5×10^7 and 1.3×10^7, respectively, in good agreement with the experimental measurement. In contrast, random distribution predicts values of k that are one order of magnitude larger.

An alternative indication on what may be the effect of the label distribution on the value of k can be obtained by comparing it with the cyclization constant, k_t, of a two-hundred-bond pyrenyl-terminated polystyrene, a chain that, on the basis of the respective values of the characteristic ratio,[11] can be considered to have a l_{eff} equal to that of PVAc. This k_t value has been interpolated from the molecular weight dependence of the data obtained by Winnik[10] in toluene, and is about 7×10^5 sec^{-1}. This leads to $k/k_t \simeq 20$. Since the viscosities of methanol and toluene are almost identical and the polidispersity effect gives only a minor change in k_t, a straightforward comparison with the values 5.2 and 52, respectively, predicted from the simple model presented here for a regular or a random distribution of the labels, would confirm a remarkable degree of regularity in the distribution of the pyrenyl groups in the PVAc-PBA sample.

A non-completely random distribution of the pyrenyl groups could be the result of the procedure used to obtain the marked polymer, consisting in three successive steps: partial saponification of PVAc up to 10% conversion, partial esterification with 1-(4-pyrenyl) butyric acid and final reacetylation of the product. The first step is well known to give rise to block sequences of hydroxyl groups[12], where the pyrenyl groups will be confined. However, this mechanism seems to be insufficient to explain the exceedingly high regularity inferred.

A reason for the small experimental k and k/k_t might originate from the failure of our assumption $k_i(h_i) = k_t(h_i)$ as a consequence of the more restricted mutual diffusion of the internal groups with respect to that of the chain ends. An evaluation of this effect could be obtained by introducing the factor τ_t/τ, given by the ratio of the rotational relaxation times of the chain ends and of the internal units, that for PVAc amounts to 0.24.[13]

In conclusion, though further experimental work on well-characterized samples, as well as a more sophisticated model for k which takes into account the non-asymptotic behavior of k_i and the non-additivity of the contribution from the different reaction sinks, are needed in order to gain more precise information, the results presented here are sufficient to indicate that the technique can be of particular interest for studying polymer motions in a range of frequency less accessible to other methods.

References

1. Vala, M. T., J. Haebig & S. A. Rice. 1965. J. Chem. Phys. **43:** 886–897.
2. Somersall, A. C. & J. E. Guillet. 1976. J. Macromol. Sci. Rev. Macromol. Chem. **14A:**

135–187.
3. MORAWETZ, H. 1977. *In* Contemporary Topics in Polymer Science. E. M. Pearce & Schaefgen, Eds. Plenum Press. New York. p. 171.
4. CUNIBERTI, C. & A. PERICO. 1977. Eur. Polym. J. **13:** 369–374.
5. PERICO, A. & C. CUNIBERTI. 1977. J. Polym. Sci. Polym. Phys. Ed. **15:** 1435–1450.
6. WEILL, G. & J. des CLOISEAUX. 1979. J. Phys. **40:** 99–105.
7. CUNIBERTI, C. & A. PERICO. 1980. Eur. Polym. J. **16:** 887–893.
8. BRANDRUP, J. & E. H. IMMERGUT, Eds. 1975. Polymer Handbook. Part **IV:** 15. Interscience. New York.
9. BIRKS, J. B., Ed. 1970. Photophysics of Aromatic Molecules. p. 357. Wiley-Interscience. London.
10. WINNIK, M. A. 1980. Macromolecules **13:** 328–335.
11. Ref. 8. Part IV: 39–41.
12. FUJII, K. 1971. J. Polym. Sci. Macromol. Rev. **5:** 431–540.
13. NORTH, A. M. & I. SOUTAR. 1972. J. C. S. Faraday I. **68:** 1101–1116.

STUDY OF CONFORMATION AND INTRAMOLECULAR MOTILITY OF POLYPEPTIDES IN SOLUTION BY A NOVEL FLUORESCENCE METHOD

E. Katchalski-Katzir,* E. Haas† and I. Z. Steinberg*

* *The Weizmann Institute of Science*
Rehovot, Israel
† *Bar-Ilan University,*
Ramat-Gan, Israel

Introduction

Linear oligomer and polymer molecules do not as a rule possess a fixed conformation in solution. Because of the thermal motion of the chain segments, the overall conformation of each molecule varies constantly, and the distance between the chain ends changes ceaselessly. The extent and rate of such movements depend on the length of the chain, the nature of the backbone, the nature of the end groups, the temperature, the viscosity of the solvent, etc. In the following, we shall adopt the concept of static flexibility to express the repertoire of molecular conformations possessed by an ensemble of linear oligomer or polymer chain molecules at any given instant. This characteristic is essentially time independent. This is distinguished from the concept of dynamic flexibility, which expresses kinetic changes in conformation in a single molecule or a chosen subgroup of molecules in the ensemble. The dynamic flexibility of the chain obviously determines the rate of thermal movements of the chain ends.

Conformational transitions in chain molecules predominantly involve rotations around bonds. However, if rotations around a single bond were the only process occurring at a given time, a large part of the chain would have to swing through the viscous medium. The probability of such a process occurring is very low. It was therefore suggested that changes in conformation in polymer chains take place by correlated transitions of a number of bonds. The motions are restricted to a short segment, while the major portion of the chain remains undisturbed.[1]

The common methods used for studying the dimensions of polymeric molecules, such as viscosity, light scattering, or ultracentrifugal sedimentation, are not applicable to oligomers of relatively small molecular weight. The conformations of such molecules are, however, of considerable interest, since they are more amenable to computational analysis by fast computers. Furthermore, since many low molecular weight oligomers, such as some of the native and synthetic peptides, show biological activity, information about their conformation in solution seems of particular interest. Stryer and Haugland[2] have suggested that distances of the order of 15–20 Å between definite sites on macromolecules can be derived from the efficiency of transfer of electronic excitation energy by the Förster mechanism[3] between suitably chosen chromophoric groups that have been attached to these sites. This is made possible by the relationship given in Equation 1, in which E denotes the efficiency of energy transfer, and r is the distance between the participating chromophores.

$$E = \frac{R_0^6}{R_0^6 + r^6} \tag{1}$$

0077-8923/81/0366-0044$01.75/0 © 1981, NYAS

R is a quantity of dimension length given by Equation 2

$$R_0^6 = \frac{9000 \ln 10\, \kappa^2 \eta_0}{128\, \pi^4 N n^4} \int_0^\infty f(\bar{\nu})\, \epsilon(\bar{\nu}) \frac{d\bar{\nu}}{\bar{\nu}^4} \qquad (2)$$

In Equation 2, $f(\bar{\nu})$ and $\epsilon(\bar{\nu})$ denote the fluorescence intensity of the donor and the extinction coefficient of the acceptor respectively, at the wave number $\bar{\nu}$; η_0 is the quantum yield of the donor in the absence of acceptor molecules; n is the index of refraction of the medium; N is Avogadro's number; and κ^2 is a dimensionless geometric factor determined by the orientation in space of the transition dipole moments of the donor and acceptor. For random orientations of the chromophores in space and short rotational relaxation times relative to the fluorescence lifetime, $\kappa^2 = 2/3$. A similar value can be assumed for many chromophores which show mixed polarization in their spectral behavior even when their rotation is restricted during the lifetime of the donor.[4] In such cases, the absorption and emission across the pertinent spectral range are not characterized by a single transition dipole moment but by a combination of two or more incoherent transition dipole moments. Stryer and Haugland[2] have demonstrated the feasibility of the method for estimating inter-chromophore distances by measuring the efficiency of energy transfer between chromophores attached to the ends of molecules of homologous proline oligopeptides, which are stiff and linear and whose lengths are known.[5]

In the following, we describe the determination of the end-to-end distance distribution for a series of homologous oligopeptide molecules that contain suitable donor and acceptor chromophores at their ends. The method employed was based on the study of the decay kinetics of the fluorescence of the donor $I(t)$. The experimental data obtained also enabled us to determine the Brownian motion of the chain ends in a series of solvents of varying viscosity.

The chromophore-bearing oligopeptides studied had the following structure:

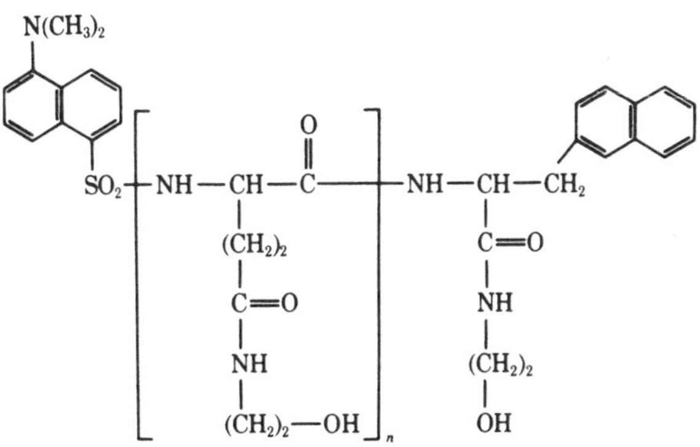

[I]

The repeating units were N_5-(2-hydroxyethyl)-L-glutamine residues; n varied from 4 to 9. A naphthalene chromophore was used as the energy donor, and a dansyl chromophore as the energy acceptor.

Several theoretical problems had to be solved before the energy transfer data could be interpreted in terms of the molecular parameters. First, the end-to-end distance in an ensemble of homogeneous flexible chain molecules is not the same; one would like therefore to design experiments from which one can derive information concerning the distribution of end-to-end distances of such chain molecules at any given instant. Secondly, under a variety of solvent conditions, the chain ends undergo translational Brownian motion relative to one another during the lifetime of the donor excited state. Such motions obviously affect the probability of nonradiative energy transfer[8,9] and should be taken into account when interpreting the data. Thirdly, in interpreting the efficiency of energy transfer in terms of distances between donors and acceptors, the problem of the orientational factor should be considered. Because of the reasons mentioned above, chromophores of relatively low polarization were chosen in our experimental studies.

Before describing our experimental results, the various theoretical aspects will be briefly reviewed. On the basis of these theoretical considerations, the results of the measurements will subsequently be analyzed and discussed.

Determination by Fluorescence Decay of the Distribution of Distances Between the Chain Ends of the Oligopeptide Molecules

The various structural conformations characterizing a given peptide in solution are "frozen" in a highly viscous medium. It is thus justifiable to assume that in such a case the distance between the donor and acceptor does not vary during the lifetime of the donor. Postulating an end-to-end distribution function, $f(r)$, for the assembly of structural conformations, one can readily show that the fluorescence decay of the donor, $I(t)$, for the total assembly, after an extremely short excitation pulse will be given by Equation 3

$$I(t) = k \int_0^{\infty} f(r) \exp\left[-\frac{t}{\tau} - \frac{t}{\tau}\left(\frac{R_0}{r}\right)^6 \right] dr \tag{3}$$

where τ is the lifetime of the donor in the absence of acceptor and k is a proportionality factor.

To illustrate the sensitivity of $I(t)$ to changes in $f(r)$, a series of hypothetical functions have been assumed, all of which are inverted parabolas with different mean values and different spread (see FIGURE 1). FIGURE 2 gives the values of $I(t)$ to be expected for the various $f(r)$ functions. The data calculated show that $I(t)$ does indeed contain relevant information concerning $f(r)$, and that it is quite sensitive to the parameters which characterize $f(r)$.

The functional forms chosen for the analysis of the experimental data described below are presented in TABLE 1. The parameters a and b were adjusted to yield the best fit between the calculated fluorescence decay curves $I_c(t)$, derived from Equation 3 in conjunction with the assumed distribution function $f(r)$, and the corresponding experimentally determined decay curves $I(t)$.

Evaluation of the Effect of Translational Brownian Motion of the Donor-Acceptor Pairs on the Fluorescence Decay Kinetics of the Donor

When the solvent viscosity is sufficiently low, conformational changes with time are expected to occur within each polymer molecule. Information about these changes is of considerable interest as it can shed light on the factors determining the dynamic flexibility of various chain molecules. In the following, a theoretical treatment is presented for the evaluation from the fluorescence decay of the donor of the variation

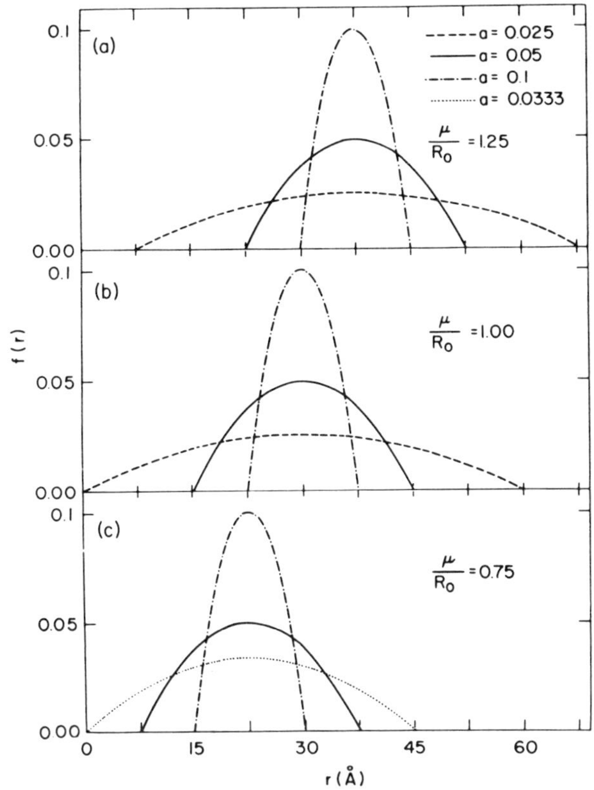

FIGURE 1. Hypothetical distributions of distances between donors and acceptors in an ensemble of donor-acceptor pairs. The distribution functions, $f(r)$, are normalized parabolas of the form $a - 16/9\ a^3(r - \mu)^2$, where a and μ are the height and the mean of the parabola, respectively, and r is the donor-acceptor distance in a pair.

with time of the distance between the ends of a flexible linear oligomer in solution containing a donor-acceptor pair at its ends (see also Reference 7).

Let $\mu(r)$ be the chemical potential of the fraction of chain molecules possessing an end-to-end distance r and let $N(r, t)\,dr$ be the number of chain molecules whose end-to-end distance is between r and $r + dr$ at time t. The equation describing the change of $N(r, t)$ with time may be derived by conceptually superimposing one end of all molecules at the origin of a coordinate system. Let $\rho(r, t)$ then be the number per unit volume of the molecules' other end as a function of r and t. Thus, $N(r, t) = 4\pi r^2\rho(r, t)$. Defining j as the flux per unit area of the free ends of the chain molecules, one obtains:

$$j = -\frac{1}{f}\frac{\partial\mu(r, t)}{\partial r}\rho(r, t) = -\frac{\rho(r, t)}{f}\frac{\partial\mu_0(r)}{\partial r} - \frac{RT}{f}\rho(r, t)\frac{\partial \ln\rho(r, t)}{\partial r} \tag{4}$$

and

$$\frac{\partial p(r, t)}{\partial t} = \frac{1}{r^2}\frac{\partial}{\partial r}(r^2 j) \tag{5}$$

where $\mu_0(r)$ is the chemical potential for $\rho = 1$ and f is a frictional coefficient. At equilibrium, $j = 0$, i.e.,

$$\frac{\partial\mu_0(r)}{\partial r} = -\frac{RT\partial\ln\rho_0(r)}{\partial r} \tag{6}$$

$\rho_0(r)$ being the number per unit volume of the free end of molecules possessing an end-to-end distance r at equilibrium. Defining an apparent diffusion coefficient D as RT/f, one obtains:

$$j = -D\rho(r,t)\frac{\partial\ln[\rho(r,t)/\rho_0(r)]}{\partial r} = -D\rho(r,t)\frac{\rho_0(r)}{\rho(r,t)}\frac{\partial[\rho(r,t)/\rho_0(r)]}{\partial r} = -D\rho_0(r)\frac{\partial\bar{\rho}(r,t)}{\partial r} \tag{7}$$

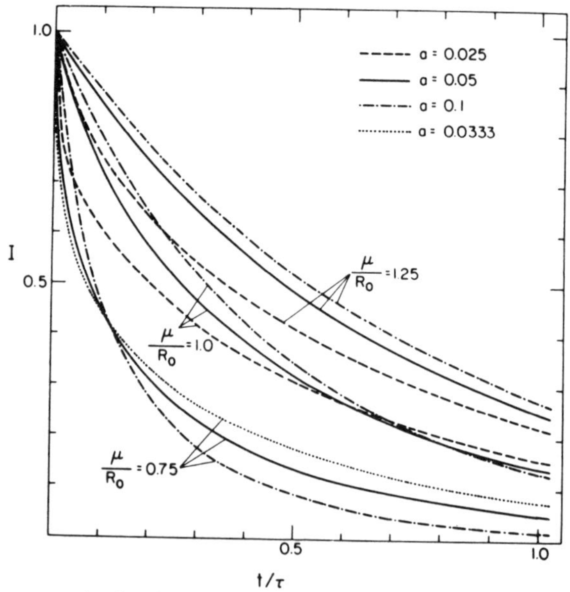

FIGURE 2. Computer-simulated donor fluorescence decay curves, $I(t)$, for ensembles of donor-acceptor pairs, the distance between which is described by parabolic distribution functions. $I(t)$ was obtained by use of Equation 3 and the distribution functions given in FIGURE 1.

TABLE 1

DISTRIBUTION FUNCTION OF END-TO-END DISTANCES USED IN THE ANALYSIS OF THE FLUORESCENCE DECAY KINETICS OF THE OLIGOMERS INVESTIGATED*

(1) $f(r) = 4\pi r^2\exp\{-a(r-b)^2\}$
(2) $f(r) = 4\pi r^2(a/\pi)\exp\{-(ar^2 + b/r^2)\}$
(3) $f(r) = 4\pi r^2(a/\pi)\exp\{-(ar^2 + b/r)\}$
(4) $f(r) = a - (16/9)a^3(r-b)^2$

* Expression (1), (2) and (3) were obtained for the end-to-end distribution of linear flexible polymers by Edwards,[15] Kumbar,[16] and Reiss.[17] Expression (4), representing an inverted parabola was used by Grinvald *et al.*[10] in simulated calculations. The parameters a and b were adjusted to fit the experimental data.

where

$$\bar{\rho}(r, t) = \rho(r, t)/\rho_0(r)$$

Thus:

$$\frac{\partial \bar{\rho}(r, t)}{\partial t} = \frac{1}{\rho_0(r) r^2} \frac{\partial}{\partial r} (r^2 D \rho_0(r) \frac{\partial \bar{\rho}(r, t)}{\partial r} \tag{8}$$

Remembering that $N(r, t) = 4\pi r^2 \rho(r, t)$ and $N_0(r) = 4\pi r^2 \rho_0(r)$, one obtains:

$$\frac{\partial \bar{N}(r, t)}{\partial t} = \frac{1}{N_0(r)} \frac{\partial}{\partial r} \left(N_0(r) D \frac{\partial \bar{N}(r, t)}{\partial r} \right) \tag{9}$$

where $\bar{N}(r, t) = N(r, t)/N_0(r)$.

Combining the effect of diffusion on $N(r, t)$ with the disappearance of electronically excited chromophores due to spontaneous decay and to energy transfer, one finally obtains:

$$\frac{\partial \bar{N}(r, t)}{\partial t} = -\frac{1}{\tau} \bar{N}(r, t) - \frac{1}{\tau} \left(\frac{R_0}{r} \right)^6 \bar{N}(r, t) + \frac{1}{N_0(r)} \frac{\partial}{\partial r} \left[N_0(r)\, D \frac{\partial \bar{N}(r, t)}{\partial r} \right] \tag{10}$$

Equation 10 has to be solved subject to the initial and boundary conditions $\bar{N}(r, t) = 0$ and $N(a, t) = 0$, i.e., there is immediate quenching of the donor when it approaches the acceptor to within some critical distance a, which is of the order of closest approach distance of the two chromophores. Once $\bar{N}(r, t)$ has been solved, the fluorescence decay curve of the donor is obtained from it according to the equation:

$$I(t) = k \int_a^{r_{max}} N(r, t)\, \mathrm{d}r = k \int_a^{r_{max}} N_0(r) \bar{N}(r, t)\, \mathrm{d}r \tag{11}$$

where k is a proportionality factor and r_{max} is the length of the fully extended chain.

The procedure followed for obtaining D from $I(t)$ was as follows: $N_0(r)$ was obtained by measuring the end-to-end distribution of distances of the chain molecules in a highly viscous solvent. τ and R_0 were obtained from independent experiments. This leaves D as the only unknown in Equation 10. By solving this equation for different values of D, one gets a set of functions $N(r, t)$, and these, when inserted into Equation 11, yield a corresponding set of hypothetical calculated functions for the fluorescence decay curve of the donor, $I_c(t)$. The D value which leads to a function $I_c(t)$ closest to the experimentally determined fluorescence decay curve $I(t)$, as determined by a least squares fit was chosen to represent the apparent end-to-end diffusion coefficient.

Materials and Methods

The chemical structure of the series of homologous peptides synthesized is shown in the Introduction. It will be designated in the following by the formula DG_nN. The peptides DG_nN (n = 4, 5, 6, 7, 8 and 9) were synthesized by the solid-phase method according to the procedure given in ref. 6. The dansyl-blocked peptides were purified by chromatography on Poropak-Q, a porous polystyrene preparation, and subsequently by gel filtration on Sephadex G-15 columns.

Each of the finally purified oligopeptide derivatives gave a single spot on thin-layer chromatography plates of cellulose, aluminium oxide, and silica gel with butanol/acetic acid/water (25:6:25 v/v) as developer. All of the peptides DG_nN

($n = 4-9$) showed the same absorption spectrum within the range studied (240–450 nm), corresponding to the sum of the absorption spectra on naphthalene and dansyl chromophores at equimolar amounts.

A peptide containing four N^5-(2-hydroxyethyl)-L-glutamine residues and a naphthyl chromophore but lacking a dansyl group (Compound II) was also prepared and used as a reference compound in the study of the fluorescence characteristics of the bound energy donor.

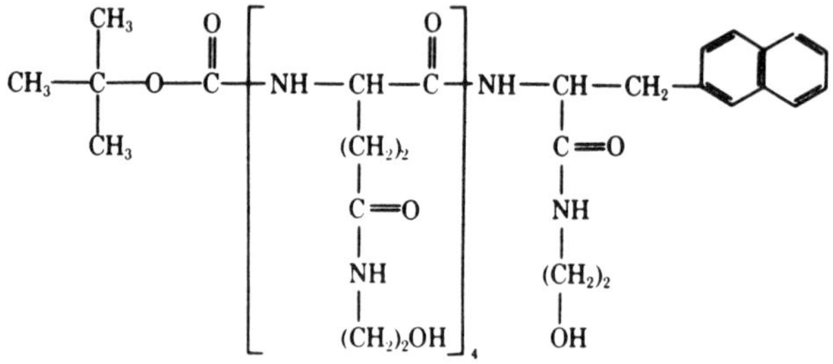

[II]

The kinetics of the fluorescence decay of the donor was measured with an instrument based on the sampling-scope technique built in our laboratory (FIGURE 3).[11,12] A deuterium flash-lamp (TRW) triggered by a Thyratron tube was used as an excitation source and a Phillips XP-1023 photomultiplier served as a signal detector. The analog pulses generated by the photomultiplier were sampled by a Tektronix ISI sampling unit and subsequently digitized and averaged in the multichannel analyzer (Hewlett-Packard 5401 A). The sampling unit as well as the multichannel analyzer were triggered and synchronized by the pulses generated by an auxiliary photomultiplier (1P28) which responded to the deuterium excitation flash lamp via a suitable light guide. Errors due to the time drift in the lamp profile were eliminated

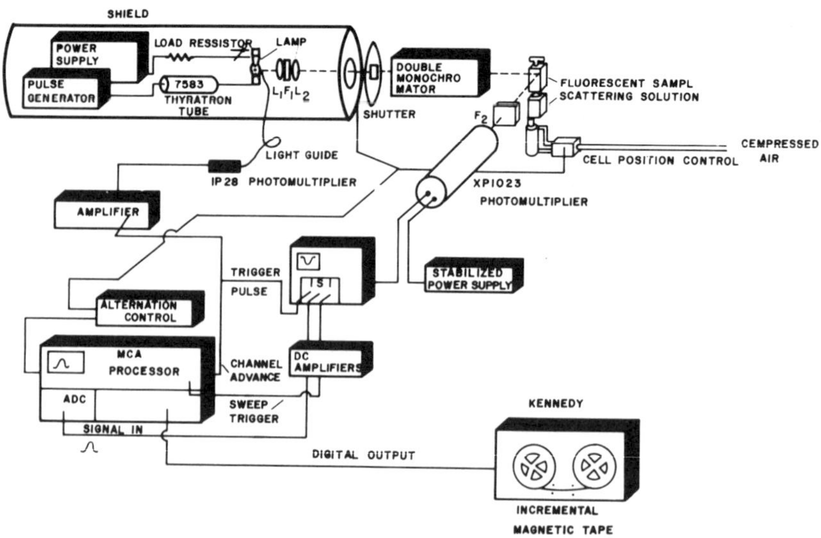

FIGURE 3. Schematic representation of the nanosecond fluorimeter based on the sampling method.

by frequent alterations in the recording of the lamp profile and the corresponding fluorescence response. Analysis of the information collected was carried out on the IBM 370/165 computer.

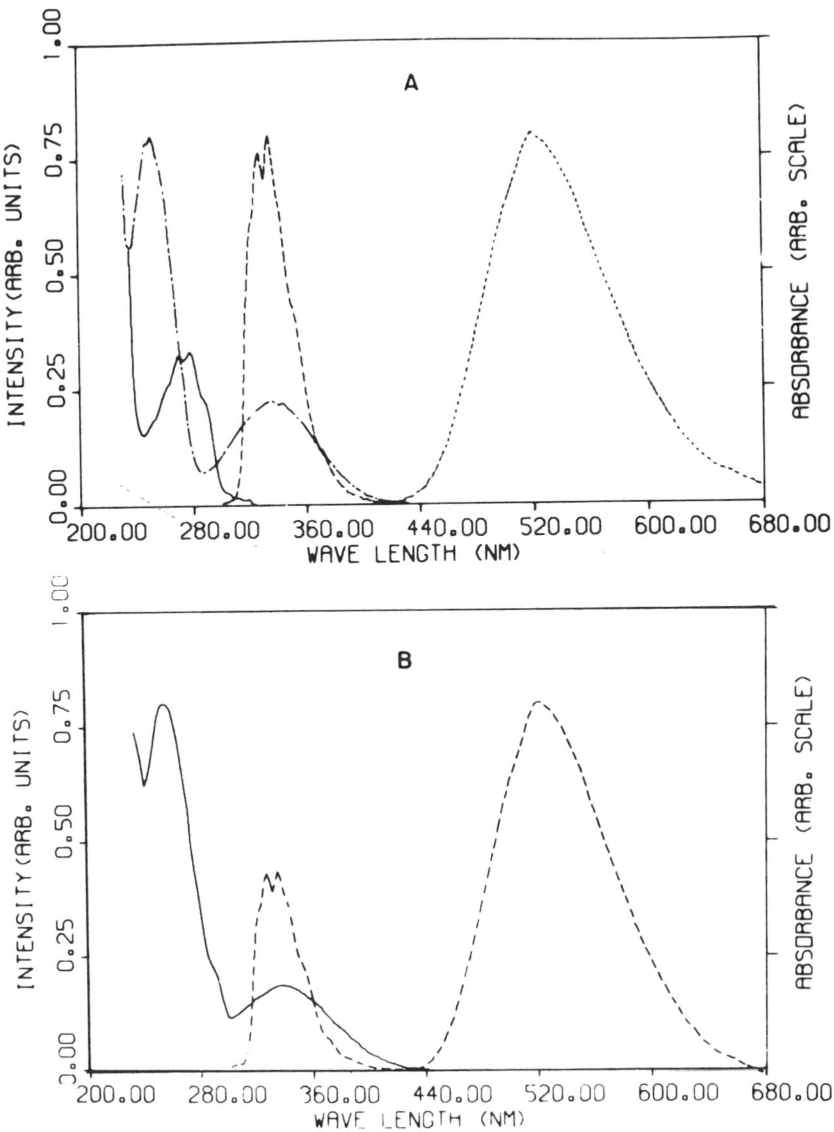

FIGURE 4. Absorption and fluorescence spectra of the chromophores present in the series of oligopeptides DG_nN (Compound I). (A) Compound II (see text): Absorption (——), fluorescence (λ_{ex} = 280 nm)(– – –). Dansylamide: absorption (–·–·–), fluorescence (λ_{ex} = 340 nm) (. . .). (B) DG_5N (Compound I, n = 5): absorption (——), fluorescence (λ_{ex} = 280 nm)(– – –). All measurements were carried out in glycerol at room temperature. The concentrations of the solutes were not determined; the absorption and fluorescence spectra are thus presented in arbitrary units.

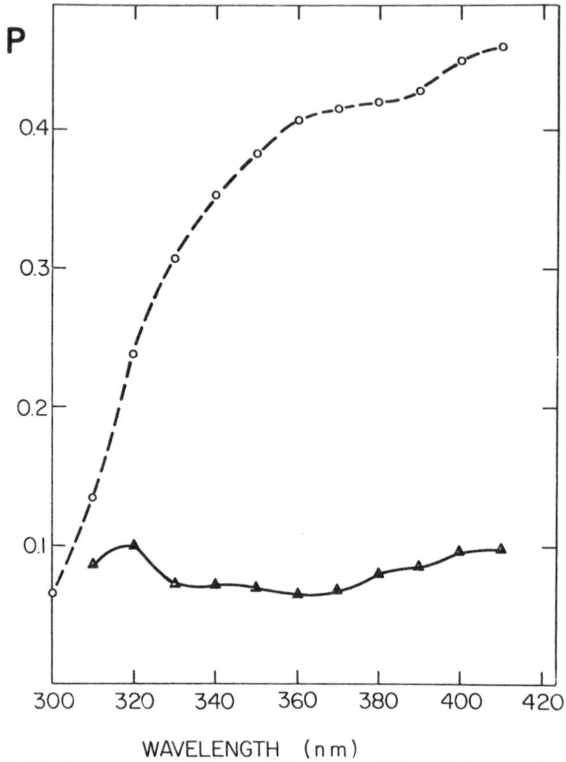

FIGURE 5. Polarization spectra of the naphthalene and dansyl chromophores of DG_8N in anhydrous glycerol: naphthalene (λ_{ex} = 290 nm)(—▲—▲—▲—), the low values for *p* were also obtained upon excitation at 270 nm to 320 nm; dansyl (λ_{em} = 500 nm)(○- -○- -○).

The experimentally obtained fluorescence decay curves, $F(t)$, were deconvoluted and fitted with the proper calculated functions by the method of least squares as described elsewhere.[13,10]

RESULTS AND DISCUSSION

Optical Characteristics of the Oligopeptides DG_nA Containing an Energy Donor, D *and an Energy Acceptor,* A, *at Their Chain Ends*

The absorption and fluorescence spectra of Compound II and of dansylamide, which are characteristic of the donor and acceptor chromophores respectively, are

FIGURE 6. Fluorescence decay of Compound II and of DG_8N (Compound I, n = 8). $g(t)$, lamp profile; $F(t)$, experimental decay of fluorescence; $F_c(t)$, calculated for the best fit using the decay function assumed. $F(t)$ and $F_c(t)$ represent the convolution integral of $I(t)$ and $I_c(t)$ with $g(t)$. The horizontal wavy lines under the fluorescence decay curves are the residuals, i.e., the differences between $F(t)$ and $F_c(t)$; the insets are the autocorrelation functions of the residuals. (A) Fluorescence decay of Compound II analyzed by fitting to monoexponential decay. RMS of fit, 0.0016. τ = 61 nsec. (B) Fluorescence decay of DG_8N analyzed by fitting to monoexponential decay. RMS of fit 0.025. (C) Experimental data as in panel B. The end-to-end distribution function given in (1), TABLE 1, was inserted in Equation 3 and convoluted with $g(t)$ to fit the experimental data. RMS of fit, 0.0019. In panels A and C, $F(t)$ and $F_c(t)$ coincide within the accuracy of the drawing.

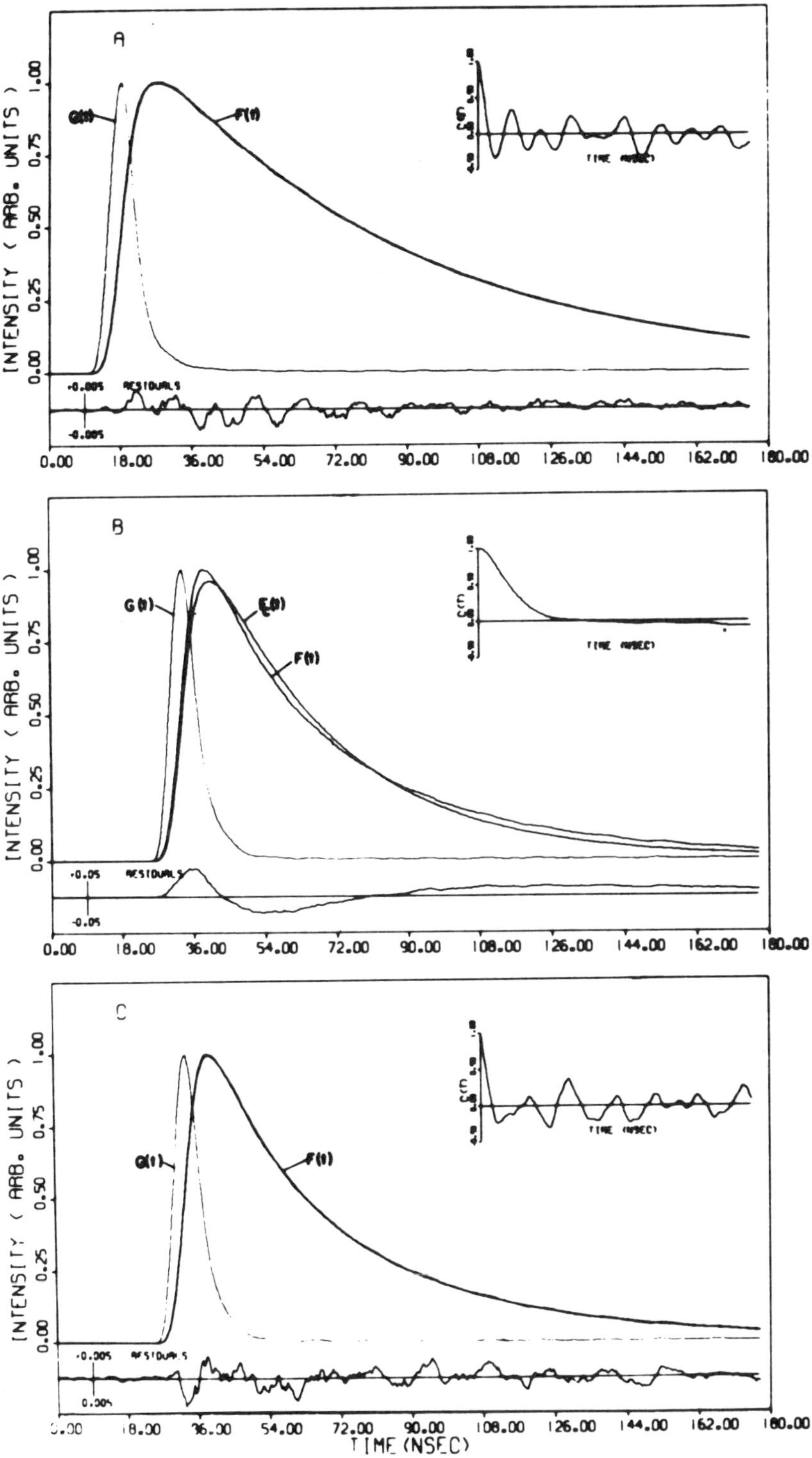
A
B
C
INTENSITY (ARB. UNITS)
0.00
0.25
0.50
0.75
1.00
G(t)
F(t)
E(t)
RESIDUALS
+0.005
-0.005
+0.05
-0.05
0.00
18.00
36.00
54.00
72.00
90.00
108.00
126.00
144.00
162.00
180.00
TIME (NSEC)

presented in FIGURE 4. The figure shows that the fluorescence spectrum of the naphthalene donor overlaps to a considerable extent with the absorption spectrum of the dansyl acceptor.

A study of the energy transfer between these two chromophores in solution[11] yielded a value of 23 Å for R_0 of this donor-acceptor pair.

The polarization spectra of the chromophores are shown in FIGURE 5. The value of p for the naphthyl chromophore is extremely low (~ 0.07), whereas that of the dansyl chromophore varies appreciably (in the range of 0.20–0.45) across the range of overlap between donor emission and acceptor absorption. By employing the theoretical treatment of the orientational factor described elsewhere,[4] it was shown that for these chromophore polarization values, the orientational factor has a negligible effect on the end-to-end distribution functions obtained by assuming a constant value of 2/3 for κ^2.

The fluorescence decay of Compound II containing only the naphthalene chromophore in glycerol is shown in FIGURE 6A. Analysis by the least-squares method of the fluorescence decay of this compound showed that it proceeds by monoexponential kinetics, as indicated by the low RMS of the fit (0.0016), and by the lack of systematic error in the fitting as reflected by the random variation of the autocorrelation of the residuals. The above findings illustrate the high accuracy of the data obtained by the fluorescence decay instrument used, and the well-behavior of the fluorescence decay of the donor under the conditions used. From the data presented, a lifetime of 61.0 nsec was obtained for the donor.

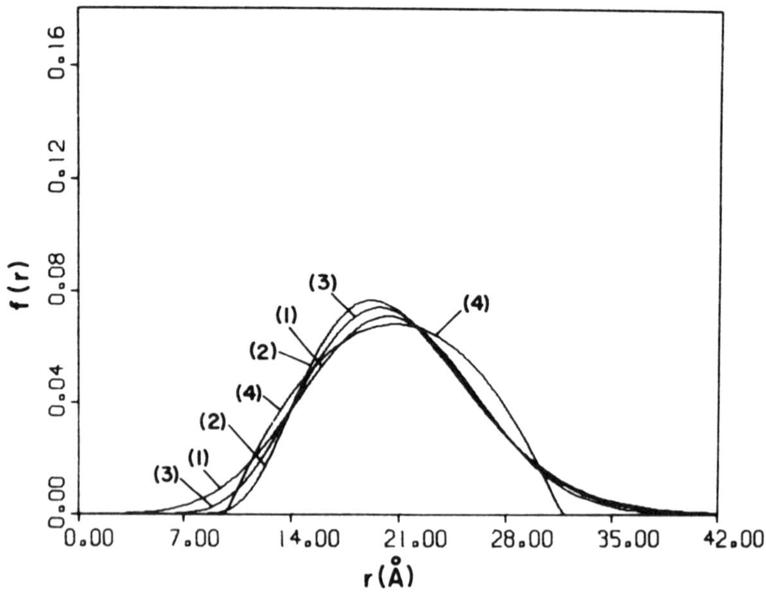

FIGURE 7. The distribution function of the distances between donor and acceptor for DG_8N (I, n = 8) in glycerol, assuming the different mathematic expressions listed in TABLE 1, and using the corresponding numerical values of a and b which were obtained on best fit with experimental data. The numbers in parentheses refer to the functions given in the Table.

End-to-End Distribution Functions for the Oligopeptides DG_nN (n = *4–9) in Glycerol, a Highly Viscous Medium*

The experiments to be described below were carried out in glycerol. Due to the high viscosity of this solvent (about 900 centipoise) one expects little Brownian motion of the chain ends relative to one another during the lifetime of the donor excited state. The results obtained for the end-to-end distribution function are thus obviously not complicated by translational diffusional effects of the chain ends.

The fluorescence decay of the donor in the peptide DG_8N (compound I, $n = 8$), containing both donor and acceptor, is presented in FIGURES 6B and C. Qualitatively similar graphs were obtained for the other peptides studied. An attempt to fit the decay kinetics of the various oligopeptide derivatives studied to a monoexponential function failed completely, the standard deviation being intolerably high. For DG_8N, for example, the RMS was $\simeq 0.025$, in comparison to RMS $\simeq 0.0016$ obtained for the donor in the absence of acceptor. This behavior indicates that the distance between donor and acceptor in each of the oligopeptides DG_nN ($n = 4–9$) is not unique and that the method of measurement used by us is sensitive to this fact.

To evaluate the distribution function of the distances between donor and acceptor attached to the ends of a given peptide, one has to search, as explained above, for a function $f(r)$ which, when inserted in Equation 3, gives an $I_c(t)$ that fits the experimental decay curve of the fluorescence of the donor. The various analytical

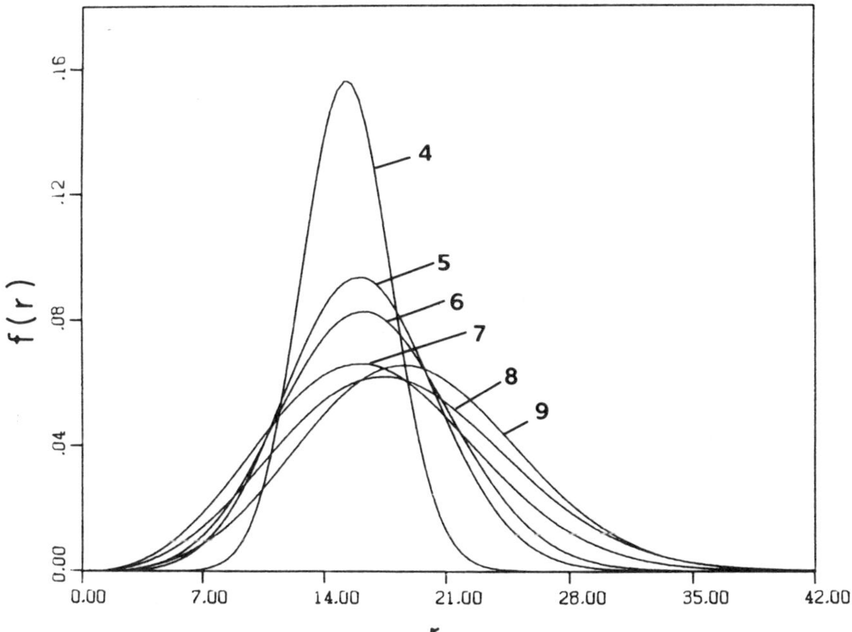

FIGURE 8. The distribution function of the distances between donor and acceptor for the series of oligopeptides 1, $n = 4, 5, 6, 7, 8$ and 9. The numbers in the figure refer to the values of n. The mathematical expression (1), TABLE 1, was used to fit the data.

distribution functions given in TABLE 1 yielded quite similar end-to-end distribution functions, as illustrated in FIGURE 7 for the peptide with $n = 8$. It should be noted that the Gaussian distribution function could not be fitted satisfactorily to the data for all of the oligopeptides studied. This finding is in accord with the conclusion of Flory[14] that $f(r)$ cannot assume a Gaussian form centered at the origin.

The end-to-end distribution functions obtained for the various peptides studied, using the analytical expression

$$f(r) = 4\pi r^2 \exp[-a(r - b)^2]$$

are presented in FIGURE 8. Similar results were obtained by using the other analytical functions presented in TABLE 1. The values of the parameters a and b used in order to yield the best fit with experimental data have been listed in TABLE 2 of Reference 6. As expected, the average end-to-end distances and the spread of the distribution of distances increase systematically with chain length. The average distances and the RMS distances for the series of oligomers studied are presented in TABLE 2.

Effect of Viscosity of the Medium on the Fluorescence Decay Kinetics of the Donor in the Oligopeptides DG_nN

Marked changes were observed in the donor fluorescence decay kinetics on reducing the solvent viscosity. The viscosity was changed by adding trifluoroethanol or methanol to glycerol. Compound II, which contains only the donor chromophore without an energy acceptor, was found to decay monoexponentially in all solvent mixtures, with a lifetime that varied only slightly with solvent composition (τ increasing from 61 nsec in glycerol to 63 nsec in a 1:9 (w/w) mixture of glycerol: trifluoroethanol). By contrast, the donor fluorescence decay in all peptides containing both energy donor and acceptor deviated markedly from monoexponential decay. Furthermore, the rates of decay (expressed, for example, by average inverse apparent lifetimes) increased markedly with decrease in solvent viscosity. The results for two of the peptides studied are illustrated in FIGURE 9. It should be noted that upon decreasing the solvent viscosity, the relative intensity of the acceptor increased in parallel with the decrease in donor fluorescence, showing that the effects observed

TABLE 2
THE MEAN AND THE ROOT MEAN SQUARE END-TO-END DISTANCES FOR THE VARIOUS OLIGOPEPTIDES STUDIED*

Peptide	$\langle r \rangle$ (Å)	$[\langle r^2 \rangle]^{1/2}$ (Å)
DG_4N	15.2	15.4
DG_5N	15.8	16.2
DG_6N	16.4	17.1
DG_7N	18.4	19.2
DG_8N	20.6	21.3
DG_9N	21.6	22.5

* The values given were derived from expression (1), TABLE 1. Similar values were obtained by making use of the other expressions given in TABLE 1.

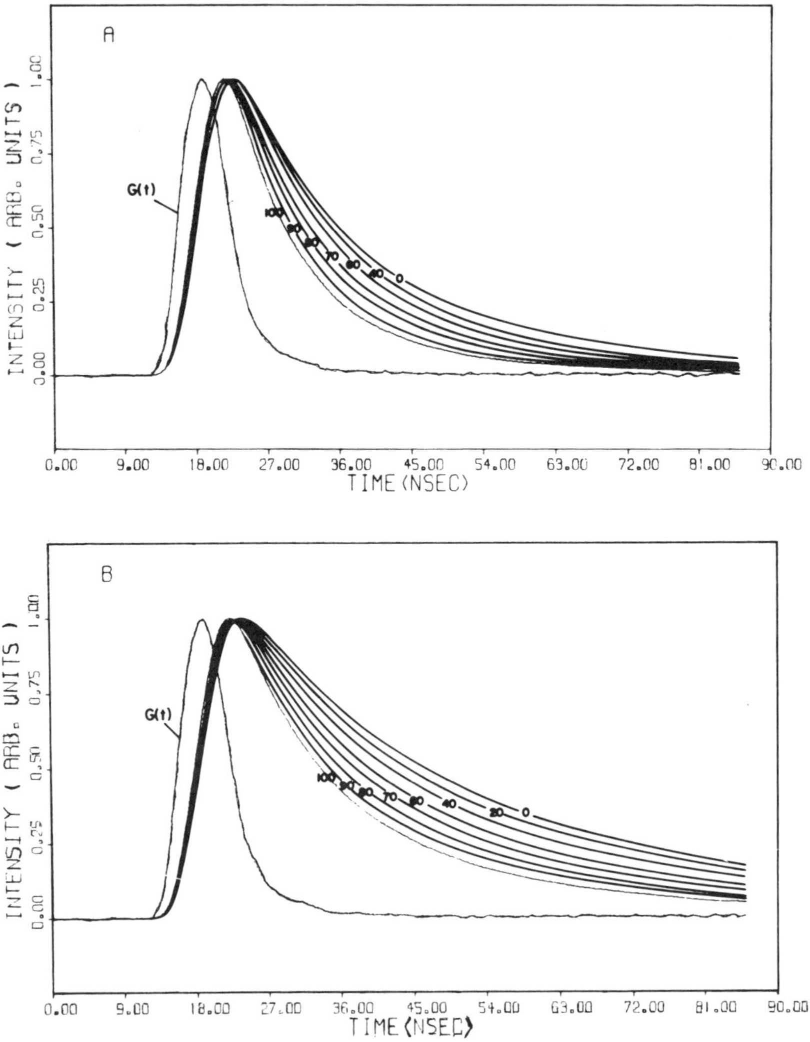

FIGURE 9. Fluorescence decay curves of the naphthalene chromophore (energy donor) of the peptides DG_5N and DG_8N in various mixtures of trifluoroethanol and glycerol. (A) DG_5N; F_3EtOH: glycerol ratios are 0, 40, 60, 70, 80, 90, 100 (% w/w). (B) DG_8N, F_3EtOH: glycerol ratios are: 0, 20, 40, 60, 70, 80, 90, 100 (% w/w).

are due to an increase in energy transfer from donor to acceptor upon reduction of the solvent viscosity.

Calculated Intramolecular Diffusion Coefficients for the Chain Ends

The fluorescence decay data obtained for the series of oligopeptides investigated in the various solvent mixtures were analyzed according to the theoretical treatment outlined above in terms of the translational diffusion of the chain ends relative to one another. In the calculations, it was assumed that the equilibrium end-to-end distance distribution in the various solvent mixtures is the same as that derived for glycerol. The justification for this is that the peptides used are polyalcohols and are highly soluble in both the alcoholic solvents used. Preferential solvation by either solvent is thus not expected to be significant. The value of R_0 used was calculated for each solvent mixture using the relevant values of the refractive index and other spectroscopic characteristics. The values of τ used were those actually measured for Compound II in the corresponding solvent mixtures.

The results obtained for the intramolecular chain-ends diffusion coefficients, D, for the peptides with n = 4, 5, 8 and 9 as a function of the viscosity of the solvent mixtures used are presented in FIGURE 10 and TABLE 3. It will be noted that for the peptide DG_5N, experiments were performed in glycerol-methanol mixtures as well as in glycerol-trifluoroethanol mixtures, yielding approximately the same values for D in solvent mixtures of the same viscosity. These results support the above assumption that there seem to be negligible solvent effects on $f(r)$ in the solvents used.

The values of D obtained in solutions of viscosities ranging between 1 cp and 200 cp fall in the range of 10^{-9}–10^{-7} cm^2/sec. This value is lower by about an order of magnitude than that of the diffusion coefficients expected for small molecules comparable in size to the chromophores attached to the ends of the oligopeptides. It is thus evident that the "internal friction" of the oligomers reduces by about an order of magnitude the diffusion rate of the ends of the oligomers, compared to the rate of free diffusion of small molecules in solution.

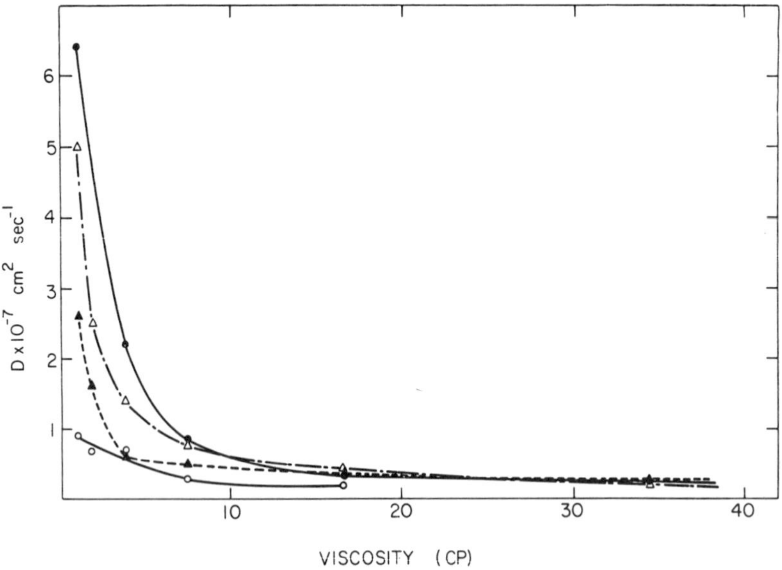

FIGURE 10. Diffusion coefficients of one molecular end relative to the other in DG_nN (n = 4, 5, 8, 9) as a function of solvent viscosity. DG_4N (-○-○-); DG_5N (-▲-▲-), DG_8N (-△-△-), DG_9N (-●-●-). Temperature, 23°C.

TABLE 3
DIFFUSION COEFFICIENTS OF CHAIN ENDS RELATIVE TO ONE ANOTHER*

% F_3EtOH in Glycerol (w/w)	Solvent Viscosity (cp)	R_o†	DG_4N		DG_5N		DG_8N		DG_9N	
			D	RMS	*D*	RMS	*D*	RMS	*D*	RMS
0	900	22.3	0.0		0.00		0.00		0.00	
20	210.30	22.6			0.01	0.0041	0.01	0.0040		
30	116.0	22.8			0.21	0.0070	0.06	0.0050		
40	66.4	23.0			0.21	0.0090	0.07	0.0049	0.14	0.0031
50	34.5	23.1			0.26	0.0062	0.20	0.0077	0.30	0.0060
60	16.6	23.4	0.2	0.0130	0.37	0.0070	0.43	0.0093	0.34	0.0100
70	7.6	23.6	0.3	0.0038	0.50	0.0036	0.76	0.0155	0.85	0.0134
80	3.8	23.8	0.7	0.0060	0.60	0.0180	1.40	0.0220	2.20	0.0220
90	1.8	24.0	0.7	0.0140	1.60	0.0160	2.50	0.0302		
100	1.0	24.3	0.9	0.0220	2.60	0.0272	5.00	0.0350	6.40	0.0276

* Obtained for the peptide series DG_nN ($n = 4,5,8,9$) in trifluoroethanol (F_3EtOH)/glycerol mixtures. The diffusion coefficients D, in 10^{-7} cm^2/sec, were computed as described in the text. The equilibrium distribution function of end-to-end distances used was $f(r) = 4\pi r^2 \exp - [a(r-b)^2]$.

† R_o was calculated assuming $\kappa^2 = 2/3$ and using the experimentally determined values for the refraction index of each solvent employed.

The dependence of D on the viscosity of the medium as presented in FIGURE 10 may be extrapolated to zero viscosity, yielding the hypothetical diffusion coefficients as determined solely by the internal friction of the oligopeptides. While the extrapolation to zero viscosity cannot be made accurately, the extrapolated values for D are probably below 10^{-6} cm^2/sec, and are smaller the shorter the oligomer. This shows that for all the oligopeptides studied, the internal friction is finite and appreciable. Furthermore, the internal friction is higher for the shorter chain molecules. It follows that the shorter chains cannot permit as much flexibility of motion at their ends as the longer chains can.

It is of interest to note that the quality of fit of the experimental decay curves to the model described by Equation 10 varies with the fluidity of the solvent. More specifically, the fit becomes poorer at low viscosities, starting roughly below 7 cp. This is reflected by the relatively high RMS yielded at best fit (see TABLE 3) and by some systematic deviations between decay curves calculated from Equation 11 and the experimental data, as reflected by the autocorrelation of the residuals. It thus seems that the model expressed by Equation 10 fits the experimental data not as satisfactorily in the low viscosity range as in the high viscosity range. This probably stems from the use of the Fick equation to describe the translational Brownian motion of the ends of the chain molecules. The Fick equation implies that every step in the translational motion in uncorrelated with the previous steps performed by the moving entity. This is not strictly the case for the motion of the ends of chain molecules, since successive steps of the chain end in a given direction are expected to make an additional step in the same direction more difficult, whereas a step in the opposite direction will be easier because of the finite time required for the conformational relaxation of the rest of the chain molecule. The concomitant deviations from Fick's equation will be relatively unimportant in viscous solvents, since the number of steps and the distance moved by the chain ends are small. In fluid solutions, the above effects apparently become significant and are reflected in the poorer fit of the data to Equation 11. The diffusion coefficients obtained in the highly fluid solvents should thus be taken as approximate values only.

CONCLUDING REMARKS

The theoretical and experimental studies of the energy transfer between chromophores attached to the ends of the series of homologous peptides synthesized, based on fluorescence decay kinetics, have proved valuable in elucidating the molecular structure of the peptides in solution. Information was obtained with regard to the end-to-end distribution of distances for an ensemble of molecules representing at equilibrium the various conformations attained by a given peptide in solution. The data accumulated for solutions of relatively low viscosity, in conjunction with those derived for a corresponding solution of high viscosity, enabled the evaluation of the dynamic flexibility of the various peptide chains studied under the experimental conditions employed. Thus, it might be expected that the study of the fluorescence decay, by the method outlined above, of donor-acceptor pairs adequately attached to peptides, proteins and other biopolymers, will enable the evaluation not only of the average distance between the donor and acceptor, but also of the conformational fluctuations and conformational changes occurring in such macromolecules under well specified conditions.

REFERENCES

1. MORAWETZ, H. 1975. Macromolecules in Solution, 2nd Ed. Wiley-Interscience. London. See also Morawetz, H. 1979. Some applications of fluorimetry to synthetic polymer studies. Science **203:** 405–410.
2. STRYER, L. & R. P. HAUGLAND. 1967. Energy transfer: A spectroscopic ruler. Proc. Natl. Acad. Sci. USA **58:** 719–726.
3. FÖRSTER, TH. 1948. Zwischenmolekulare Energiewanderung und Fluoreszenz. Ann. Phys. **2:** 55–75.
4. HAAS, E., E. KATCHALSKI-KATZIR & I. Z. STEINBERG. 1978. Effect of the orientation of donor and acceptor on the probability of energy transfer involving electronic transitions of mixed polarization. Biochemistry. **17:** 5064–5070.
5. STEINBERG, I. Z., W. F. HARRINGTON, A. BERGER, M. SELA & E. KATCHALSKI. 1960. The configurational changes of poly-*L*-proline in solution. J. Am. Chem. Soc. **82:** 5263–5279.
6. HASS, E., M. WILCHEK, E. KATCHALSKI-KATZIR & I. Z. STEINBERG. 1975. Distribution of end-to-end distances of oligopeptides in solution as estimated by energy transfer. Proc. Natl. Acad. Sci. USA **72:** 1807–1811.
7. HAAS, E., E. KATCHALSKI-KATZIR & I. Z. STEINBERG. 1978. Brownian motion of the ends of oligopeptide chains in solution as estimated by energy transfer between the chain ends. Biopolymers. **17:** 11–31.
8. ELKANA, Y., J. FEITELSON & E. KATCHALSKI. 1968. Effect of diffusion on transfer of electronic excitation energy. J. Chem. Phys. **48:** 2399–2404.
9. STEINBERG, I. Z. & E. KATCHALSKI. 1968. Theoretical analysis of the role of diffusion in chemical reactions, fluorescence quenching and nonradiative energy transfer. J. Chem. Phys. **48:** 2404–2410.
10. GRINVALD, A., E. HAAS & I. Z. STEINBERG. 1972. Evaluation of the distribution of distances between energy donor and acceptors by fluorescence decay. Proc. Natl. Acad. Sci. USA **69:** 2273–2277.
11. HAZAN, G. 1973. Ph.D. Thesis. Submitted to the Feinberg Graduate School, Weizmann Institute of Science, Rehovot, Israel.
12. HAZAN, G., A. GRINVALD, M. MAYTAL & I. Z. STEINBERG. 1974. An improvement of nanosecond fluorimeters to overcome drift problems. Rev. Sci. Instrum. **45:** 1602–1604.
13. GRINVALD, A. & I. Z. STEINBERG. 1974. On the analysis of fluorescence decay kinetics by the method of least squares. Anal. Biochem. **59:** 583–598.
14. FLORY, P. J. 1973. Moments of the end-to-end vector of a chain molecule, its persistence and distribution. Proc. Natl. Acad. Sci. USA **70:** 1819–1823.
15. EDWARDS, S. F. 1965. The statistical mechanics of polymers with excluded volume. Proc. Phys. Soc. (London) **85:** 613–624.
16. KUMBAR, V. A. 1973. Comparative study of distribution functions for the excluded volume problem. J. Macromol. Sci. Chem. **A7:** 461–474.
17. REISS, H. 1967. Variation principle treatment of the excluded problem. J. Chem. Phys. **47:** 186–194.

TRIPLET PROBE STUDY ON THE RATES OF DIFFUSION-CONTROLLED REACTIONS IN SOLUTION

Itaru Mita

Institute of Space and Aeronautical Science
University of Tokyo
Tokyo, Japan

INTRODUCTION

Diffusion-controlled reactions are observed much more frequently in polymer systems than in the usual organic system because, on the one hand, polymers, as reactants, diffuse more slowly than small molecules in any medium, and, on the other hand, any reactant diffuses more slowly in media containing polymers than in usual solvents. But the diffusion-controlled reactions of the reactive groups attached to polymers are not as simple as those of small molecules. In the case of small molecules, the translational diffusion of a molecule as a whole and that of the reacting group are the same and the second-order rate constant k_2^D for the formation of an encounter pair can be expressed by the famous Smoluchowski equation

$$k_2^D = 4\pi RD \qquad \textbf{(1)}$$

where R is the radius of reaction sphere and D is the sum of the diffusion coefficients of two reactants. When the reactive group is a part of a flexible polymer molecule, however, the situation becomes rather complicated as the polymer chain itself has a large degree of freedom of motion. The diffusion of a reactive segment and that of the center of gravity of the polymer are no longer the same and furthermore, intermolecular access of the reactive groups may be affected by the interpenetration characteristics of two randomly coiled polymer molecules. In other words, the reaction rate may be strongly influenced by the chain dynamics and thermodynamics of the polymer, which are controlled by the nonreactive part of the polymer molecule.

These considerations have long been the subject of discussions on the termination reaction of radical polymerization,[1-3] which is a typical diffusion-controlled reaction. But no accurate rate constants for radical termination reactions as a function of the degree of polymerization of polymers, solvent, etc., have been reported, because of experimental difficulties inherent in the indirect method of extracting them from kinetic data and the polydispersity of the polymeric radicals. More direct methods for obtaining the rate constants for the diffusion-controlled reactions with substantially monodisperse polymers have been needed. The use of triplet probes seems to answer that need. If a reactive group to be excited to triplet state is attached to the chain end of a polymer and its quenching group to another polymer, their quenching reaction can be thought to be a typical diffusion-controlled reaction of polymer chain ends and the rate of the reaction can be followed directly by proper photophysical means.

0077-8923/81/0366-0062$01.75/0

The diffusion-controlled reaction between two reactive groups bound to the chain ends of a flexible polymer molecule is another subject of interest. Its rate constant k_1^D may reflect directly the conformational chain dynamics of large scale. Several theories dealing with the reaction[4-5] have been advanced already, but experimental data are not numerous even for low molecular weight polymers and no experiments have been reported for high molecular weight polymers ($P > 5 \times 10^2$). The first data of k_1^D for high polymers determined in the author's laboratory are presented and discussed here for the case of a reaction between two triplet probes attached to each of both chain ends of a polymer molecule.

The true meaning of the kinetic rate constants for diffusion-controlled reactions is understood only on the basis of the knowledge of the chain dynamics and thermodynamics affecting the diffusion and collision process of the reactive groups. On the contrary, the experimental determination of the rate constants for the diffusion-controlled reactions contributes considerably to insights on the movement of the flexible polymer chain in solution.

TRIPLET PROBES

The methods using singlet probes such as excimer formation and fluorescence deporalization have been shown to be useful for the studies of whole chain dynamics of oligomers and local motions of polymers in solution. But the lifetimes of the singlet state (10^{-7}–10^{-9} sec) probes are too short to be applied successfully to the study of a phenomenon such as intermacromolecular chain end collisions of high molecular weight polymers. On the other hand, the lifetimes of the triplet state in solution are of the order of 10^{-6}–10^{-3} sec and they fit in well with the time scale of the above phenomenon.

Suppose that the Stern Volmer's relation is used to obtain the bimolecular quenching rate constant either by the dynamic lifetime (τ) measurement or by measuring the stationary luminescent intensity (I).

$$\frac{\tau_0}{\tau}\left(or\ \frac{I_0}{I}\right) = 1+ \tau_0 k_q\,[Q] \qquad \textbf{(2)}$$

Then the concentration of quencher $[Q]$ should be of the order of or higher than $(k_q\tau_0)^{-1}$ to get reliable data. With k_q of the order of $10^8 M^{-1}\text{sec}^{-1}$ for polymers in dilute solution, $[Q]$ is of the order of 1 M or higher for singlet quenching with $\tau_0 = 10^{-8}$ sec. This means that the base molar concentration $[M]$ of the polymer with $P = 10^3$ should exceed $10^3 M$ ($[M]=P[Q]$), an impossible concentration. When a triplet probe with $\tau_0 = 10^{-4}$ sec is used, however, $[M]$ is allowed to be of the order of 10^{-1}M, a reasonable concentration for the study in dilute solution.

The relation of the lifetime of luminescence to the rate of intramolecular diffusion-controlled reactions is straightforward. The first-order decay lifetime of the reactions $(1/k_1^D)$ for a high polymer is of the order of 10^{-4}–10^{-5} sec which is just comparable to the lifetimes of triplet probes.

The limitation of concentration is imposed not only on the polymer-carrying quencher group but also on the polymer-carrying triplet group. Excepting the case of a very high intensity incident light, the fraction of the ground-state reactant converted to triplet is very small. For the solution to be dilute, however, the concentration of the ground-state reactant group should not be higher than 10^{-4}–10^{-5}M. This situation precludes the possibility of resorting to the triplet-triplet absorption in the dynamic lifetime measurement and also the stationary intensity measurement of phosphorescence, because the phosphorescence is usually very weak in solution in addition to the fact that few compounds phosphoresce in fluid medium.

Thus, in the present work, the dynamic lifetime measurement of luminescences from laser-pulse-excited triplets such as phosphorescence and delayed fluorescence is used to monitor the diffusion-controlled reactions of triplets: phosphorescence for the study on intermolecular reactions between triplet and quencher and delayed fluorescence for the study on intramolecular triplet-triplet annihilation reactions.

Another advantage of the use of triplet probes is that the radius of reaction sphere or the distance of triplet energy transfer is not very large and is comparable to those of usual chemical reactions for which the reacting groups must come into contact.

Benzil ($E_T = 223.4\ kJ/\text{mol}$) is one of those rare compounds that phosphoresce in fluid solution (quantum yield $\phi_{ph} = 0.01$) and its triplet state with a lifetime of 7×10^{-5} sec (in benzene)[6] is quenched by anthracene ($E_T = 175.7\ kJ/\text{mol}$) at a diffusion-controlled rate because of a large difference in their triplet energy levels (E_T). Triplet anthracene has a much longer lifetime (1 msec)[7] in solution and it reacts with another triplet anthracene resulting in singlet anthracene which then fluoresces (*P*-type delayed fluorescence).

These triplet probes were incorporated into substantially monodisperse polystyrenes with different degrees of polymerization *P*. Benzil end-capped polystyrene PS-COO-CH_2-p-C_6H_4-CO-CO-C_6H_5 (PS-B) was synthesized by the reaction of poly-

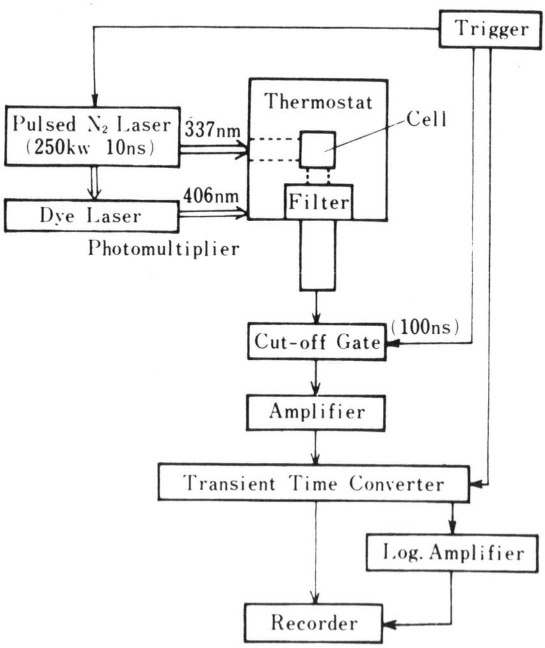

FIGURE 1. Schematic diagram of the apparatus.

styrenecarbonyl chloride derived from living polystyryl monoanion with hydroxymethylbenzil prepared by four steps. Anthracene end-capped polystyrene PS-$CH_2C_{14}H_9$ (PS-A) was synthesized by the direct reaction of living polystyryl monoanion with 9-chloromethylanthracene. Both end-capped α,ω-dianthrylpolystyrene (A-PS-A) and di(polystyryl)anthracene (PS-A-PS), which contains an anthracene group at the center of the polystyrene chain, were prepared by using a polystyryl dianion and 9-10-dichloromethylanthracene respectively. The polydispersity parameters $\bar{M}_w/\bar{M}_n$ of the polystyrene derivatives thus obtained are less than 1.1 for most samples.

A schematic diagram of the apparatus for measuring the decay of luminescence is shown in FIGURE 1. A pulsed light (406 nm) of dye laser, induced by nitrogen laser (250 kW, 10 nsec), was used to excite benzil for the quenching studies and the light of the nitrogen laser itself (337 nm) to excite anthracene for the T-T annihilation studies. To avoid the effect of very strong direct fluorescence, a cut-off gate circuit operating for 100 nsec after pulse irradiation was put before the amplifier.

Thoroughly degassed polymer solution in spectrosol grade solvent was irradiated. In the case of quenching experiments, the linear decay of logarithmic intensity of phosphorescence against time provides the lifetimes τ and τ_0 in the presence and absence of quencher polymer, respectively. Typical decay curves are shown in FIGURE 2[8]. The quenching rate constant was calculated according to Equation 2. In the case of intramolecular T-T annihilation, more complex analysis of the data is required. This case will be discussed in a later section of this paper. Triplet anthracene was found to be quenched slowly by polylstyrene both inter- and intramolecularly.[7,9]

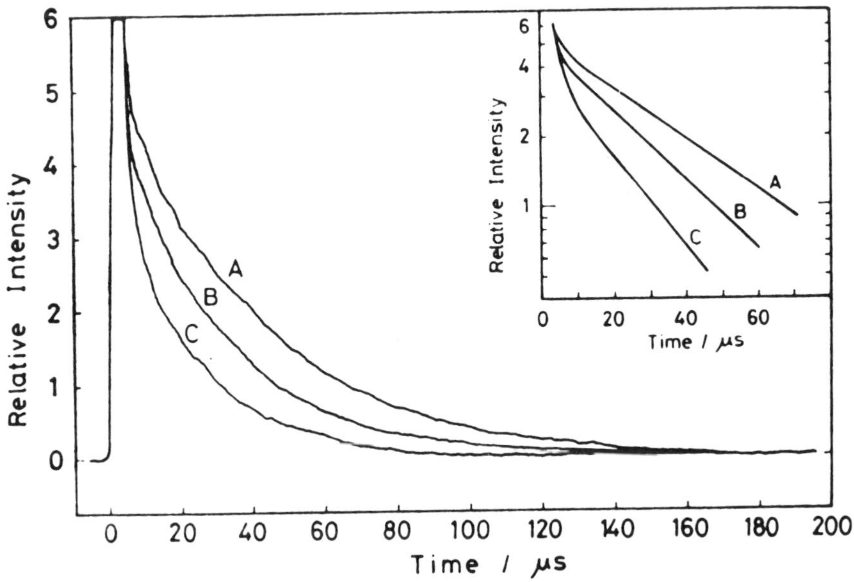

FIGURE 2. Phosphorescence decay curves of *PS60-B* in the absence and presence (*B*, *C*) of *PS71-A* in benzene at 30° *C*, and their semilogarithmic plots (insert). Concentration of *PS71-A*: *B*:1.04 × 10^{-5}*M*; *C*:2.14 × 10^{-5} *M*.

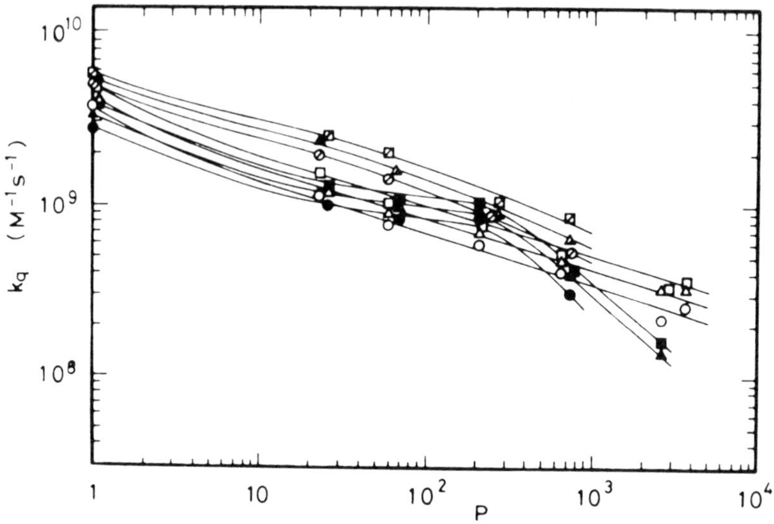

FIGURE 3. Dependence of k_q on the degree of polymerization P in benzene (○, △, □), cyclohexane (●, ▲, ■), and butanone (⊘, ◬, ⊠). Plots for $P = 1$ correspond to the reaction of benzil with 9-methylanthracene. Temperature: (○, ●, ⊘) 20° C; (△, ▲, ◬) 30° C; (□, ■, ⊠) 40° C.

This reduces considerably the lifetime of triplet anthracene bound to polystyrene. Triplet benzil is quenched by polystyrene with a similar rate constant to that of triplet anthracene ($k_2 \simeq 10^3 M^{-1}sec^{-1}$). But the effect is not important because the lifetime of triplet benzil is more than one order shorter than that of triplet anthracene.

Intermolecular Reactions in Dilute Solution

Polymer-Polymer Reactions

The rate constants k_q for quenching of PS-B by PS-A measured in several solvents are shown as functions of degree of polymerization P in FIGURE 3.[10] The reaction of benzil with 9-methylanthracene is taken as the reaction for $P = 1$. The magnitude of k_q for polymer-polymer reactions lies in the range of 2×10^9–1×10^8 $M^{-1}sec^{-1}$ and seems a little larger than those known for termination of polystyryl radicals. This may be attributed to the larger reaction sphere (ca. 1.3 nm) and smaller steric hindrance in triplet quenching than those of radical reactions. The rate constants k_q in butanone are larger than the corresponding k_q in benzene partly as a result of the lower viscosity of butanone. The activation energies E_q obtained show some scatter (9–12 kJ/mol in benzene for example), but they are regarded as constant, independent of P, and almost equal to the temperature dependence of the viscosity of solvent (10.6, 11.9 and 8.4 kJ/mol for benzene, cyclohexane, and butanone, respectively).

Almost straight lines are obtained in benzene for k_q as a function of P on a log-log scale. The exponent for the relation $k_q = A \times P^{-n}$ is about 0.3 for the range of P = 23–3900. This fact clearly denies the possibility that the reaction rate is controlled solely by the translational diffusion of polymer coils. The translational diffusion coefficient of the polymer molecule D_p as well as that of segment D_s is proportional to absolute temperature and inversely proportional to solvent viscosity η_0. Consequently, when k_q is multipled by η_0/T, the resultant reduced rate constant $k_q\eta_0/T$ is independent of η_0 and T regardless of whether the rates are controlled by the translational diffusion step of the polymer molecule or by that of the segment carrying the active group.

The reduced rate constants $k_q\eta_0/T$ are shown in FIGURE 4 against P. Their values at different T in each solvent come into a single normalized curve characterizing the effect of P and solvent-polymer interaction or excluded volume effect. In the range

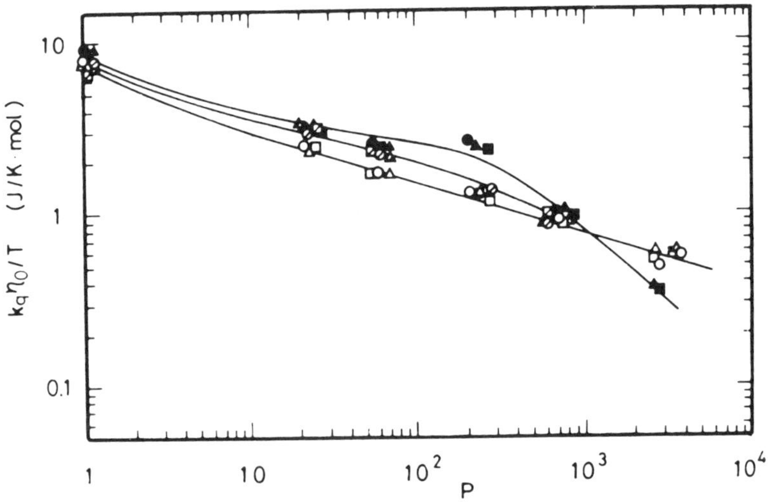

FIGURE 4. Reduced-rate constant $k_q\eta_0/T$ against P for polymer-polymer reactions in various solvents. Symbols are the same as indicated in FIGURE 3.

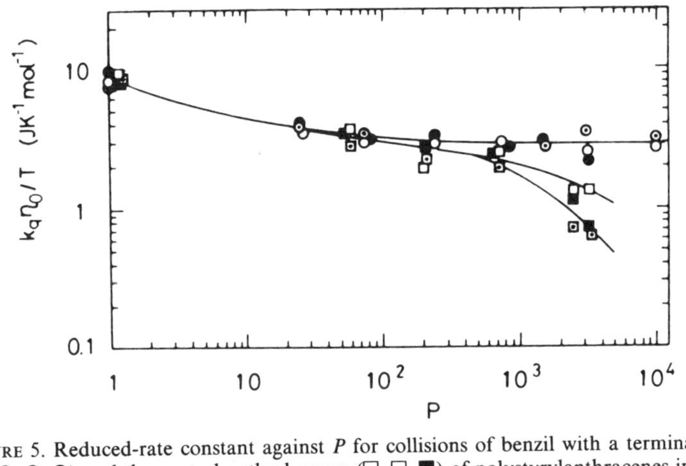

FIGURE 5. Reduced-rate constant against P for collisions of benzil with a terminal anthryl group (○, ○, ●) and the central anthryl group (□, □, ■) of polystyrylanthracenes in benzene (○, □), butanone (○, □), and cyclohexane (●, ■).

of $P < 2 \times 10^2$, the poorer the solvent, the larger becomes the reduced rate constant for the same P. This tendency is in accord with the prediction of several theories which take into consideration the thermodynamic equilibrium potential of coiled polymers; i.e., that the repulsion of polymer molecules against each other is stronger in better solvent and this makes the access of polymer, hence the access of the reactive groups, difficult. The effect can be called thermodynamic excluded volume effect. Even the experimental P dependencies agree well with those calculated from the theory presented by the authors. It assumes that the reaction rate is controlled by the diffusion of the reactive segment and that the latter is a function of the position of the segment in the polymer coil.[3]

Beyond $P > 5 \times 10^2$, the reduced rate constants in poor solvent, especially in cyclohexane, decrease markedly with P and become smaller than the coresponding values in benzene at the same P. This behavior cannot be anticipated from any existing theory. But the fact that the reduced rates are smaller in poorer solvent at $P > 10^3$ has been reported by several authors for a radical termination reaction[11,12] and also for an activation-controlled reaction.[13] In a high P region, non-equilibrium or kinetic excluded volume effect seems operative. As the kinetic excluded volume effect is thought to come from the topological hindrance of the polymer chain against the approach of the reactive sites to each other, its effect would be naturally affected by the segment density, especially at high P.

The quenching rate constants k_q depend slightly on the polymer concentration. Consequently, for a more rigorous discussion, this fact should be considered. Most experiments described above were made for concentrations of less than 1%. Even when the data in this dilute concentration range are extrapolated to infinite dilution by using some empirical relation or by using the theoretical relation reported,[14] the tendency of the reduced rate constants in poor solvent to decrease sharply at $N > 10^3$ does not change at all.[15]

Polymer-Small Molecule Reaction

Quenching reactions of triplet benzil by the anthracene group attached to the chain end of polystyrene (PS-A) and that incorporated at the center of the polymer

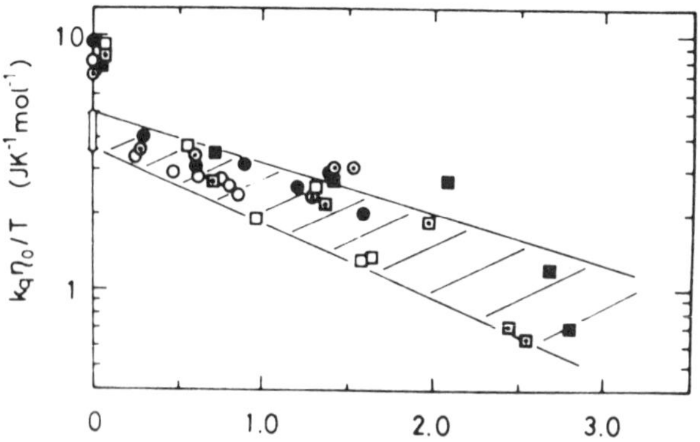

FIGURE 6. Reduced-rate constants against ν. Symbols are the same as in FIGURE 5.

(PS-A-PS) have been studied.[16] As in the case of polymer-polymer reactions, the activation energies of the quenching rate constants k_q had similar values to those of solvent viscosity in any solvent studied. Thus, for discussions of factors other than viscosity, the reduced rate constants $k_q\eta_0/T$ can be used. They are shown as functions of P of polystyrene moiety in FIGURE 5. The reaction for $P = 1$ is represented by the reaction of triplet benzil with 9,10-dimethylanthracene, which is found to have the same rate constant as 9-methylanthracene within experimental error.

The reduced rate constants for *PS-A* are the same in three solvents, benzene, butanone and cyclohexane, showing that in this case the thermodynamic excluded volume effect, observed in polymer-polymer reaction, is not operative and the viscosity is the only factor influencing k_q. The reduced rate constants decrease with increasing P of *PS-A* up to $P = 10^2$ to attain a constant value which is about 1/2 to 1/3 of the value for the model reaction ($P = 1$). In Smoluchowski's equation, R does not change with P. However, D, which is the sum of two diffusion coefficients, D_B (benzil) and D_A (anthryl group), decreases from $D \simeq 2D_B$ at $P = 1$ to $D = D_B$ at large

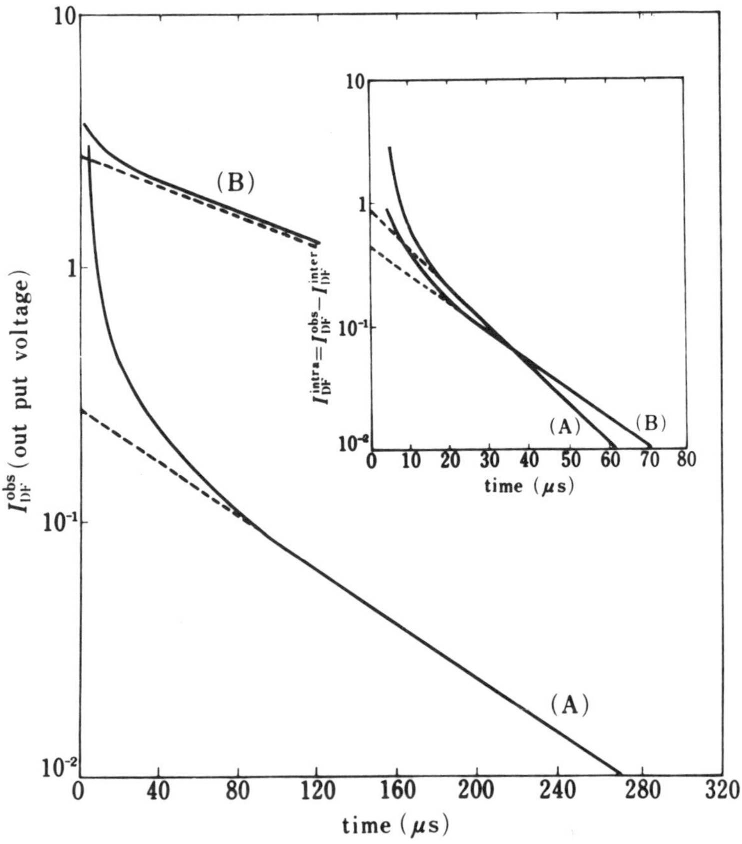

FIGURE 7. Smoothed semilogarithmic decay curves of delayed fluorescence of *A-PS-A* ($P = 280$) in benzene (A) and in butanone (B) at 30°C. Insert: semilogarithmic decay curves of subtracted delayed fluorescence.

P because D_A is nearly equal to D_B for $P = 1$ while it becomes more than one order smaller and neglegible at large P.

The reduced rate constants for the reaction of benzil with the central anthryl group in the polystyrene chain (*PS-A-PS*) shows a sigmoidal dependence on P and they become a little smaller than the corresponding k_q for the terminal anthryl group, especially in poor solvents at large P. As only the diffusion constant of a small molecule contributes to the reaction rate at large P, the results observed show that the diffusive characteristics of triplet benzil, until reaching the collision site, is influenced by the position of the anthryl group in the polymer or, more rigorously speaking, by the distribution of inert polymer segments (segment density) around the anthryl group, which carries one polystyryl chain (*PS-A*) or two polystyryl chains with varying P. To evaluate the effect of the segment density distribution, the number of collisions ν of a benzil molecule with chain segments during its hypothetical linear course of traverse from outside the polymer coil to the anthryl group was calculated. When the reduced rate constants are plotted against ν (FIGURE 6), most of the plots are included in the shaded area. This suggests the proportionality of $k_q\eta_0/T$ to $\exp(-C\nu)$ or by a relation determined by a single parameter, regardless of the position of the anthryl group or the quality of solvent. The extrapolation to $\nu = 0$ gives a value which is just half the value for $P = 1$ where $D_A = D_B$. The value of C is 0.5–0.7. This means that k_q is reduced to about one half for every collision with a polymer segment or, in other words, the apparent diffusion coefficient inside the polymer coil is much smaller than that outside the coil.

INTRAMOLECULAR REACTIONS IN DILUTE SOLUTION

The intramacromolecular diffision-controlled reaction between the two chain end groups of a flexible polymer molecule has been studied with the system of a triplet-triplet annihilation reaction of α,ω-dianthrylpolystyrene.[17] All the experiments were

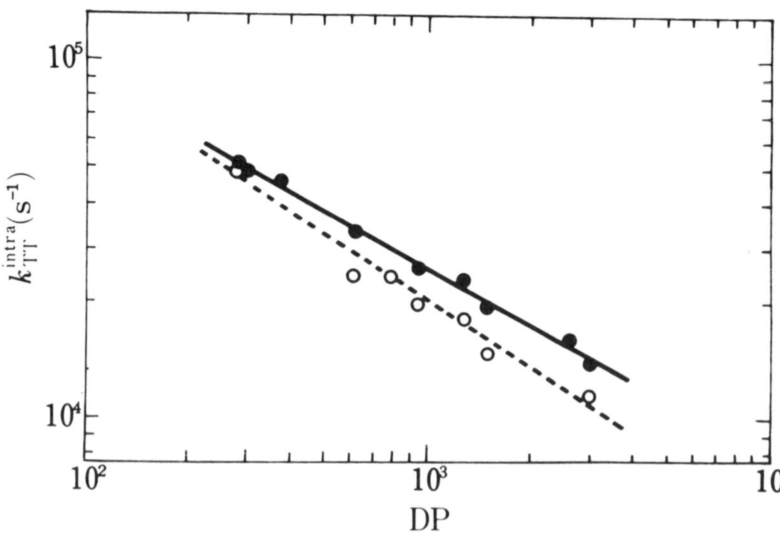

FIGURE 8. Dependence of k_{TT}^{intra} on P in benzene (●) and in butanone (○) at 30°C.

made at polymer concentrations of less than 1%. The typical decay curves of the delayed fluorescence derived from triplet anthracene groups excited by nitrogen laser-pulse are shown in FIGURE 7. They are composed of three stages, each having a different mechanism as the main cause of the decay of the triplet. The first stage reflects mainly second-order decay of the triplet due to intermolecular *T-T* annihilation (k_{TT}^{inter}) and the third one corresponds to first-order decay of the triplet due to thermal deactivation (k_d). Of course, this decay is always monitored by the delayed fluorescence due to the second-order *T-T* annihilation reaction. (slope of log (intensity) vs. time is $-2k_d$) But at this stage of decay, the contribution to the decay of the triplet by *T-T* annihilation is disregarded ($k_{TT}^{inter} \cdot [^3A^*] << k_d$). The delayed fluorescence at the second stage represents mainly first-order intramolecular *T-T* annihilation (k_{TT}^{intra}) with some contribution from the two other mechanisms. So, when the linearly extrapolated values from the third stage of decay in the semilogarithmic scale are subtracted from the observed ones and the resultant delayed fluorescence intensity is again plotted in a semilogarithmic scale against time, a linear decay line is obtained in the range of the second stage as shown in FIGURE 7 and the first-order decay constant k_{TT}^{intra} is determined from the slope. It should be noted that k_d obtained at the third stage depends on the polymer concentration because triplet anthracene is, as already stated, quenched slowly by polystyrene.

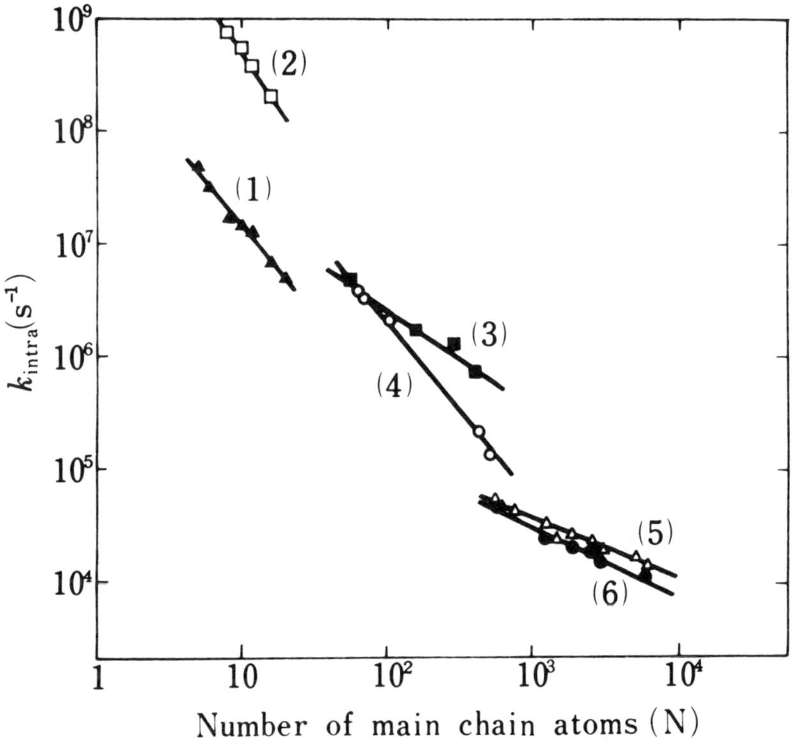

FIGURE 9. Rate constants for intramolecular chain-end reactions against a number of main chain atoms *N* by different authers: (1) ref.21, (2) ref.18, (3) ref.20, (4) ref.19, (5) current work in benzene, (6) current work in butanone.

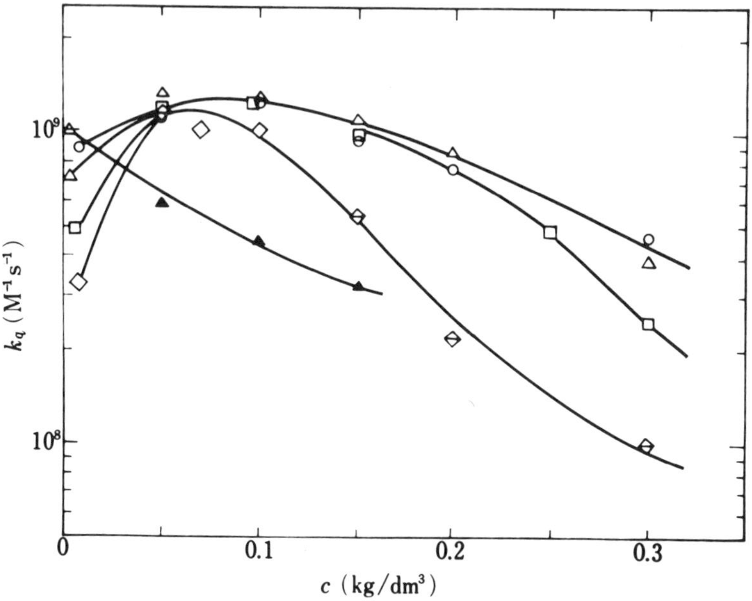

FIGURE 10. Dependence of k_q at 30°C on the polymer concentration in benzene (○, △, □, ◇, ⊖) and in cyclohexane (▲). The combinations of polymers are *PS*60*B*-*PS*71*A*-*PS*98 (○) *PS*340*B*(*PS*210*B*)-*PS*400*A*-*PS*410 (△, ▲), *PS*640*B*-*PS*800*A*-*PS*1000 (□), and *PS*2900*B*-(*PS*1200*B*)-*PS*3900*A* (*PS*3300*A*)-*PS*3800 (◇, ⊖).

The rate constants for intramolecular *T*-*T* annihilation $k_{TT}{}^{intra}$ thus obtained at 30°C are plotted against *P* in FIGURE 8. They are of the order of 10^4 sec^{-1} and, as expected, they decrease with increasing degree of polymerization. The exponent *n* in the relation $k_{TT}{}^{intra} = B \times P^{-n}$ is 0.53 in benzene and 0.60 in butanone for the range of $P = 280$–3000. The activation energies could not be determined accurately because of large experimental error. But it was confirmed that they do not exceed by much those of solvent viscosity.

The exponents *n* obtained in the present work are very small as compared with those in the low *P* region reported by other authors with the use of excimer formation: $n = 1.5$ (oligomethylene),[18] 1.35 (polystyrene, $P\,2.5 \times 10^2$),[19] and 1.0 (polyoxyethylene, $P\,4 \times 10^2$).[20] They are summarized in FIGURE 9 along with the rate constants for the electron exchange reaction between naphthalene anion radical and naphthalene.[21] The meaning of this rate constant is not the same as that for the other reactions because the electron exchange reaction can be repeated without deactivation. As the polymers (flexibility), solvents (viscosity and expansion) and reactions (reaction diameter and steric hindrance) are not the same for these experiments, a direct comparison of the absolute rate constants may not be adequate. As for the slopes, however, there is a net trend that the exponents decrease with *P*. So our experimental results are not inconsistent with other experiments from this point of view, but they are clearly contradictory to several theories which predict 1.5–2 for *n*.[4,5] More experiments and reconsideration of theories seem necessary.

Intermolecular Reactions in Semi-Dilute Solution

Preliminary results of the quenching reactions of *PS-B* by *PS-A* in a semi-dilute solution of inert polystyrene[22] and in gel[23] are described briefly.

The quenching rate constants k_q in the presence of inert linear polystyrene with P similar to that of probe polymers have been measured in the concentration range of up to 30% mainly in benzene. Up to 5%, k_q increases with increasing concentration in benzene while it decreases in cyclohexane. The behaviors just correspond to the increase and decrease in diffusion coefficient of polystyrene in each respective solvent (see Fig. 10)[24]

For C = 5 − 10%, k_q does not depend any more on P and it begins to decrease again with further increase in C at a critical concentration which appears earlier for the polymer with higher P. The critical concentration is considered to correspond to the onset of so-called gel effect in radical polymerization.[25,26] The exponent of concentration dependence of k_q after the critical point is −1.7–1.9. This value seems to agree with the prediction $D_{self} \alpha C^{-1.75}$ by DeGennes' dynamic scaling theory based on a reptation model.[27,28]

For the study of quenching reactions in gel, crosslinked polystyrenes (radically polymerized with divinylbenzene) swelled to equilibrium in benzene were used. The quenching rate constant for benzil-anthracene system (small molecule—small molecule reaction) is not affected at all by the presence of crosslinks. It decreases monotonically with increasing polymer concentration (reciprocal of degree of swelling) and fits in well with the curve of k_q against C for the linear polymer. On the other hand, when polymeric probes are used, the ratios of k_q in gel and in linear polymer solution at the same concentratiopn, $(k_q)_{gel}/(k_q)_{linear}$ are no longer unity and they are smaller for larger P of the probe polymer, e.g., the ratio is about 0.2 for $P \simeq 10^{-3}$ at $C \simeq 0.2$.

References

1. North, A.M. 1968 Progress in High Polymers, Vol. 2. J.C. Robb & F.W. Peaker, Eds. 95. Illife. London.
2. North, A.M. (1974) Reactivity, Mechanism and Structure in Polymer Chemistry. A.D. Jenkins & A. Ledwith, Eds.: 142. Wiley. London.
3. Horie, K., I. Mita & H. Kambe. 1973. Polym. J. **4:** 341.
4. Wilemski, G. & M. Fixman. 1970. J. Chem. Phys. **60:** 866, 878.
5. Doi, M. 1975. Chem. Phys. **9:** 455.
6. Horie, K. & I. Mita. 1976. Polym. J. **8:** 227.
7. Ushiki, H., K. Horie, A. Okamoto & I. Mita. 1979. Polym. J. **11:** 691.
8. Horie, K. & I. Mita. 1977. Polym. J. **9:** 201.
9. Ushiki, H., K. Horie, A. Okamoto & I. Mita. 1980. Polym. J. **12:** 35.
10. Horie, K. & I. Mita. 1978. Macromolecules **11:** 1175.
11. Cameron, G. G. & J. Cameron. 1973. Polymer **14:** 107.
12. Kiwi, J. & W. Schnabel. 1976. Macromolecules **9:** 468.
13. Morawetz, H., J.R. Cho & P.J. Gans. 1973. Macromolecules **6:** 624.
14. Mahabadi, H.KH. & K.F. O'Driscoll. 1977. Macromolecules **10:** 55.
15. Horie, K. & I. Mita. 1980. Report to Institute of Space and Aeronautical Science. University of Tokyo. **45:** 87.
16. Horie, K., K. Tomomune & I. Mita. 1979. Polym. J. **11:** 539.
17. Ushiki, H., K. Horie, A. Okamoto & I. Mita. 1981. Polymer J. **13:** In Press.
18. Yamamoto, Y., K. Goshiki, T. Kanaya & Y. Nishijima. 1977. Chem. Phys. Lett. **56:** 333.
19. Winnik, M.A., A.E.C. Redpath & D.H. Richards. 1980. Macromolecule **13:** 311.

20. Cuniberti, C. & A. Perico. 1977. Eur. Polym. J. **13:** 369.
21. Shimada, K. & M. Szwarc. 1975. J. Am. Chem. Soc. **97:** 3313.
22. Mita, I., K. Horie & M. Takeda. 1979. Preprints Macro-Mainz. p. 297.
23. Mita, I., K. Horie & M. Masuda. 1980. Presented at 7th Discussion Conference on Polymer Network.
24. Rehage, G., O. Ernst & J. Fuhrmann. 1970. Discuss Faraday Soc. **49:** 208.
25. Turnner, D. T. 1977. Macromolecules **10:** 221.
26. Lee, H.B. & D.T. Turnner. 1977. Macromolecules **10:** 226, 231.
27. De Gennes, P.G. 1976. Macromolecules **9:** 587, 594.
28. De Gennes, P.G. 1979. Nature **282:** 367.

THE DYNAMICS OF POLYMER CYCLIZATION 3. EXCLUDED VOLUME EFFECTS IN THE END-TO-END CYCLIZATION OF POLYSTYRENE PROBED BY INTRAMOLECULAR PYRENE EXCIMER FORMATION

A.E.C. Redpath and Mitchell A. Winnik

Lash Miller Chemical Laboratories
Department of Chemistry, and Erindale College
University of Toronto
Toronto, Canada M5S 1A1

INTRODUCTION

Since the early days of this century, chemists have been fascinated by the idea that flexible chain molecules might cyclize. The discovery that musks, important components in the fabrication of perfumes, were macrocyclic hydrocarbons, gave impetus to the pioneering efforts of Carothers[1] and Ziegler[2] to synthesize these molecules. This in turn prompted others[3,4] to address the question of what factors affected end-to-end chain cyclization.

In 1934, Kuhn published his now classic paper elucidating the random-walk-like nature of polymer conformation.[5] His work suggested a Gaussian distribution function for the end-to-end distance of the chain and predicted that the cyclization probability $W(0)$ could decrease as $N^{-3/2}$, where N is the number of bonds in the chain. In this context, Jacobsen and Stockmayer[6] related the equilibrium constant for end-to-end cyclization (K_{cy}) to $W(0)$ and demonstrated how K_{cy} might be obtained from ring-chain equilibrium polymerization studies.[7] They suggested that if the cyclized polymer could reopen in N different ways, the K_{cy} would decrease as $N^{-5/2}$. Much more recently, Brown and Slusarczuk[8] measured K_{cy} by isolating the macrocycles produced in dimethyl siloxane polymerization and provided the major piece of evidence that K_{cy} does in fact decrease, for long chains, as $N^{-5/2}$. Flory and Semlyen[9] subsequently calculated K_{cy} in this system in terms of the rotational isomeric state model. Semlyen has followed up this work with extensive experimental and computational studies, which he recently reviewed.[7] While the siloxanes provide the only solid experimental evidence on this point, it is generally believed that in θ-solvents, where polymer chains are Gaussian, that $K_{cy} \sim N^{-5/2}$ for chains that cyclize and can reopen at any monomer-monomer bond, and that $K_{cy} \sim N^{-3/2}$ for chains which cyclize and reopen at one unique bond.

A new chapter in this history was opened in 1973 and 1974 when Wilemski and Fixman[10] published their theoretical investigations of the dynamics of end-to-end cyclization in polymers. Their analyses are couched in terms of two models for the chain. The first, the harmonic spring (HS) model considers the chain as a harmonic spring connecting two terminal pendant groups whose adjacency defines the cyclized state. The second description is developed in terms of the Rouse-Zimm ball-and-spring model where the chain is taken to be a series of beads (friction centers where

0077-8923/81/0366-0075$01.75/0 © 1981, NYAS

the hydrodynamic interactions occur) separated by harmonic springs. As in the HS model, cyclization is defined by the approach of the two terminal beads to within a critical distance of one another.

The Wilemski-Fixman papers have had a major influence on the field of polymer dynamics. Doi[11] has published a series of papers examining some of the fine points of the HS model and its description of cyclization dynamics. In addition, he has undertaken a computer simulation of cyclization dynamics using a model which resembles the free-draining Rouse-Zimm chain.[11d] Perico and Cuniberti[12] have explored the hydrodynamic aspects of the Rouse-Zimm model for partially draining finite chains in a θ-solvent. Parameterizing the hydrodynamic interactions from intrinsic viscosity measurements,[12b] they predict values for the rate constant k_{cy} for dynamically controlled cyclization. Another, simplified description of cyclization has been presented by Brereton and Rusli.[13]

There are several classes of experiments which lend themselves to the measurement of the rates of cyclization of polymer chain ends. A theoretician would consider these boundary conditions on the initial state of the reaction. One such experiment might begin with cleavage of a single bond in a macrocycle and observe the re-encounter of the dangling chain ends. A second type of experiment might subject a conformationally relaxed polymer to an excitation pulse that would selectively excite a group on one end to an electronically excited state. If the groups on the chain ends were properly chosen, approach of the two ends of the chain to within a critical distance a would lead to an observable event. This event could be a chemical reaction, fluorescence quenching, or the generation of a new emission.

Although examples of the former class of experiments are known, the latter has the attraction that it is amenable to study by a number of contemporary techniques which are themselves the subject of considerable current interest. For example, the phenomenon of non-radiative (Förster) energy transfer permits one to study the approach of well-chosen terminal substituents on a polymer chain to within distances a that are large with respect to the molecular diameters of those substituents. Katchalski-Katzir[14] and his co-workers have carried out pioneering work in the application of this methodology to the study of polypeptide dynamics. Using this technique, one could in principle examine predictions of the theory which suggest that k_{cy} should depend upon a.

When chains are very long, the time for end-to-end cyclization may be much longer than that of fluorescence from even the longest-lived singlet electronic excited states of aromatic chromophores. Techniques based upon triplet states of these chromophores have a special appeal. One such technique involves triplet-triplet annihilation, leading to delayed fluorescence when a chain containing two terminal substituents, each excited to its triplet state, cyclizes. Mita and his group[15] have begun important studies along these lines with anthracene end-capped polystyrene. The technique is a difficult one. The most difficult problem appears to be in producing a large fraction of excited polymers with both ends excited to the triplet excited state.

Intramolecular excimer formation between pyrenes attached to the ends of a polymer chain had a special appeal to us for measuring the dynamics of chain cyclization.[16,17] First, it is necessary to excite only one chain end of only a small fraction of the sample in order to observe the excimer. Second, excimer emission occurs far to the red of the localized pyrene fluorescence, making its detection relatively unambiguous.[18] Third, excimer kinetics have a particularly simple form, which we felt would facilitate data analysis; excimer kinetics have been well-studied for bimolecular reactions.[18] Intramolecular excimer formation is an active area. Here it appears that excimer kinetics only become complicated when the chain connecting the two terminal chromophores is sufficiently short and sufficiently stiff to restrict the mutual orientation of the end-groups when they are in proximity.

Pioneering studies using pyrene excimer formation to detect dynamically controlled end-to-end cyclization in poly(ethylene oxide) were carried out by Cuniberti and Perico.[17] Their results were qualitative. It is unfortunate that they faced two very difficult problems—incomplete functionalization of the chain ends, and lack of access to state-of-the-art photochemical instrumentation for measuring fluorescence lifetimes.

We began our studies with pyrene-terminated polystyrene, following in spirit the approach of Cuniberti and Perico.[17] Beginning with bis-HO-terminated polystyrene synthesized by Dr. D.H. Richards in England and by one of us (MAW) in Strasbourg, we attached pyrenes quantitatively to both ends as esters of 4(1′-pyrene)-butyric acid to yield the polymer.

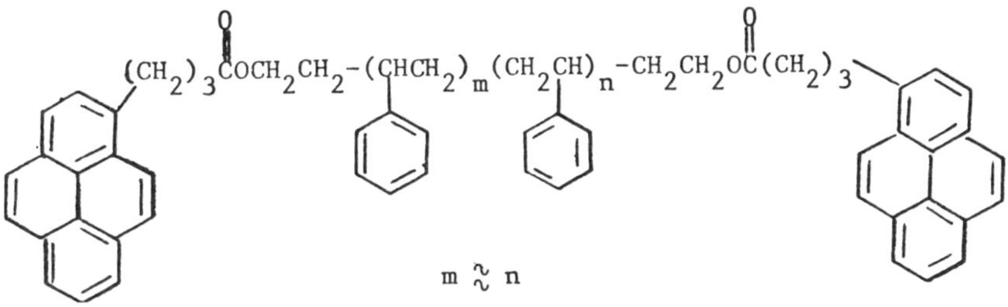

Our initial studies[19] demonstrated (i) that intramolecular excimer formation could be observed without complication from bimolecular reactions for molecular weights up to at least $\bar{M}_n = 30000$; (ii) the rate constant k_{cy} is a strong function of molecular weight, decreasing rapidly as the chains get longer; (iii) weak excimer emission for long chains suggested that the techniques might be limited to chains of less than 1000 to 2000 bonds [$\bar{M}_n \leqslant 50000$ to 100000 for polystyrene]; (iv) because of the sensitivity of k_{cy} to chain length N, polymer samples for quantitative work would have to be highly fractionated and the results corrected for the distribution of molecular weights in the sample.

This paper describes our progress to date, with individual polymer samples sufficiently fractionated that meaningful quantitative data could be obtained for both the dynamics of cyclization and the equilibrium constant for end-to-end cyclization.

Experimental

Fluorescence spectra were measured on a Hitachi-Perkin Elmer MPF-44B spectrometer. Some spectra were rerun by Dr. L. Costa at the Eastman-Kodak Research Laboratories, Rochester, New York. These served as a helpful and useful check on the accuracy of our I_E/I_M intensity measurements, especially for very long polymers. UV spectra were run on a Cary 210 UV-visible absorption spectrometer. Molecular weight analysis was carried out by gel permeation chromatography using a Waters Associates high pressure liquid chromatograph in conjunction with three microstyragel columns of pore size 10^4 Å, 10^3 Å, and 500 Å respectively. These were calibrated with "monodisperse" polystyrene standards purchased from Pressure Chemical Corporation, Pittsburgh, Pa.

Fluorescence decay curves were measured by the time-correlated single photon counting technique.[20] For all fluorescence measurements, the sample temperature was controlled by circulating water through the sample block. The temperature of the circulating water was controlled by a Lauda model K2R circulating bath with the block temperature remaining constant to within ±0.1 °C. Samples for fluorescence measurements were weighed out on a microbalance and diluted to 10.0 ml. These solutions, approximately 0.7 to 3.0×10^{-6} *M*, were placed in 12.7 mm O.D. pyrex tubes and degassed by five successive freeze-pump-thaw cycles, then sealed under a vacuum of better than 1×10^{-5} Torr.

Toluene for all measurements was purified by refluxing over sodium metal (under nitrogen) for several hours, followed by fractional distillation through a 20 cm × 1 cm jacketed column packed with glass helices. Cyclohexane was stirred over sulfuric acid, under nitrogen. The sulfuric acid was changed daily. When the sulfuric acid layer remained clear, the cyclohexane was decanted, washed successively with water and aqueous sodium bicarbonate, dried over $CaSO_4$ and distilled from sodium metal.

Bis-HO-terminated polystyrene was synthesized by standard methods using potassium naphthalene as the initiator and ethylene oxide to quench the living polymer ends. These syntheses were carried out in the laboratory of Dr. Paul Rempp at the Centre de Recherches sur les Macromolecules in Strasbourg, France, with the helpful assistance of Mr. Schuster. These polymers had polydispersities $[\bar{M}_w/\bar{M}_n]$ ranging from 1.2 to 1.6. Attempts to prepare very low molecular weight polymer gave material with polydispersities slightly above 2.

New techniques were developed based upon silica gel column chromatography to fractionate these polymers. These chromatography columns were run by Ms. Katherine Paton in our laboratory. A complete description will be published elsewhere.[21] In summary, we can say that in this way we could prepare 100 to 200 mg fractions of varying molecular weight with polydispersities in the range of 1.09 to 1.16. After attachment of the pyrene end-groups as esters of pyrenebutyric acid, the polymers were further purified by preparative gel permeation chromatography. This gave a material of usefully low polydispersity ($\bar{M}_w/\bar{M}_n = 1.05$ to 1.13) and free of fluorescent low molecular weight impurities.

Results and Discussion

The overall focus of this research is to explore several aspects of end-to-end cyclization dynamics of polymers in dilute solution. Our approach is to synthesize polymers—in this case polystyrene—to which specific chromophores are attached to the chain termini. We carry out kinetic studies of the photochemical and photophysical interactions between these chromophores after specifically exciting one of them to an electronically excited state with light of an appropriate wavelength. From these data, we attempt to extract the rate constant for interaction between the terminal chromophores and to relate it to the behavior of k_{cy}, the rate constant for dynamically controlled end-to-end cyclization for a polymer of unique molecular weight.

The fluorescence spectrum of pyrene-end-capped polystyrene in dilute toluene solution is shown in FIGURE 1. At short wavelengths, we observe a structured pyrene-like fluorescence. We call the intensity of this emission I_M. In addition, a broad, structureless emission centered at 495 nm, is observed. This emission is due to the pyrene excimer, an excited, sandwich-shaped complex formed from an excited pyrene and a ground state pyrene upon intramolecular, end-to-end cyclization of the polymer chain. We call the intensity of the excimer emission I_E. It is clear from FIGURE 1 that the intensity ratio I_E/I_M is sensitive to the length of the polymer separating the chain ends.

These excimers are formed in a dynamic process reflecting the Brownian motion of the isolated polymer chains in solution. In the ground electronic state pyrenes interact only repulsively.[18] Thus before irradiation, the polymer is in its conformationally relaxed random configuration. A negligible fraction of these chains have their ends in proximity. After excitation with light of low intensity and of an appropriate wavelength, one pyrene on each of a very tiny fraction of the polymer chains becomes excited. As the chains in the subset of the polymer sample undergo conformational changes, their chain-ends sometimes encounter one another. Excimer formation occurs between the pyrene substituents. Binding in the excimer is thought to be sufficiently strong ($\Delta H_o \approx -8$ kcal/mol)[18] that the excimers emit their green fluorescence before they can dissociate back to the open form of the polymer. In steady-state fluorescence measurements, (FIG. 1), the intensity ratio I_E/I_M is sensitive to the dynamics of chain cyclization. In fluorescence decay measurements, one has direct means of observing the time profile of excimer formation and decay.

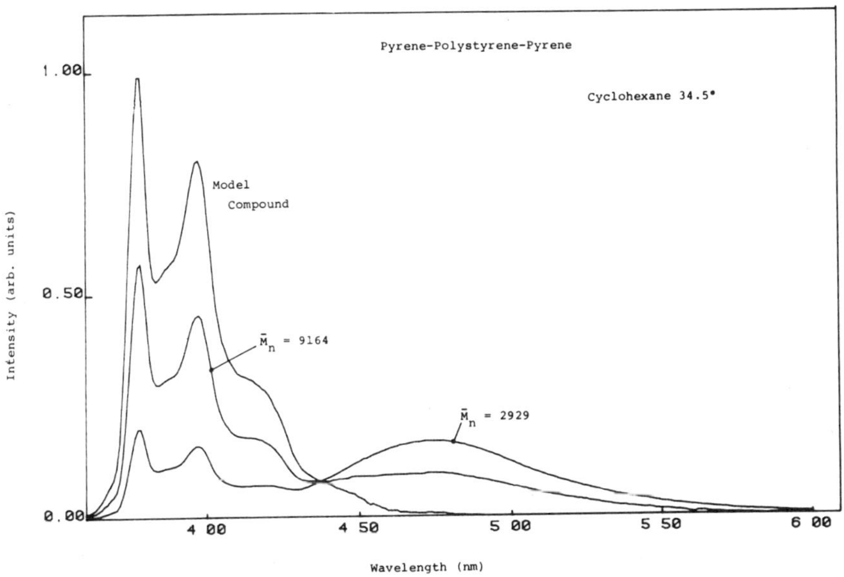

FIGURE 1. Fluorescence spectra of the polymers pyrene-polystyrene-pyrene in degassed cyclohexane at 34.5°C at 2×10^{-6} M. The "model compound" is **2**, the methyl ester of pyrenebutyric acid. Note that for the lowest molecular weight polymer, the integrated excimer emission I_E is more intense than the locally excited pyrene fluorescence. The spectra are uncorrected.

The meaningful values for testing theories of end-to-end cyclization dynamics are those of the rate constants for intramolecular excimer formation between the pyrenes attached to the ends of the polymer molecules. Rate constants in kinetic studies are always obtained by reference to a hypothetical reaction mechanism. This means that conclusions drawn from interpretation of rate constants are no more secure than the assumed mechanism itself. In quantitative studies of molecular conformation and dynamics, based upon reaction kinetics, it is essential to demonstrate that the rate law predicted by the reaction mechanism applies to the system being studied.

The Mechanism of Excimer Formation

We assume that intramolecular excimer formation between the pyrenes (P) attached to the two ends of our polystyrene chains follows the mechanism in Scheme I below. Irradiation of the sample with light of an appropriate wavelength (typically 347 nm) causes one pyrene on a small number of the polymer chains to become electronically excited. This locally excited group can decay back to the ground state by giving off light (fluorescence) or heat, with k_M describing the sum of these rates; or it can form a pyrene excimer intramolecularly, with a rate described by $\langle k_1 \rangle$, through the process of end-to-end cyclization. We write $\langle k_1 \rangle$ in brackets because it is predicted to be sensitive to chain length; hence its value represents an average over the molecular weight distribution in the sample. It will be convenient, following Birks, to define $X = k_M + \langle k_1 \rangle$.

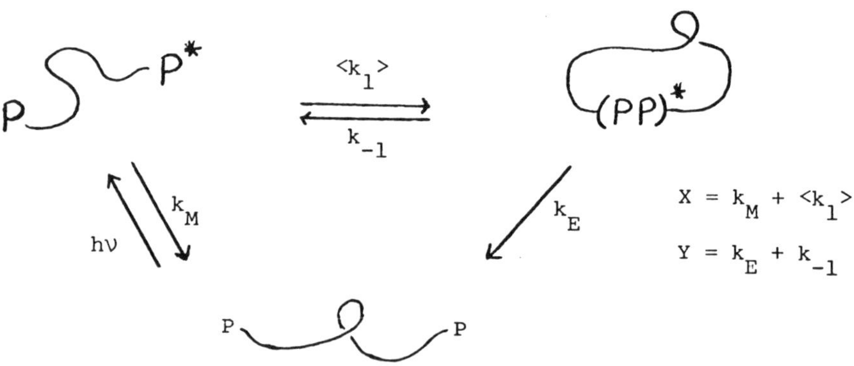

Scheme I.

According to this scheme, the excimer, once formed, can dissociate back to the locally excited pyrene by ring opening described by k_{-1}, or it can decay either by fluorescence or without radiation, back to the ground state, with k_E describing the sum of these rates. We define $Y = k_E + k_{-1}$ as the sum of all decay paths of the excimer.

Bimolecular reactions of P^* can also lead to excimer formation. At the concentrations we employed in our experiments [7×10^{-7} *M* to 3×10^{-6} *M*], these contributions are negligible.

Scheme I predicts that the locally excited pyrene excited state, after excitation with an infinitely short pulse of light, will decay as the sum of two exponential terms.[18] Its fluorescence intensity $I_M(t)$ will decay according to

$$I_M(t) = a_1\exp(-\lambda_1 t) + a_2\exp(-\lambda_2 t) \tag{1}$$

The excimer fluorescence $I_E(t)$ is described by a difference of the same two exponential terms:[18]

$$I_E(t) = a_3[\exp(-\lambda_1 t) - \exp(-\lambda_2 t)] \quad \textbf{(2)}$$

The decay constants λ_1, λ_2 are related to the rate constants in the scheme via the complicated expression[18]

$$\lambda_1, \lambda_2 = 1/2\{(X + Y) \mp [(Y - X)^2 - 4 <k_1> k_{-1}]^{1/2}\} \quad \textbf{(3)}$$

To obtain all four rate constants in the scheme, one needs four measurable quantities. From the decay studies, one has in principle three independent values: λ_1, λ_2, and (a_2/a_1). This problem can be resolved by reference to an appropriately chosen model compound. Its exponential fluorescence decay time τ_M can be assumed to give k_M according to $\tau_M^{-1} = k_M$ when measured under the same conditions as λ_1 and λ_2. In our experiments, we use the methyl ester (**2**) of pyrene butyric acid (**1**) as the appropriate model compound.

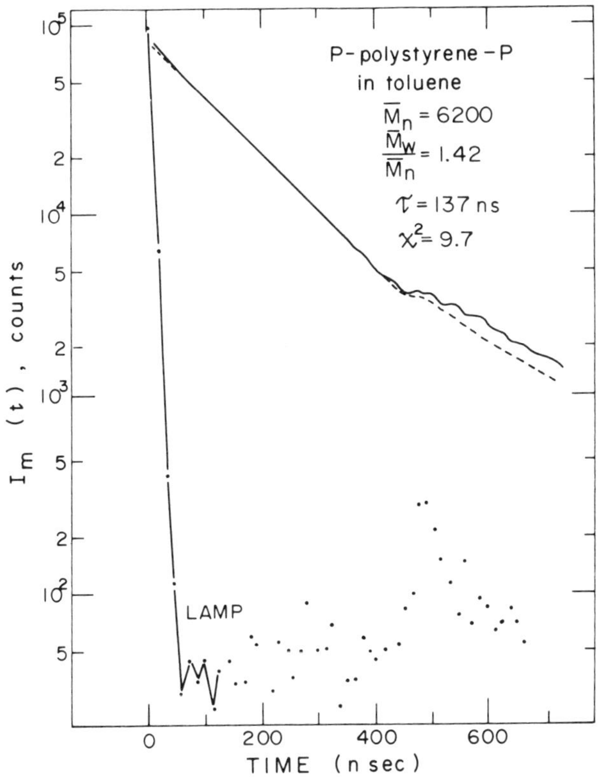

FIGURE 2. Semilogarithmic plot of log $I_M(t)$ vs time for (a) Pyrene-polystyrene-pyrene of $\bar{M}_n = 6400$, $\bar{M}_w/\bar{M}_n = 1.3$, solid line; (b) Lamp decay profile, ·····; (c) An exponential decay convoluted with the decay profile of the pulsed lamp, ----. Note deviations of $I_M(t)$ from the exponential decay at both short and long times.

Irreversible Excimer Formation

It is commonly believed that because bimolecular pyrene excimer formation in solution is essentially irreversible at room temperature, intramolecular excimer formation, as described by Scheme I, will also be irreversible at room temperature. This is tantamount to assuming that k_{-1} is negligibly small, which according to Equation 3 means that the pyrene locally excited state I_M(t) should decay exponentially with a decay time $\lambda_1 = \langle k_1 \rangle + k_M$. Under these conditions $\langle k_1 \rangle$ can be determined from

$$\langle k_1 \rangle = \lambda_1 - k_M \quad \textbf{(4)}$$

with the understanding that when the chains are very long, $\lambda_1 \approx k_M$, and one will know $\langle k_1 \rangle$ with very poor precision.

This prediction that $I_M(t)$ decays exponentially is an experimentally testable hypothesis. As FIGURE 2 indicates, for polymers of modestly narrow polydispersity ($\bar{M}_w/\bar{M}_n$ = 1.2 to 1.4), substantial deviations from exponential behavior can be

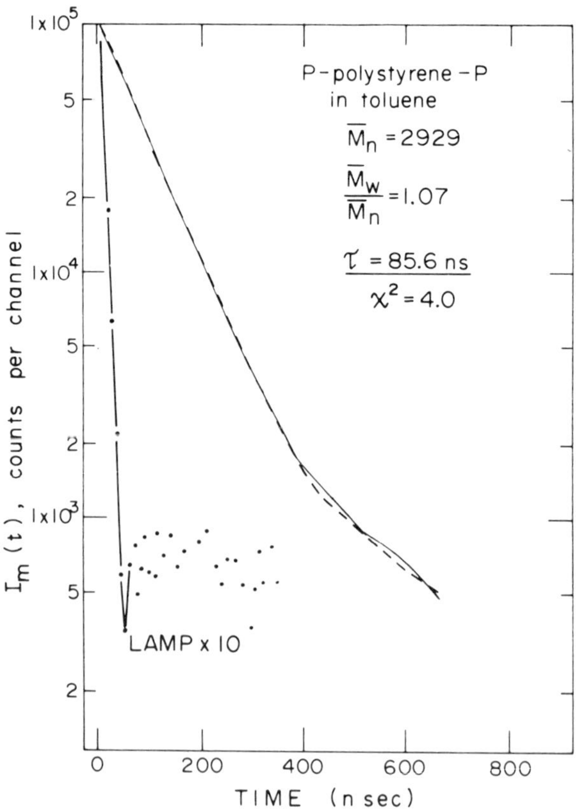

FIGURE 3. Semilogarithmic plot of log $I_M(t)$ vs time for (a) Pyrene-polystyrene-pyrene of $\bar{M}_n$ = 2929, $\bar{M}_w/\bar{M}_n$ = 1.07, solid line; (b) Lamp decay profile, · · · ; (c) An exponential decay convoluted with the decay profile of the pulsed lamp, ---.

observed both at short times and at long times in the fluorescence decay curve. Note that the theoretical exponential decay in the semilog plot in FIGURE 2 is not a straight line. The exciting pulse has a finite width (4 nsec at half maximum) and a long, very weak tail. The theoretical exponential decay represents a convolution of an exponential decay with the excitation profile of the lamp pulse.

There are at least two possible explanations for these deviations from exponential behavior of $I_M(t)$ in toluene at 22°C. First, the mechanism in Scheme I may be incorrect, or the assumption that k_{-1} is negligibly small may not be true. Second, it may be that $<k_1>$ is so sensitive to chain length that short chains in the sample have much shorter decay times than long chains in the sample. This dilemma can be resolved by further fractionation of the polymer samples.

FIGURE 3 shows a decay curve for a polystyrene sample of $\bar{M}_n$ = 2929 with ($\bar{M}_w/\bar{M}_n$ = 1.07). Its fluorescence decay is exponential for more than two decades (>99%, >5 lifetimes). The statistical deviation χ^2 = 2.0 is also reasonable for a polymer sample of narrow polydispersity, indicating that at this temperature, the amount of the fast component in the fluorescence decay (a_2) has become negligibly small. Under these conditions, reversal of excimer formation can be neglected, and $<k_1>$ can be calculated from Equation 4.

For polystyrene samples of higher molecular weight in toluene at 22°C, the decay curves are decreasingly sensitive to the polydispersity as the chain length increases. An exponential decay was obtained for the sample of $\bar{M}_n$ = 46,000 where we had difficulty fractionating the polymer further than ($\bar{M}_w/\bar{M}_n$ = 1.26). These data are collected in TABLE 1.

It is worth stressing that 1% or less of polymer containing only one pyrene substituent will make a noticeable contribution to the decay curve in certain instances. Since these molecules should have values of λ_1^{-1} = 193 ns, when the chains are sufficiently short that λ_1^{-1} for the rest of the sample is less than 140 to 150 ns, this impurity will appear as a long-lived tail in the third decade of the emission decay. Such a weak tail appears in the decay curves of three samples where $\bar{M}_n$ = 2929,

TABLE 1
MOLECULAR WEIGHTS AND POLYDISPERSITIES OF PYRENE-POLYSTYRENE-PYRENE SAMPLES, AND THEIR DECAY TIMES IN TOLUENE AT 22°C

$\bar{M}_n$	$\bar{M}_w/\bar{M}_n$	λ_1^{-1} (ns)	χ^2	I_E/I_M
2929	1.07	84.2*	0.9	1.39
3905	1.08	101.8*	1.1	.994
4489	1.07	115.4	7.2	.782
6565	1.06	138.3	1.2	.465
9164	1.13	156.6	1.4	.249
15590	1.12	178.3	2.1	.096
19726	1.05	183.5	4.6	.072
22700	1.13	186.2	3.3	.046
25268	1.22	183.0	0.9	.027
33594	1.09	191.6	3.1	.020
46140	1.20	192.4	1.3	.011
99640	1.18	193.2	3.4	.0035
Model (2)	—	193.3	2.1	—

* Derived from fitting the excimer curve $I_E(t)$ to Equation 2.

3905, and 4489. In the last example, its effect is to raise the χ^2 in the data analysis. In the first two examples, small refinements in the value of λ_1 can be obtained by fitting the excimer curve $I_E(t)$ to Equation 2.

Reversible Excimer Formation in Cyclohexane

Initial experiments in cyclohexane at 34.5°C, the θ-temperature, suggested that here, too, pyrene fluorescence decays could be described by a single exponential function. The distinction between single and double exponential decay functions is sometimes subtle. We were alerted to the possibility of deviations from exponential behavior in the decay functions by a systematic increase in χ^2 in the fit to our data with samples of decreasing molecular weight. In the first sample that we examined thoroughly in terms of a two-exponential fit [$\bar{M}_n = 3905$], we found values of λ_1 and λ_2 in $I_M(t)$ essentially identical to those obtained from analysis of the excimer $I_E(t)$. Here, clearly, k_{-1} is no longer negligible. In cyclohexane at 34.5°C, the data were treated in terms of Equations 1 and 2. For the polymers of $\bar{M}_n < 10^4$ there were clear indications of double exponential decays. Furthermore, values of λ_1 and λ_2 obtained from analysis of the pyrene locally excited state decay were always within experimental error of those obtained by analysis of the excimer emission, lending further support for the mechanism in Scheme I. Nonetheless, it was not always an easy matter to obtain the best values for these decay times. When a_2/a_1 is larger than 2.0 or less than 0.1, it is very difficult to obtain reliable values of the λ's. Even within that range, considerable care and judgment had to be exercised to have confidence in these values.

We found that when excimer emission was substantial, the $I_E(t)$ curves gave the most precise values of λ_1 and λ_2. Using these values as estimates, the non-linear least squares routine gave some scatter in the λ_1, λ_2 values in the fit to $I_M(t)$, but consistent values of a_2/a_1. For longer chains, in which λ_1 was much longer than λ_2, $I_M(t)$ gave

TABLE 2
DECAY TIMES AND COEFFICIENTS FOR DECAY OF THE LOCALLY EXCITED STATE AND EXCIMER FOR THE FLUORESCENCE OF PYRENE-TERMINATED POLYSTYRENE IN CYCLOHEXANE AT 34.5°C

$\bar{M}_n$	λ_1^{-1}(ns)	λ_2^{-1}(ns)	a_2/a_1	χ^2	I_E/I_M*
2929	83.3	39.3	1.594	2.9	3.95
3905	88.6	39.7	.521	1.3	2.69
4489	99.4	44.2	.362	1.4	2.30
6565	120.0	40.3	.112	1.3	1.37
9164	151.9	49.0	.099	1.0	.786
15590	190.7	—	—	1.4	.318
19726	197.4	—	—	1.2	.250
25268	209.7	—	—	1.8	.139
33594	226.9	—	—	1.0	.073
46140	224.9	—	—	1.3	.045
99640	235.9	—	—	2.1	.013
Model (2)	241.3	—	—	1.2	—

* Derived from uncorrected spectra. + λ_1, λ_2 values shown are taken from the decay of the locally excited state $I_M(t)$. In calculating $\langle k_1 \rangle$, both $I_M(t)$ and $I_E(t)$ results were used. See text.

consistent values of both λ_1 and a_2/a_1, with significant error in λ_2; whereas the excimer emission $I_E(t)$ gave consistent values of both λ_1 and λ_2. These data are collected in TABLE 2.

Beyond a critical length in cyclohexane at 34.5°C, a_2/a_1 became so small that the curves became indistinguishable from single exponential decays. This is understandable in terms of Equation 3, since the factors that control the form of the decay curve are the relative magnitudes of $(Y - X)$ and $4\langle k_1 \rangle k_{-1}$. Thus even if k_{-1} is small, its product with $\langle k_1 \rangle$ may not be small compared to the difference between Y and X if $\langle k_1 \rangle$ is large. For long chains, $\langle k_1 \rangle$ is also small (see below), and the second term indeed becomes negligible.

Rate Constants for End-to-End Cyclization

Absolute values for $\langle k_1 \rangle$, the rate constant for end-to-end cyclization, were obtained from fluorescence decay studies. Equation 4 was employed when $I_M(t)$ was exponential. When $I_M(t)$ was fit to the sum of two exponentials, the data were fit to Equation 3 plus the expression[18]

$$\frac{a_2}{a_1} = \frac{X - \lambda_1}{\lambda_2 - X} \tag{5}$$

Relative values of $\langle k_1 \rangle$ can be obtained from steady-state fluorescence measurements of the polymer samples. For very long polymers, a very weak excimer emission can be detected and its intensity measured. Consequently, one can use Equation 6 to calculate relative values of $\langle k_1 \rangle$ for those samples having $\lambda_1 \approx k_M$.

$$\frac{(I_E/I_M)_1}{(I_E/I_M)_2} = \frac{\langle k_1 \rangle_1}{\langle k_1 \rangle_2} \tag{6}$$

Cyclohexane at the θ-Temperature

Values of $\langle k_1 \rangle$ obtained from both fluorescence decay measurements and steady-state fluorescence studies are presented in FIGURE 4 as a plot of log $\langle k_1 \rangle$ vs log $\bar{N}$, where $\bar{N}$ is the mean number of bonds in the polymer sample. In this instance, $\bar{N}$ is obtained from $\bar{M}_n$ (determined by gpc) by subtracting the mass of the two pyrene substituents and adding the 15 bonds associated with the short spacer chains at each end of the polymer and the head-to-head polystyrene —CH_2—CH_2— bond formed during initiation of the anionic polymerization. It is remarkable to note that values of $\langle k_1 \rangle$ decrease by two orders of magnitude as the polystyrene molecular weight increases from 3000 to 100,000.

The linearity of the log-log plot in FIGURE 4 offers good support for the Wilemski-Fixman theory[10,12] that in θ-solvents the dependence of k_{cy} upon $\bar{N}$ is characterized by a single critical exponent γ. The value of γ, however, cannot be obtained simply from the slope of the line in FIGURE 4. The problem with the data in FIGURE 4 is that $\langle k_1 \rangle$ and $\bar{N}$ (i.e., $\bar{M}_n$) represent different averages over the molecular weight distribution in each sample.

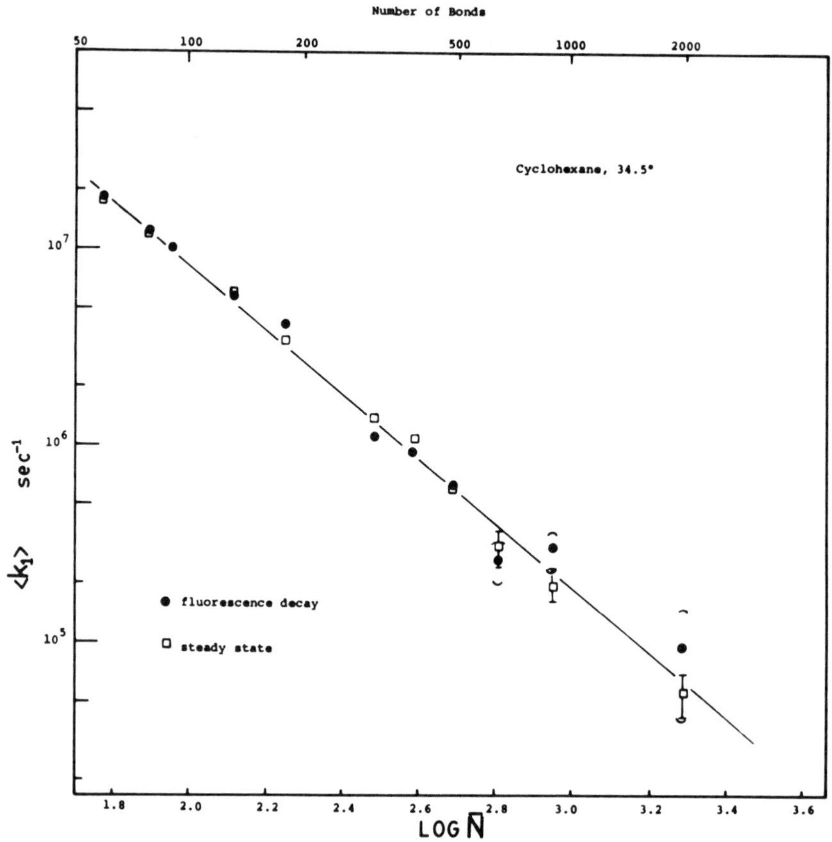

FIGURE 4. Logarithmic plot of $<k_1>$ vs mean chain length $\bar{N}$ (mean number of bonds per polymer) for experiments in cyclohexane at 34.5°C: (●) from fluorescence decay measurements; (□) from steady state fluorescence measurements.

To obtain the "correct" value of γ, one must renormalize either the $<k_1>$ values or the $\bar{N}$ values so that they correspond to the same average of the chain length distribution in the samples. This can be done, since, from gpc measurements, we know the mass distribution in each sample corresponding to each $<k_1>$ value. By assuming that for each discrete chain length, $k_{cy} = AN^{-\gamma}$, we can write

$$<k_1> = \sum_i k_{cy,i}\, \chi_i \tag{7}$$

$$<k_1> = A \sum_i \chi_i\, N_i^{-\gamma} \tag{8}$$

where in Equation 8, χ_i refers to the mole fraction of chains of length N_i.

A least squares fit to Equation 8 gives a best-fit value of $\gamma = 1.62 \pm 0.1$. The slope in FIGURE 4 is approximately −1.52. The difference arises from the larger polydispersity of the samples with $\bar{M}_n$ = 46140 and 99640 relative to the lower molecular weight samples, and the fact that it is the short chain components of each sample that make the major contribution to excimer formation.

TABLE 3
RATE CONSTANTS FOR EXCIMER FORMATION AND EXCIMER DISSOCIATION FOR PYRENE-POLYSTYRENE-PYRENE IN CYCLOHEXANE AT 34.5°C

$\bar{M}_n$	10^{-7} $\langle k_1 \rangle$ (sec^{-1})	10^{-6} k_{-1} (sec^{-1})	K_{eq}	10^{-7} k_e (sec^{-1})
2939	1.80	2.1±0.4	8.5	1.71
3905	1.21	2.8±0.4	4.4	1.74
4489	1.02	2.8±0.4	3.6	1.73
6565	.57	2.6±0.8	2.3	1.75
9164	.42	3.1±0.8	1.3	1.70

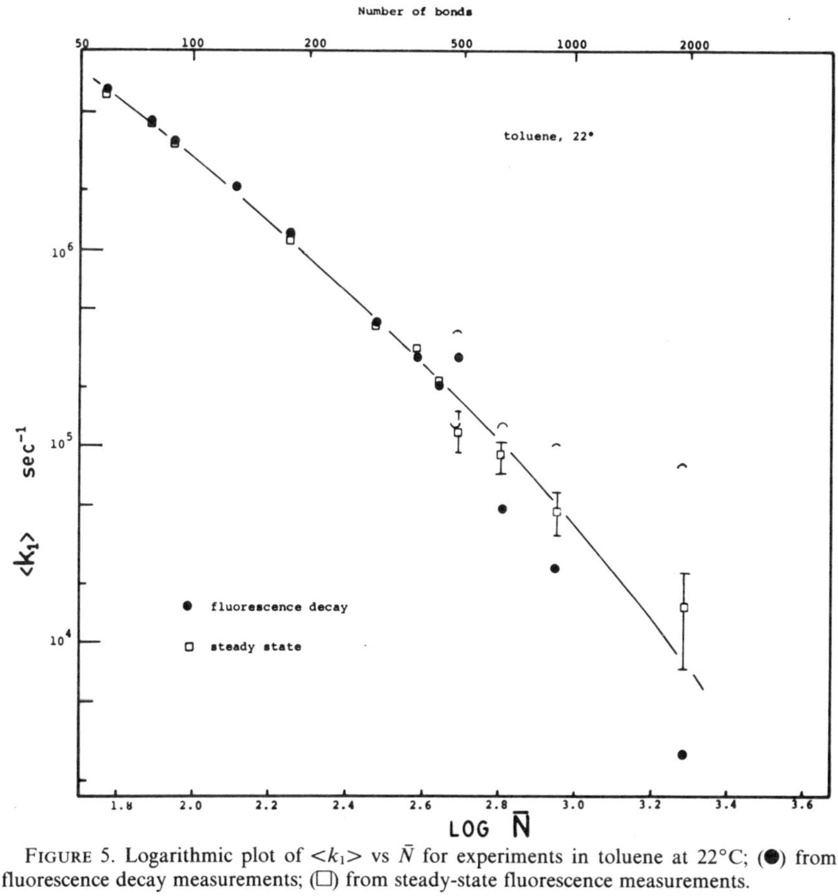

FIGURE 5. Logarithmic plot of $<k_1>$ vs $\bar{N}$ for experiments in toluene at 22°C; (●) from fluorescence decay measurements; (□) from steady-state fluorescence measurements.

The Equilibrium Constant K_{cy} for End-to-End Cyclization

Five polystyrene samples in cyclohexane at 34.5°C give sufficient data that all four rate constants in Scheme I can be evaluated. These are the polymers of $\bar{M}_n$ = 3000 to 9000. The values of the rate constants are collected in TABLE 3.

The most important result to be noted in TABLE 3 is that, for these chain lengths, k_{-1} is independent of chain length. Thus it is sensitive only to interactions localized about the excimer and is insensitive to the configurational properties of the polymer chain. This conclusion should not be extrapolated to very short chains, where cyclization to form the excimer can only occur from high energy conformations of the chain.

It is an important question in contemporary polymer theory as to whether *dynamic* and *equilibrium* properties of polymer chains can be described by the same critical exponent.[22] Hence, since k_{-1} is insensitive to chain length, k_{cy} and K_{cy} must have the same dependence on chain length.

Toluene at Room Temperature (22° C)

In spite of the fact that toluene at 22°C is less viscous than cyclohexane at 34.5°C, end-to-end cyclization is slower. Values of $\langle k_1 \rangle$ from both steady-state and fluorescence decay measurements are shown in FIGURE 5. For polymers of low molecular weight, very similar values are obtained from both types of experiments. For $\bar{M}_n \geqslant$ 25000, λ_1 is so close to k_M that the standard deviations in $\langle k_1 \rangle$ values calculated from Equation 4 are enormous.

There are two very striking features of the data in FIGURE 5. First, values of $\langle k_1 \rangle$ decrease by nearly three orders of magnitude over a molecular weight range of $\bar{M}_n$ = 3000 to 100,000. Second, the plot of log $\langle k_1 \rangle$ vs log $\bar{N}$ is curved downward. In toluene, a good solvent for polystyrene at 22°C, the dependence of $\langle k_1 \rangle$ upon N cannot be described by a single critical exponent.

In order to carry out a more detailed interpretation, we constructed a first derivative plot of our data, FIGURE 6. The effects of scatter in the data were minimized by taking as our $\langle k_1 \rangle$ values the mean of the values obtained from steady-state and decay measurements. The derivative plot was then constructed using a five-point running average over our data. The dashed horizontal lines in FIGURE 6 give a measure of the range in log $\bar{N}$ associated with each $d \log \langle k_1 \rangle / d \log \bar{N}$ value. These data can be fit with a straight line which suggests that log $\langle k_1 \rangle$ has a downward curving parabolic dependence upon log $\bar{N}$, with the slope decreasing from -1.0 at $\bar{N}$ = 50 to -2.4 at $\bar{N}$ = 1000.

In FIGURE 6, we also show the result of a similar treatment of our data in cyclohexane at 34.5°C. Here, the first derivative plot is virtually flat (within experimental error), emphasizing that in a θ-solvent, $k_{cy} \sim N^{-\gamma}$.

Excluded-Volume and End-to-End Cyclization

FIGURE 7 summarizes our data for the cyclization rate constant for polystyrene in toluene at 22°C and in cyclohexane at the θ-point. In spite of a higher solvent viscosity, the chains of a similar length cyclize four to ten times faster in cyclohexane than in toluene. The simplest interpretation of this result is that in a good solvent, the mean dimensions of the polymer are larger, and the chain ends are on the average

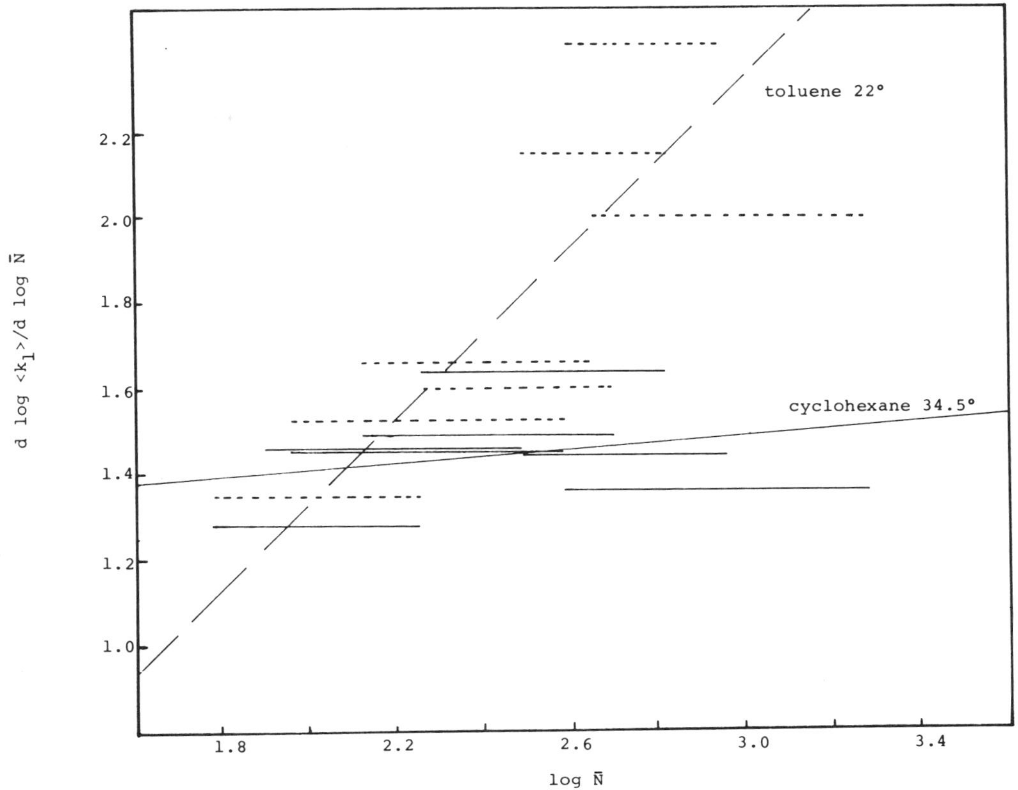

FIGURE 6. First derivative plot of $d \log <k_1>/d \log \bar{N}$ vs $\log \bar{N}$ from data taken from FIGURE 7. To minimize scatter in the data, a five-point running average was taken of the data. The horizontal lines [(– – – –) toluene, and (——) cyclohexane] indicate the range of $\bar{N}$ associated with each data point.

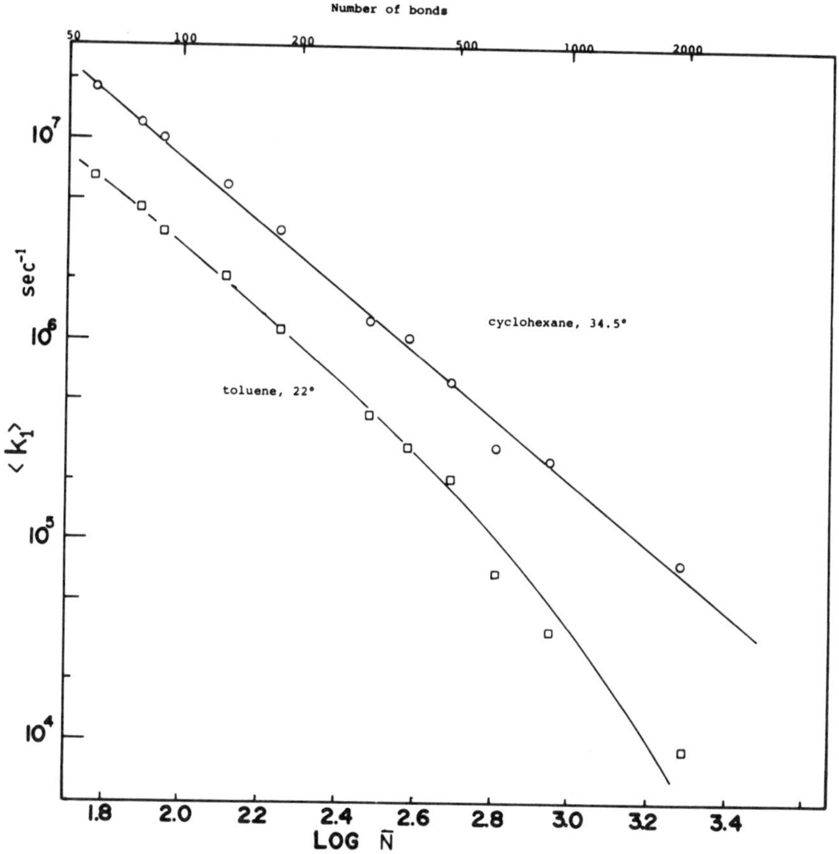

FIGURE 7. Logarithmic plot of $<k_1>$ vs $\bar{N}$ for pyrene-polystyrene-pyrene for experiments in (○) cyclohexane at 34.5°C, and (□) in toluene at 22°C. These data points are the arithmetic mean of the fluorescence decay and steady-state values shown in FIGURE 4 for cyclohexane and FIGURE 5 for toluene.

further apart. In a θ-solvent, the polymer chains are much more compact. The chain ends are on the average closer together, and the time necessary for them to find one another is much reduced. This is illustrated below:

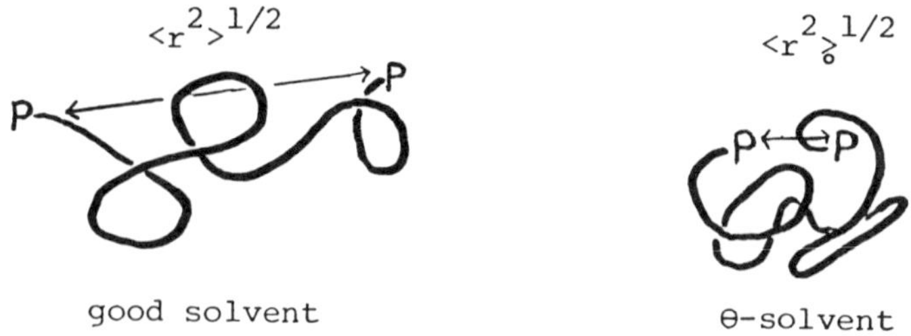

Since intrinsic viscosity $[\eta]$ is also a measure of molecular dimensions, it was of interest to compare the behavior of $<k_1>$ with that of $[\eta]$. A plot of log $<k_1>$ vs log

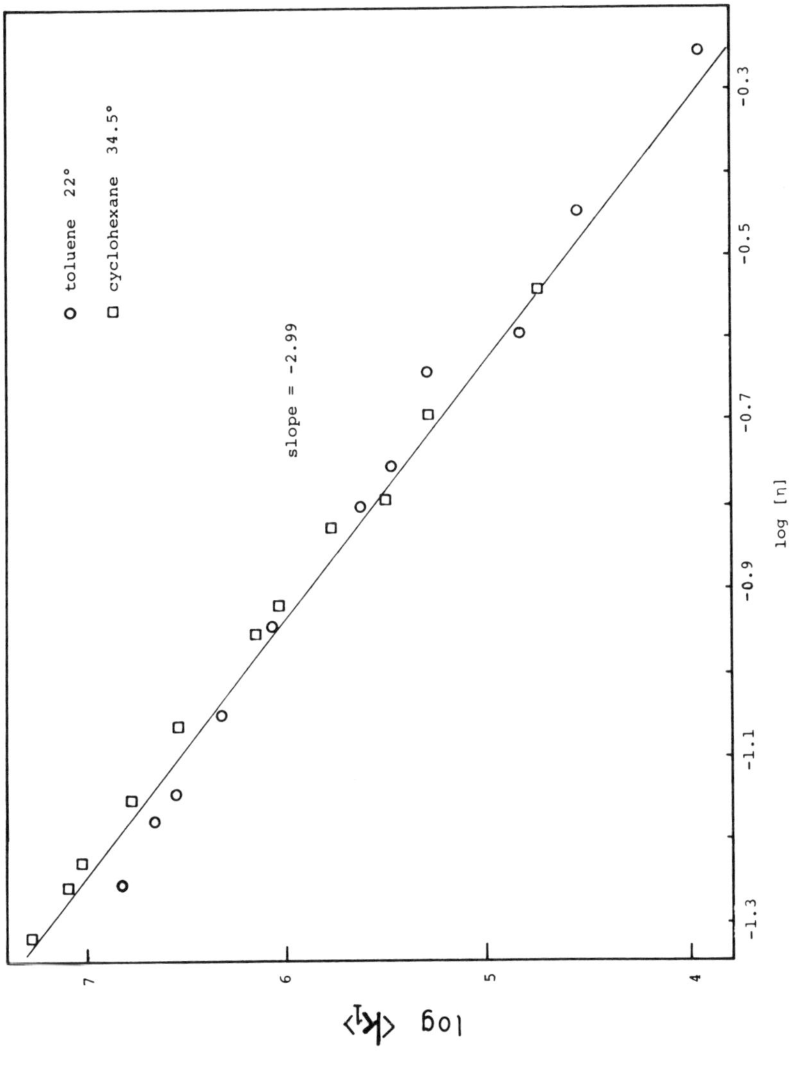

FIGURE 8. Logarithmic plot of $<k_1>$ vs the intrinsic viscosity $[\eta]$ for experiments (○) in cyclohexane at 34.5°C, and (□) in toluene at 22°. The $[\eta]$ data are at 20°C and are taken from Berry, G. C. 1967. J. Chem. Phys. **46:** 1338.

[η], FIGURE 8, using literature values of the latter, was linear with a slope of −2.99. In this plot, the differences in the behavior of $<k_1>$ and [η] in cyclohexane and toluene disappear.

This rather satisfying result supports the suggestion of Perico and Cuniberti[12b] that the same aspects of mean dimensions and chain-solvent hydrodynamic interactions enter into the rate constant for cyclization dynamics as into the intrinsic viscosity. Furthermore, the plot in FIGURE 8 emphasizes that $<k_1>$ is more sensitive than [η] to chain length by a factor of 3 in the exponent of $\bar{N}$.

ACKNOWLEDGMENTS

The authors are grateful for financial support from NSERC Canada through its Operating and Strategic Grant programs. Acknowledgment is made to the Donors of the Petroleum Research Fund, administered by the American Chemical Society, for partial support of this research.

We thank Dr. P. Rempp at the Center of Macromolecular Research, (CRM), Strasbourg, France for his hospitality. It was in his laboratories that one of us (MAW), with the able assistance of Mr. Schuster, prepared the polymers used in this study. Among the people whose discussion contributed to our understanding, we would especially like to cite, Professors Weil (CRM, Strasbourg), Morawetz (Brooklyn, NY), and Stockmayer [Dartmouth, NH], as well as Dr. H. Kohl and Dr. L. Costa at the Eastman-Kodak Research laboratories in Rochester, NY, and Dr. S.-T. Cheung in our research group.

REFERENCES

1. CAROTHERS, W. H. J. Am. Chem. Soc., **54:** 1557 (1932); **55:** 5023 (1933); **58:** 654 (1936).
2. ZIEGLER, K., & W. HECHELHAMMER. 1937. Annalen. **528:** 114.
3. STOLL, M., & A. ROUVÉ. Helv. Chim. Acta. **17:** 1283, 1289 (1934); **18:** 1087 (1935).
4. BENNET, G. M. 1941. Trans. Faraday Soc. **37:** 794.
5. KUHN, W. 1934. Kolloid Z. **68:** 2.
6. JACOBSEN, H. & W. H. STOCKMAYER. 1950. J. Chem. Phys. **18:** 1600.
7. SEMLYEN, J. A. 1976. Adv. Polym. Sci. **21:** 41.
8. BROWN, J. F. & G. M. SLUSARCZUK. 1965. J. Am. Chem. Soc. **87:** 931.
9. FLORY, P. J. & J. A. SEMLYEN. 1966. J. Am. Chem. Soc. **88:** 3209.
10. (a) WILEMSKI, G. & M. FIXMAN. J. Chem. Phys. **58:** 5009 (1973); **60:** 866, 878 (1974); (b) FIXMAN, M. J. Chem. Phys. **69:** 1527, 1538 (1978).
11. (a) DOI, M. Chem. Phys. **9:** 455 (1975); **11:** 107, 111 (1975); (b) SANAGAWA, S. & M. DOI. Polym. J. **7:** 604 (1975); **8:** 239 (1976); (c) SAKATA, M. & M. DOI, Polym. J. **8:** 409 (1976).
12. (a) PERICO, A., P. PIAGGIO & C. CUNIBERTI. 1975. J. Chem. Phys. **62:** 4911; (b) PERICO, A. & C. CUNIBERTI. 1977. J. Polym. Sci. Polym. Phys. Ed. **15:** 1435.
13. BRERETON, M. G. & A. RUSLI. 1976. Polymer. **17:** 395.
14. (a) HAAS, E., E. KATCHALSKI-KATZIR & I. Z. STEINBERG. 1978. Biopolymers **17:** 11; (b) SISIDO, M., Y. IMANISHI & T. HIGASHIMURA. 1979. Macromolecules **12:** 975.
15. USHIKI, H., K. HORIE, A. OKAMOTO & I. MITA. Submitted for publication.
16. ZACHARIASSE, K. & W. KUHNLE. 1976. Z. Phys. Chem. N.F. **101:** 267.
17. CUNIBERTI, C. & A. PERICO. 1977. Eur. Polym. J. **13:** 369.
18. (a) BIRKS, J. B. 1970. Photophysics of Aromatic Molecules, J. Wiley & Sons, New York. Chapter 7; (b) BIRKS, J. B. 1975. Rep. Prog. Phys. **38:** 903.
19. (a) WINNIK, M. A., A. E. C. REDPATH & D. H. RICHARDS. 1980. Macromolecules. **13:** 535. (b) REDPATH, A. E. C., & M. A. WINNIK. 1980. J. Am. Chem. Soc. **102:**.
20. WARE, W. R. 1971. *In* Creation and Detection of the Excited State. Vol. 1, Chapter 5. A. Lamola, Ed. Marcel Dekker. New York.
21. PATON, K. & M. A. WINNIK. To be published.

EXCIMER FORMATION IN POLY- AND DICHROMOPHORIC MOLECULES IN SOLUTION

F. C. De Schryver, K. Demeyer, M. van der Auweraer, and E. Quanten

Department of Chemistry
University of Leuven
B-3030 Heverlee, Belgium

INTRODUCTION

Intermolecular excimer formation, first observed for organic molecules by Förster,[1] can be described by the following scheme:

$$\begin{array}{ccc} M + M^* & \underset{k_{-a}}{\overset{k_a}{\rightleftarrows}} & \text{Excimer } (E) \\ k_M \big\downarrow & & \big\downarrow k_E \end{array}$$

Scheme I

where

- k_M equals the sum of the radiative and non-radiative rate constants of the excited state M^* in absence of complex formation
- k_E equals the sum of the radiative and non-radiative rate constants of the excimer in absence of its dissociation
- k_a equals the rate constant of complex formation
- k_{-a} equals the rate constant for complex dissociation.

This scheme can be quantitatively formulated by the following equations:[2]

$$\ln \frac{\phi_{FE}}{\phi_{FM}} = \frac{k_{FE}}{k_{FM}} \cdot \frac{k_a[M]}{k_{-a} + k_E} \tag{1}$$

where

ϕ_{FE} is the quantum yield of excimer fluorescence
ϕ_{FM} is the quantum yield of monomer fluorescence
k_{FE} is the radiative rate constant for fluorescence of E
k_{FM} is the radiative rate constant for fluorescence of M

and by

$$I_m(t) = A_1 e^{-\lambda_1 t} + A_2 e^{-\lambda_2 t} \tag{2}$$

$$I_E(t) = A_3(e^{-\lambda_1 t} - e^{-\lambda_2 t}) \tag{3}$$

where $I_m(t)$ and $I_E(t)$ represent respectively the fluorescence intensity of M^* and E^* as a function of time and where λ_1, λ_2 and A are parameters describing the decay. On the basis of these parameters, the rate constants of the different processes can be determined.

0077-8923/81/0366-0093 $01.75/0 © 1981, NYAS

As a function of temperature, two regions can be discussed assuming that k_{FE} and k_{FM} are either temperature independent or have the same temperature dependence (e.g., refractive index). The first region would be one in which $k_{-a} \ll k_E$. In this region, the slope of the plot of ln ϕ_{FE}/ϕ_{FM} as a function of $1/T$ yields a difference in activation energies between formation and decay of the excimer. If $k_M + k_a \gg k_E + k_{-a}$, it can be stated that the monomer intensity decays as one exponential and the excimer intensity decays as the difference of two exponentials. In the second temperature region, in which $k_{-a} \gg k_E$, the slope of the same plot yields information on the binding energy of the complex; monomer and excimer fluorescence intensity decay respectively as a sum and a difference of two exponentials.

Intramolecular excimer formation in di- and polychromophoric systems has been the subject of active research since the first observations by Hirayama[3] and has been recently reviewed.[4] Fluorescence data obtained using stationary and non-stationary methods were, even recently,[5] interpreted on the basis of a scheme which is the monomolecular equivalent of Scheme I. However, some results related to the fluorescence intensity of excimer and monomer components in bichromophoric systems and reported in the literature as early as 1966[6] indicated that conformational aspects[7] were important in intramolecular excimer formation. It has also been observed that the tacticity of polystyrene[8] influenced the excimer over monomer fluorescence ratio. These results were still analyzed on the basis of the classical scheme. Recently, the applicability of this scheme to intramolecular excimer formation was questioned.[9] Furthermore, we were able to establish the role of ground state conformations on intramolecular exciplex formation in α-aryl-ω-*N*-dialkylalkanes.[10]

In the present contribution, the fluorescence properties of poly-2-vinylnaphthalene will be discussed and compared with the racemic (2) and *meso* (1) 1,1′-di-2-naphthyldiethylether.

```
CH3—CH—O—CH—CH3
     |      |
2-naphthyl  2-naphthyl
```

Results and Discussion

Absorption and Steady State Fluorescence Measurements of Poly-2-Vinylnaphthalene in 2-Methyltetrahydrofuran

The absorption spectrum of a $10^{-4} M$ solution of poly-2-vinylnaphthalene (P2VN) in MTHF is within experimental error superimposable on the absorption spectrum of 2-methylnaphthalene. Upon excitation at 285 nm at room temperature, emission from the naphthyl chromophore as well as excimer emission with a maximum at 399 nm is observed. Intramolecular excimer formation in 1,3-di-2-naththylpropane has been reported.[11]

The ratio of excimer over monomer intensity is, however, dependent on the excitation wavelength as seen in Figure 1 and the excitation spectra clearly show a similar dependence when analyzed below 395 nm (Figure 2a). Above this wavelength, the excitation spectra are identical with the absorption spectra. Below this wavelength, a broadening of the L_A band is observed. Upon lowering the temperature, this broadening decreases, see Figure 2b. The emission spectrum of P2VN in MTHF

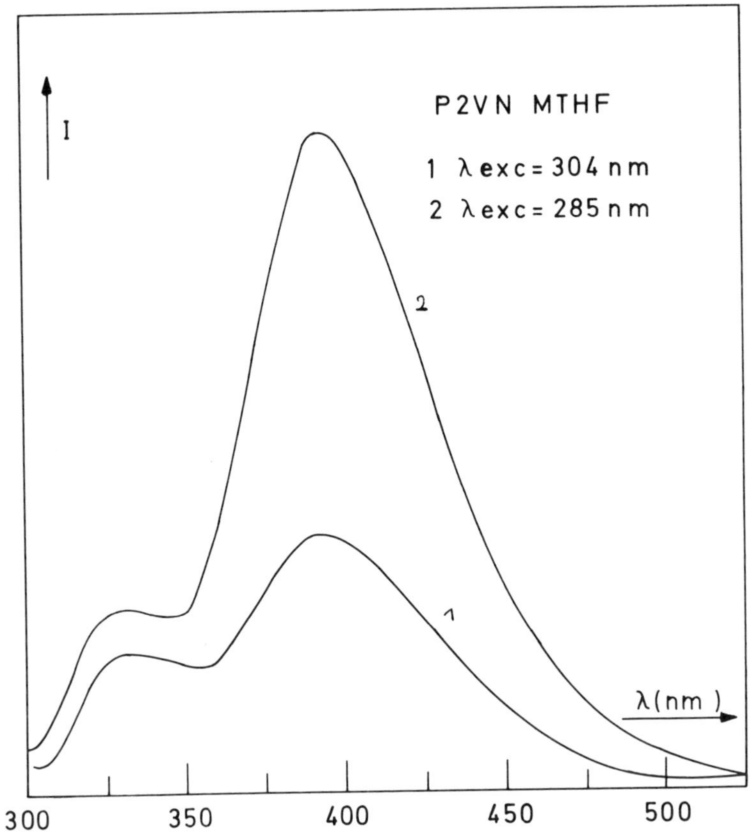

FIGURE 1. Excitation wavelength dependence of the excimer over monomer intensity in P2VN in MTHF at room temperature.

as a function of temperature is presented in FIGURE 3. Going from 208 K to 260 K, the intensity of the monomer band decreases and the excimer band increases, above 290 K, the monomer and excimer band both decrease in intensity. It should be further noticed that the maximum of the excimer emission undergoes a hypsochromic shift and that the FWHH broadens towards the hypsochromic side upon increasing the temperature.

Fluorescence Decay of P2VN in MTHF

Analysis in the monomer region at 330 nm should, if the above mentioned scheme is valid, be single exponential in the low temperature region and biexponential in the high temperature region. The decay analyzed at 330 nm is, at 230 K as well as at 298 K, (FIGURE 4) multiexponential. The correlation function for the decay reported in FIGURE 4, which is analyzed as two exponential, shows clear evidence for a poor fit. Analysis in the excimer region at 420 nm should result in a decay which can be

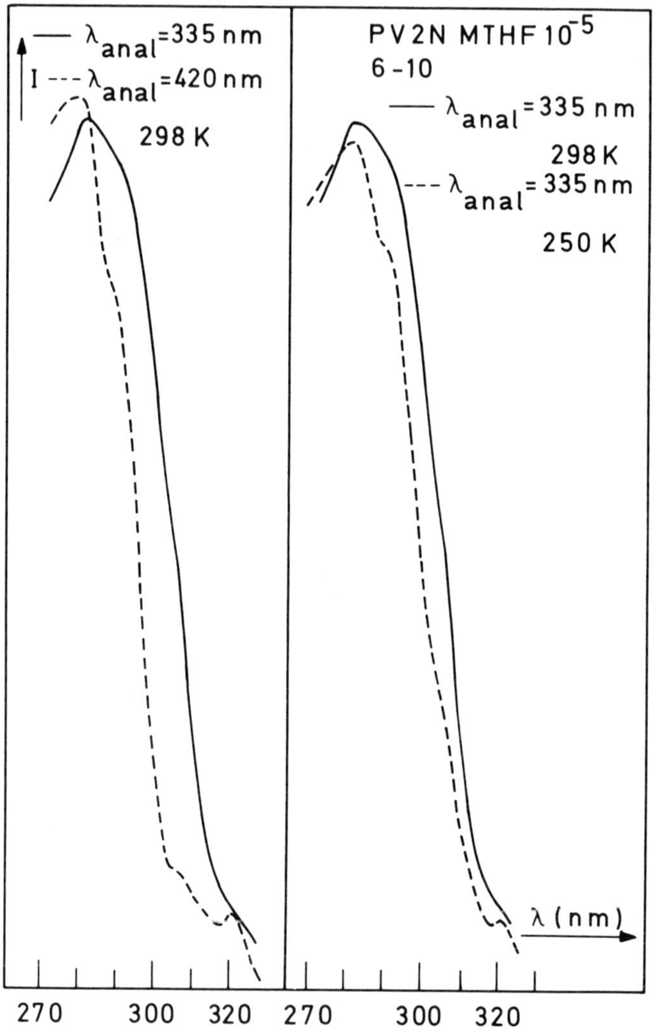

FIGURE 2. Excitation spectra of a 10^{-5} M solution of P2VN in MTHF
(*left*) at room temperature --- analysis 420 nm
—— analysis 335 nm
(*right*) at different temperatures --- analysis 335 nm, 250 K
—— analysis 335 nm, 298 K.

described as the difference of two exponentials. The decay analyzed from 420 nm to 460 nm is, at all temperatures, a triexponential function for which the growing-in can only be resolved below 273 K. At room temperature (FIGURE 5), the decay of the excimer intensity can be described as the sum of two exponentials. This clearly indicates that excimer formation in P2VN cannot be described by the monomolecular equivalent of Scheme I.[9a]

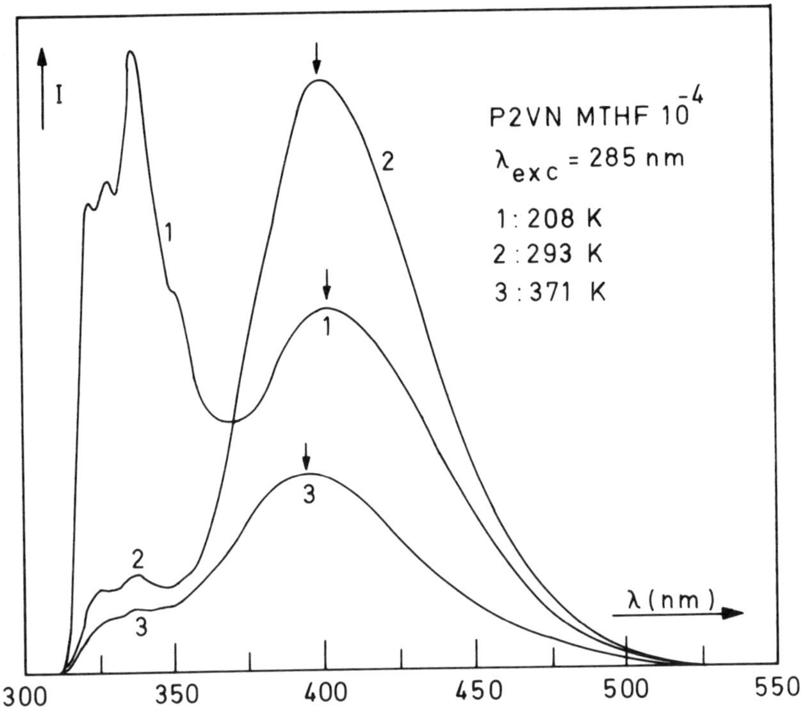

FIGURE 3. Temperature dependence of the fluorescence emission of P2VN in MTHF. (1) 208 K; (2) 293 K; (3) 371 K.

Time-Resolved Spectra of P2VN in MTHF at Room Temperature

To gain further insight, time-resolved spectra were recorded as a function of the time window and the excitation wavelength. The time-resolved spectra taken in the time window between 3.5 ns and 5 ns with 304 nm and 285 nm, respectively, as excitation wavelengths (FIGURE 6) clearly show the difference in relative intensity of monomer and excimer emission. At long times (ΔT 250 nsec δt 600 nsec), only excimer emission is observed indicating that even at 320 nm some residual excimer intensity remains, but that at room temperature (FIGURE 7), dissociation of the long-living excimer to the monomer excited state can be neglected. This could invalidate some of the conclusions drawn previously[9a] concerning the return of the long-living excimer.

As suggested before, the experimental evidence does indicate the presence of more than one type of local environment which do not interconvert. Possible candidates are the iso- and syndiotactic sequences in the polymer. To investigate this point, model systems were studied.

Fluorescence Spectra of 1,1'-di-2-naphthyldiethylether (D2NEE)

As model systems for the syndiotactic and isotactic sequences in P2VN, the racemic D2NEE and *meso*-D2NEE were synthesized. The fluorescence spectra in isooctane

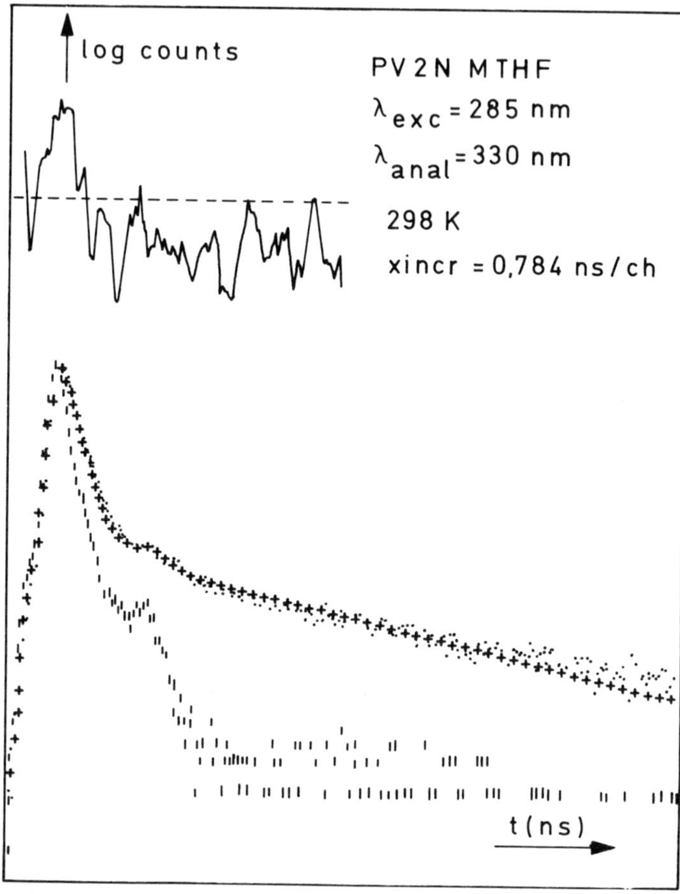

FIGURE 4. $I_m(t)$ of P2VN in MTHF at 298 K analyzed at 350 nm as a sum of two exponentials.

are presented in FIGURE 8 a+b. The relative intensity of excimer emission is substantially smaller in *rac*-D2NEE than in *meso*-D2NEE as previously observed for 2,4-diphenylpentane.[6,7a] Furthermore, the excimer over monomer ratio of *meso*-D2NEE but not of the *rac*-D2NEE is excitation wavelength dependent (FIGURE 8b). This dependency is found also in the excitation spectra where it is seen that the deviation occurs in the region below 380 nm. This observation substantiates the observation in P2VN which is due to the isotactic sequences in the polymer.

At 77 K, the emission spectra of both isomers are, within experimental error, identical and no excimer emission is observed. In isopentane between 200 K and 240 K, no excimer emission is observed for *rac*-D2NEE while an iso-emissive point is present from 240 K up to 298 K. For *meso*-D2NEE the intensity of excimer over monomer fluorescence decreases from 200 K upwards.

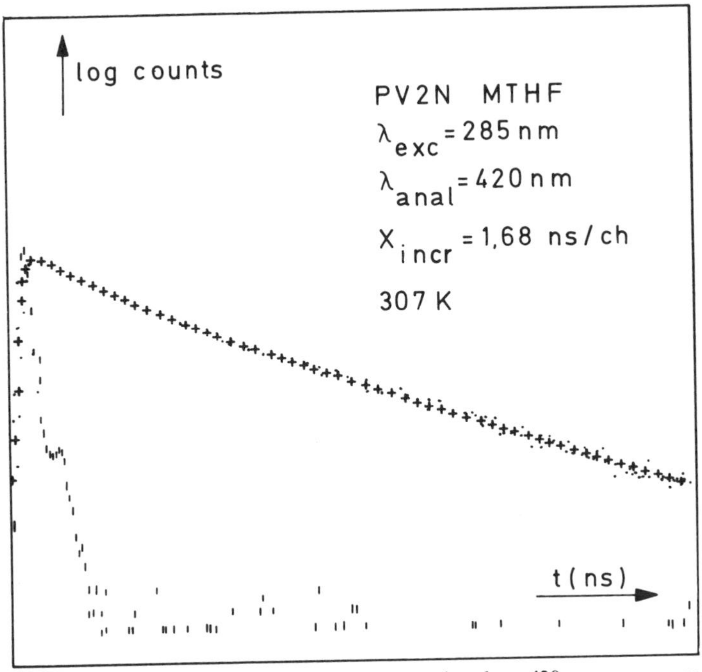

FIGURE 5. $I_E(t)$ of P2VN in MTHF at 298 K analyzed at 420 nm as a sum of two exponentials.

The maximum of the excimer emission upon excitation at 304 nm shifts from 400 nm at 173 K to 385 nm at 350 K. The excimer band shows a shoulder at 385 nm at 298 K. A dual emission in the excimer region has been observed previously for polyvinylcarbazole[12] and can also clearly be seen in the temperature dependent fluorescence spectrum of *meso*-1,1′-di-1-naphthyl-diethyl-ether (FIGURE 9) in isooctane.

Fluorescence Decay of D2NEE

The time dependence of the fluorescence intensity of the *meso*-D2NEE analyzed at 430 nm shows a growing in and a two exponential decay only at temperatures below 190 K. Above this temperature, it can be analyzed as a sum of two exponentials. At 298 K (FIGURE 10), λ_1^{-1} equals 28 nsec and λ_2^{-1} equals 10 nsec, the ratio of the pre-exponentials equals 0.3 indicating a contribution of one tenth of the emission by the short-living component. Analysis at 320 nm is difficult due to the low intensity and the spectral overlap with the excimer band.

From this observation, it can be deduced that the biexponential decay of the excimer in P2VN is due to the contribution of isotactic sequences to the total excimer emission. The time dependence of the emission of racemic D2NEE analyzed at 420

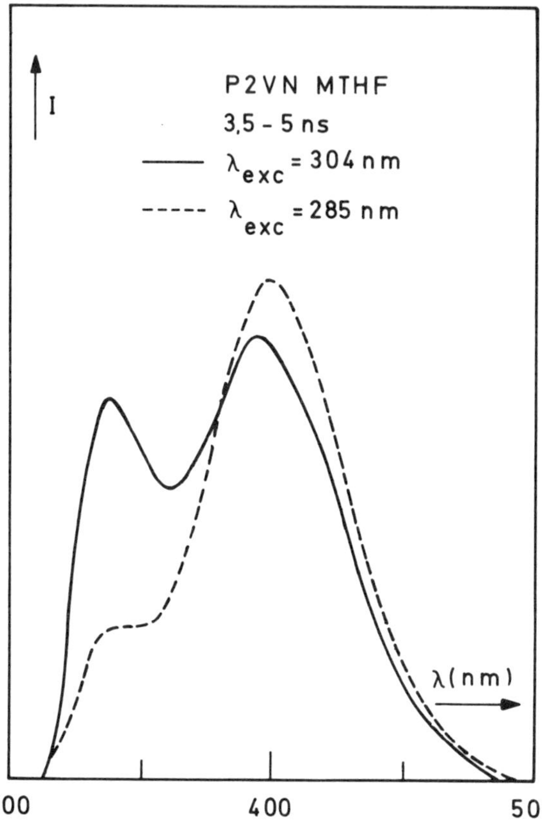

FIGURE 6. Time-resolved spectrum of P2VN in MTHF at room temperature Δt 3.5 ns; δt 2 ns.

nm could, over the whole temperature range from 240 K to 350 K, accurately be described as the difference of two exponentials. The decay parameters λ obtained in this analysis are reported in FIGURE 11. At 258 K, the "growing-in" of the excimer equals 19 ns and its lifetime equals 33.5 ns. At 320 K, the "growing-in" equals 5.2 ns and the decay parameter equals 25.2 ns. Analyzed at 330 nm, the time dependence can be described as a single exponential between 200 and 240 K where no excimer is formed and the measured lifetime equals that of the model compound 1,2-naphthylethyl methylether. Between 240 K and 277 K, the decay is analyzed as the sum of two exponentials of which the short-living component corresponds with the "growing-in" of the excimer (FIGURE 11) while the long-living component is slightly shorter than the excimer lifetime.

Above 277 K, the decay can be analyzed as the sum of two exponentials. The obtained parameters from this curve fitting are reported in FIGURE 11. The short-living component in the monomer region deviates and is longer-lived than the short-living component in the excimer region; the long-living component in the monomer region is close to the value of the long-living decay parameter in the excimer region.

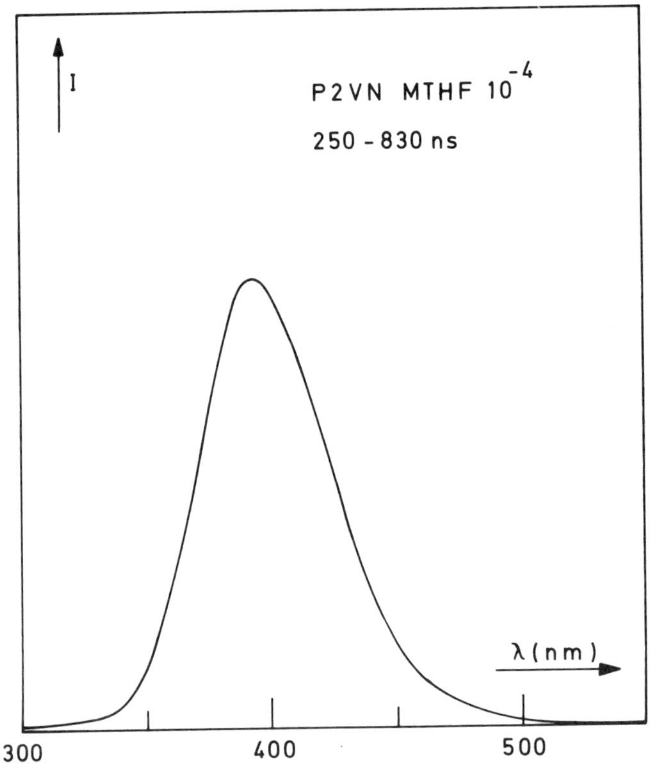

FIGURE 7. Time-resolved spectrum of P2VN in MTHF at room temperature Δt 250 ns; δt 600 ns.

DISCUSSION

Although one should keep in mind that the systems were studied in different solvents, which could affect a number of parameters, and that rotational barriers around an oxygen atom are different from those around a methylene group, general trends can be discussed.

The above reported results indicate that excimer formation in poly-chromophoric systems such as poly-2-vinylnaphthalene can be looked at as a superposition of excimer forming processes from different conformational sites. Furthermore, the efficiency of excimer formation at the different conformational sites differs substantially as can be seen in the residual naphthyl chromophore emission in racemic and *meso*-D2NEE. This means that in a polychromophoric system, the emission recorded in the monomer region relates to a certain extent to other sites more than the emission recorded in the excimer region.

If one now considers the different conformational possibilities within the *meso* and racemic compound (FIGURE 12a) and if the conformational distribution in D2NEE is similar to that reported for 2,4 diphenylpentane[7a,b,13] and for 2,4 di-2-naphthylpen-

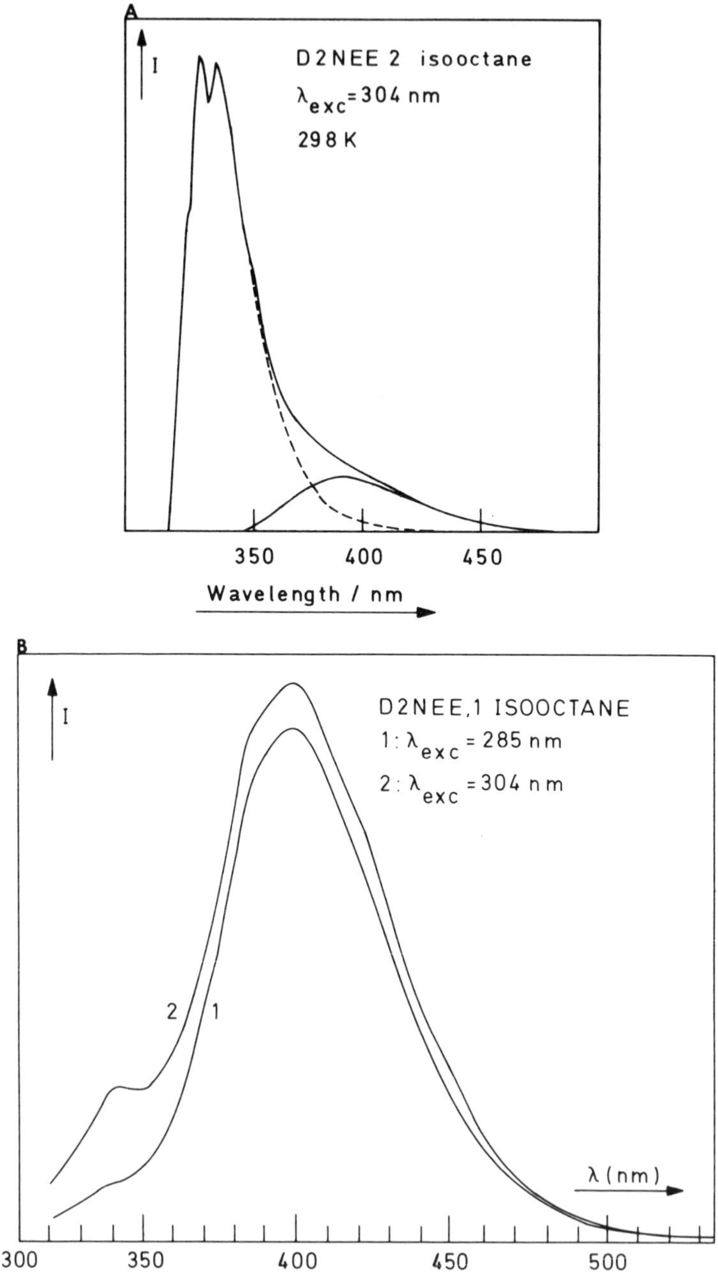

FIGURE 8. Fluorescence spectrum of (A) *rac*-D2NEE in isooctane at 298 K. Excitation at 304 nm. (B) *meso*-D2NEE in isooctane at 298 K. (1) Excitation at 285 nm and (2) excitation at 304 nm.

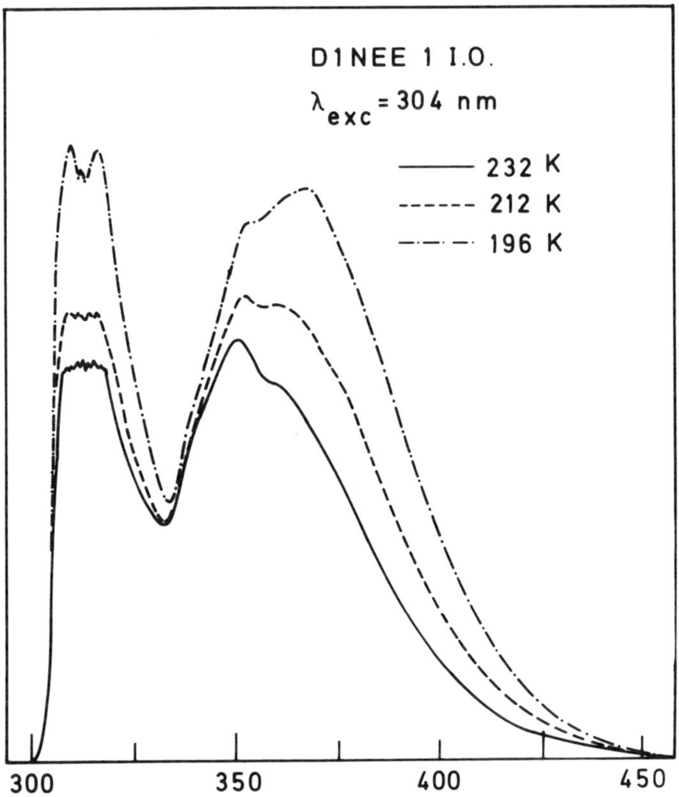

FIGURE 9. Temperature dependent emission spectrum of *meso*-1,1′-di-1-naphthyl-diethylether in isooctane.

tane,[7c] then it can be understood that the *meso*-D2NEE, with an almost exclusive population in a *TG* conformation, forms the *TT* excimer via a fast step with a low barrier. The observation of a dual decay, if related to two types of excimers differing in properties, could be due to the relative orientation in the ground state of the naphthyl groups prior to excimer formation. This rotation around the CH-naphthyl band is hindered and could be slower than the excimer formation.

The formation of the $\bar{G}\bar{G}$ excimer, as another candidate to explain the dual decay, cannot be excluded but is less likely on the basis of energetic considerations. This relative orientation could also be the origin of the wavelength dependence of the *meso*-compound. Considering the racemic D2NEE, the observed results could be interpreted on the basis of one of the following schemes.

If one considers only the temperature region below 278 K, the obtained results could be analyzed within a classical scheme. However this would mean that the enthalpy of formation is only slightly exothermic if not endothermic. The values obtained for τ_E, the true excimer lifetime, corresponds to the shorter living component

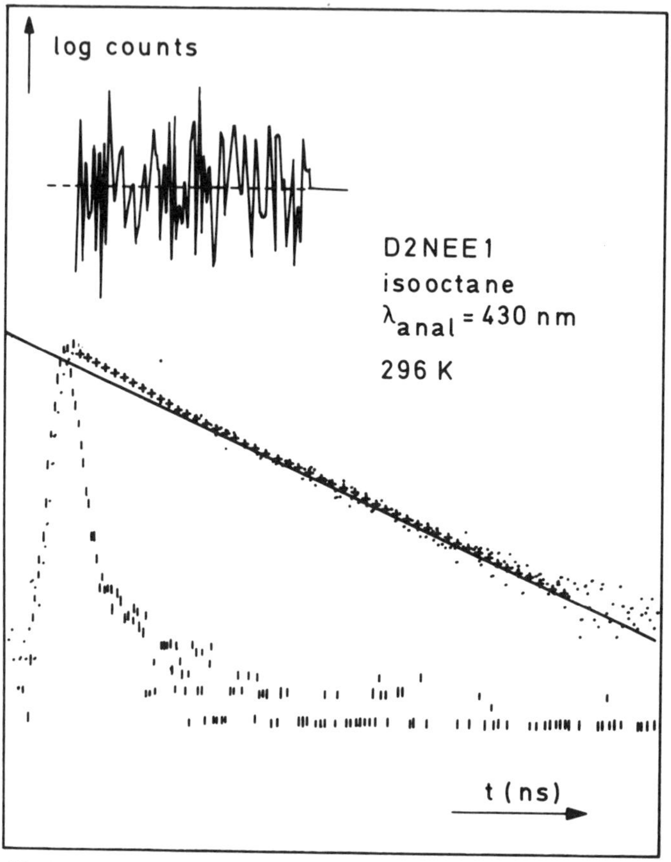

FIGURE 10. $I_E(t)$ of *meso*-D2NEE in isooctane at 298 K analyzed as the sum of two exponentials.

in the excimer region. In the temperature region above 278 K, the classical scheme cannot be applied in view of the observed deviation in the decay parameters.

An alternative interpretation could be based on the conformational distribution within the racemic D2NEE. In this case, the two exponential decay in the monomer region does not relate to excimer return but to the establishment of an equilibrium between the *GG* and *TT* conformations and to the formation of the $T\bar{G}$ excimer from *TT* (Scheme II).

$$\begin{array}{ccccc} GG & \rightleftarrows & TT & & \\ \downarrow\uparrow k_7 & & \downarrow\uparrow k_7 & & \\ GG^* & \underset{k_4}{\overset{k_3}{\rightleftharpoons}} & TT^* & \underset{k_9}{\rightarrow} & \bar{G}T^* \overset{\tau_E^{-1}}{\rightarrow} \end{array}$$

Scheme II

Within this scheme, the long-living component in the excimer region corresponds to the "growing-in" of the excimer while the short-living component in the excimer region corresponds to the excimer lifetime.

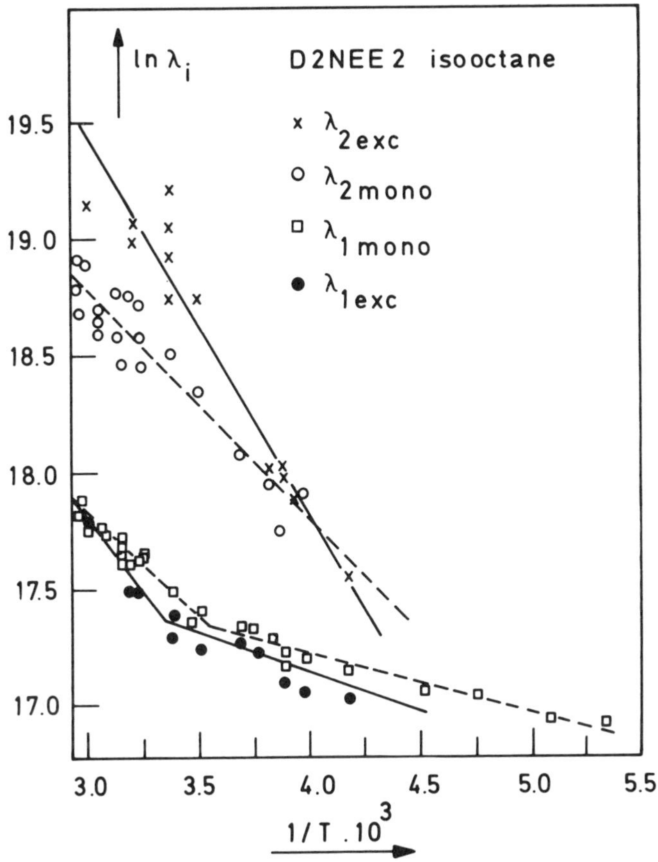

FIGURE 11. Decay parameters λ plotted as ln λ as a function of $1/T$.

On the basis of this scheme, the decay in the monomer region can be described by Equation 4:

$$A_1 \exp -\lambda_1 t + A_2 \exp -\lambda_2 t \tag{4}$$

and in the excimer region by Equation 5:

$$PR_1 \exp -\lambda_1 t + PR_2 \exp -\lambda_2 t + PR_E \exp -\lambda_2 t \tag{5}$$

where $PR_E = -(PR_1 + PR_2)$.

Analysis of the data reported in FIGURE 11 within this scheme gives the following rate constants for the different processes at 258 K:

TABLE 1
RATE CONSTANTS k_3, k_4, k_9 FOR RACEMIC D2NEE AT 258 K

k_3	0.8 10^7 sec^{-1}
k_4	0.8 10^7 sec
k_9	0.2 10^8 sec^{-1}

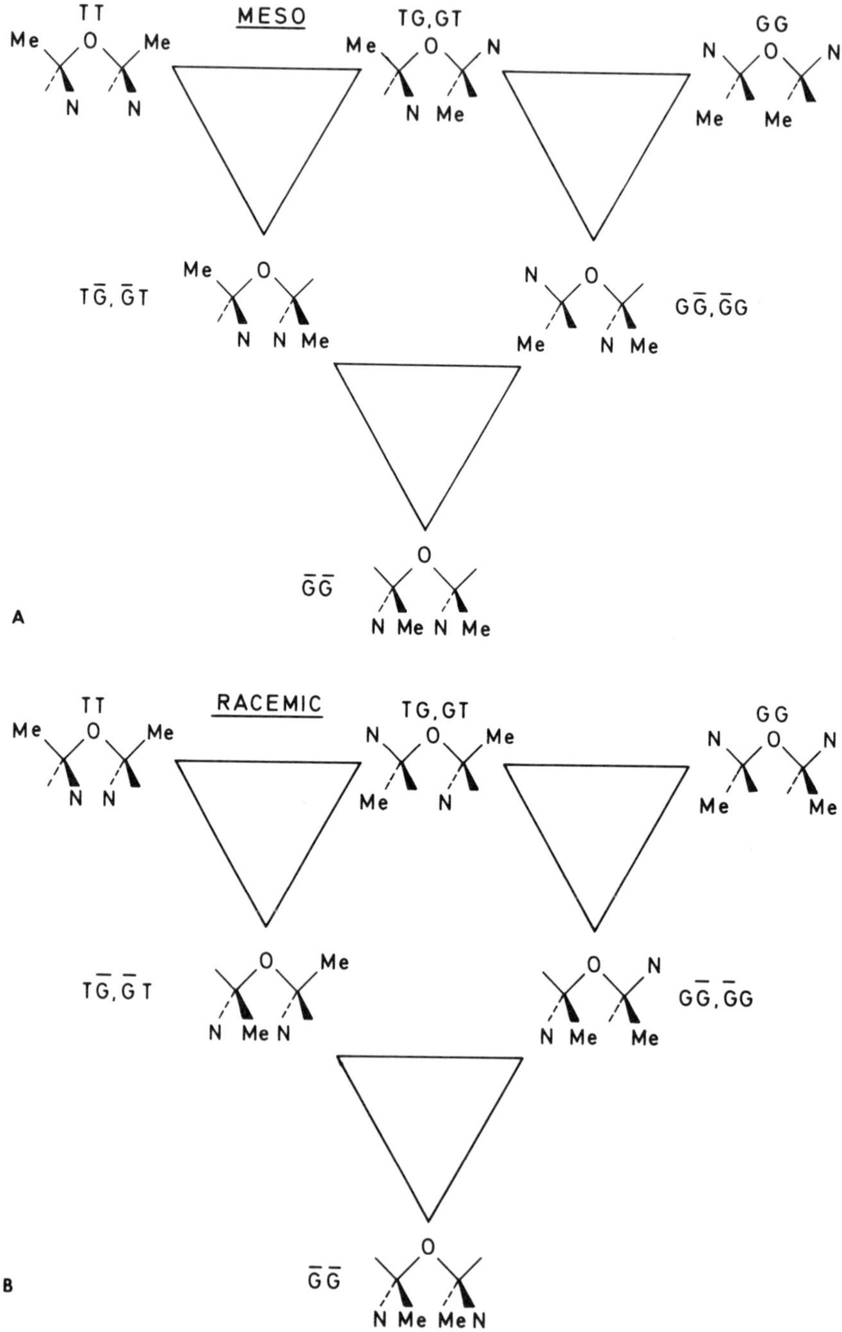

FIGURE 12. Conformational possibilities and transitions in, respectively, (A) *meso* and (B) racemic D2NEE.

From the temperature dependence, within an Arrhenius formalism, activation energies for the k_3 and k_{9-1} could be determined and are found to be 3.4 and 2.8 kcal mole respectively, which is close to the value for a rotation around an oxygen carbon bond. Analysis within other schemes, including return from the excimer *GT* and taking into account conformational transitions, can at present not be excluded and are under investigation on the above mentioned and other appropriate bichromophoric systems. However, it is clear that the classical scheme cannot as such be applied in all poly- and bichromophoric systems.

Summary

Excimer formation in poly-2-vinylnaphthalene in MTHF is studied using stationary and non-stationary techniques in a large temperature domain. On the basis of the time dependence of the intensity of monomer and excimer fluorescence and on the basis of time-resolved emission spectra, it could be shown that intramolecular excimer formation in such a system does not follow the established kinetic scheme for intermolecular excimer formation. Comparison with diasteromeric model compounds leads to the conclusion that the emission properties depend on the conformational distribution of the polymer.

Acknowledgments

The authors are indebted to the Ministry of Scientific Programmation, the Belgian National Science Foundation and the University Research Fund for financial aid to the laboratory. The Belgian National Science Foundation and the IWONL are thanked for predoctoral fellowships to MvdA and KD respectively. FDS is indebted to NATO for a grant facilitating exchange of information with Dr. K. Zachariasse and Prof. M. Winnik.

References

1. Forster, Th., K. Kasper. 1954. Z. Phys. Chem. (Frankfurt), **1:** 275.
2. Birks, J. B. 1970. Photophysics of Aromatic Molecules. Wiley. New York. Chapter 7.
3. Hirayama, F. 1965. J. Chem. Phys. **42:** 3163.
4. (a) Nishijima, Y. 1973. J. Macromol. Sci. Phys. **B8**(3–4): 389. (b) Klöpffer, W. 1973. Organic Molecular Photophysics. J. B. Birks, Ed. Wiley. London, p. 357. (c) Somersall, A. C., J. E. Guillet. 1975. J. Macromol. Sci. Revs. Macromol. Chem. **C13**(2):135. (d) De Schryver, F. C., N. Boens, J. Put. 1977. Adv. Photochem. **10:** 359. (e) Beavan, S. W., J. S. Hargreaves, D. Phillips. 1979. Adv. Photochem. **11:** 207.
5. Ghiggino, K., R. Wright, D. Phillips. 1978. J. Polym. Sci. Polym. Phys. Ed. **16:** 1499.
6. Longworth, J., F. A. Bovey. 1966. Biopolymers **4:** 1115
7. (a) Bokobza, L., B. Jasse, L. Monnerie. 1977. Eur. Polym. J. **13:** 921. (b) Bokobza, L., B. Jasse, L. Monnerie. Eur. Polym. J. In press. (c) Nishijima, Y., M. Yamamoto. 1979. Polymer Reprints **20:** 391. (d) Morawetz, H. IUPAC Symposium Macromolecular Chemistry. Mainz. In press. (e) Harrah, L. A. 1972. J. Chim. Phys. **56:** 383. (f) Frank, C. W., L. A. Harrah. 1974. J. Chim. Phys. **61:** 1526.
8. (a) David, C., N. Putman de Lavareille, G. Geuskens. 1974. Eur. Polym. J. **10:** 617. (b) David, C., L. Lempereur, G. Geuskens. 1974. Eur. Polym. J. **10:** 1181. (c) Ishii, T., T. Handa, S. Matsunaga. 1977. Makromol. Chem. **178:** 2351.

9. (a) De Schryver, F. C. 1979. Symposium on Photochemistry in Polymers. Mulhouse. Demeyer, K., M. Van Der Auweraer, L. Aerts, F. C. De Schryver. 1980. J. Chim. Phys. **77:** 493. (b) Phillips, D., A. J. Roberts, I. Soutar. 1980. J. Polym. Sci., Polym. Lett. Ed. **18:** 123. (c) Zachariasse, K. Personal communication.
10. (a) Van Der Auweraer, M., A. Gilbert, F. C. De Schryver. 1980. J. Am. Chem. Soc. **102:** 6107. (b) Van Der Auweraer, M., A. Gilbert, F. C. De Schryver. 1980. Nouv. J. Chim. **4:** 153. (c) Meeus, F., M. Van Der Auweraer, F. C. De Schryver. 1980. Chem. Phys. Lett. **74:** 218.
11. (a) Chandross, E. A., C. J. Dempster. 1970. J. Am. Chem. Soc. **92:** 3586. (b) Zachariasse, K. A., W. Kühnle, A. Weller. 1978. Chem. Phys. Lett. **59**(3): 375.
12. (a) Johnson, G. E. 1975. J. Chem. Phys. **62:** 4697. (b) Hoyle, C. E., T. L. Nemzek, A. Mar, J. E. Guillet. 1978. Macromolecules **11:** 429. (c) Ghiggino, K. P., R. D. Wright, D. Phillips. 1978. Eur. Polym. J. **14:** 567.
13. Yoon, D. Y., P. J. Flory, P. R. Sundararajan. 1975. Macromolecules **8:** 776.

MULTICOMPONENT FLUORESCENCE DECAY IN VINYL AROMATIC POLYMERS AND COPOLYMERS

A. J. Roberts, D. V. O'Connor, and D. Phillips*

Davy Faraday Research Laboratory
The Royal Institute of Great Britain
London W1X 4BS, England

INTRODUCTION

The observation of fluorescence from a dimer formed by the interaction of an electronically excited state with a ground state identical molecule ("excimer") is widespread. Excimer formation is particularly facile in aromatic molecules, and many synthetic vinyl polymers with aromatic groups pendant to the vinyl chain exhibit this behavior. For free aromatic moieties in dilute solution, "excimer" kinetics are well established, and the simplest kinetic scheme capable of explaining observed behavior is represented by FIGURE 1.

Analysis of this scheme shows that the fluorescence decay of the uncomplexed aromatic species $i_M(t)$ (referred to here as monomer) is the sum of two exponential terms (Equation 1), whereas that of the excimer, $i_D(t)$, will be the difference of the same two exponential terms. (Equation 2).

$$i_M(t) = A_1e^{-\lambda_1 t} + A_2e^{-\lambda_2 t} \tag{1}$$

$$i_D(t) = A_3(e^{-\lambda_1 t} - e^{-\lambda_2 t}) \tag{2}$$

where

$$\begin{aligned}\lambda_1 + \lambda_2 &= \tau_1^{-1} + \tau_2^{-1} \\ &= k_{FM} + k_{IM} + k_{DM}[M] + k_{FD} + k_{ID} + k_{MD}\end{aligned}$$

These kinetics have been established in many small molecule systems, with accurate evaluation of individual rate constants possible when concentration dependence is investigated. There have been many apparently successful attempts to fit these simple kinetics to synthetic polymer systems in which excimer fluorescence has been observed. However, recognizing the heterogeneous nature of such polymers, with pendant aromatic groups distributed randomly along the polymer chains, it is perhaps surprising that such simple schemes are successful. Intuitively, one might distinguish between two types of monomer species, that which has near neighboring aromatic groups in such configurations as to enable excimer formation through facile exciton diffusion and small-range conformational changes, and that which would be largely isolated, forming excimers only through long-range interactions. In such cases, it is easy to show (see below) that additional terms to those given in Equations 1 and 2 will be of importance in the fluorescence decay of polymer systems. Attempts to observe these experimentally require a time-resolved fluorescence apparatus capable

* To whom correspondence should be addressed.

0077-8923/81/0366-0109$01.75/0 © 1981, NYAS

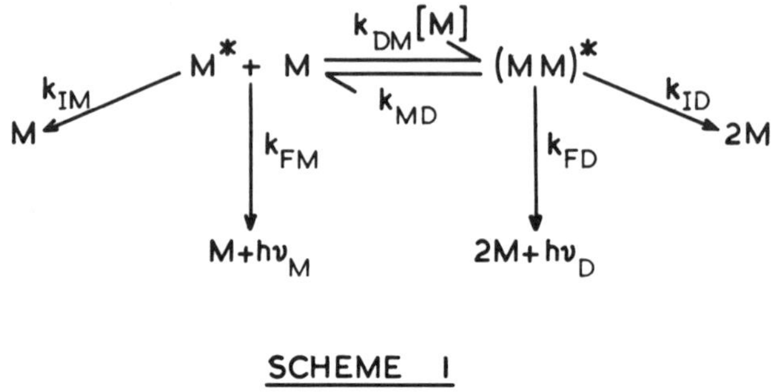

FIGURE 1. Kinetic scheme for excimer formation and decay after Birks.

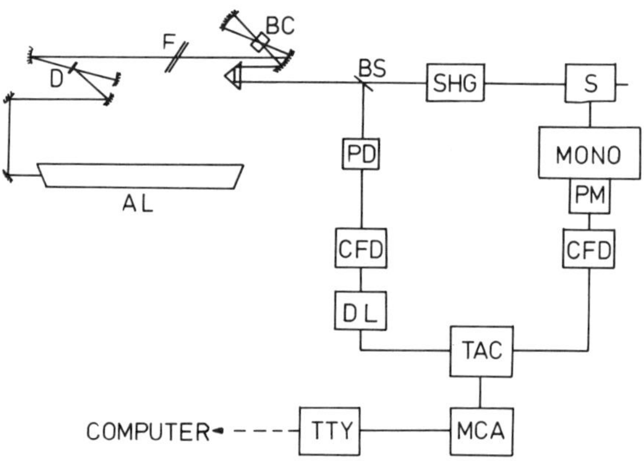

FIGURE 2. Block diagram of time-resolved emission spectrometer: AL—argon ion laser; D—dye jet; F—birefringent filter; BC—Bragg Cell; BS—beam splitter; SHG—frequency doubling crystal; S—sample; MONO—monochromator; PM—photomultiplier tube; CFD—constant fraction discriminator; PD—photodiode; DL—variable delay; TAC—time-to-pulse height converter; MCA—multichannel analyzer; TTY—teletype.

of superior time and spectral resolution, and particularly good signal-to-noise characteristics, and computational curve-fitting routines which can be proved to yield realistic results when used to analyze such experimental decay curves. We describe here an apparatus and curve-fitting routines that we can show to be capable of yielding accurate results when used to analyze multicomponent fluorescence decay, and apply these techniques to the study of excimer formation and decay in the following polymers: poly (1-vinylnaphthalene) and copolymers with methyl methacrylate and methyl acrylate; poly (acenaphthylene); and poly (*N*-vinyl carbazole).

EXPERIMENTAL

A block diagram of the time-resolved fluorescence spectrometer used for the measurements reported here is shown in FIGURE 2.

A 4 W argon ion laser (Spectra-Physics model 166) is used to excite a jet stream dye laser (Coherent Radiation Ltd, model 590). With the dye used (Rhodamine 6G in ethylene glycol), laser output is obtained over the range 570 nm–640 nm, wavelength selection being achieved with a 2-plate birefringent filter. The normal dye-laser output mirror was removed and, in its place, a modified Spectra-Physics (model 366) cavity dumper installed. In this device, the laser cavity is folded and the beam focused to a waist at the center of a quartz flat (Bragg cell). A piezoelectric transducer, situated on a non-optical face, is used to convert a pulsed rf signal into an acoustic wave which propagates across the cell causing regions of rarefaction and compression and, hence, periodic changes in the index of refraction. Thus, as the sound wave traverses the beam, the intracavity laser light is deviated from its normal path and out of the cavity. With careful alignment, pulses as short as 10 ns FWHM (Full width at half maximum intensity) could be obtained at repetition rates from single shot to 5 MHz. Pulse peak powers were over 50 times greater than the average output power, enabling relatively efficient second harmonic generation. The pulse profile was invariant over hours of operation and the interpulse separation was stable to 1 part in 10^5. To provide sample excitation in the wavelength range 290–305 nm, these pulses were "frequency doubled" using a 90° phase-matched temperature and angle tuned ADA crystal. For the measurements reported here, sample excitation was at 293 nm and a pulse repetition rate of 1 MHz was used.

The sample was contained in a lcm^2 quartz cuvette and the fluorescence was monitored, at 90° to the direction of excitation, through a lm grating monocromator by a high gain photomultiplier tube (Philips 56DUVP). A polarizer could be inserted before the monochromator to allow corrections to be made for time-dependent depolarization effects.[1] Conventional Ortec single photon counting detection equipment was used,[2] the data being stored in a multichannel analyzer (Northern NS 600) and output to an ICL 2970 computer for analysis. The ratio of fluorescence photons recorded to excitation pulse rate was kept below 0.5%. Even so, decay curves with greater than 30000 counts in the maximum channel, and lnm fluorescence resolution, could be accumulated in under 10 mins. This system has been shown to be capable of resolving sub-nanosecond fluorescence decays and multicomponent emissions with accuracy.[3]

By setting upper- and lower-level voltage discriminators on the output signal from the time-to-pulse height converter, and by scanning the monochromator, the fluorescence spectrum of only those photons emitted a certain time after excitation could be measured. Such time-resolved spectra have proved invaluable in resolving complex fluorescence emissions without the need for the assumption of a functional form for the fluorescence decay.[4–7]

Materials

The preparation and purification of the 1-vinyl naphthalene and its copolymers,[8] of poly (acenaphthylene),[9] and poly (*N*-vinyl carbazole)[10] have been described elsewhere. The former polymers were prepared by the research group of Dr. I. Soutar, Heriot-Watt University, and the latter by that of Professor A. Ledwith, University of Liverpool.

The polymers were dissolved in tetrahydrofuran (THF) which was purified by refluxing over and fractional distillation from Li AlH_4 immediately prior to use. Solutions were degassed by repeated freeze-pump-thaw cycles.

Data Analysis

Although deconvolution is a well-defined mathematical procedure, its application to fluorescence decay curves is attended with numerous difficulties owing to the counting errors and instrumental distortions that accompany single-photon counting data. It is now generally accepted that least squares iterative reconvolution is the most satisfactory method of analyzing nanosecond decay data,[11-13] and it is the method we prefer.

In its simplest form, deconvolution involves solution of the equation

$$F(t_i) = \int_0^{t_i} G(t')P(t_i - t')\,dt' \tag{3}$$

in which

$F(t)$ is the measured decay cure

$G(t)$ is the true decay function

and

$P(t)$ is the measured excitation profile, also known as the instrument response function.

In the least squares fitting procedure, $G(t)$, assumed to be of a specific functional form with adjustable parameters, is convoluted according to Equation 3 with the measured function, $P(t)$. The calculated curve $C(t)$ is compared with the measured curve $F(t)$ and the reduced chi-square, χ_ν^2, is calculated.

$$\chi_v^2 = \frac{\sum_{i=1}^n W_i^2(F(t) - C(t_i))^2}{n-p} = \frac{\sum_{i=1}^n (r_i)^2}{n-p} \tag{4}$$

W_i is the weighting factor for the ith data point ($= 1/\sqrt{F(t_i)}$), n is the number of channels over which fitting is performed, and p is the number of adjustable parameters in the fitting function. The values of the adjustable parameters are varied until a minimum value of χ_ν^2 is obtained. If the minimum value of χ_ν^2 is close to one, the function assumed for $G(t)$ is taken to be correct and the final values of the adjustable parameters are assumed to have real physical significance.

For single exponential decay functions with reasonably long decay times (>2 ns), deconvolution poses few difficulties. Double exponential decays, for which $G(t)$ contains 4 variable parameters, are much more difficult to analyze correctly and a number of tests have been proposed to ensure that the fitting procedure does not mask distortions in the data or the presence of a further component. The most common of these tests is a plot of the weighted residuals, which should be randomly distributed about zero. Also of use is a plot of the autocorrelation function of the residuals.[14]

While the analysis of single and double exponential decays may now be regarded as fairly routine, the successful deconvolution of triple exponential functions is fraught with difficulties. Correlation among the six variable parameters in $G(t)$ cannot be avoided and unless strict precautions are taken, the decay function recovered from deconvolution may have little resemblance to the true fluorescence decay.[15] It is, of course, necessary to have data that are as statistically accurate and as free from instrumental artifacts as possible. A systematic study is at present under way to try to ascertain the number of counts that must be collected in order to establish triple exponential functions with satisfactory accuracy. For the present we can say that, for most of the decays we have studied, 30,000 counts in the channel of maximum counts is a lower limit. Fortunately, with our laser excitation source such high data counts can be achieved in a matter of minutes. We cannot, however, avoid the difficulty introduced by having to collect the pump profile at the excitation wavelength. We do on occasion correct for the variation of the time response of the photomultiplier with wavelength by shifting the zero time of the pump profile with respect to the decay curve time scale, but since the procedure introduces a further variable parameter into the fitting routine, we choose not to follow it in analysis of triple exponential decays. It should be noted that the error resulting from this source is negligible compared to the shortest decay time that is extracted.

Even when data are statistically very accurate and free from instrumental distortions, triple exponential analysis may be meaningless. For instance, in a trial test with simulated data, the function

$$G(t) = \textstyle\sum_{j=1}^{3} a_j e^{-t}/\tau_j \qquad \textbf{(5)}$$

with $a_1 = 0.3$, $\tau_1 = 1.5$, $a_2 = 0.2$, $\tau_2 = 5.0$, $a_3 = 0.1$, $\tau_3 = 10.0$ was convoluted with a real pump profile of width 7 ns. Gaussian noise was then added. Deconvolution using our triple exponential routine failed to recover the correct decay parameters although the χ_υ^2 value was quite satisfactory (1.03). It must therefore be admitted that triple exponential analysis is not possible for certain combinations of decay parameters. However, the three lifetimes that are of interest in the present work are quite well separated and two of them exceed the width of the excitation pulse. Tests with simulated data demonstrated the ability of the routine to recover the correct parameters as well as showing the inability of a two-component function to fit three-component data of the type encountered in these polymer systems. As an illustration, FIGURE 3(a) shows a two-component fit to the three-component decay generated by convolution of a real laser pulse profile with a $G(t)$ in which $a_1 = 0.25$ $\tau_1 = 2.5$, $a_2 = 0.07$, $\tau_2 = 10.00$, $a_3 = 0.025$, $\tau_3 = 40.0$. The plots of residuals (FIGURE 3(b)) and autocorrelation of residuals (FIGURE 3(c)) are clearly diagnostic of an inappropriate fitting function. The fit to a three-component decay is shown in FIGURE 4(a) with residuals and autocorrelation of residuals plotted in FIGURES 4(b) and 4(c). All plots are satisfactory, and in fact, the parameters recovered with this analysis were $a_1 = 0.25$, $\tau_1 = 2.54$, $a_2 = 0.07$, $\tau_2 = 9.72$, $a_3 = 0.026$, $\tau_3 = 39.44$, in good agreement with the correct values. Similar tests with many other trial functions lead to the conclusion that, with the pumping pulse profiles obtained from the present excitation source and decay functions containing decay constants of the magnitude met in the polymer systems under investigation, triple exponential analysis produces meaningful results if statistical accuracy is high and instrumental artifacts are maintained at a minimum. (A small amount of stray light, for instance, would render the analysis meaningless).

In order to increase the level of confidence in our analysis, we calculate a number of other statistics in addition to χ_ν^2. These include the standard deviation of the weighted residuals, the mean residual, a skewness of fit parameter,[16] and, especially,

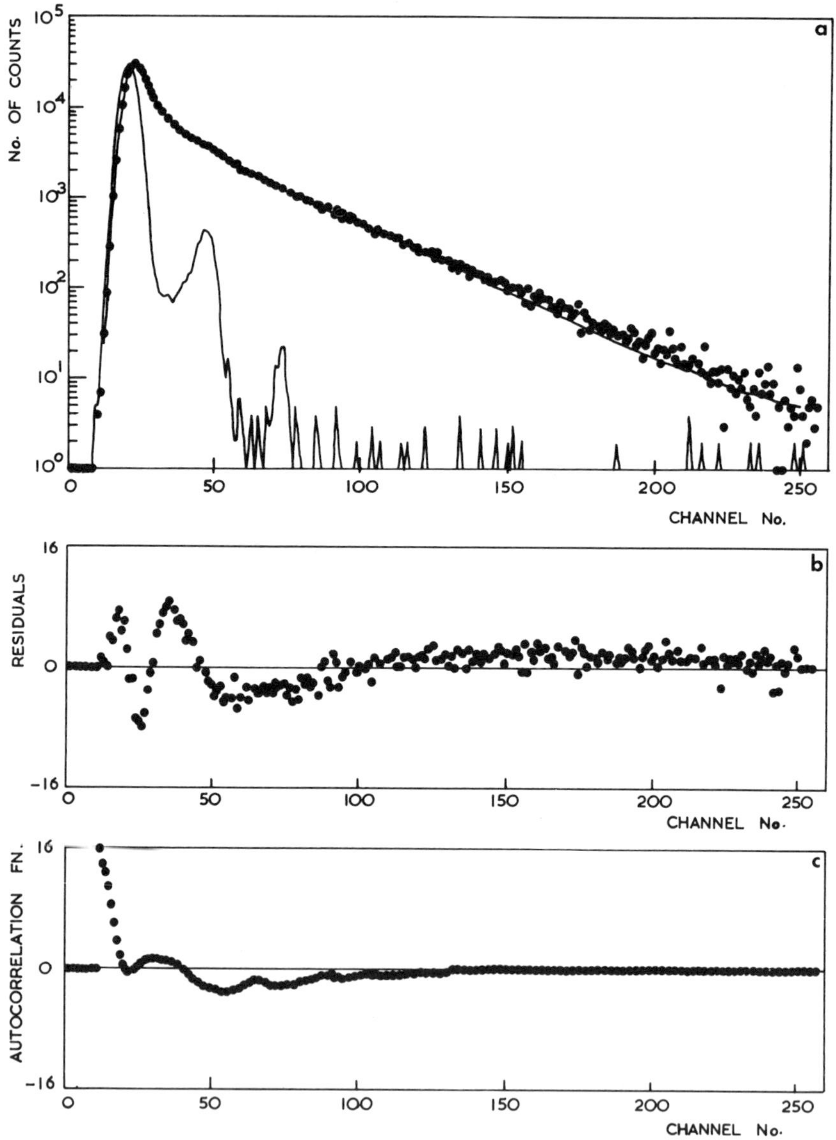

FIGURE 3. Plots of (a) two-component fit to simulated three-component fluorescence decay (see text); (b) weighted residuals; (c) autocorrelation function.

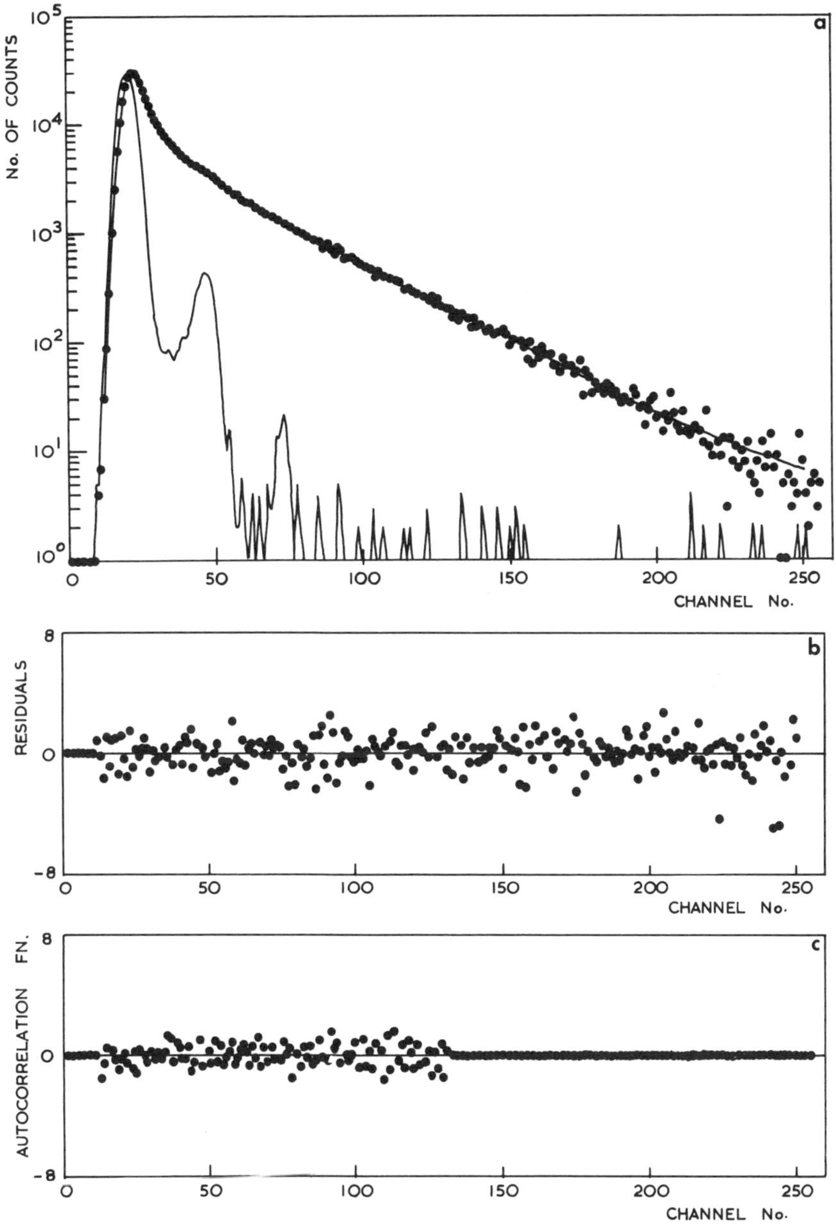

FIGURE 4. Plots of (a) three-component fit to the same simulated three-component fluorescence decay as in FIGURE 3; (b) weighted residuals; (c) autocorrelation function.

the Durbin-Watson factor, DB.[17,18] By using DB, calculated according to

$$DB = \frac{\sum_{i=2}^{n} (r_i - r_{i-1})^2}{\sum_{i=1}^{n} r_i^2} \quad \textbf{(6)}$$

and the table of values in Reference 18, one has an additional criterion concerning the suitability of the chosen fitting function.

Applications to Fluorescence of Synthetic Polymers

Poly (1-vinyl naphthalene) and its Copolymers with Methyl Acrylate[19] *and Methyl Methacrylate*

This work, carried out jointly with Dr. Ian Soutar, Heriot-Watt University, has been described in part by him in an earlier paper in this volume. Time-resolved spectra (Figure 5) reveal only two types of spectrally distinct fluorescent sites in these polymers, but decay characteristics require three exponential terms for successful fitting of monomer decay functions monitored at 330 nm and shown in Table 1. There are two kinetic schemes which could account for these results, shown in Schemes 2 and 3, which are modifications of the Birks Scheme 1, and in both of which two types of monomer unit, M_1^* are distinguishable.

Scheme 2 (Figure 6)

In this photophysical scheme it is proposed that M_1^*, and D^* interact by the generally accepted exciton diffusion mechanism. M_2^* is considered to be an isolated naphthalene chromophore which can transfer energy into M_1^* with a transfer rate characterized by the rate coefficient k_t. Reverse transfer from M_1^* to M_2^* is considered unimportant since exciton diffusion is expected to be very efficient within sequences of naphthalene chromophores within the chain comprising the M_1^* sites. In view of the reduced lifetime of M_1^* relative to M_2^* and of the delocalized nature of the energy within extended chromophore sequences which increases the effective separation of M_1^* and M_2^*, M_1^* to M_2^* energy transfer by Foster or Dexter mechanisms is diminished relative to the M_2^* to M_1^* process.

Scheme 3 (Figure 7)

An alternative proposal considers M_2^* to be an isolated naphthalene chromophore which can form excimers through a long-range interaction involving segmental diffusion to bring the M_2^* species into close proximity with a ground state chromophore. Such interactions have been cited as the dominant mechanistic routes to excimer formation in polymers such as poly (naphthyl methacrylate).

Assuming the emissions from M_1^* and M_2^* to be spectrally indistinguishable and that

$$k_M = k_{FM} + k_{IM}$$

is identical for each species, then the decay profiles $i_M(t)$ and $i_D(t)$ recorded for

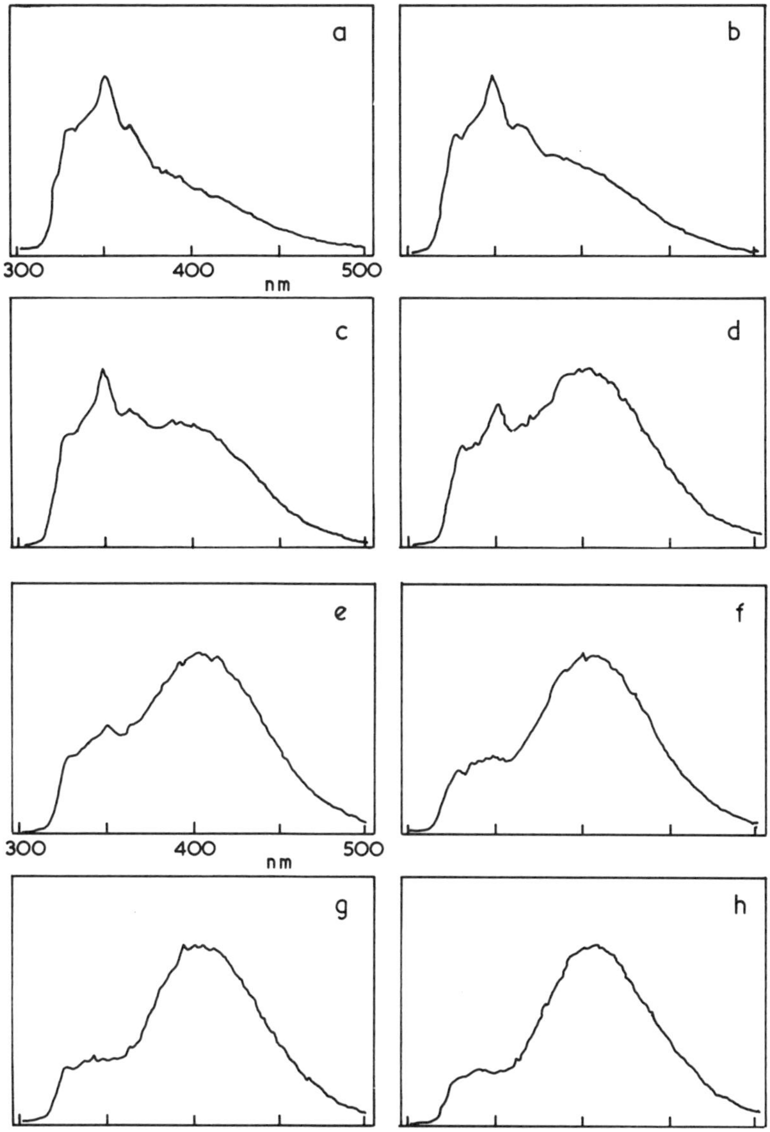

FIGURE 5. Time-resolved spectra in poly (vinyl naphthalene)-poly (methyl methacrylate) copolymer. (Sample 5) Delays: (a) 0 ns; (b) 3.2 ns; (c) 6.4 ns; (d) 12.8 ns; (e) 19.2 ns; (f) 25.6 ns; (g) 32.0 ns, and (h) 38.0 ns.

monomer and excimer respecitvely may be derived as

$$i_M(t) = A_1 \exp(-\lambda_1 t) + A_2 \exp(-\lambda_2 t) + A_3 \exp(-\lambda_3 t) \tag{7}$$

and

$$i_D(t) = A_4 \exp(-\lambda_1 t) + A_5 \exp(-\lambda_2 t) + A_6 \exp(-\lambda_3 t) \tag{8}$$

TABLE 1
TRIPLE EXPONENTIAL DECAY DATA FOR 1-VINYL NAPHTHALENE METHYL METHACRYLATE COPOLYMERS

Sample	f_n*	A_1	τ_1/ns	A_2	τ_2/ns	A_3	τ_3/ns
1	0.17	0.005	73.55	0.023	12.65	0.112	40.06
2	0.27	0.011	67.60	0.041	15.31	0.092	34.17
3	0.38	0.027	59.13	0.057	11.88	0.074	26.44
4	0.47	0.021	59.50	0.075	8.46	0.066	23.09
5	0.58	0.025	53.34	0.100	6.41	0.069	19.52
6	0.66	0.025	53.24	0.115	4.18	0.065	16.36
7	0.75	0.032	48.83	0.149	3.57	0.072	14.41
8	0.83	0.029	47.07	0.257	1.87	0.072	13.52

* f_n is mole fraction of naphthalene polymer.

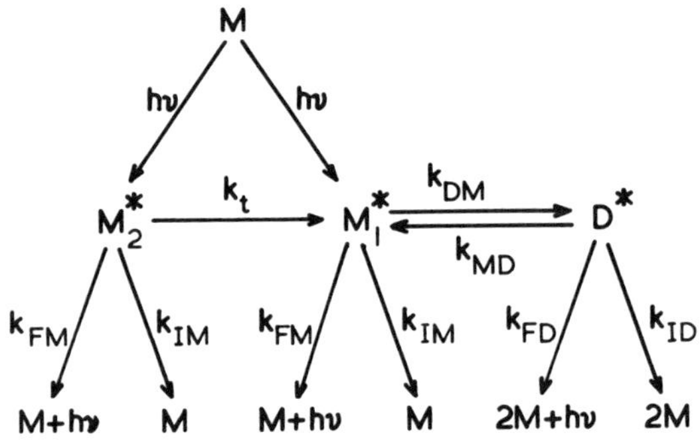

SCHEME 2

FIGURE 6. Possible scheme for excimer kinetics in copolymers of vinyl naphthalene.

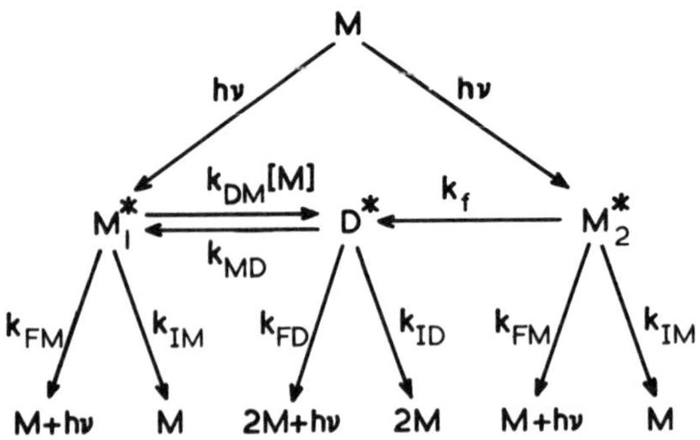

SCHEME 3

FIGURE 7. Alternative scheme for excimer kinetics in copolymers of vinyl naphthalene.

where

$$\lambda_{1,2} = \tfrac{1}{2}[(X + Y) \pm \{(Y - X)^2 + 4k_{MD}k_{DM}[M]\}^{1/2}]$$

and

$$X = k_M + k_{DM}[M];$$

$$Y = k_D + k_{MD}$$

where

$$k_D = k_{FD} + k_{ID};$$

and

$$\lambda_1 + \lambda_2 = k_M + k_{DM}[M] + k_D + k_{MD};$$

and

$$\lambda_3 = k_t + k_M \quad \text{(for Scheme (2))}$$

or

$$\lambda_3 = k_f + k_M \quad \text{(for Scheme (3))}$$

The empirical triple exponential fit to $i_{330}(t)$ is consistent with the expression for $i_M(t)$ in Equation 7 derived from the kinetic schemes above. It will be noted that the expression for $i_M(t)$ contains terms involving $k_{DM}[M]$, which represents a concentration term. We have discussed at length elsewhere the necessity of carrying out studies on copolymer systems of varying composition in order to eliminate the concentration dependence from the rate constant k_{DM}, and the methods used for evaluation of individual rate constants.[8] Such a procedure yields a true rate coefficient k_{DM} modified by a factor α which is not concentration dependent, and other rate-coefficients given in TABLE 2, in which several points are noteworthy:

TABLE 2
KINETIC PARAMETERS FOR VINYLNAPHTHALENE COPOLYMER SYSTEMS

Rate Parameter	Value/10^7 S^{-1}	
	Methyl Acrylate* Comonomer	Methyl Methacrylate† Comonomer
k_M‡	1.5	1.3
	2.0	2.15
	1.0	1.0
αk_{DM}	45	12.05
$k_{MD} + k_D$	14.3	6.35
k_D	2.35	2.2
k_{MD}	12.0	4.2
k_t	11.4	7.8

* Derived from a complete study on a series of copolymers of vinyl naphthalene and methyl acrylate.

† Derived from a complete study on a series of copolymers of vinyl napthalene and methyl methacrylate.

‡ Values quoted are obtained by three different methods.

(1) Rate parameters for the different copolymers which are internal deactivation rates of monomer (k_M) and excimer (k_D) agree well.
(2) Rate parameters which might be expected to depend significantly on the flexibility of the polymer chains show the anticipated dependence, in that αk_{DM}, and k_{MD} are much greater in the more flexible acrylate copolymers. k_t shows a similar dependence although recent temperature studies show only a weak temperature dependence of k_t. It is thus still not possible on experimental grounds to differentiate between Schemes 2 and 3.

Poly (acenaphthylene)[6,9]

Although in this polymer excimer formation cannot occur via nearest neighbor interactions, it was again found to be necessary to model the kinetics using triple exponential decays.

Poly (N-vinyl) carbazole

This polymer is unique in that in the homopolymer, no monomer emission is observed, but two spectrally distinct excimer sites are distinguishable at λ max 420 nm, and λ max 380 nm.[20] Although these two sites interconvert,[10] there is no isoemissive point in the steady-state fluorescence spectra, (FIGURE 8), and the kinetics must therefore be complex. Time-resolved studies have confirmed this point, and have led to the proposal of a Scheme 4 below in which excimer sites E_2 and E_3 are the excimers emitting at 420 and 380 nm respectively, and E_1, also fluorescing at 380 nm, is the subject of some debate, Tagawa *et al.*[21] favouring a "relaxed" monomeric group, although there is no spectral evidence for this, while Roberts *et al.*[10] prefer a structure similar to E_3, but in a configuration such that it may form E_2. Both E_1 and E_3 are populated on a time-scale of less than 10 ps.[8]

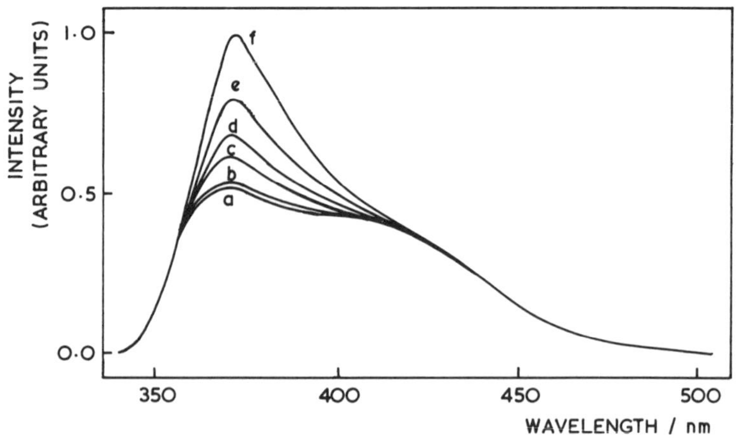

FIGURE 8. Steady-state fluorescence spectra of poly (N-vinyl carbazole) as a function of temperature. (a) 295 K; (b) 275 K; (c) 260 K; (d) 240 K; (e) 220 K, and (f) 200 K.

Three decay parameters can be measured as shown in TABLE 3 and identified approximately with the rate constants shown in Scheme 4 as follows:

$$\tau_1^{-1} = k_2 + k_4$$

$$\tau_2^{-1} = k_5 + k_6$$

$$\tau_3^{-1} = k_7$$

Since E_1 and E_3 are spectrally indistinct, it is fair to assume that the internal deactivating rate constants k_2 and k_7 are similar. The strong temperature dependence of τ_1 must therefore be associated with k_4, since τ_3 varies little with temperature. An Arrhenius plot of $\ln(\tau_1^{-1} - \tau_3^{-1})$ (FIGURE 10) yields an activation energy of 3.5 kJ mol^{-1} for the conversion of E_1 to E_2. The relative contributions of the E_1 and E_3 species to the fluorescence centred at 380 nm can be estimated from the time-resolved decay curves and data in TABLE 3 and the steady-state data in FIGURE 8. Results are shown in FIGURE 11, from which it is seen that Φ_1 is reduced at lower temperatures whereas Φ_3 is increased dramatically. This is only consistent with the invariance of τ_3 with temperature if the quantum yield Φ_3 changes as a result of the branching

TABLE 3
DECAY DATA FOR FLUORESCENCE OF POLY N-VINYLCARBAZOLE AT 370NM FITTED TO A TRIPLE EXPONENTIAL SCHEME

Temp/K	A_1	τ_1/ns	A_2	τ_2/ns	A_3	τ_3/ns	X^2_v	DB
295	0.229	2.43	0.053	35.69	0.045	10.43	1.17	1.93
275	0.195	2.82	0.036	36.05	0.062	10.50	1.16	1.82
260	0.187	2.88	0.031	35.04	0.069	10.55	1.10	1.83
250	0.168	3.11	0.027	35.25	0.081	10.75	1.04	1.97
240	0.147	3.49	0.025	35.28	0.089	11.11	1.08	1.87
230	0.118	3.44	0.026	33.92	0.106	11.00	0.93	1.96
220	0.088	3.70	0.017	34.20	0.102	11.75	1.11	2.19
210	0.084	4.14	0.015	34.87	0.114	12.89	1.14	1.88
200	0.075	4.25	0.013	35.21	0.125	13.57	1.24	2.14

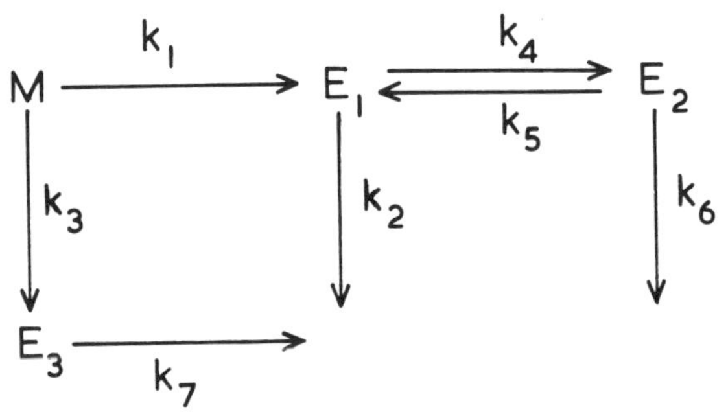

SCHEME 4

FIGURE 9. Kinetic scheme for excimer formation and decay in poly (N-vinyl carbazole).

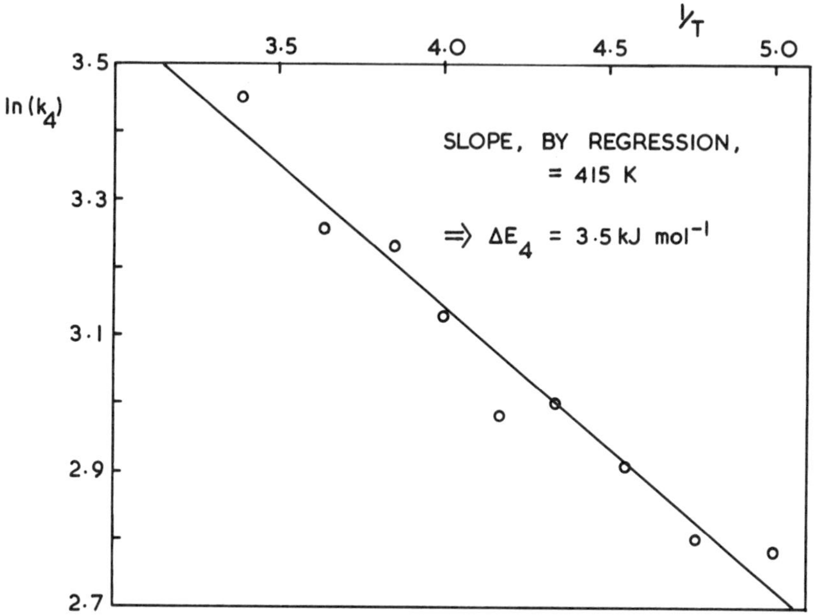

FIGURE 10. Arrhenius plot of $\ln(\tau_1^{-1} - \tau_3^{-1})$ vs. $1/T$ (see text) for poly (N-vinyl carbazole).

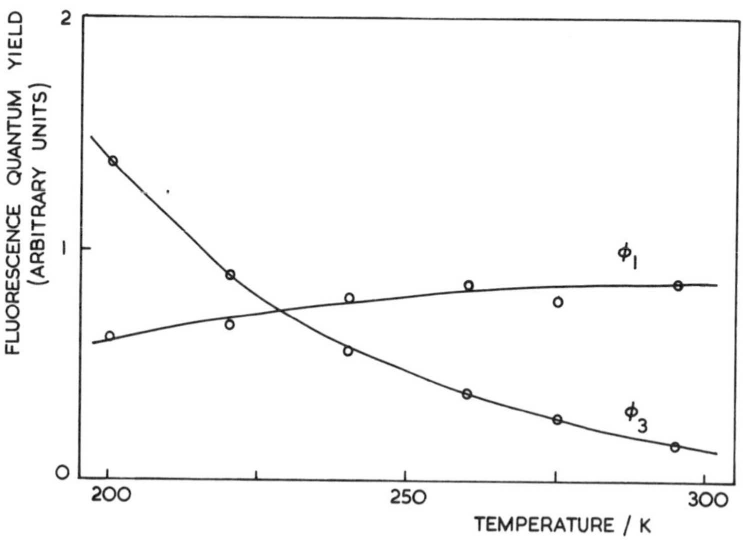

FIGURE 11. Plots of quantum yield of fluorescence from E_1 (Φ_1) and E_3(Φ_3) sites (Scheme 4) as a function of temperature.

ratio $k_3/(k_1 + k_3)$ changing with temperature. If E_1 and E_3 are "preformed" dimer sites, which is consistent with their population on a time-scale of less than 10 ps, evidently E_3 is favored at low temperatures.

Conclusions

In all polymers studied to date in dilute solution, modeling of fluorescence decay characteristics requires the use of three exponential terms. This is demonstrably not an experimental artifact, since simpler model systems can be well fitted to single and dual exponential decays under similar excitation conditions, and thus appears to be a general feature of vinyl polymers with pendant aromatic groups. Moreover, the three decays are explicable with plausible physical models.

Summary

The kinetics derived for excimer formation in small molecule systems are inappropriate for successful modeling of fluorescence decays of monomeric aromatic moieties and excimer sites in vinyl polymers with pendant aromatic groups. In particular, the decay of the monomer fluorescence requires three exponential functions for successful fitting, and these can be rationalized in terms of two types of monomer unit, one which can partake in facile and reversible excimer formation, the other which behaves essentially as an isolated chromophore, forming excimers only by long-range interactions.

Acknowledgments

We are grateful to The Science Research Council, The Royal Society, and U.S. Army Office of Scientific Research, Europe, for financial support.

References

1. Tao, T. 1969. Biopolymers. **8:** 609–632.
2. Hackett, P. A. & D. Phillips. 1974. J. Phys. Chem. **78:** 671–678.
3. Ghiggino, K. P., A. J. Roberts & D. Phillips. 1980. J. Phys. E. **13:** 446–450.
4. Meech, S. R., D. V. O'Connor, A. J. Roberts & D. Phillips. On the construction of nanosecond time-resolved emission spectra. Photochem. Photobiol. Submitted.
5. Costa, S. M. de B., M. J. Prieto, K. Ghiggino, D. Phillips & A. J. Roberts. 1980. J. Photochem. **12:** 11–16.
6. Phillips, D., A. J. Roberts & I. Soutar. 1980. J. Polym. Sci. Polym. Lett. Ed. **18:** 123.
7. Hoyle, C. E., T. L. Nemzek, A. Mar & J. E. Guillet. 1978. Macromolecules **11:** 429–431.
8. Phillips, D., A. J. Roberts & I. Soutar. Transient decay studies of photophysical processes in aromatic polymers I. Multiexponential fluorescence decays in copolymers of 1-vinylnaphthalene and methyl methacrylate. J. Polym. Sci. Polym. Phys. Submitted.
9. Phillips, D., A. J. Roberts & I. Soutar. Transient decay studies of photophysical processes in aromatic polymers III. Poly (acenaphthylene). Eur. Polym. J. Submitted.
10. Roberts, A. J., C. G. Cureton & D. Phillips. Time-resolved studies of poly (N-vinyl carbazole) using pulsed laser excitation. Chem. Phys. Lett. In press.

11. KNIGHT, A. G. & B. K. SELINGER. 1971. Spectrochim. Acta **27A:** 1223.
12. MCKINNON, A. G., A. G. SZABO & D. R. MILLER. 1977. J. Phys. Chem. **81:** 1564.
13. O'CONNOR, D. V., W. R. WARE & J. C. ANDRE. 1979. J. Phys. Chem. **83:** 1333.
14. EASTER, J. H., R. P. DE TOMA & L. BRAND. 1976. Biophys. J. **16:** 571.
15. GRINVALD, A. & I. Z. STEINBERG. 1974. Anal. Biochem. **59:** 583.
16. IRWIN, D. J. G. & A. E. LIVINGSTON. 1974. Comput. Phys. Commun. **7:** 95.
17. DURBIN, J. & G. S. WATSON. 1950. Biometrika **37:** 409.
18. DURBIN, J. & G. S. WATSON. 1951. Biometrika **38:** 159.
19. PHILLIPS, D., A. J. ROBERTS & I. SOUTAR. Transient decay studies of photophysical processes in aromatic polymers II. Copolymers of 1-vinyl naphthalene and methyl acrylate. Polymer. Submitted.
20. JOHNSON, G. E. 1975. J. Chem. Phys. **62:** 4697.

PART II. LUMINESCENCE FROM SOLUTIONS OF BIOLOGICAL MACROMOLECULES

LIGHT SCATTERING BY CHROMOPHORES AT THEIR ABSORPTION BANDS

Izchak Z. Steinberg and Jacob Anglister

Department of Chemical Physics
Weizmann Institute of Science
Rehovot, Israel

INTRODUCTION

Light energy that is absorbed by molecules at their absorption bands can be dissipated by a variety of mechanisms, both radiative and non-radiative. The yields of the various processes which occur following light absorption depend as a rule on the individual properties of the absorbing molecules. There is one process of dissipation, however, which inherently and invariably accompanies light absorption, and its magnitude bears a fixed relation to the molar extinction coefficient, but is otherwise independent of the individual properties of the absorbing molecules (except for some simple characteristics such as the dipolar or multipolar nature of the transition involved and the assumption that the chromophore is small in dimension relative to the light wavelength). This process is the elastic scattering of light. The association of elastic light scattering with light absorption is intuitively evident in the classical picture of the absorption process, which is assumed to take place by forced oscillations of molecular oscillating dipoles under the influence of the incident beam. Such oscillations also radiate light of the same frequency as, and at a fixed phase relative to, the incident light, the radiated light propagating in all directions in space. To demonstrate that the scattering is quantitatively related to the intensity of the absorption and not to other individual properties of the absorbing molecules, the following considerations may be made. From the superposition principle of electromagnetic fields, it follows that the attenuation of the incident beam by the intervening absorbing medium may, and should, be viewed as due to the superposition of the original incident beam with electromagnetic fields produced by the medium having the same frequency as the original beam but having a phase difference of π relative to it, so as to attenuate its amplitude. Given the dipolar nature attributed to the oscillators and the molecular dimensions of the absorbing entities, as well as their random distribution in solution, their contribution to the electromagnetic fields in the forward direction, as manifested by the attenuation of the incident beam, must be accompanied by energy-carrying electromagnetic fields in other directions, i.e., by scattered light, in a fixed quantitative relation to the absorption intensity. Similar considerations apply to the quantum-mechanical view of the above processes.

In the following, the theoretical background of the elastic light scattering accompanying light absorption will be discussed. Obviously, one should include in the elastic scattering also the contribution due to the real part of the refractive index,

0077-8923/81/0366-0125$01.75/0 © 1981, NYAS

which is expected to be enhanced upon approaching an absorption band. The experimental results to be presented on the light scattered by a few carotenoid pigments at their visible absorption bands indicate that the elastic scattering is the dominant scattering mechanism at the blue part of the absorption band. In this spectral range, fair agreement has been found between the experimental results and theoretical predictions. However, at the red part of the absorption bands, the observed scattered intensity significantly exceeds the levels theoretically expected for elastic scattering as evaluated from the absorption intensity.

THEORETICAL CONSIDERATIONS

The contribution of the refractive index increment of solute molecules to the intensity of Rayleigh scattering is well known. As stated above, light scattering at absorption bands cannot be described exclusively by the real part of the refractive index; the absorption coefficient also has a marked contribution to the scattering intensity. There are various ways of evaluating the latter contribution, and for the sake of simplicity we shall present a rather intuitive approach along the lines suggested by Kauzmann.[1]

When an absorbing medium is placed in the path of a light beam, the radiative electromagnetic fields following the medium are of course smaller than those of the incident beam. By the superposition principle of electromagnetic fields, we can view this attenuation as due to a contribution of electromagnetic fields by the medium; these fields are of opposite direction to those of the incident light beam. Let us consider the absorption by a thin layer of thickness Δz of absorbing medium perpendicular to the light beam. The attenuation ΔI of the light intensity (associated with an attenuation of magnitude E_s in the electric field) is given by:

$$\frac{\Delta I}{I_0} = -2 \cdot 3\epsilon \frac{1000N}{N_A} \Delta z \tag{1}$$

and

$$\frac{\Delta I}{I_0} = \frac{2E_s}{E_0} \tag{2}$$

where I_0 and E_0 are the incident light intensity and the amplitude of the associated electric field, respectively, ϵ is the molar extinction coefficient, N is the number of solute molecules per milliliter, and N_A is Avogadro's number. Equation 2 follows from the quadratic dependence of I on E and the small values of $\Delta I/I_o$ and E_s/E_0 (due to the assumed small value of Δz). Hence, one obtains

$$\frac{E_s}{E_0} = -\frac{2 \cdot 3}{2} \times 10^3 \epsilon \frac{N}{N_A} \Delta z \tag{3}$$

Let us restrict the following discussion to molecules which have random orientations and which exhibit dipolar transitions at the absorption band under discussion. Following the approach used in Reference 1 (pages 600–608) used for the evaluation of the forward scattering of molecules due to their in-phase polarizability, one can obtain an expression for E_s for the case under discussion (compare with Reference

1, Equation E.34),

$$\frac{E_s}{E_0} = -(4\pi^2 N/\lambda)\Delta z(\alpha'/3)/n \tag{4}$$

where λ is the wavelength in vacuum of the incident light, α' is the out-of-phase polarizability of the absorbing molecules along the direction of their transition dipole moment, and n is the real part of the refractive index of the medium. Whereas Equation 4 has been derived in the same way in which Equation E.34 of Reference 1 was, a few differences may be noted between the forms of the equations, which follow from the difference in the problems treated. (a) The phase angle was dropped in Equation 4, since E_s due to light absorption is π out of phase to E_0, which is signified by the minus sign in front of the right-hand side of this equation; (b) In the present case, the molecules are assumed to be optically extremely anisotropic, i.e., dipolar. Due to the presumed random orientations of the absorbing molecules, the effective value of the out-of-phase polarizability is $\alpha' \cdot \langle \cos^2\theta \rangle = \alpha'/3$ (θ being the angle between the transition dipole moment and the direction of polarization of the incident beam); (c) The factor $1/n$ which appears in Equation 4, but is absent in Equation E.34 of Reference 1, is due to the fact that in the latter case the calculations were performed for molecules in vacuum. [For molecules embedded in a medium of refractive index n, the phase angle $\cos\omega\left(t - \frac{r}{c}\right)$ of Equation E-32d (Reference 1) should be changed to $\cos\omega\left(t - \frac{r}{c/n}\right)$. The integration step following this equation will thus introduce an additional factor of $1/n$ in the result].

Combining Equations 3 and 4, one obtains an explicit expression for α':

$$\alpha' = \frac{3 \times 2{\cdot}3 \times 10^3 \epsilon\lambda n}{8\pi^2 N_A} \tag{5}$$

Combining the above expression for α' and that for the total energy, ω, radiated by a dipolar oscillator,

$$\omega = 16\pi^4 m_0^2 c/3\lambda^4 \tag{6}$$

and remembering that $m_0 = \alpha' E_0 \cos\theta$, $I_0 = c|E_0|^2/8\pi$, and that $\langle \cos^2\theta \rangle = 1/3$, one obtains,

$$\tau = \frac{2 \times 2.3^2 \times 10^6 \epsilon^2 n^2}{\lambda^2 N_A^2} \cdot N = \frac{2\pi \times 2.3^2 \times 10^3 \epsilon^2 n^2 C}{\lambda^2 N_A} = \frac{32\pi^3 \times 10^3 n^2 C}{\lambda^4 N_A}\left(\frac{\partial k}{\partial C}\right)^2 \tag{7}$$

where τ the turbility, is the total light scattered by 1 cm^3 of solution of molarity C, and k is the imaginary part of the refractive index, given by $k = 2.3\epsilon C\lambda/4\pi$.

It is often convenient to express the scattering data by the Rayleigh ratio, which can be deduced from the above to yield

$$R_{v,V+H}(90°) = \frac{9600\pi^2 n^2 C}{\lambda^4 N_A} \cdot \left(\frac{\partial k}{\partial C}\right)^2 \tag{8}$$

where $R_{v,V+H}$ is the Rayleigh ratio for vertically polarized incident light and the sum of vertically and horizontally polarized scattered light. To include the contribution of the real part of the refractive index of the solute molecules to the scattered light, one should substitute the expression $\left[\left(\frac{\partial n}{\partial C}\right)^2 + \left(\frac{\partial k}{\partial C}\right)^2\right]$ for $\left(\frac{\partial k}{\partial C}\right)^2$ in Equation 7 and 8.

The problem of light scattering by molecules at absorption bands was treated theoretically by Miller[2] by a different approach. In this treatment, the results of the Mie theory for light scattering were extrapolated to particle size that is small compared to the wavelengths of the incident light. The results obtained by Miller differ from Equation 7 by a factor of 3, the reason being that Miller's treatment applies to isotropic particles whose optical properties do not depend on orientation. Obviously, this does not apply to non-degenerate dipolar transitions, which are the most likely to be met in practical cases. Equations 7 and 8 do indeed describe such systems.

It is of interest to note that the light scattering at absorption bands varies with ϵ^2. If the system tested consists of a heterogeneous mixture of molecules with different ϵ values, the measured extinction coefficient will be an average of ϵ, $<\epsilon>$, whereas the measured Rayleigh scattering will be related to the average of ϵ^2, which will differ from $<\epsilon>$. Deviations from the behavior described by Equation 7 and 8 are thus expected to be observed if $<\epsilon>^2$ is used in these expressions.

The real part of the refractive index increment due to the solute molecules will, as stated, contribute additionally to the intensity of the scattered light. The real part is related to the whole absorption spectrum of the molecules under discussion by the Kronig-Kramers relationship:

$$n(\lambda_0) = n_0(\lambda_0) + \frac{2 \cdot 3\lambda_0^2 C}{2\pi^2} \int_0^\infty \frac{\epsilon(\lambda)}{\lambda_0^2 - \lambda^2}\, d\lambda \qquad \textbf{(9)}$$

where $n(\lambda_0)$ and $n_0(\lambda_0)$ are the refractive indices of the solution and solvent, respectively, at wavelength λ_0. The contributions of remote bands fall off markedly as their λ differ more from λ_0. Thus, any given absorption band will have a dominant effect on n at the spectral region of this band, both as regards the magnitude and the polarization properties of n.

The highly anisotropic nature of the scattering molecules at absorption bands will of course result in appreciable depolarization of the scattered light. For dipolar transitions, the depolarization ratio, defined as the ratio of depolarized to polarized intensities of the scattered light, is expected to be 1/3. This was indeed confirmed by experiments carried out on a number of chromophores, except at the blue edge of the bands, where $\partial n/\partial C$ may assume even negative values and where contributions from other bands are not negligible.[3]

Experimental

Method of Measurement

For practical reasons, light absorption by the solute puts an upper limit on the concentrations of chromophores that can be used in light scattering experiments at absorption bands. It was found impractical to exceed the optical density of the solution beyond about one. Since scattering varies linearly with C but quadratically with ϵ, dyes scatter more at a given optical density of the solution if their ϵ has a higher value. For dyes of molar extinction coefficients of 10^4 or $10,^5$ it was thus impossible to exceed molar concentrations of about 10^{-4} or 10^{-5}, respectively. Under such circumstances, the polarized scattered light intensity due to the dye is expected to be 1 or 2 orders of magnitude less than the polarized scattered intensity due to the solvent. However, due to the extreme optical anisotropy of the solute dye molecules, their depolarized scattered light is relatively high and by proper choice of solvents which exhibit low depolarized scattering (say, less than 2% of the total), the contribution of the dye to the depolarized scattering is significant. The polarized scattered light, which is due predominantly to the solvent, was exploited to be used as an internal calibration measure to correct for the attenuation by absorption of the incident beam and scattered beam. This attenuation is difficult to evaluate by computation since it depends on the geometric factors of the sample cell and its relation to the incident beam and detection system. Since the solution is isotropic, the attenuation is identical for the polarized and depolarized scattered light, and the former can serve as a standard for the latter.

The information about the light scattering was thus extracted from the data on the depolarization ratio of the solution, $\rho_{v,sn}$, and of the solvent, $\rho_{v,st}$.

$$\rho_{v,sn} = \frac{I_{\perp,d} + I_{\perp,st}}{I_{\parallel,d} + I_{\parallel,st}} \quad \textbf{(10)}$$

where $I_\perp$ and $I_\parallel$ are, respectively, the depolarized and polarized scattered intensities, and the subscripts d, st and sn refer to the dye, solvent and solution, respectively. Upon multiplying both sides of Equation 10 by $(I_{\parallel,d} + I_{\parallel,st})/I_{\perp,st}$ and rearrangement, one obtains,

$$\rho_{v,sn}\left(\frac{1}{\rho_{v,d}} \cdot \frac{I_{\perp,d}}{I_{\perp,st}} + \frac{1}{\rho_{v,st}}\right) = \frac{I_{\perp,d}}{I_{\perp,st}} + 1 \quad \textbf{(11)}$$

From the above it follows that,

$$\frac{I_{\perp,d}}{I_{\perp,st}} = \frac{(\rho_{v,sn}/\rho_{v,st}) - 1}{1 - (\rho_{v,sn}/\rho_{v,d})} \quad \textbf{(12)}$$

From Equation 11, one can obtain the Rayleigh ratio, $(R_{v,H})_d$, for the depolarized scattering by the dye (the incident light is assumed to be vertically polarized) if the Rayleigh ratio for total scattering by the solvent, $(R_{v,V+H})_{st}$, is known:

$$\frac{(R_{v,H})_d}{(R_{v,V+H})_{st}} = \frac{\rho_{v,sn} - \rho_{v,st}}{1 - (\rho_{v,sn}/\rho_{v,d})} \cdot \frac{1}{1 + \rho_{v,st}} \quad \textbf{(13)}$$

Thus, the depolarized scattering intensity of the dye [Equation 12] or its Rayleigh ratio [Equation 13] may be evaluated by the measurement of $\rho_{v,st}$ and $\rho_{v,sn}$. The absolute depolarized scattering intensity of the solvent, $I_{\perp,st}$, or its Rayleigh ratio, should be known, information which can be obtained from the published literature in conjunction with the measurements on $\rho_{v,st}$. In addition, a knowledge of $\rho_{v,d}$ is required, which was measured[3] to be very close to 1/3 (except at the blue edge of the absorption band, where $\frac{\partial n}{\partial C}$ may assume negative values[3]), in agreement with the theoretical expectations for dipolar radiating transitions.

Instrumentation

The light source for the light-scattering measurements was an Argon-ion laser (Spectra-Physics Model 164), which has ten lasing lines in the visible range of the spectrum, with polarization values better than 50:1. In some of the experiments, we used a dye laser (Spectra-Physics, Model 375) pumped by an Argon-ion laser (Spectra-Physics, Model 171-019). The laser dye used was coumarin 6. Two circular apertures of 2mm diameter were placed in the path of the incident laser beam. The scattered light was detected at 90° to both the direction of propagation and plane of polarization of the incident light. Three circular apertures of 3mm diameter, spaced 6cm apart, were placed in the direction of the detected scattered light in order to define a small acceptance angle. This is necessary in order to prevent errors in the measurement of the depolarized scattering, especially when the depolarization ratio is small.[4] The scattered light was modulated by an elasto-optic light modulator (Hinds, Model PEM-3) placed between two crossed Glan-Thompson polarizers of extinction ratio better than 10^{-5}. The front polarizer defines the plane of polarization of the component of the scattered light to be measured. The modulated scattered light was then focused on the entrance slit of a Jarrel-Ash 0.25m double monochromator (Model 82-410), set to pass a spectral bandwidth of 3.2 nm centered at the wavelength of the incident beam. The monochromator serves to reject Raman, resonance Raman, possible traces of fluorescence (although non-fluorescent dyes were chosen), etc. The light was detected by an EMI photomultiplier (Model 9789. Model 9863B was used in the measurements carried out with the dye laser), cooled to 2°C, the output of which was amplified by a phase-sensitive amplifier (Ithaco, Dynatrac 391-A), which received a reference signal from the light-modulator. The readings were corrected for the different transmission of the monochromator to light of different polarization. From measurements performed by setting the polarizer in front of the modulator parallel and perpendicular to the direction of polarization of the incident beam the values of ρ_v could be evaluated.

The intensity of the incident laser beam was set not to exceed 300 mW, and measurements were carried out as quickly as possible (in less than 60 sec) in order to minimize the danger of possible bleaching of the dye or the occurrence of turbulence caused by local heating of the solution.

The spectrophotometric cells used in the light scattering experiments were cleaned by the method of Dubin *et al.*[5] and the solutions were cleared of particulate matter by filtering through 0.22μ millipore filters. The absorption spectra of the pigment solutions were measured on a Carl-Zeiss Spectrophotometer (Model PMQ II).

All measurements were carried out at room temperature.

Materials

Lycopene (all-*trans*) was prepared from tomato paste devoid of artificial coloring material by the method of Davis[6]. β-Carotene, of purity better than 95%, was a product of Koch-Light, and was recrystallized from petrol ether. Bixin, an Annato extract, was a product of Tokyo Kasei, and was purified by recrystallization from ethyl acetate.

All solvents used were spectrograde (Fluka, BDH or Merck). Filtered nitrogen was bubbled through the pigment solutions for 15 min before measurements.

RESULTS

The pigments studied (lycopene, β-carotene, and bixin) were chosen because they have intense absorption bands in the visible (the ϵ at the peak of absorption exceeding $10^5 M^{-1} cm^{-1}$), and because they are practically non-fluorescent. Furthermore, their visible absorption bands overlap with at least some of the Argon-ion laser lines.

The scattering intensities of the pigments studied were measured as described above and compared with the theoretically expected values for the Rayleigh scattering from the systems studied. The out-of-phase contribution expected from theory was calculated by use of Equation 8. The expected in-phase contribution was calculated from the refractive-index increment as evaluated by the Kronig-Kramers relationship (Equation 9), using the light absorption data in the spectral range 250-600 nm. It is believed that contributions to n from far absorption bands outside this absorption range will be relatively small in the range of the very intense absorption band in the visible and will be approximately constant in this range. The integral in Equation 9 was evaluated by fitting by computer the absorption spectrum in the range of 250-600 nm to cubic spline functions, the integral thus being analytic for each spline if the borders of the spline do not coincide with λ_0. In the following, we present both the real and imaginary parts of the refractive index for each of the pigments studied, as well as their contributions to the Rayleigh ratio. The estimated phase lags of the oscillating dipoles calculated by $arctg(\partial k/\partial C/(\partial n/\partial C)$ are also included, although we have no experimental data about this quantity. It is, however, a measurable quantity, as was illustrated by Drexhage.[7]

Lycopene (Scheme I)

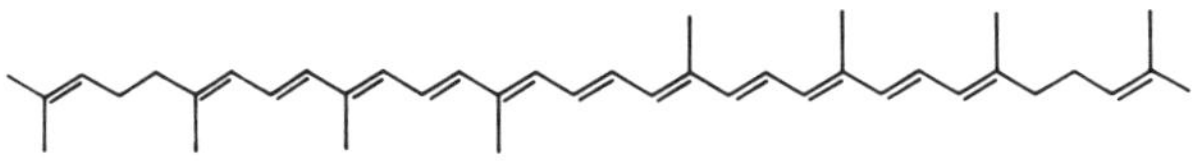

I

Some results on the scattering of light by lycopene in carbon tetrachloride solution at the visible absorption band were presented previously.[8] They have since been extended by use of a dye laser pumped by an Argon-ion laser. The results are presented in FIGURE 1. The light scattering by lycopene in cyclohexane solution is described in FIGURE 2. In the latter solvent, the absorption spectrum is shifted by about 8nm to the blue (compared to the spectrum in CCl_4) and ϵ at the peak is 1.81 $\times 10^5 M^{-1} cm^{-1}$, at 475 nm (compared to $1.6 \times 10^5 M^{-1} cm^{-1}$ at 483 nm, in CCl_4). The Rayleigh scattering intensities expected from theory are also presented in the above figures. As can be seen, good agreement is obtained between the measured and calculated values for the light scattering at the blue side of the absorption band. In contrast, at the red part of the absorption band, the measured values exceed appreciably (by up to 70%) the calculated values. It is of interest to note that on passing the absorption band towards the red the theoretical and experimental result tend to converge again. In this spectral range, the relative contribution of the real part of the refractive index to the scattering is large.

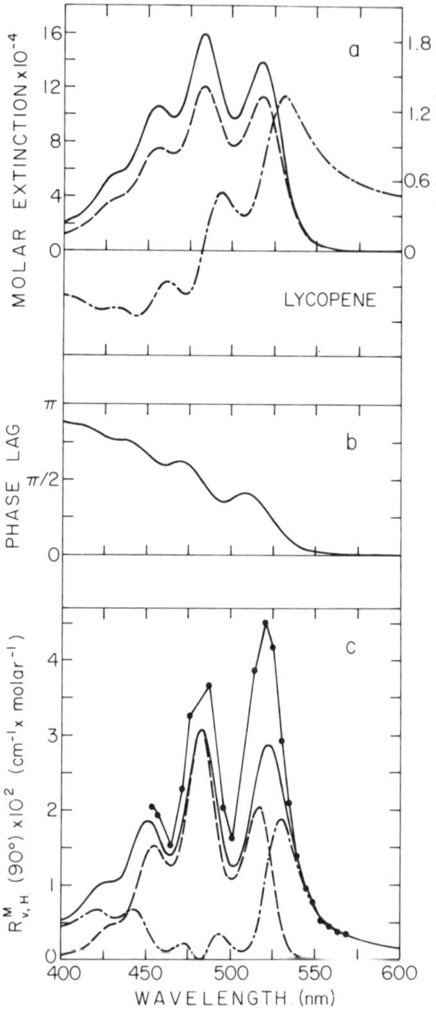

FIGURE 1. Spectroscopic and scattering data for lycopene in CCl_4. (a) ——, the molar extinction of lycopene in CCl_4; –·–, $\partial n/\partial C$; – – – $\partial k/\partial C$. (b) The phase lag of the scattering dipole. (c) ●, the experimental Rayleigh ratio, $R^{M}_{v,H}(90°)$, for 1M concentration of lycopene; –·–, the calculated contribution of $\partial n/\partial C$ to the scattered light; – – –, the calculated contribution of $\partial k/\partial C$; ——, the sum of the two calculated contributions. (The experimental results in the wavelength range 520-570 nm were obtained by using a coumarin 6 dye laser).

β-Carotene (Scheme II)

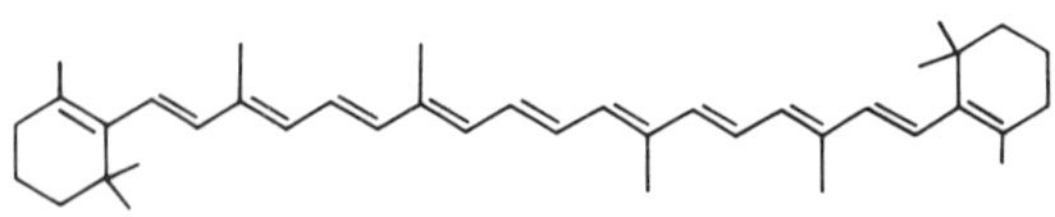

II

Light scattering measurements on β-carotene were carried out in CCl_4 and cyclohexane solutions. The absorption spectrum in cyclohexane is shifted by 9nm to the blue relative to that in CCl_4, and ϵ at the peak is appreciably higher in the former

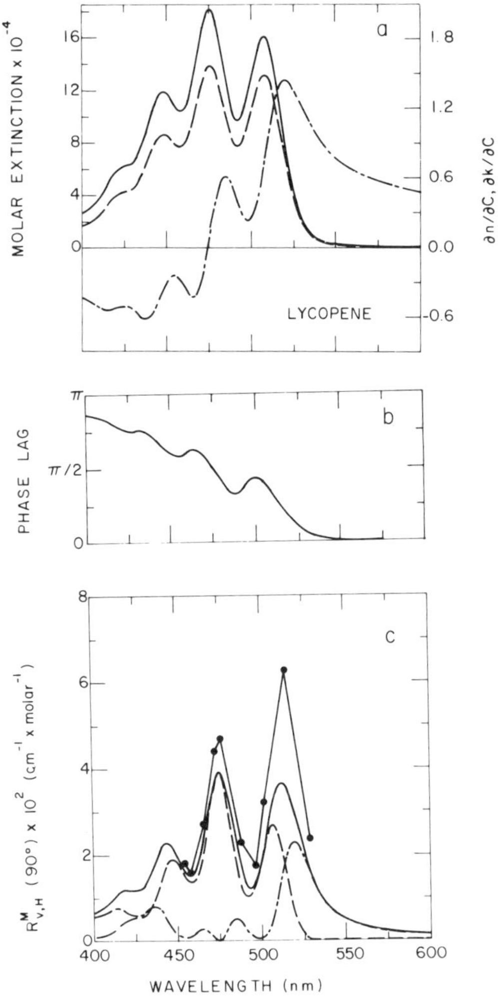

FIGURE 2. Spectroscopic and scattering data for lycopene in cyclohexane. (a) ——, the molar extinction of lycopene in cyclohexane; –·–, $\partial n/\partial C$; – – –, $\partial k/\partial C$. (b) The phase lag of the scattering dipole. (c) ●, the experimental Rayleigh ratio $R^{M}_{v,H}(90°)$ for 1M concentration of lycopene; –·–, the calculation contribution of $\partial n/\partial C$ to the scattered intensity; – – –, the calculated contribution of $\partial k/\partial C$ to the scattered intensity; ——, the sum of the calculated contributions.

solvent ($1.35 \times 10^5 M^{-1}cm^{-1}$) than in the latter ($1.16 \times 10^5 M^{-1}cm^{-1}$). The shift of the spectrum to the blue in cyclohexane permitted the use of the longest wavelength laser line for a measurement of light scattering at the extreme of the red edge of the absorption band. The results of both the measured intensities of light scattered by β-carotene and the theoretically expected values are presented in FIGURE 3 for CCl_4 solution and in FIGURE 4 for cyclohexane solution. The spectra of $\frac{\partial n}{\partial C}$ and $\frac{\partial k}{\partial C}$, as well as their individual contributions to the Rayleigh scattering are also included in these figures. The agreement between the measured and theoretical values in the case of β-carotene follows the same trends as in the case of lycopene, although on the whole it is poorer. At the blue part of the absorption band, the measured scattering is by 20% to 50% higher than the theoretical values but at the red part it is much larger, being about threefold higher at the red local peak. Upon crossing the

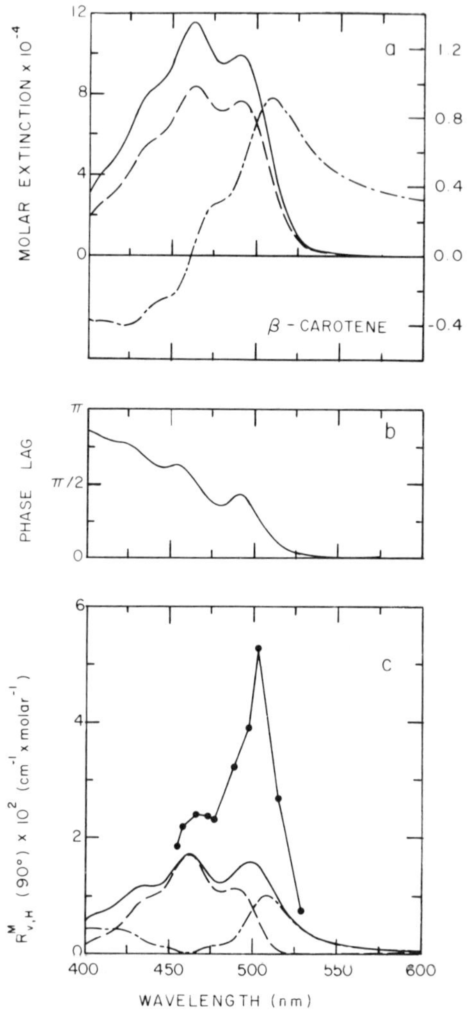

FIGURE 3. Spectroscopic and scattering data for β-carotene in CCl_4. (a) ——, the molar extinction of β-carotene in CCl_4; –·–, $\partial n/\partial C$; – – –, $\partial k/\partial C$. (b) The phase lag of the scattering dipole. (c) ● the experimental Rayleigh ratio $R^M_{v,H}(90°)$ for 1M concentration of β-carotene; –·–, the calculated contribution of $\partial n/\partial C$ to the scattered intensity; – – –, the calculated contribution of $\partial k/\partial C$ to the scattered intensity; ——, the sum of the two calculated contributions.

absorption band towards the red the agreement between theory and experiment is nearly perfect (see FIGURE 4, at $\lambda = 528.7$ nm).

Bixin (Scheme III)

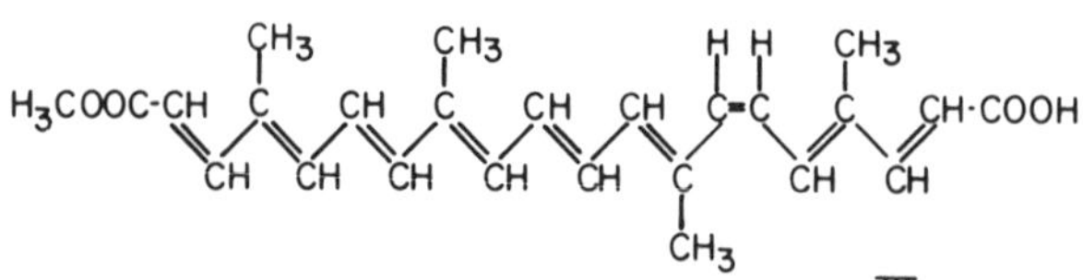

III

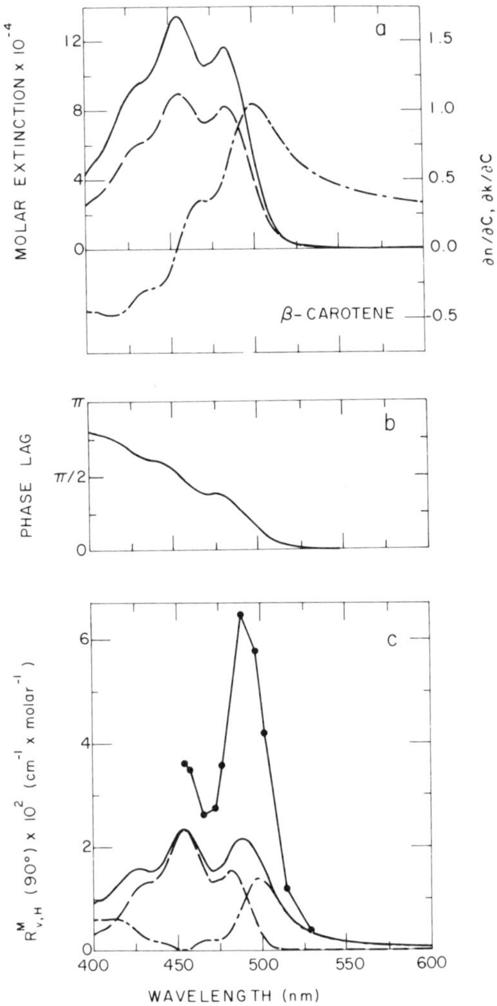

FIGURE 4. Spectroscopic and scattering data for β-carotene in cyclohexane. (a) ——, the molar extinction of β-carotene in cyclohexane; –·–, $\partial n/\partial C$; – – –, $\partial k/\partial C$. (b) The phase lag of the scattering dipole. (c) ●, the experimental Rayleigh ratio $R^M_{v,H}(90°)$ for 1M concentration of β-carotene; –·–, the calculated contribution of $\partial n/\partial C$ to the scattered intensity; – – –, the calculated contribution of $\partial k/\partial C$ to the scattered intensity; ——, the sum of the two calculated contributions.

Light scattering by bixin was studied in CCl_4 and methanol solutions. The absorption spectrum is shifted in methanol by 15nm to the blue compared to that in CCl_4 (see FIGURE 5a and 6a). The light scattering intensities are presented in FIGURE 5c and 6c, which include also the theoretically calculated values as well as the contributions to the intensity by the in-phase and out-of-phase polarizabilities. The trends are similar to those presented above for the other carotenoids, except that the effect of solvent is more pronounced. In the blue part of the absorption band, the measured intensities exceed those expected theoretically for Rayleigh scattering by about 25%. The deviations are much more pronounced at the red end of the spectrum and are markedly different for the two solvents used. In this range of the spectrum, the experimental values exceed the theoretical ones by as much as 3.38-fold in CCl_4, whereas in methanol the deviation is at most about 70%.

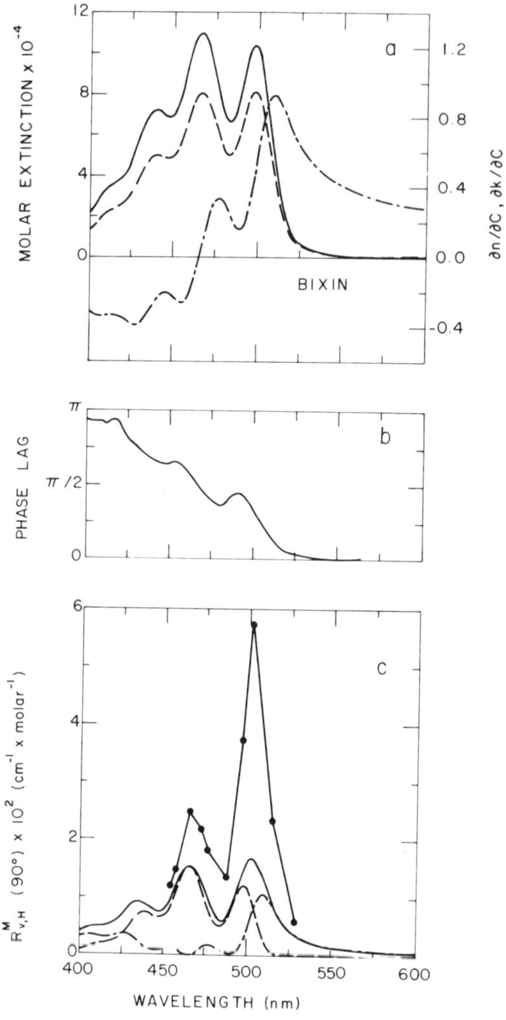

FIGURE 5. Spectroscopic and scattering data for bixin in CCl_4. (a) ——, the molar extinction of bixin in CCl_4; –·–, $\partial n/\partial C$; – – –, $\partial k/\partial C$. (b) The phase lag of the scattering dipole, (c) ●, the experimental Rayleigh ratio $R^M_{v,H}(90°)$ for 1M concentration of bixin; –·–, the calculated contribution of $\partial n/\partial C$ to the scattered intensity; – – –, the calculated contribution of $\partial k/\partial C$ to the scattered intensity, ——, the sum of the two calculated contributions.

DISCUSSION

The light energy lost from an incident beam upon its interaction with an absorbing molecule can be dissipated in a variety of radiative or non-radiative processes, such as fluorescence, Raman scattering, internal conversion (and hence, heat), and photochemical conversion. As shown above, the light absorption process is invariably associated with an elastic scattering process, whatever the nature and properties of the absorbing molecules. This scattering process accounts for a small part of the absorbed energy; the balance can channel through one or more of the other processes, depending on the nature of the molecules involved and possibly on the experimental conditions (solvent, temperature, etc.). The amount of light energy scattered elasti-

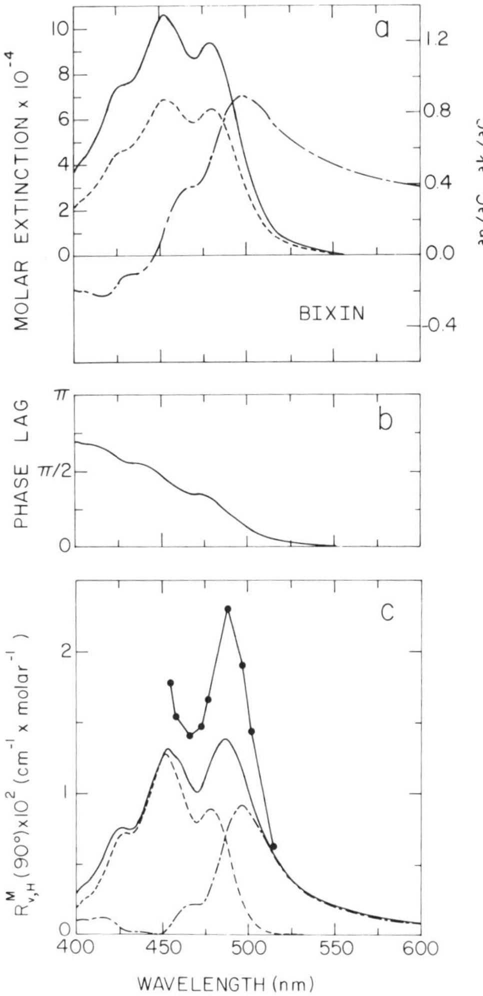

FIGURE 6. Spectroscopic and scattering data for bixin in methanol. (a) ——, the molar extinction of bixin in methanol; –·–, $\partial n/\partial C$; – – –, $\partial k/\partial C$. (b) The phase lag of the scattering dipole. (c) ●, The experimental Rayleigh ratio, $R^M_{v,H}(90°)$ for 1M concentration of bixin; –·–, the calculated contribution of $\partial n/\partial C$ to the scattered intensity; – – –, the calculated contribution of $\partial k/\partial C$ to the scattered intensity; ——, the sum of the two calculated contributions.

cally depends quadratically on ϵ, whereas the amount absorbed depends linearly on ϵ for a given local light intensity. The fraction of absorbed light which is scattered elastically thus increases proportionally with ϵ. For an ϵ of $10^5 M^{-1}cm^{-1}$, the fraction of the light absorbed which is scattered elastically is about 2×10^{-6} (at $\lambda = 514$Å), a small number indeed. The other emissive and non-emissive processes thus involve orders of magnitude more energy than the elastic scattering process.

The light scattered by three carotenoids at the wavelength of the incident beam was measured in a few solvents. In no case was the scattering intensity found to be smaller than that calculated theoretically for the elastic scattering as discussed above. This is in accord with the statement that the elastic scattering is inherently associated with the light absorption process. In a few cases (e.g., lycopene, at the blue end of its visible absorption band; see FIGURE 1c and 2c), the agreement between theory and experiment is quite good. Obviously, in these cases no radiative process besides the elastic scattering occurs at the wavelength of the incident beam. In other cases, marked discrepancies were noted between theory and experiment, the measured intensities sometimes exceeding those expected for elastic scattering by more than threefold (see, for example, the red region of the absorption band of β-carotene; FIGURE 3c and 4c). It should be noted, however, that large as these deviations are, they represent only a tiny fraction of the absorbed light.

The information presently available does not permit the identification of the radiative mechanism responsible for the light scattering which occurs in addition to the elastic scattering at the red end of the absorption bands. Although the pigments used are considered to be non-fluorescent, the occurrence of low intensities of fluorescence cannot be excluded at this point. It should be stressed, however, that the theoretical possibility exists that all of the observed light scattering is due to elastic scattering and that the discrepancy observed between experiment and theory (in those circumstances where it is observed) is apparent rather than real. As was explained in the theoretical section above, if the scattering solution contains a mixture of molecules which have different ϵ values, the observed extinction coefficient of the mixture will be an average of ϵ (i.e., $< \epsilon >$), whereas the scattering intensity will depend on the average of ϵ^2 (i.e., $< \epsilon^2 >$). Since $< \epsilon^2 >$ is greater than $< \epsilon >^2$, when ϵ does not have a unique value, the expected scattered light intensity from a heterogeneous mixture of molecules (with respect to their extinction coefficients) will exceed that calculated from Equation 7 and 8 upon inserting $< \epsilon >$ or k in these equations. The heterogeneity in ϵ among different molecules may be momentary, i.e., a molecule may assume a given extinction coefficient at a given instance depending on the conformation or mode of solvation at that instance. Different molecules, even if they are of the same kind, may thus have different ϵ values at a given instance. A special source for heterogeneity in ϵ may occur at the red edge of the absorption band, where transitions occur from thermally excited molecules in the electronic ground state. Since not all molecules are equally thermally excited at any given instance, the distribution of thermal excitation obeying a Boltzmann distribution, a heterogeneity in ϵ at the red edge of the absorption band seems to be very likely. This will explain the especially high discrepancies between theory and experiment invariably observed at the red part of the absorption bands (FIGURES 1–6). It may be noted that the above discrepancies cannot be ascribed in the cases studied to aggregation of the dye molecules, since the experimentally determined Rayleigh ratio varied linearly with solute concentration.

Equations 7 and 8, which express the intensity of light scattering as a function of ϵ, were derived for dipolar transitions. This does not seem to present any limitations from the practical point of view, since only intense absorption bands are likely to be amenable to experimental study of their contribution to light scattering, and these are associated with dipolar transitions. It is of interest to note that the scattering intensity as expressed by Equations 7 and 8 includes the quasi-elastic scattering due to translational Brownian motion but not that due to rotational Brownian motion. The reason for this is simple to see: at 0° to the incident beam, translational motion does not cause a spectral broadening in the forward scattered light; this scattered light thus contributes to ϵ by interfering with the incident light; in contrast, light which is spectrally broadened by rotational motion cannot interfere with the incident beam and thus cannot contribute to ϵ.

In conclusion, it may be pointed out that besides its intrinsic interest, light scattering from chromophores at their absorption bands may be of biochemical and biophysical interest. It may thus help us to understand properties of light scattered from stained cells, liposomes, etc. Light scattered from chromophoric ligands bound to macromolecules (such as proteins or nucleic acids) may permit the determination of the number of ligands bound per macromolecule, since the scattered intensity is expected to be proportional to this number for a given concentration of ligand molecule, provided the macromolecule is small compared to λ. Finally, the degree of orientational order in bound chromophoric ligands when many of them are attached to macromolecules, liposomes, etc., may be evaluated from the depolarization ratio of the scattered light, completely isotropic organization of the bound ligands leading to negligible depolarized scattering.

Summary

The refractive index of a chromophore undergoes marked changes in its magnitude at its absorption bands. These changes are expected to affect the intensity of the Rayleigh scattering by the chromophore. Theory predicts that the extinction coefficient (i.e., the imaginary part of the refractive index) has an additional contribution to the Rayleigh scattering which is of comparable magnitude to that of the real part of the refractive index. At absorption bands, the scattering entities are optically highly anisotropic, a factor which contributes additionally to the scattering intensity as expressed by the Cabannes factor. This anisotropy also leads to marked depolarization of the scattered light. Light scattering experiments were performed on three carotenoid pigments in a few solvents. Reasonable agreement was found between theory and experiment at the blue part of the absorption bands, but marked discrepancies were observed at the red part. The latter deviations may be due to noncoherent mechanisms of scattering which contribute in addition to the Rayleigh scattering, or to instantaneous heterogeneity in the absorption intensity of different molecules.

References

1. Kauzmann, W. 1957. Quantum Chemistry. Academic Press. New York.
2. Miller, G. A. 1978. J. Phys. Chem. **82:** 616–618.
3. Anglister, J. & I. Z. Steinberg. 1981. J. Chem. Phys. Accepted for publication.
4. Farinato, R. S. & R. L. Rowell. 1976. J. Chem. Phys. **65:** 593–595.
5. Dubin, S. B., N. A. Clark & G. B. Benedeck. 1971. J. Chem. Phys. **54:** 5158–5164.
6. Davis, W. B. 1949. Anal. Chem. **21:** 1226–1228.
7. Drexhage, K. H. 1974. *In* Progress in Optics. E. Wolf, Ed. Vol. **12** 65–232. North Holland.
8. Anglister, J. & I. Z. Steinberg. 1979. Chem. Phys. Lett. **65:** 50–54.

FLUOROPHORES IN POLAR MEDIA: SPECTRAL EFFECTS OF THE LANGEVIN DISTRIBUTION OF ELECTROSTATIC INTERACTIONS

Robert B. Macgregor, Jr. and Gregorio Weber

Roger Adams Laboratory
University of Illinois
Urbana, Illinois 61801

Introduction

The models employed for the analysis of the dependence of the fluorescence spectrum upon the dielectric properties of the solvent[1-5] consider the latter as a continuum which causes changes in the electronic energy levels through the interaction of the fluorophore with the reaction field that it induces upon its dielectric environment. In models of this kind, the change in energy of the excited state with respect to an unperturbed state is given by an expression of the form:

$$\frac{hc}{\mu - \mu^*} \Delta\bar{\nu} = (R_{\text{or}} - R^*_{\text{or}})a^{-3} \qquad \textbf{(1)}$$

where $\Delta\bar{\nu}$ is the difference in the average wavenumber of emission between unperturbed and dielectrically perturbed states; μ^* and μ, the dipole moments of the fluorophore in fluorescent and ground state, respectively, R^*_{or} and R_{or}, the orientational reaction fields corresponding to them, and a, the radius of the spherical cavity scooped in the dielectric to accommodate the fluorophore. The dimensions of $R_{\text{or}}a^{-3}$ are charge per unit area corresponding to the density of neutralizing charge over the surface of the cavity. The largest spectral differences between the fluorescence emitted by the same fluorophore in an apolar solvent (e.g. cyclohexane) and a strongly polar one (e.g. water) do not exceed 5,000 cm^{-1}. From molecular dimensions, $\mu^* - \mu$ cannot be larger than some 30 Debye units so that the last equation indicates that in the limit, the surface density of charge appearing on a sphere of 4 Å radius would approach the value of a single electronic charge. This suggests that computation of the explicit interactions of a small number of solvent dipoles with the fluorophore would furnish a sufficient approximation, and one much more physically realistic than the commonly employed continuum models. Further such explicit models should permit straightforward extension to include the purely kinetic effects of temperature and viscosity upon the spectrum. We consider below the details of such a model that we are presently exploring by computation and experiment.

Our aim in developing this model is not simply the improvement in the agreement between the observed spectral displacements and the prediction from recognizable solvent and fluorophore parameters, as has been repeatedly tried in the past. From such previous attempts, as well as from many observations, it is already quite clear that a valid computation of the effects must include the time-dependent changes in interactions peculiar to each fluorophore and solvent and is unlikely to be achieved

0077-8923/81/0366-0140$01.75/0 © 1981, NYAS

by methods other than those of molecular dynamics. Our present purpose is to provide a quantitative framework for the study and understanding of some previously unobserved or little understood phenomena: the heterogeneous character of the emission that leads to both "blue edge" and "red edge" excitation effects and the spectral shifts observed in the high and low temperature regimes in which the solvent viscosity becomes irrelevant.

THEORY

Langevin Distribution of Monopole-Dipole Interactions

The assumption of a point dipole of strength μ or μ^*, placed at the center of the fluorophore does not appear as a plausible starting point. We prefer to identify in the fluorophore a multipole distribution, with each monopole interacting at short range with a single solvent dipole. In some fluorophores at least, there will be a single positive and a single negative monopole that arise because of the existence of two unique electron donor and acceptor groups attached to an aromatic, electronically "indifferent" ring system. 2 dimethylamino(6 propionyl)naphthalene (PRODAN) is one such molecule in which the origin and position of the monopole charges is not in doubt.[6] Further, the same monopoles, though with different strength, are present in the ground and excited states, thus eliminating the additional complications that would arise in cases of more complex charge distributions.[7] The scheme of the monopole—dipole interactions for this simple case is shown in FIGURE 1. The fluorophore charges are separated by a distance ℓ, known with all accuracy from X-ray crystallographic data. If we associate with the fluorophore a dipole moment μ

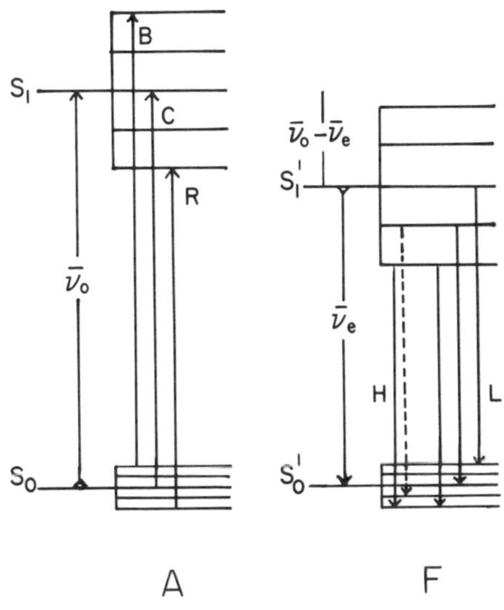

FIGURE 1.

(ground state) or μ^* (fluorescent state) the corresponding monopole charges would be:

$$q = \mu/\ell$$
$$q^* = m\mu/\ell \quad (2)$$

respectively, with

$$m = \mu^*/\mu$$

Each such charge will interact with a single solvent dipole of strength μ_s placed at minimal distance a from it, as it would be difficult or impossible to interact strongly with more than one dipole. The distance a is computed as the sum of the van der Waals radia of the atoms carrying the monopole and the oppositely charged end of the dipole plus the interatomic distance of the dipole ends in the solvent molecule. In alcohols, for example, a would be about 3.1 Å. The energy of interaction of dipole and monopole is given by:

$$u(\phi) = u \cos \phi$$
$$u^*(\phi) = u^* \cos \phi \quad (3)$$
$$u = -\mu_s\mu/\ell a^2$$
$$u^* = mu$$

where ϕ is the angle of the dipole axis with the direction giving maximal interaction energy. As both ends are equivalent, we have for the average orientational energy

$$\langle u \rangle = 2u \overline{\cos \phi};\ \langle u^* \rangle = \cdots \quad (4)$$

The average value of $\cos \phi$ will depend upon the disorienting effect of temperature. We shall take the Boltzman average for this case as determined by the alignment of the dipole in the field of the monopole according to the Langevin distribution,[8] giving

$$\langle u \rangle = 2u\pounds(u/kT)$$
$$\langle u^* \rangle = 2u^*\pounds(u^*/kT) \quad (5)$$

where

$$\pounds(u/kT) = \text{ctanh}\left(\frac{u}{kT}\right) - \frac{kT}{u}$$
$$\pounds(u^*/kT) = \cdots \quad (6)$$

The level from which absorption takes place has an average energy

$$S_0 = S_0^{\,0} + 2u\pounds(u/kT) \quad (7)$$

where $S_0^{\,0}$ is the zero vibrational level of the ground state in the absence of permanent dipole interactions. Because of Franck-Condon restrictions, the energy of the state reached in absorption is altered by an amount $u^*\pounds(u/kT)$ since the dipole moment of the fluorophore has instantly changed to μ^*, but the dipole distribution around it is still that corresponding to μ. Accordingly the lowest level of the excited state has, immediately after excitation, an energy

$$S_1 = S_1^{\,0} + 2u^*\pounds(u/kT) \quad (8)$$

where S_1^0 is the zero vibrational level of the excited state in the absence of permanent dipole interactions.

If the temperature and viscosity of the solvent are such as to permit rearrangement of the solvent cage during the fluorescent lifetime, the transition $\pounds(u/kT) \rightarrow \pounds(u^*/kT)$ can take place. Its effect is a further change of the energy of the fluorescent state to the level

$$S_1' = S_1^0 + 2u^*\pounds(u^*/kT) \quad \textbf{(9)}$$

Emission takes place from this state to a ground level given by

$$S_0' = S_0^0 + 2u\pounds(u^*/kT) \quad \textbf{(10)}$$

If $\mu^* > \mu$, the usual case, this state lies evidently *below* the starting ground state, corresponding as it does to a stronger interaction with solvent dipoles. The final step in the return to the original ground state is the conversion of the excess electronic energy: $u[\pounds(u^*/kT) - \pounds(u/kT)]$ into entropy with a corresponding rise in the electronic energy level. The energy of the $0 - 0'$ absorption transition is given by

$$E_A = S_1^0 - S_0^0 + 2(u^* - u)\pounds(u/kT) \quad \textbf{(11)}$$

The energy of the $0' - 0$ emission is

$$E_F = S_1^0 - S_0^0 + 2(u^* - u)\pounds(u/kT) \quad \textbf{(12)}$$

and the increase in the Stokes shift owing to dipole interactions is given by

$$\Delta E_F = 2(u^* - u)[\pounds(u^*/kT) - \pounds(u^*/kT)] \quad \textbf{(13)}$$

We assume in the following that Equations 10 to 13 apply also to the average energy of absorption or emission in a broad electronic band. The last of the equations above is valid provided the solvent shell can undergo during the fluorescent lifetime the changes necessary to satisfy the equilibrium distribution of solvent dipoles interacting with μ^*. Considering this rearrangement as a simple two-state process,[9] in general it will be accomplished during the excited state to a fractional extent

$$\beta = \tau/(\tau + \rho_s) \quad \textbf{(14)}$$

where τ is the fluorescence lifetime and ρ_s the correlation time for the solvent rearrangement. From the last equation, we obtain as the general expression for the spectral shift of the fluorescence spectrum with respect to a state unperturbed by permanent dipole interactions:

$$\Delta E_F = 2(u^* - u)[(1 - \beta)\pounds(u/kT) - \beta\pounds(u^*/kT)] \quad \textbf{(15)}$$

Distribution of the Spectral Displacements

It will be noticed that u/k, or u^*/k have dimensions of temperature which we designate as θ and θ^* respectively. These are defined as characteristic temperatures at which thermal energy and optimal electrostatic interaction energy become equal, and are given by

$$\theta = \mu\mu_s/a^2\ell k \, ; \, \theta^* = m\theta \quad \textbf{(16)}$$

It is seen that the set of parameters μ or μ^*, μ_s, a and ℓ contract to a single one, θ or

θ^*, which alone determines the distribution of the perturbations of the energy levels of S_0 or S_1 about their mean values. The distribution of $u(\phi)$ values applicable to our case must be evidently a binary distribution over the two monopoles. For simplicity, we shall adopt a model of discontinuous energy states which may be expected to yield reasonable results if $\Delta\bar{\nu} = u/hc$ is no larger than σ, the half-width of the unperturbed absorption or fluorescence band. We assume that there are only three permissible directions for dipole alignment: opposing the monopole field with fractional population f_-, parallel to it with fraction f_+, and neutral to it with fraction f_0. The Langevin distribution is approximated by setting:

$$\begin{aligned} f_- &= [1 + (\alpha - 1)\exp\alpha]/D \\ f_0 &= [\exp\alpha + \exp(-\alpha) - 2]/D \\ f_+ &= [1 - (\alpha + 1)\exp(-\alpha)]/D \qquad \mathbf{(17)} \\ D &= \alpha[\exp\alpha - \exp(-\alpha)] \\ \alpha &= \theta/T \end{aligned}$$

It will be noticed that as $\alpha \to 0, f_+, f_- \to 1/4, f_0 \to 1/2$. If $\alpha \gg 1$,

$$\begin{aligned} f_- &\to 1 - \alpha^{-1} \\ f_0 &\to \alpha^{-1} \qquad \mathbf{(18)} \\ f_+ &\to \alpha^{-1}\exp(-\alpha) \end{aligned}$$

and also that

$$f_- - f_+ = \pounds(\alpha)$$

Each of the two monopole-dipole interactions of the fluorophore belongs to one of these three energy states. If they are independent, they combine to yield a population of fluorophores of five energy states with fractional occcupancies,

$$\begin{aligned} f(1) &= f_-^2 \\ f(2) &= 2f_-f_0 \\ f(3) &= f_0^2 + 2f_-f_+ \qquad \mathbf{(19)} \\ f(4) &= 2f_0f_+ \\ f(5) &= f_+^2 \end{aligned}$$

The disposition of the five levels in absorption and emission for the case $m = 4$ is shown in FIGURE 2.

Effects upon the Absorption

We assume in the following that the various transitions within an electronic band involving simultaneous vibrational changes are affected by the electrostatic interactions in the same manner as the pure electronic transition. Accordingly, the absorption

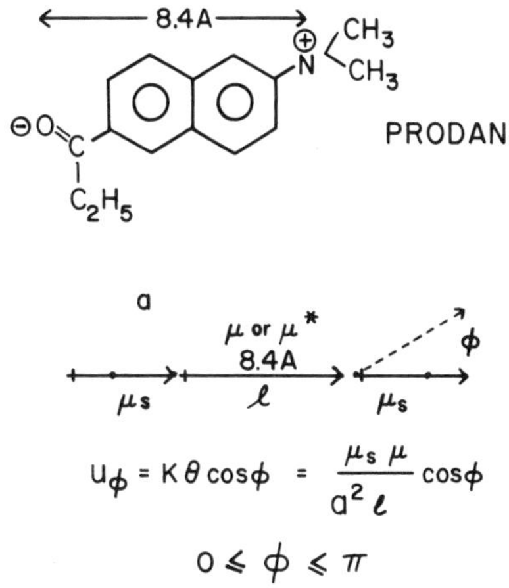

FIGURE 2. Arbitrary splitting of ground and fluorescent states of a fluorophore into five levels by electrostatic interactions with solvent. A—absorption; F,L—Fluorescence, when the population of S_1 levels is determined by ground state equilibrium ($\beta = 0$); F,H—Fluorescence when S_1 population distribution is determined by equilibrium in the excited state ($\beta = 1$).

starting in each of these five states has a wavenumber $\bar{\nu}_j$ of most probable transition given by

$$\bar{\nu}_j = \bar{\nu}_0 + (3 - j)\langle \Delta\bar{\nu} \rangle \qquad 1 \leqslant j \leqslant 5 \tag{20}$$

where ν_0 corresponds to the unperturbed absorption and

$$\langle \Delta\bar{\nu} \rangle = -\frac{k\theta}{hc}(m - 1) \tag{21}$$

If the unperturbed absorption band is a Gaussian band centered at $\bar{\nu}_0$, with absorbance given by

$$A(\bar{\nu}) = A_0 \exp - [(\bar{\nu} - \bar{\nu}_0)^2/1.21\sigma^2] \tag{22}$$

it follows from Equation 10 that the absorption in the presence of dipole interactions is

$$A(\bar{\nu}) = \sum A_j(\bar{\nu}) \qquad A_j(\bar{\nu}) = f(j) \exp - [(\bar{\nu} - \bar{\nu}_j)^2/1.21\sigma^2] \tag{23}$$

Equation 23 can be extended to include transitions to S_2 and higher levels. It requires specification of the heights, A', and width σ' of the Gaussian bands, as well as the dipole strength ratio m' of each state. The latter fixes the values of $\bar{\nu}_j$ for each absorption band. If $m > 1$ replacement of a nonpolar solvent by a polar one produces a red shift and a broadening of the absorption band. Increase in temperature results in a systematic blue shift arising from two separate causes: the thermal agitation that disorders the dipole orientations and the increase in distance a between monopole and dipole that results from thermal expansion of the solvent. The latter has the effect of making θ a slowly varying function of temperature, and from Equation 16 one finds

$$\frac{d\theta}{dT} = -\frac{2}{3}\gamma\theta$$

where γ is the coefficient of thermal expansion of the solvent. From the last equation and Equations 11 and 21

$$\frac{d}{dT}\Delta\bar{\nu} = \frac{k}{hc}(m-1)[1 - \alpha^2 \operatorname{csch}^2\alpha + \frac{2}{3}\gamma T(\alpha \operatorname{ctanh}\alpha - \alpha^2\operatorname{csch}^2\alpha)] \quad \textbf{(24)}$$

With values that, to a first approximation, apply to PRODAN in a fluid solvent with $\mu_s = 2D$, $\theta = 381$ K; $m = 8$; $\gamma = 0.0011$ K^{-1} and $T = 298$ K, the blue shift with temperature is predicted to be about 5 cm^{-1} K^{-1}. Some 60% of this is due to increasing thermal disorientation and the remainder to the solvent expansion. FIGURE 3 shows theoretical absorption spectra. The parameters above have been used to characterize the $S_0 - S_1$ absorption. These and the additional parameters for $S_0 - S_2$ absorption, given in the legend, are intended to provide a reasonable representation of the polarity dependence of the two lower frequency bands of PRODAN.

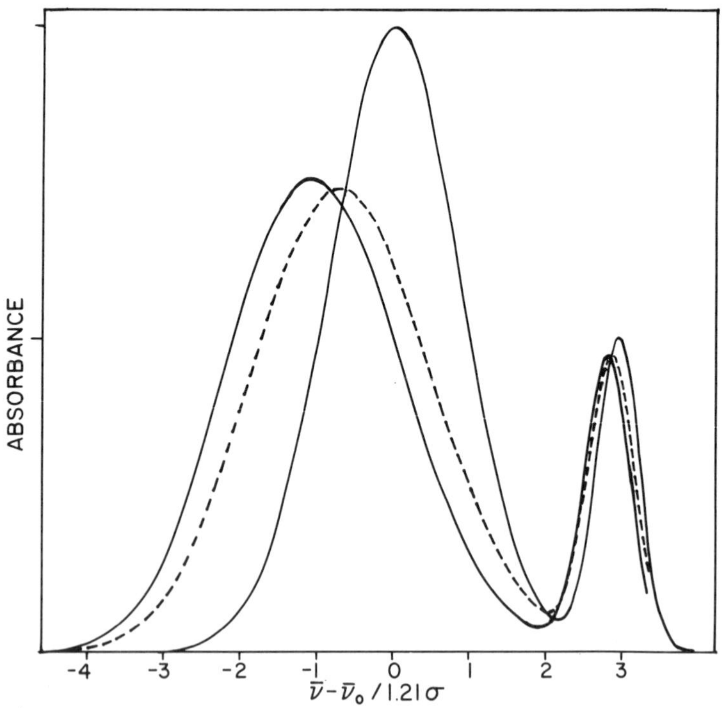

FIGURE 3. Absorption spectra expected for a fluorophore in a polar solvent at 200 K —— and 300 K ----, and in an ideal apolar solvent. The parameters A, σ, ν_0 and m for the $S_0 - S_1$ absorption are given in text. For $S_0 - S_2$: $A'/A = 1/2$; $\sigma'/\sigma = 0.34$; $m' = 2$; $(\bar{\nu}_0{}^1 - \nu_0)/\sigma = 2.95$. The units of the abscissa are half-widths of the $S_0 - S_1$ unperturbed absorption band.

Mean Value of the Fluorescence Shift with Temperature

It is easy to recognize that three different temperature regimes are predicted by Equation 15.

1. At low temperatures, defined by $\tau \ll \rho_s$ or $\beta \rightarrow 0$, the fluorescence shift is given by

$$\Delta E_F = 2(u^* - u)\pounds(u/kT) \tag{25}$$

2. At high temperatures, defined by the condition $\tau \gg \rho_s$ or $\beta \rightarrow 1$

$$\Delta E_F = 2(u^* - u)\pounds(u^*/kT) \tag{26}$$

3. At intermediate temperatures, when τ and ρ_s are of comparable magnitude, one regime goes progressively into the other. For a fluorescence lifetime of a few nanoseconds in a medium with a temperature coefficient of viscosity of approximately 10% per degree, this transition takes place over a span of approximately 35 degrees below to 35 degrees above a midtemperature T_0 defined by the condition

$$\rho_0 \equiv \rho_s(T_0) = \tau \tag{27}$$

The intermediate temperature regime, in cases in which μ^* is larger than μ consists in a red shift with temperature that has been much studied both for simple fluorophores in solution and for fluorophores complexed with polymers and proteins.[9-14]

Here we shall concern ourselves primarily with the low and high temperature regimes, but to describe the effects over the whole temperature range, it is necessary to introduce the variation of ρ_s with the temperature and viscosity of the solvent. This variation is that of the ratio T/η, and over the limited range of temperatures in which the change in ρ_s is of any interest we can set,

$$\begin{aligned} \rho_s &= \rho_0 \exp \zeta(T_0 - T) \\ \beta &= [1 + \exp \zeta(T_0 - T)]^{-1} \end{aligned} \tag{28}$$

where ζ is the thermal coefficient of the ratio T/η in the significant range about T_0. Plots of the average wave number of emission against the temperature are shown in FIGURE 4 for the range of temperature of 20 to 400 K. T_0 was assumed to be 235 K and $\zeta = 0.085$ K^{-1}, values that apply to propylene glycol. In the high temperature regime, in which the solvent is fluid, liquids like acetone or alcohol have similar values of γ in the vicinity of 1.10^{-3} K^{-1}. In the low temperature regime, when the solvent has become vitrified γ is much smaller and we have adopted a value of 1/5 of the high temperature figure. These plots show clearly the three temperature regimes: low temperature blue shift, intermediate red-shift relaxation region, and high temperature blue shift.

From comparison of Equations 11 and 16, it would follow that Equation 24 describes the fluorescence blue shift at high temperatures if α is replaced by $\alpha^* = m\theta/T$. However, Equation 24 presupposes that the relative increase in a with temperature is that of the intermolecular distance in the solvent, an assumption only valid if the energy of the electrostatic interactions is no greater than the thermal energy. If θ^*/T is much larger than unity, the increase in a will be much smaller than expected from the thermal expansion of the solvent. On this basis, we expect that the fluorescence blue shift with temperature is best approximated by the first

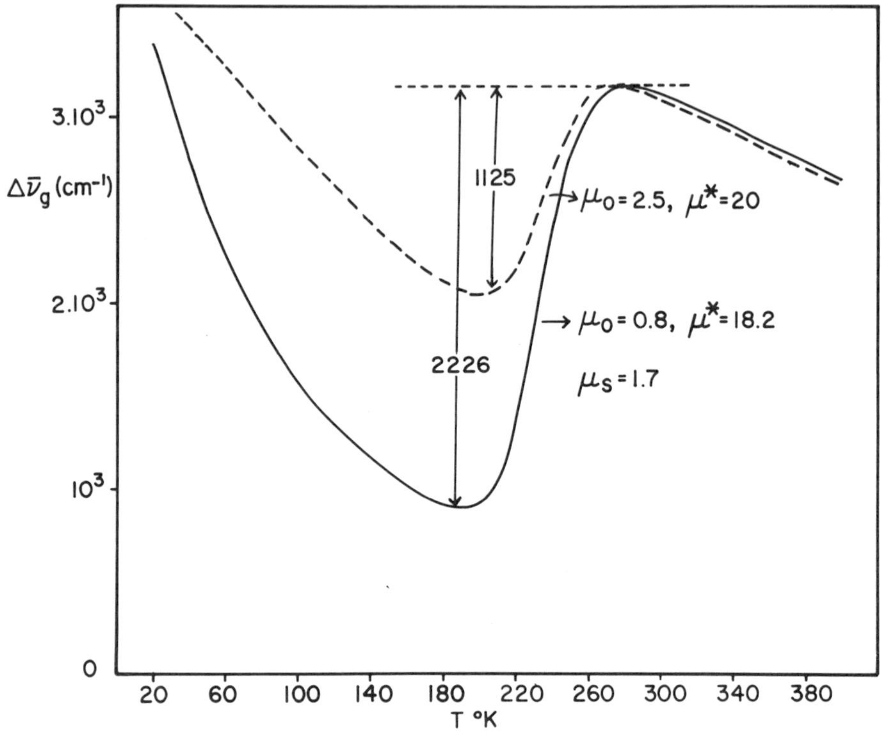

FIGURE 4. Plots of the center of mass of the fluorescence emission versus temperature calculated for two values of μ_0 and $m = 8$. The figure shows the three temperature regimes and illustrates the sensitivity of the relaxation red shift to the value of the ground state dipole of the fluorophore.

term in Equation 24, and has a value approximately equal to that of the absorption, or about 5 cm^{-1} K^{-1}.

Dependence of the Fluorescence Spectrum upon the Wavelength of the Excitation

If the fluorescence emission in a nonpolar medium is a Gaussian band centered at $\bar{\nu}_e$, the spectral distribution is given by the equation

$$F(\bar{\nu}) = F_0 \exp - [(\bar{\nu} - \bar{\nu}_e)^2/1.21\sigma^2] \quad \textbf{(29)}$$

$\bar{\nu}_e$ differs from $\bar{\nu}_0$ by the energy losses in the excited state unconnected with solvent dipolar interaction, and $\bar{\nu}_0 - \bar{\nu}_e$ is the Stokes shift observed in a nonpolar medium. In a polar medium,

$$F(\bar{\nu}) = F_0 \sum_{j=1}^{5} f(j) \exp - [(\bar{\nu} - \bar{\nu}_j)^2/1.21\sigma^2] \quad \textbf{(30)}$$

where

$$\bar{\nu}_j = \bar{\nu}_e + (3 - j)\langle\Delta\bar{\nu}\rangle \qquad \textbf{(31)}$$

with $\langle\Delta\nu\rangle$ given by Equation 21. The distribution of $f(j)$ values in Equation 30 varies greatly depending upon the value of β. In the high temperature regime, solvent reorientation is accomplished in a fraction of the lifetime τ, and the distribution, determined then by the parameter α^*, is practically independent of the wavelength of the excitation: whatever the $f(j)$ distribution at the time of the excitation, the same equilibrium distribution in the excited state is reached before emission. In the temperature regime in which solvent reorientation is minimal during the fluorescence lifetime ($\beta \rightarrow 0$), the distribution of $f(j)$ values is that present at excitation, and depends upon the wavenumber of the excitation $\bar{\nu}$, according to

$$f(j) = A_j(\bar{\nu}) / \sum_{j=1}^{5} A_j(\bar{\nu}) \qquad \textbf{(32)}$$

$A_j(\bar{\nu})$ being given by Equation 23.

FIGURE 5 shows the fluorescence spectra that would be expected when excitation in the low temperature regime is carried out at the wavelength of maximum

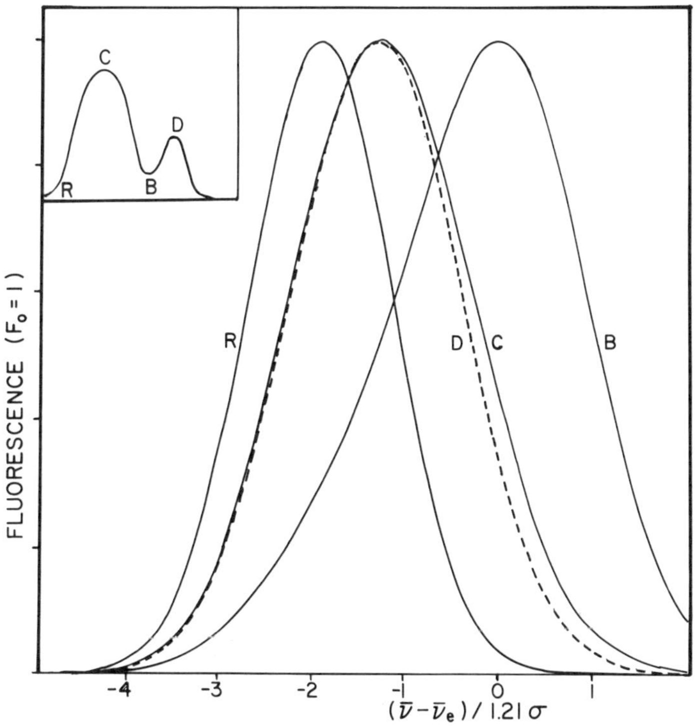

FIGURE 5. Computed fluorescence spectra in a frozen polar solvent. Parameters as in FIGURE 3. Excitation at the places in the absorption shown in the inset.

absorption (C) and at the two wavelengths, respectively B and R, at which absorption has fallen to 1/8 of the maximum. The spectra largely preserve the original shape but differ both in average wavenumber and half-width, the emission becoming narrower as it shifts to the red with the wavelength of the excitation. Excitation at R results in a red shift with respect to the emission excited at C on the order of $\langle \Delta \nu \rangle$ or approximately 1200 cm^{-1} for a fluorophore like PRODAN dissolved in propylene glycol or glycerol ($\mu \sim 2D$). Excitation at B produces effects that greatly depend upon the overlap of the first two absorption transitions. If this is negligible, a blue shift with respect to the spectrum excited at C, of the order of $\langle \Delta \nu \rangle$ and some broadening may be expected. If absorption band overlap is appreciable, it involves red-shifted levels of $S_0 - S_2$, and blue shifted levels of $S_0 - S_1$, with the former adding a red-shifted fluorescence that results in a much broadened emission band. In fact, an analysis of the emission excited in the region where the two transitions contribute appreciably can be used to estimate the polarity of the S_2 state.

EXPERIMENTAL

In this first explicit application of the Langevin distribution of electrostatic interactions to the absorption and emission of polar fluorophores, we describe experiments that demonstrate the blue shift of the fluorescence in the high temperature regime and the large dependence of the fluorescence spectrum upon the excitation wavelength in the temperature regime where the solvent reorientation is slow.

The absorption corresponding to the $S_0 - S_1$ transition is expected to exhibit a red shift of magnitude $\langle \Delta \bar{\nu} \rangle$ when a strongly polar solvent, like an alcohol, is compared to a truly nonpolar one, like cyclohexane. TABLE 1 shows the absorption maxima and half-width of the band for PRODAN in cyclohexane, ethanol, and propylene glycol. The shifts in the polar solvents are respectively 1,220 cm^{-1} and 2,200 cm^{-1} against a predicted value of 2,000 cm^{-1} from an ideal Gaussian absorption in a nonpolar medium. The broadening of the band in the alcohols is 21%, almost coincident with the predicted broadening. Evidently, although other factors besides heterogeneous electrostatic interactions may be involved, this single cause is sufficient to account for the magnitude of the shift and broadening of the $S_0 - S_1$ absorption of PRODAN in the polar solvents.

TABLE 1
ABSORPTION OF PRODAN IN CYCLOHEXANE AND ALCOHOLS

	$S_0 - S_1$		$S_0 - S_2$	
	$\bar{\nu}_{max}$	$\Delta\bar{\nu}_{1/2}$	$\bar{\nu}_{max}$	$\Delta\bar{\nu}_{1/2}$
Cyclohexane	29,154	2,221	35,724	750
Ethanol	27,932	2,695	35,587	1,000
Propylene Glycol	27,037	2,721	35,460	1,000

All values in cm^{-1}. $S_0 - S_2$ is strongly overlapped by $S_0 - S_3$, and the values of $\Delta\bar{\nu}_{1/2}$ for this band are therefore only approximate.

Fluorescence Observations

Fluorescence emission spectra were acquired using an updated version of the instrument described by Jameson *et al.*[14]; improvements included a holographic grating with low parasitic light from SLM Instruments and replacement of the EMI 6256S phototube by a Hammamatsu R 928. The improved sensitivity in the 600–800 nm range greatly reduces the magnitude of the corrections required to convert the technical into the molecular fluorescence spectrum. Temperatures were measured with a copper-constantan thermocouple placed in the sample chamber within 2 mm of the cuvette. PRODAN was synthesized as described by Weber & Farris[6] and 3 times crystallized from ethanol. Spectral purity was demonstrated by high pressure liquid chromatography. Spectral quality solvents (purissimum grade) were used without further treatment, except for propylene glycol (Fluka, purissimum grade) which was distilled once. The mean wave number of the fluorescence was numerically computed by means of the equation

$$\bar{\nu}_g = \sum \bar{\nu}\, F(\bar{\nu}) / \sum F(\bar{\nu})$$

where $F(\bar{\nu})$ is the photon count and the summation covers the range of intensities significantly different from the solvent background. The spectral bandwidth $\Delta\bar{\nu}_{1/2}$ was computed as the first moment of the spectral distribution about ν_g.

FIGURE 6 shows a plot of $\bar{\nu}_g$ against T for PRODAN in propylene glycol excited at 365 nm. T covers the range of temperatures over which the transition from the relaxation regime (red shift) to the high temperature regime (blue shift) takes place. The former has been extensively studied and the latter has been occasionally noticed,[16] but to our knowledge, never systematically discussed. This blue shift (see inset of figure) is about 5 cm^{-1} K^{-1}, in reasonable agreement with the prediction from Equation 24. The high temperature blue shifts observed for other solvents are collected in TABLE 2. It is evident that while the magnitude of the shifts is that predicted from the Langevin distribution, there are variations among solvents with very similar dipole moment which must arise from causes not explicitly considered in the theory. Such could be interactions with more than one solvent dipole, induced dipole effects and in the case of associated solvents like alcohols or water, propagation of effects to neighboring solvent molecules. All these causes must be of lesser importance as the theory is sufficient to predict the correct magnitude of the effects. From the quantities quoted in TABLE 2, it could be erroneously concluded that it is difficult to give a simple qualitative demonstration of the high temperature blue shift and that this may only be demonstrated by accurate photometric measurements. This is not so; in solvents like ethanol or acetone that remain fluid over a large span of temperatures below room temperature, a simple visual comparison of the color of the fluorescence of two solutions, one at room temperature and the other cooled in an alcohol-solid CO_2 bath, convincingly demonstrates the high temperature blue shift.

Excitation Wavelength Dependence of the Fluorescence in Very Viscous Solvents

FIGURE 7 shows the plot of photon counts vs. wavenumber for the fluorescence of solutions of PRODAN in glycerol at −41°C (232 K) and at 4°C (277 K) excited at wavelengths corresponding to B, C and R of FIGURE 5. TABLE 3 below summarizes

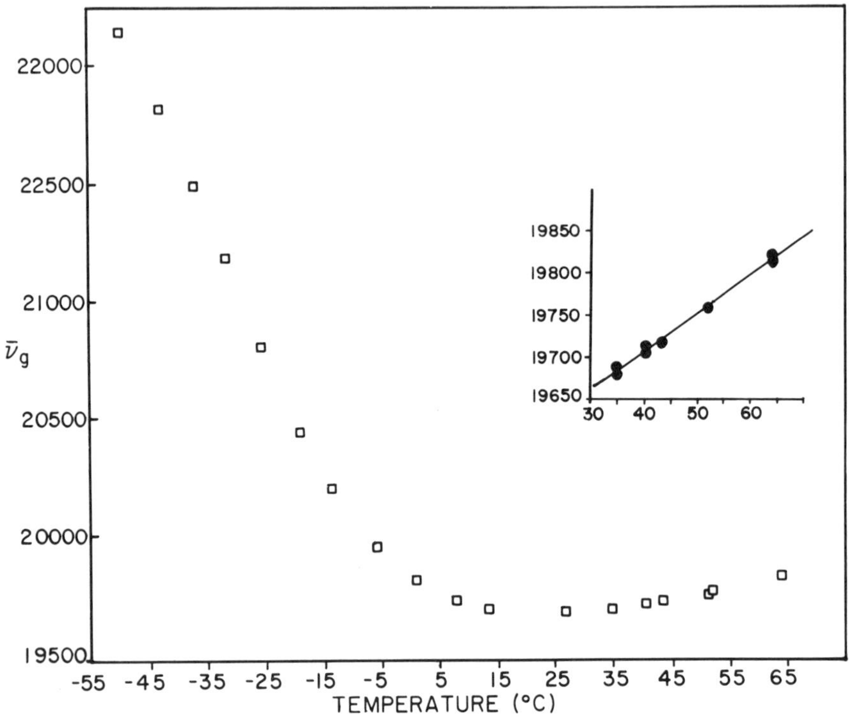

FIGURE 6. Experimental values of $\bar{\nu}_g$ (Equation 33) for PRODAN in propylene glycol excited at 365 nm in the temperature interval − 60 to +65°C, showing the transition from relaxation red shift to high temperature blue shift. Inset: Expanded scale plot of the latter.

TABLE 2
FLUORESCENCE BLUE-SHIFT WITH TEMPERATURE; PRODAN IN FLUID SOLVENTS

	$d\bar{\nu}_g/dT$ ($cm^{-1}K^{-1}$)	μ (D)
Ethanol	6.0	1.69
Ethylene Glycol	4.5	2.28
Propylene Glycol	5.0	2.77
Acetone	3.7	2.88
Acetonitrile	5.6	3.5
Dimethyl phthalate	1.4	2.8

For the five aliphatics, the average is 5 ± 0.8 $cm^{-1}K^{-1}$. The low dimethyl phthalate value may arise from the large induced effects.

the values of $\bar{\nu}_g$ and $\Delta\bar{\nu}_{1/2}$. The constancy of the fluorescence spectrum in the high temperature regime and its very large dependence upon the excitation wavelength in the low temperature regime are dramatically shown. The blue shift upon excitation on the spectral region where the two lower frequency absorption bands overlap is

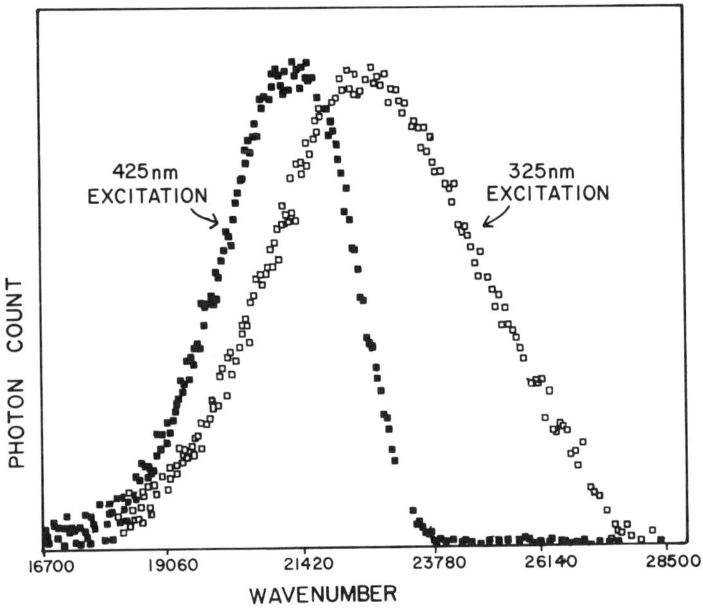

FIGURE 7. Dependence of the Fluorescence Spectrum of PRODAN in glycerol at −41°C upon the wavelength of excitation. □ Excitation at 325 nm. ■ Excitation at 425 nm.

TABLE 3
DEPENDENCE OF FLUORESCENCE-EMISSION OF PRODAN IN GLYCEROL UPON EXCITATION WAVELENGTH

λ_{exc}	$\bar{\nu}_g$	$\Delta\bar{\nu}_{1/2}$
	−41°C	
325 nm	23056	2457
385 nm	21910	1742
425 nm	21108	1475
	3.9°C	
325 nm	19934	1804
385 nm	19869	1877
425 nm	19630	1663

accompanied by appreciable band broadening, as expected. The progressive red shift observed in the fluorescence spectrum of PRODAN as the excitation is moved towards the red results in greatly diminished overlap between the red-edge excited fluorescence and the absorption spectrum of the whole population. This is evidently sufficient to explain a failure of energy transfer among the PRODAN molecules upon red edge excitation. This phenomenon has been observed with many fluorophores,[17] some with excited states of little polar character and therefore the value of this red shift of the emission as a general explanation for the failure of energy transfer upon red edge excitation requires much further investigation.

References

1. Bayliss, N. S. & E. G. McRae. 1954. J. Phys. Chem. **58:** 1002.
2. Ooshika, Y. 1954. J. Phys. Soc. Jpn. **9:** 594.
3. Mataga, N., Y. Kaifu & M. Koizumi. 1956. Bull. Chem. Soc. Jpn. **29:** 465.
4. Lippert, E., Z. Elekrochem. 1957 **61:** 962.
5. Bilot, L. & A. Kawski. 1962. Z. Naturforsch. **17a:** 621.
6. Weber, G. & F. J. Farris. 1979. Biochemistry **18:** 3075.
7. Marcus, R. A. 1963. J. Chem. Phys. **38:** 335.
8. Debye, P. 1929. *In* Polar Molecules, p. 27. Dover, N. Y.
9. Bakshiev, N. G. 1964. Opt. Spectrosc. **16:** 446.
10. Ware, W. R., P. Chow & S. K. Lee 1968. Chem. Phys. Lett. **2:** 356.
11. Anufrieva, E. V., M. V. Vol'khenstein & V. A. Samokish. 1970. Dokl. Biophys. **195:** 159.
12. DeToma, R. P. & L. Brand. 1977. Chem. Phys. Lett. **47:** 231.
13. Gafni, A., R. P. DeToma, R. E. Manrow & L. Brand. 1977. Biophys. J. **17:** 155.
14. Weber, G. & J. R. Lakowicz. 1973. Chem. Phys. Lett. **22:** 419.
15. Jameson, D. M., R. D. Spencer, & G. Weber. 1976. Rev. Sci. Instrum. **47:** 1034.
16. Lippert, E., W. Luder & F. Moll. 1959. Spectrochim. Acta **10:** 858.
17. Weber, G. & M. Shinkitzky. 1970. Proc. Natl. Acad. Sci. USA **65:** 823.

ORIENTATIONAL EFFECTS IN INTRA- AND INTERMOLECULAR LONG RANGE EXCITATION ENERGY TRANSFER

J. Eisinger and W. E. Blumberg

Bell Laboratories
Murray Hill, New Jersey 07974

R. E. Dale

Paterson Laboratories, Christie Hospital
and Holt Radium Institute
Manchester M20 9BX England

Introduction

Long-range (Förster) excitation energy transfer (EET) has found numerous applications to structural problems in molecular biology and polymer science in recent years.[1,2] The development of improved experimental techniques, particularly for the measurement of time-resolved fluorescence and depolarization,[3] and advances in methods of analyzing the data obtained in such experiments[4–6] can be expected to lead to a growing number of attempts to solve similar problems by EET.

The present contribution focuses on the orientational dependence of the rate and efficiency of EET, which will be seen to depend on the mutual donor and acceptor orientations as well as on the extent of orientational relaxation of these fluorophores. We have attempted to present our results in a form which will allow the experimenter to analyze data graphically, once a reasonable model for the system under study has been established.

It would of course be impossible to present in this space protocols for the many and varied experimental situations which are encountered in biological systems. We will therefore confine this discussion to two types of EET experiments which appear to be of the greatest current interest. These are

(1) The determination of a fixed, generally intramolecular, separation between intrinsic or specifically bound donor (D) and acceptor (A) fluorophores and
(2) The estimation of the extent of randomicity of the distribution of D-A separations from the efficiency of EET between donors and acceptors on surfaces, generally obtainable from intermolecular EET experiments.

The determination of the distance between a donor and an acceptor bound at specific sites on a protein is an example of the first of these cases, while the measurement of an (apparent) local concentration of membrane-bound acceptors, illustrates the second.

0077-8923/81/0366-0155 $01.75/0 © 1981, NYAS

Excitation Energy Transfer and Polarization

It is by now well established that for *D-A* separations greater than a few angstroms, the Förster or dipole-dipole mechanism of EET[7,8] is the only one of importance. It follows that the rate (k_T) of EET between a *D*, *A* pair has the same angular and distance dependence as the square of the dipole-dipole interaction energy. The energy transfer rate may be expressed by

$$k_T = (3/2)\kappa^2(1/\tau_0)(R_0/R) \quad (1)$$

where R_0 is the so-called critical Förster distance, which depends on the energy overlap between the donor emission and acceptor absorption spectra, κ^2 is the orientation factor for dipole-dipole interaction (Fig. 1), and R is the *D-A* separation. At a separation R_0, k_T is equal to the donor de-excitation $\tau/_0$ rate in the absence of acceptors if κ^2 has its dynamic average value of 2/3.

R_0 can usually be evaluated from spectroscopic data with good accuracy, and in most applications of energy transfer, the experimenter seeks to determine the distance R or its distribution. To do so it is, according to Equation 1, necessary to obtain some estimate of the orientation factor

$$\kappa^2 = (\cos\theta_T - 3\cos\theta_D\cos\theta_A)^2 \quad (2)$$

which can in principle have any value between 0 and 4, although in a random assembly, values at or near the lower limit occur with high frequency (Fig. 2). In a real situation, a considerable amount of angular averaging of κ^2 may be expected.[5] As will be demonstrated later, the extent of angular averaging is indeed of paramount importance in the interpretation of a given energy transfer efficiency in terms of intramolecular separations.

Some authors have made use of κ^2 distributions (cf. Figure 2) to assess an error in a determination of R caused by a particular choice of κ^2. At best, such considerations have statistical validity for cases in which ***D*** and ***A*** have a large number of entirely independent possible orientations on a common substrate. It is, however, inappropriate to use the κ^2 distribution to draw conclusions about the κ^2 value in a case for which some particular *D*, *A* geometry obtains, since the experimenter has no way of knowing if his particular case is a "probable" or an "improbable" one.

The most promising strategy for delimiting the possible values of the orientation factor makes use of the results of depolarization measurements to quantitate the model illustrated in Figure 3a. The upper part of the figure depicts the sequence of events which occurs in the course of a typical energy transfer experiment: the donor is excited, its transition moment direction $\boldsymbol{D}_i$ reorients to $\boldsymbol{D}_j$, transfer occurs to the acceptor with orientation $\boldsymbol{A}_i$, which then reorients to the orientation $\boldsymbol{A}_j$.

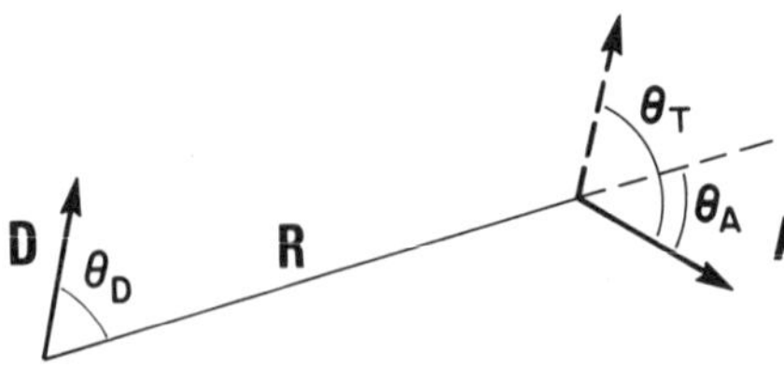

Figure 1. The diagram illustrates the angular parameters needed in the definition of κ^2. ***D*** and ***A*** represent the transition dipole moments of the donor and acceptor and R is the separation vector between them.

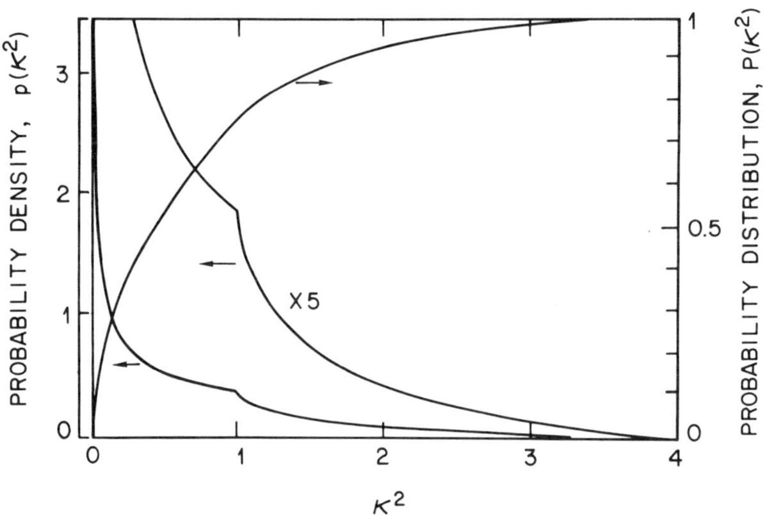

FIGURE 2. The probability density $p(\kappa^2)$ as a function of κ^2, where κ^2 is the orientation factor between two transition dipoles, each of which can adopt any orientation, independently of the other. $P(\kappa^2)$ is the probability distribution for κ^2 being between zero and κ^2. The probability of κ^2 being less than 0.1, for instance, is seen to be 0.25 for this model.

For application to an ensemble of identical *D*, *A* pairs attached to identical substrates, the analysis of this model is greatly facilitated by expressing the reorientational averaging which occurs in terms of depolarization factors.

When the direction of the transition moment of a fluorophore is changed by an angle θ (over random azimuths), the emission anisotropy, initially r_i, is related to its final value, r_f, by

$$r_f = dr_i \tag{3}$$

in which *d*, the depolarization factor, is given, according to Perrin's law of isotropic depolarization,[9] by

$$d = (3/2)\cos^2\theta - (1/2). \tag{4}$$

It can moreover be shown[10] that, for a series of *n* independent depolarizing events, the overall depolarization is related to the individual ones by

$$d = \prod_{i=1}^{n} d_i \tag{5}$$

where d_i is the depolarization factor for the ith step.

Now if the orientation of a transition moment is defined by a unit vector $\boldsymbol{M}_i$, its reorientation, say, $\boldsymbol{M}_i \rightarrow \boldsymbol{M}_j$, is independent of any intermediate steps in the reorientation process. If, in particular, the vectors $\boldsymbol{M}_i$ are distributed with axial symmetry about the axis $\boldsymbol{M}^x$, this depolarization is equivalent to the two-step path

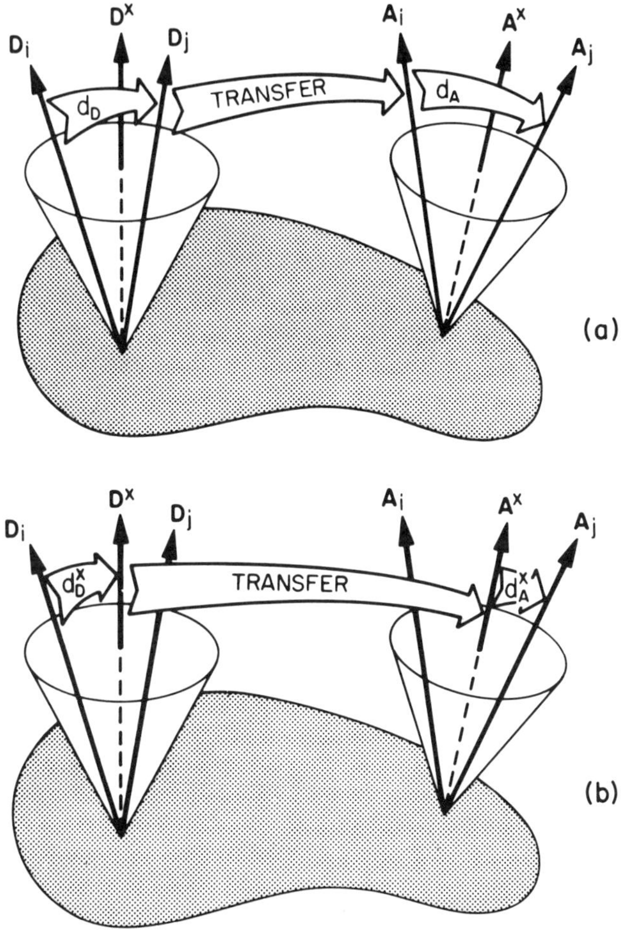

FIGURE 3. (a) Schematic representation of the depolarizing steps following the absorption of excitation energy by the donor: donor depolarization, transfer depolarization and acceptor depolarization. (b) An equivalent representation to (a), in terms of axial depolarization of $\boldsymbol{D}$ and $\boldsymbol{A}$ and transfer between the symmetry axes of the donor and acceptor transition dipole distribution ($\boldsymbol{D}^x$ and $\boldsymbol{A}^x$).

$\boldsymbol{M}_i \rightarrow \boldsymbol{M}^x$ followed by $\boldsymbol{M}^x \rightarrow \boldsymbol{M}_j$. It is then convenient to define an average depolarization factor $\langle d \rangle$ and an average axial depolarization factor $\langle d^x \rangle$:

$$\langle d \rangle = (3/2)\langle(\boldsymbol{M}_i \cdot \boldsymbol{M}_j)^2\rangle_{i,j} - (1/2) \quad (6)$$

$$\langle d^x \rangle = (3/2)\langle(\boldsymbol{M}_i \cdot \boldsymbol{M}^x)^2\rangle - (1/2)$$
$$= 3/2\langle(\boldsymbol{M}_j \cdot \boldsymbol{M}^x)^2\rangle - (1/2). \quad (7)$$

It is then immediately evident from Equation 5 that the observed depolarization factor $\langle d \rangle$ is related to $\langle d^x \rangle$ by

$$\langle d \rangle = \langle d^x \rangle^2. \quad (8)$$

The reorientational averaging which is indicated by the angular brackets is assumed to belong to the dynamic regime, as is discussed in detail elsewhere,[5] i.e., it is assumed that reorientational relaxation is fast compared to the fluorescence and transfer rates.

Orientation Factors for Intramolecular Excitation Energy Transfer

The events discussed above are described pictorially in Figure 3, which represents a donor and acceptor pair, D and A, with limited orientational freedom about the axes $\boldsymbol{D}^x$ and $\boldsymbol{A}^x$, attached to an immobile substrate and separated by a fixed distance. By considering EET between them from the point of view of the preceding section, it is readily shown[5] that the observed average depolarization may be written:

$$\langle d_T \rangle = \langle d_D{}^x \rangle d_T{}^x \langle d_A{}^x \rangle \quad \textbf{(9)}$$

where

$$d_T{}^x = (3/2)\cos^2\Theta_T - (1/2) \quad \textbf{(10)}$$

is the depolarization factor which corresponds to the reorientation $\boldsymbol{D}^x \rightarrow \boldsymbol{A}^x$, and Θ_T is the angle between $\boldsymbol{D}^x$ and $\boldsymbol{A}^x$.

Depolarization measurements generally permit the evaluation of $\langle d_D \rangle$, $\langle d_A \rangle$ and $\langle d_T \rangle$ (see Figure 3a). If the possible ambiguity with respect to sign in deriving $\langle d^x \rangle$ from $\langle d \rangle$ is ignored for the moment, such experiments yield a minimum of *three* angular parameters related to the model of Figure 3: Θ_T, and the angular widths of the $\boldsymbol{D}$ and $\boldsymbol{A}$ orientational distributions. $\langle \kappa^2 \rangle$ can be seen, however, to depend on at least *five* angular parameters: the three angles appearing in Equation 1, plus at least two more (depending on the orientational distribution model invoked, see, e.g., Reference 5), *those* characterizing the widths of the (axially symmetrical) $\boldsymbol{D}$ and $\boldsymbol{A}$ distributions.* Thus, while polarization experiments cannot, except in special situations, yield *unique* values for $\langle \kappa^2 \rangle$, they can provide *bounds* for $\langle \kappa^2 \rangle$ which are consistent with the measured depolarization factors. This is accomplished by expressing $\langle \kappa^2 \rangle$ in terms of the angular parameters which define the model of Figure 3:[5]

$$\begin{aligned}\langle \kappa^2 \rangle = \kappa^{x2} \langle d_D{}^x \rangle \langle d_A{}^x \rangle &+ (1/3)(1 - \langle d_D{}^x \rangle) + (1/3)(1 - \langle d_A{}^x \rangle) \\ &+ \cos^2\Theta_D \langle d_D{}^x \rangle (1 - \langle d_A{}^x \rangle) + \cos^2\Theta_A \langle d_A{}^x \rangle (1 - \langle d_D{}^x \rangle) \quad \textbf{(11)}\end{aligned}$$

where

$$\kappa^x = \cos\Theta_T - 3\cos\Theta_D \cos\Theta_A \quad \textbf{(12)}$$

is the orientation factor for the symmetry axes $\boldsymbol{D}^x$ and $\boldsymbol{A}^x$. Θ_D and Θ_A are, by analogy to Figure 1, the angles between the separation vector $\boldsymbol{R}$ and $\boldsymbol{D}^x$ and $\boldsymbol{A}^x$, respectively.

A more general, but more cumbersome, vector formulation of Equation 11, including cases in which the $\boldsymbol{D}$ and $\boldsymbol{A}$ distributions lack axial symmetry, has also been carried through.[11]

We have constructed contour maps for obtaining minimum and maximum values of the orientation factor, $\langle \kappa^2 \rangle_{\min}$ and $\langle \kappa^2 \rangle_{\max}$, for particular values of $\langle d_D{}^x \rangle$, $\langle d_A{}^x \rangle$

* It is important to note, however, that the analysis which follows does not require detailed knowledge of the orientational distributions and is independent of them as long as axial symmetry obtains to a good approximation. The appropriate $\langle d^x \rangle$ values are entirely sufficient to characterize the distributions in this context.

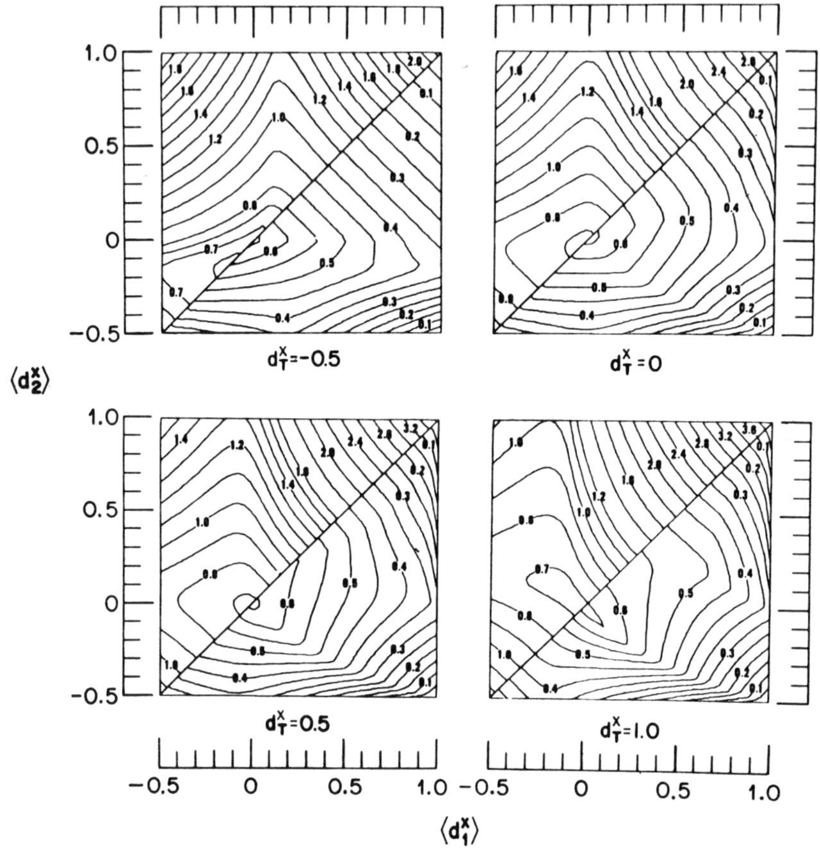

FIGURE 4. Contour maps for determining upper and lower limits for $\langle \kappa^2 \rangle$, the dynamically averaged orientation factor, as a function of the axial depolarization factors $\langle d_1^x \rangle$ and $\langle d_2^x \rangle$, where the subscripts may refer to the donor or acceptor interchangeably. Minimum and maximum levels appear in the lower right and upper left regions respectively of each contour map. The four maps are for the d_T^x values indicated on the figures. Additional maps for intermediate values of d_T^x appear in Reference 5.

and d_T^x which appear in Equations 11 and 9. To this end, we made a computer search for minimum and maximum values of $\langle \kappa^2 \rangle$ as functions of $\langle d_D^x \rangle$ and $\langle d_A^x \rangle$ over all geometries consistent with given d_T^x values. These solutions are plotted in a series of contour maps of $\langle \kappa^2 \rangle_{min}$ and $\langle \kappa^2 \rangle_{max}$ as functions of $\langle d_1^x \rangle$ and $\langle d_2^x \rangle$, as illustrated in FIGURE 4. The subscripts 1 and 2 are used to denote D and A interchangeably, since Equation 11 is symmetric in D and A and thus invariant to an exchange between them. As a result, the contour maps are symmetric about their diagonals $\langle d_1^x \rangle = \langle d_2^x \rangle$. For a given d_T^x, this allows solutions for both $\langle \kappa^2 \rangle_{min}$ and $\langle \kappa^2 \rangle_{max}$ to be displayed in a single square contour plot like those presented in FIGURE 4.

In the absence of an experimental determination of $\langle d_T \rangle$, d_T^x cannot be determined and a somewhat wider range of possible $\langle \kappa^2 \rangle$ values is obtained from the single contour plot shown in FIGURE 5. Contour plots for d_T^x values in intervals of 0.1 have also been published.[5] When $\langle d \rangle$ is less than 0.25, the sign of $\langle d^x \rangle$ may be degenerate. When using the contour plots in such cases, both axial depolarization factors should be checked to determine the absolute minimum and maximum values of $\langle \kappa^2 \rangle$.

It may be instructive to examine in some detail the use of such contour plots in the determination of the *D-A* separation *R*. If other sources of error (e.g., imperfectly dynamic orientational averaging, uncertainty in probe attachment site) are negligible, the uncertainty in $\langle \kappa^2 \rangle$ introduces a negligible error in ***R*** only when both ***D*** and ***A*** have nearly isotropic (or pseudo-isotropic) distributions. In this case depolarization is complete, $\langle d \rangle$ and therefore $\langle d^x \rangle$, become vanishingly small for both donor and acceptor, and the $\langle \kappa^2 \rangle$ value is indeed close to 2/3. Since R_0 is proportional to $(\langle \kappa^2 \rangle)^{1/6}$, a factor of two between the minimum and maximum values of $\langle \kappa^2 \rangle$ produces an uncertainty of only some ±6% in *R*. FIGURE 4 shows that such precision is readily obtainable for most situations in which the two axial depolarization factors are in the range ±0.4, corresponding to $\langle d \rangle$ values of less than 0.16. Serious uncertainties in *R* obtain when both fluorophores are relatively immobile, with axial depolarization factors near unity, e.g. $\langle d^x \rangle > 0.8$.

Consider for example the case for which $\langle d_D{}^x \rangle, \langle d_A{}^x \rangle \geq 0$, for an arbitrary value $\langle d_T \rangle$. It has been shown[5] that $\langle \kappa^2 \rangle_{\min}$ and $\langle \kappa^2 \rangle_{\max}$ are then given by

$$\langle \kappa^2 \rangle_{\min} = (2/3)[1 - (1/2)(\langle d_D{}^x \rangle + \langle d_A{}^x \rangle)] \quad \textbf{(13)}$$

$$\langle \kappa^2 \rangle_{\max} = (2/3)[1 + \langle d_D{}^x \rangle + \langle d_A{}^x \rangle + 3\langle d_D{}^x \rangle \langle d_A{}^x \rangle]. \quad \textbf{(14)}$$

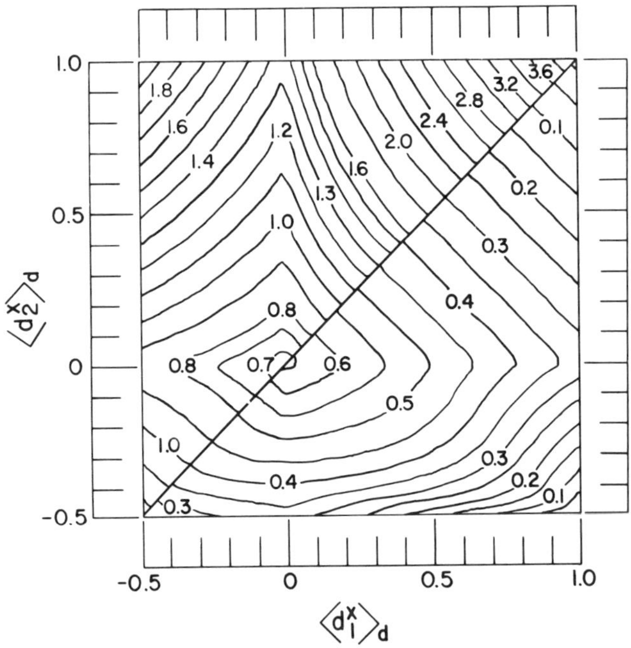

FIGURE 5. Contour maps for determining upper and lower bounds of $\langle \kappa^2 \rangle$, as in FIGURE 4, but independently of $d_T{}^x$. These are the appropriate limits when $d_T{}^x$ is unknown.

The condition, $\langle d_D^x \rangle$, $\langle d_A^x \rangle > 0.8$, therefore corresponds to a factor of more than 22.6 between the upper and lower bounds of $\langle \kappa^2 \rangle$, or an error greater than ±25% in R, and this error is not greatly reduced when $\langle d_T \rangle$ is also measured. This limit still corresponds to relatively small measured depolarization factors, $\langle d_D \rangle = \langle d_A \rangle = 0.64$. When $\langle d_D \rangle$ and $\langle d_A \rangle$ exceed 0.8, so that $\langle d^x \rangle > 0.9$, the error in the determination of R increases to more than ±45%.

For a fluorophore adsorbed to a protein, values of $\langle d^x \rangle$ at least as high as 0.82 can be found in the literature.[12] At the opposite extreme, one can derive $\langle d^x \rangle$ values as low as (+)0.23 for a fluorophore attached to a substrate by a presumably quite flexible 3-carbon aliphatic sidechain.[13] A donor-acceptor combination characterized by these two depolarizations will have a $\langle \kappa^2 \rangle_{max}/\langle \kappa^2 \rangle_{min}$ ratio of about 5.5, as can be seen either from Equations 13 and 14 or from FIGURE 5, and the corresponding uncertainty in R would be about ±16%. This is a substantial error, even though one of the fluorophores comes close to being isotropically averaged. If both the donor and the acceptor have axial depolarization factors as large as about 0.9 (corresponding to $r_0 \simeq 0.32$, which would by no means constitute an unexpectedly high value), the resulting uncertainty in R approaches a factor of 2, as indicated above.

The extent of angular averaging which corresponds to particular $\langle d^x \rangle$ or $\langle d \rangle$ values can be gauged by postulating or inferring a specific angular distribution function. Thus, for random orientation within the volume of a cone with half-angle Ψ,[4,5]

$$\langle d^x \rangle = (1/2)\cos\Psi(1 + \cos\Psi). \qquad \textbf{(15)}$$

In the two cases considered above, the cones would have half-angles of 29° and 70°, respectively. Several other angular distribution functions have been considered elsewhere.[5]

INTERMOLECULAR ENERGY TRANSFER FOR SURFACE DISTRIBUTIONS

In this section, we will consider EET among a distribution of donors and acceptors which are confined to planes or to spherical surfaces. Examples of such assemblies of interest to biologists include surface-labeled spherical macromolecules (e.g., proteins) and detergent micelles. Phospholipid bilayer vesicles containing fluorophores adsorbed at the surface or into the bilayer interior also provide such a distribution. Planar distributions may be viewed as the limiting cases approached by spheres of large radius and are appropriate to oriented planar bilayers as well as to cell membranes whose curvature is small compared to that of a sphere with a radius on the order of the Förster distance.

EET experiments for such systems are particularly useful for determining local acceptor densities around donor sites. These may then be compared with the density calculated for uniform (random) surface distributions.[14,15] Such experiments can therefore shed light on the surface topographical relationships between specifically labeled membrane components, such as antibody binding sites, hormone receptors and ion transport proteins.

While the present discussion confines itself to cases in which both donors and acceptors are located on the same spherical surfaces, extensions to more general cases can be envisaged. The critical dependence of the EET efficiency for such systems on the sphere diameters becomes evident from the results presented below.

We base our calculations of EET efficiency on the following model: donors and acceptors, considered as point dipoles, are randomly distributed over the surface of a sphere with diameter D. Transfer between fluorophores on different spheres, as well as donor-donor (self-) transfer within a single sphere, is considered negligible. All donors have equivalent depolarization factors ($\langle\langle d_D\rangle\rangle$), as do all acceptors ($\langle\langle d_A\rangle\rangle$). The acceptors have an average surface concentration c, and the critical Förster concentration c_0 corresponds to one acceptor per disc of radius R_0, i.e., $c_0 = 1/\pi N' R_0^2$, where N' is Avogadro's number per mmole.

Planar Distributions

The decay of the excited donor population may be calculated in analogy to Förster's classical treatment in three dimensions,[16] and the limiting case for the plane has long been known:[17]

$$\bar{\rho}(t) = \exp\{-[(t/\tau_0) + (c/c_0)(t/\tau_0)^{1/3}\Gamma(2/3)]\} \qquad \textbf{(16)}$$

where the bar indicates an average over all acceptor distributions, c is the acceptor concentration in mmoles per cm^2 and τ_0 is the donor decay time in the absence of acceptors.

In this system, the donor quantum yield relative to its yield in the absence of acceptors has recently been expressed analytically[18] as

$$\langle\Phi\rangle/\Phi_0 = \int_0^\infty \bar{\rho}(t)\,dt \Big/ \int_0^\infty \rho_0(t)\,dt \qquad \textbf{(17a)}$$

$$= \sum_{j=0}^{\infty} \{(-[c/c_0])\Gamma[2/3])^j\,\Gamma[(j/3)+1]/j!\} \qquad \textbf{(17b)}$$

where ρ_0 is the donor decay function in the absence of acceptors. The efficiency of two-dimensional transfer is then given by

$$\langle T\rangle = 1 - (\langle\Phi\rangle/\Phi_0). \qquad \textbf{(18)}$$

Spherical Surface Distributions

It has been pointed out[6] that the spherical case can be carried through in an equivalent manner, with due regard to the Poisson distribution of the numbers of acceptors per sphere. It gives rise to integrals identical in form to the ones which yield Equation 16 for the planar case, but since their upper limits depend on the sphere diameter, they do not lead to a convenient analytical solution.

For the sake of completeness, and to aid our understanding of orientational effects in the case of spherical surfaces of arbitrary size, we outline in the following section an analysis of this problem, together with computer-derived graphical solutions for the transfer efficiency.

For spheres of diameter D and surface area $a = \pi D^2$ which harbor one or more donors and k acceptors, the Poisson (Smoluchowski) distribution of k is given[19] by

$$P_k = (\nu^k/k!)e^{-\nu} \qquad \textbf{(19)}$$

with

$$\sum_{k=0}^{\infty} P_k = 1 \tag{20}$$

where ν is the average number of acceptors per sphere given by

$$\nu = aN'c. \tag{21}$$

For such a sphere, one obtains, following Förster's approach[16]

$$\rho(t) = \exp[-(t/\tau_0)] \prod_{j=1}^{k} \int_0^D \exp[-(t/\tau_0)(R_0/R)^6]\omega(R_j)dR_j \tag{22}$$

$$= \exp[-(t/\tau_0)][J(t)]^k \tag{23}$$

where $\omega(R_j)$ is the probability of finding an acceptor at a distance between R_j and $R_j + dR_j$ from the donor. On averaging $\rho(t)$ over the distribution of Equation 19 one obtains

$$\bar{\rho}(t) = \sum_{k=0}^{\infty} \exp[-(t/\tau_0)][J(t)]^k(\nu^k/k!)e^{-\nu} = \exp[-(t/\tau_0)] \exp\{-\nu[1 - J(t)]\}. \tag{24}$$

The integral $J(t)$ is conveniently expressed in terms of θ, the angular separation between D and A (cf. FIGURE 7):

$$J(t) = \frac{1}{2}\int_0^{\pi} \exp\{-8(t/\tau_0)[R_0^2/D^2(1 - \cos\theta)]^3\}\sin\theta\, d\theta. \tag{25}$$

Writing $\alpha = 1 - \cos\theta$ and $x = t/\tau_0$ and defining a concentration c_s (in mmoles per cm^2) as

$$c_s = 1/\pi N' D^2, \tag{26}$$

corresponding to an average of one acceptor per sphere, Equation 25 becomes

$$J(x) = \frac{1}{2}\int_0^{2} \exp[-8(c_s/c_0)\alpha^{-3}x]d\alpha. \tag{27}$$

Since $\nu = c/c_s$, the donor decay function Equation 24 and ultimately the EET efficiency, $\langle T \rangle$, can thus be expressed in terms of the dimensionless parameters c/c_s and c/c_0. As will be discussed below, the orientational dependence of $\bar{\rho}(t)$ and $\langle T \rangle$ is contained in R_0, i.e., the appropriate orientation factor is assumed to be independent of R.

By substituting Equation 27 into Equation 24 and performing the double integration over α and x (i.e., over θ and t) numerically, then using Equations 17a and 18, EET efficiencies were calculated as functions of (c/c_0) for a range of spheres sizes expressed in terms of $(D/R_0)^2 = (c_0/c_s)$. These results are shown as functions of acceptor concentration and sphere diameter in FIGURE 6 where they are compared to the planar case, $(D/R_0) \rightarrow \infty$.

It can be seen from FIGURE 6 that $\langle T \rangle$ becomes vanishingly small as the sphere diameter becomes much less than R_0, as would be expected from the vanishing probability of finding any acceptor at all on the sphere's surface. As the sphere

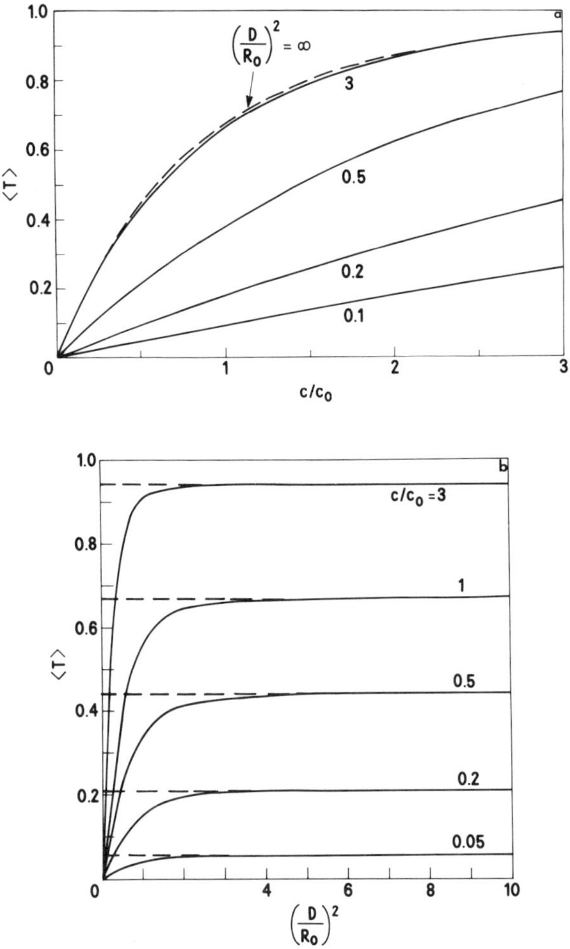

FIGURE 6. The average EET efficiency between donors and acceptors distributed randomly over the surface of a sphere, as a function of (a) the acceptor concentration and (b) the sphere diameter. The orientation factor κ^2 for each *D*, *A* pair is assumed to be the same. These results may be compared with those for planar *D*, *A* distributions, which are indicated by the dashed curves.

diameter increases, $\langle T \rangle$ increases rapidly until D/R_0 is of the order of unity, where it begins to approach the limiting value for a planar distribution ($D/R_0 \rightarrow \infty$). For values of D/R_0 greater than about 1.5, the sphere's curvature plays a negligible role.

While this discussion assumes only that the κ^2 for each *D*, *A* pair is uncorrelated with the *D-A* separation, it is clear that, for finite spheres, the only experimentally approachable case of this kind is the one in which the fluorophore orientations are distributed isotropically.

Orientational Effects in Energy Transfer on Spherical Surfaces

We finally wish to consider the effects of donor and acceptor orientational distributions on EET between them. Only axially symmetrical distributions are considered, and the symmetry axis is assumed to lie along the sphere's radius. This model is illustrated in Figure 7.

The dynamically averaged orientation factor for a particular D, A pair with angular separation θ is then obtained from Equations 11 and 12 as

$$\langle \kappa^2(\theta) \rangle = (1/4)(3 - \cos\theta)^2 \langle d_D{}^x \rangle \langle d_A{}^x \rangle + (1/3)(1 - \langle d_D{}^x \rangle) + (1/3)(1 - \langle d_A{}^x \rangle) + (1/2)(1 - \cos\theta)[\langle d_D{}^x \rangle (1 - \langle d_A{}^x \rangle) + \langle d_A{}^x \rangle (1 - \langle d_D{}^x \rangle)]. \quad \mathbf{(28)}$$

In the plane, $\theta \rightarrow 0$ and Equation 28 reduces to

$$\langle \kappa^2 \rangle = \langle d_D{}^x \rangle \langle d_A{}^x \rangle + (1/3)(1 - \langle d_D{}^x \rangle) + (1/3)(1 - \langle d_A{}^x \rangle). \quad \mathbf{(29)}$$

The axial depolarization factors are obtained in the usual way from a time-resolved depolarization measurement[4,20] as

$$\langle d^x \rangle = (r_\infty/0.4)^{1/2} \quad \mathbf{(30)}$$

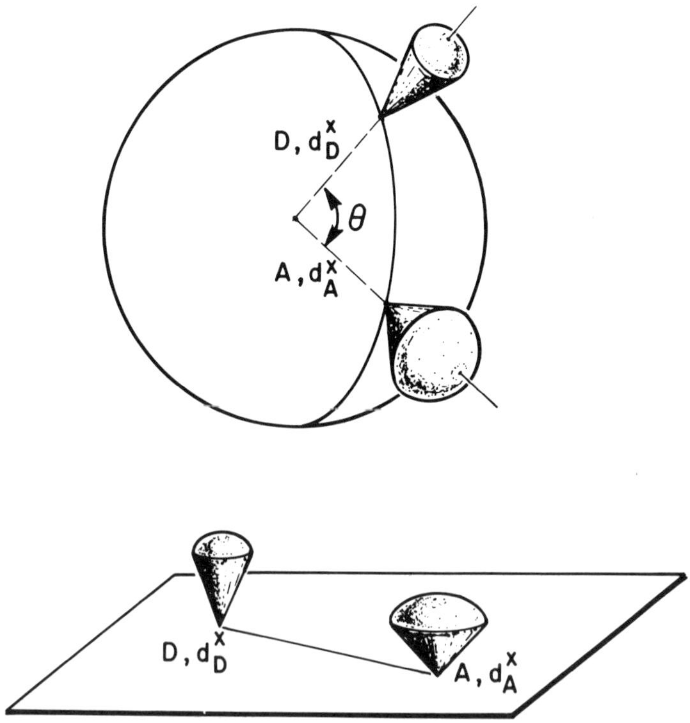

Figure 7. Schematic representation for the model used to calculate the orientational dependence of the EET efficiency among donors and acceptors distributed on planes and spherical surfaces. The orientational distributions of the D and A transition moments have axial symmetry and are characterized by the axial depolarization factors $\langle d_D{}^x \rangle$ and $\langle d_A{}^x \rangle$.

where r_∞ is the limiting emission anisotropy at long times and where possible degeneracy of the sign of $\langle d^x \rangle$ is taken into consideration when $r_\infty \leq 0.1$. The assumption that the anisotropy decay to r_∞ is fast compared with the transfer time (dynamic averaging regime) will be well approximated, at least for low transfer efficiencies and/or small degrees of reorientational freedom (r_∞ high).

For fluorophore distributions confined to a plane, the orientational effects are taken into account by the appropriate choice of κ^2 in R_0, so that c_0 in Equations 16 and 17 would be replaced by $c_0/[(3/2)\kappa^2]^{1/3}$. For the dynamically averaged cases considered here, all D, A pairs are orientationally identical and the $\langle \kappa^2 \rangle$ which enters in this scaling of c_0 is the appropriate dynamic average value given by Equation 29.

For spherical surfaces, on the other hand, there exists in general a *statically* averaged component of the orientational dependence (cf. Equation 28). If the orientational distributions of $\boldsymbol{D}$ and $\boldsymbol{A}$ are isotropic, c_0 can be defined with $\langle \kappa^2 \rangle = 2/3$ as discussed for the planar case above. If not, the orientation factor is clearly correlated with the D-A separation and cannot be treated independently. It is then convenient to define a new critical separation, R_0', and a corresponding concentration, c_0', which are independent of the orientation factor:

$$R_0' = (3/2)^{1/6} R_0 \qquad \textbf{(31)}$$

$$c_0' = (2/3)^{1/3} c_0 \qquad \textbf{(32)}$$

which effectively contain κ^2 values of unity.

In order to illustrate the orientational effects of EET for spherical surface distributions, we present a number of limiting cases in which three types of (axially symmetrical) transition moments for $\boldsymbol{D}$ and $\boldsymbol{A}$ are considered (cf. FIGURE 8): linear transitions perpendicular to the plane or radial on a sphere (L), planar transitions whose normal is aligned the same way (P), and isotropic transitions (I). The latter two cases are of course equivalent to dynamically averaged linear transitions in a plane tangential to the surface and in three dimensions, respectively.

The values of $\langle \kappa^2(\theta) \rangle$ derived from Equation 28 after setting $\langle d^x \rangle$ equal to 1, $-1/2$ and 0 for the L, P and I cases, respectively, are given in TABLE 1. The table also contains $\langle \kappa^2 \rangle$ values for the various corresponding cases of planar D, A distributions.

For each model considered in TABLE 1, the appropriate value of the orientation factor was substituted into Equation 25 as $\langle \kappa^2(\theta) \rangle R_0'^6$ for R_0^6. Transfer efficiencies were then computed by numerical integration as described in the previous section. The integrations were performed for a series of concentrations and sphere sizes, defined by the dimensionless parameters c/c_0' and $(D/R_0')^2$, respectively, and the results are shown in FIGURES 9-12 and 14.

It can be seen that, for EET between radial linear transition moments on a sphere ($L \leftrightarrows L$, FIGURE 9) and also for $L \rightleftarrows I$ transfer on a sphere (FIGURE 10), the transfer efficiency may significantly exceed that for the same acceptor concentration in a planar distribution when the sphere diameter is of the order of R_0' or somewhat greater. For larger spheres, the transfer efficiencies asymptotically approach those for planar fluorophore distributions.

On the other hand the EET efficiencies for transfer of the type $P \rightleftarrows P$ (FIGURE 11) and $P \rightleftarrows I$ (FIGURE 12) on a sphere are always lower than those for planar distributions with the same acceptor concentrations and approach the limiting in-

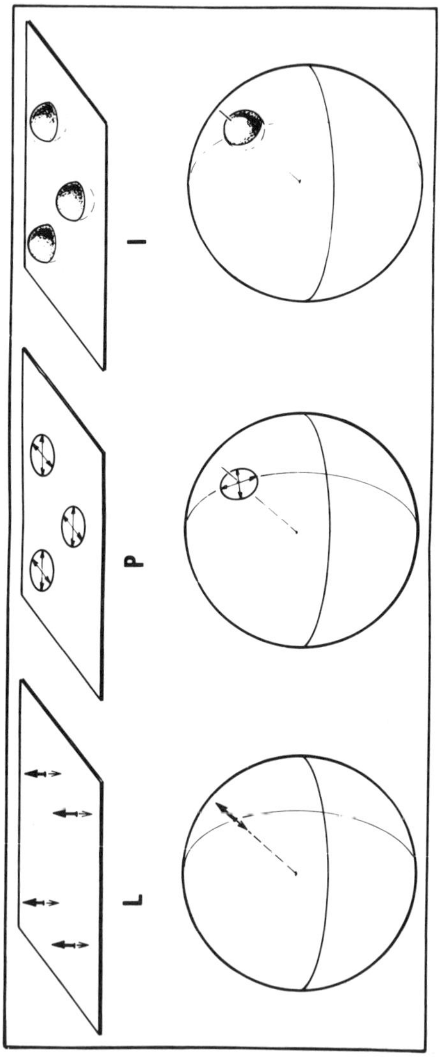

FIGURE 8. Schematic representation of the three limiting cases of the transition moment distributions considered in the present discussson: linear (L), planar (P) and isotropic (I).

TABLE 1

THE AVERAGE ORIENTATION FACTORS FOR DONOR-ACCEPTOR PAIRS IN A PLANE AND ON THE SURFACE OF A SPHERE, WHERE *D* AND *A* ARE SEPARATED BY AN ANGLE θ. *L*, *P* AND *I* CHARACTERIZE THE AXIALLY SYMMETRICAL TRANSITION MOMENT DISTRIBUTIONS OF *D* AND *A* (cf. FIGURE 7).

	$\langle\kappa^2\rangle$	
D → A	D,A IN PLANE	D,A ON SPHERE
I → I	2/3	2/3
L → L	1	$\frac{1}{4}(3-\cos\theta)^2$
P → P	5/4	$\frac{1}{16}(13+6\cos\theta+\cos^2\theta)$
L → P P → L	0	$\frac{1}{8}\sin^2\theta$
L → I I → L	1/3	$\frac{1}{6}(5-3\cos\theta)$
P → I I → P	5/6	$\frac{1}{12}(7+3\cos\theta)$

plane efficiencies more slowly than is the case for transfer of the type $I \rightleftarrows I$ (FIGURE 6). These effects are related to the manner in which $\langle\kappa^2(\theta)\rangle$ varies with θ, the angular separation between donor and acceptor, which is shown in FIGURE 13: $\langle\kappa^2(\theta)\rangle$ is seen to increase monotonically as θ goes from 0 to π for the *L,L* and *L,I* combinations and decreases monotonically for the *P,P* and *P,I* cases.

The $L \rightleftarrows P$ transfer case, which is illustrated in FIGURE 14, is of particular interest because it is characterized by a vanishing orientation factor for planar distributions and hence *no* transfer occurs in that case (see TABLE 1). For donors and acceptors on a sufficiently small sphere, on the other hand, high transfer efficiencies are to be expected. For a given R_0' and acceptor concentration there exists a sphere diameter for which energy transfer is maximally efficient. The dependence of this diameter on acceptor concentration is weak, with corresponding D/R_0' values lying between 1 and 1.2 for c/c_0' values up to 3. This case is less exotic than might at first be supposed: the rod-like fluorophore diphenylhexatriene (DPH) in some lipid bilayers can approximate the *L* model quite closely,[21,22] while the *P* model may be well approximated for example by the long aliphatic chain indocarbocyanine derivative diI_{18}.[23]

In a real experimental situation, the fluorophores on a sphere have transition moment distributions which at best usually only approximate the *L*, *P* and *I* cases

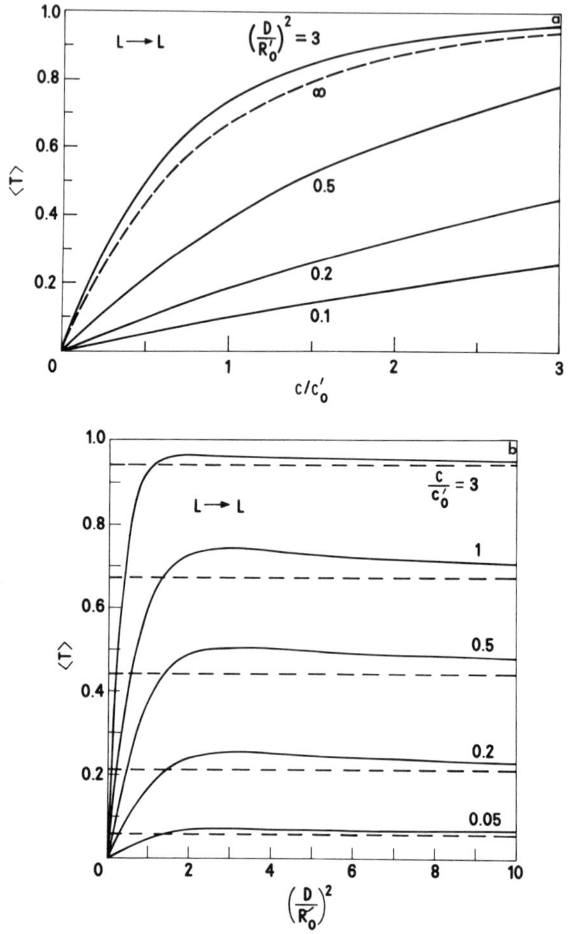

FIGURE 9. The average EET efficiency, $\langle T \rangle$, as a function of (a) acceptor concentration and (b) sphere size and with linear transition moments for both $\boldsymbol{D}$ and $\boldsymbol{A}$, and (transfer type $L \leftrightarrows L$ cf. TABLE 1), with $\boldsymbol{D}$ and $\boldsymbol{A}$ randomly distributed over the surface of a sphere. The solutions for the planar case ($D \rightarrow \infty$) are shown by dashed lines.

considered above. It is then necessary to determine the axial depolarization factors $\langle d_D^x \rangle$ and $\langle d_A^x \rangle$ and substitute them into Equation 28 for $\langle \kappa^2(\theta) \rangle$. The average transfer efficiency $\langle T \rangle$ may then be evaluated by numerical integration as described above for particular limiting cases.

CONCLUSION

We have attempted to illustrate the importance of taking orientational effects into account whenever EET experimentation is employed to measure intramolecular distances or to obtain information about distributions of donors and acceptors. In the former application, the best (indeed, the only) practical method for estimating

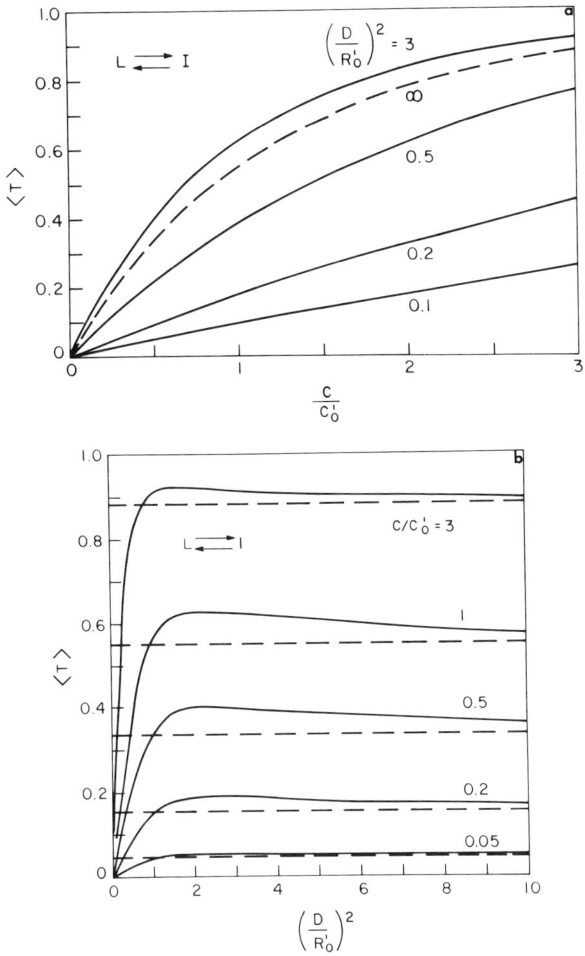

FIGURE 10. $\langle T \rangle$ as a function of (a) acceptor concentration and (b) sphere size, as in FIGURE 9, but for transfer of the type $L \rightleftarrows I$. (cf. TABLE 1).

the uncertainty in the derived separation due to the orientation factor is to use depolarization measurements to provide upper and lower bounds for $\langle \kappa^2 \rangle$ and hence for R. In the latter case, such measurements provide *all* the information necessary to allow for the orientational effects within the limits of the model considered. Whenever possible, time-resolved fluorometry should be employed to determine the required depolarization factors.

SUMMARY

The orientational dependence of long-range excitation energy transfer is reviewed for intramolecular donor-acceptor pairs and also examined for intermolecular energy

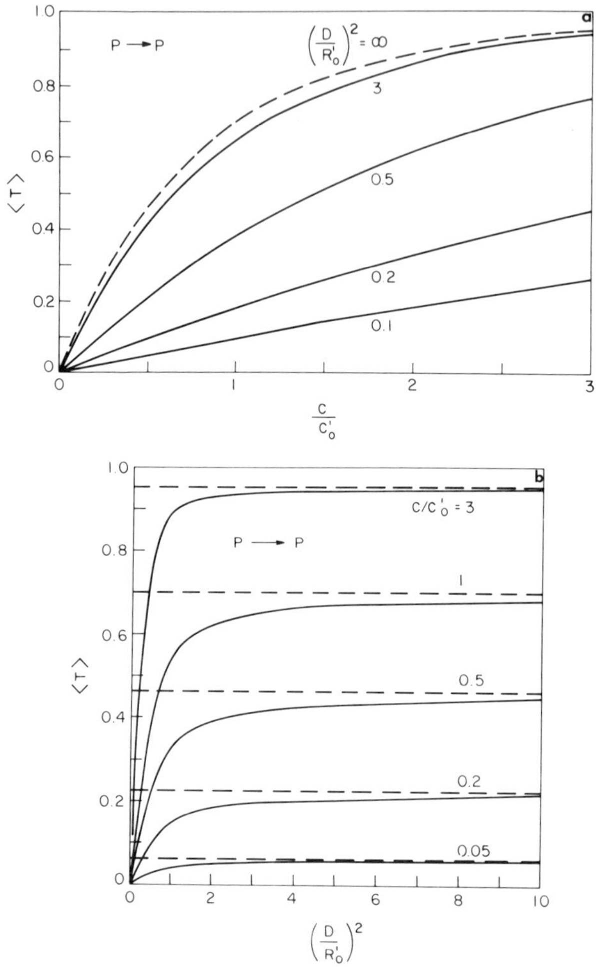

FIGURE 11. ⟨T⟩ as a function of (a) acceptor concentration and (b) sphere size, as in FIGURE 9, but for transfer of the type $P \leftrightarrows P$ (cf. TABLE 1).

transfer between donors and acceptors distributed randomly over a spherical surface. It is shown that the energy transfer efficiency has an appreciable dependence on the orientations and reorientational motion of donors and acceptors in both situations.

The results of fluorescence depolarization experiments may be used to provide estimates for the magnitude of these orientational effects and protocols for doing so are provided. For fluorophores on spherical surfaces, both the general case is discussed and the limiting cases in which the dynamically averaged transition moments approach linear, planar and isotropic distributions. The dependence of the dynamically averaged transfer efficiency on sphere size and on acceptor concentration was calculated and is compared with that for planar distributions of fluorophores.

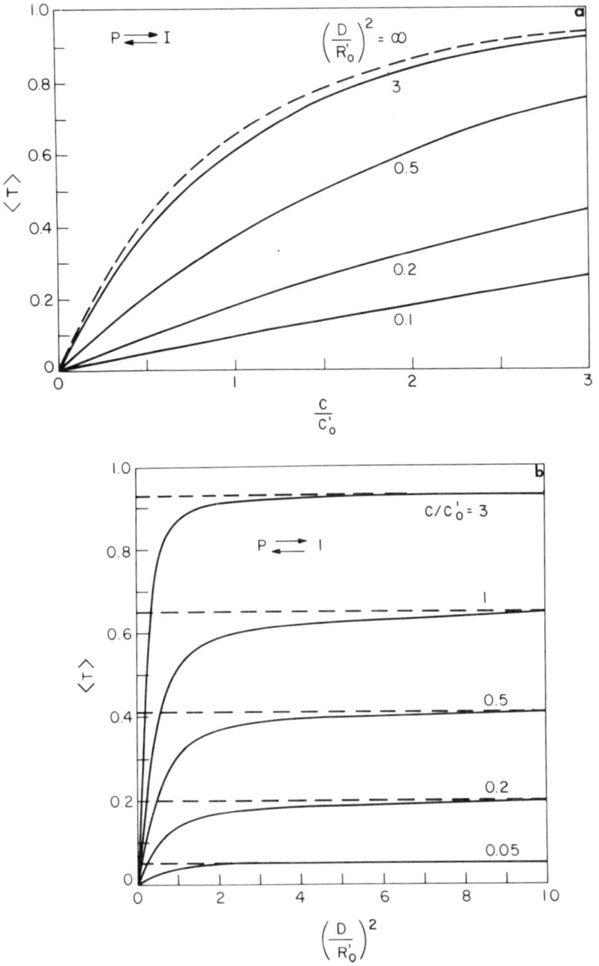

FIGURE 12. ⟨T⟩ as a function of (a) acceptor concentration and (b) sphere size, as in FIGURE 9, but for transfer of the type $P \rightleftarrows I$ (cf. TABLE 1).

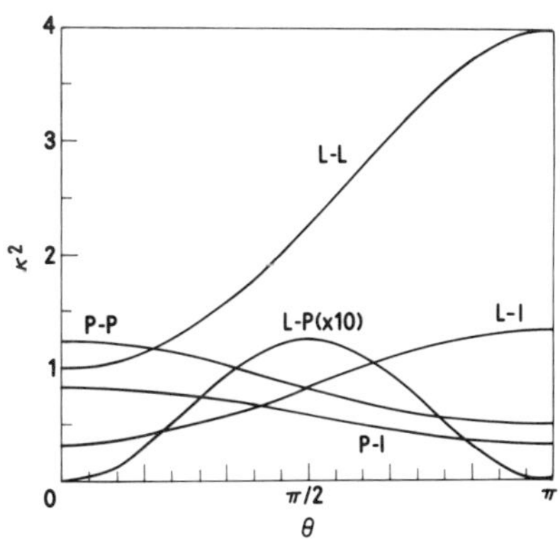

FIGURE 13. The orientation factor κ^2 for a D, A pair on the surface of a sphere, as a function of the angular separation θ. Various combinations of the limiting donor and acceptor transition moment distributions (L, P and I) are considered.

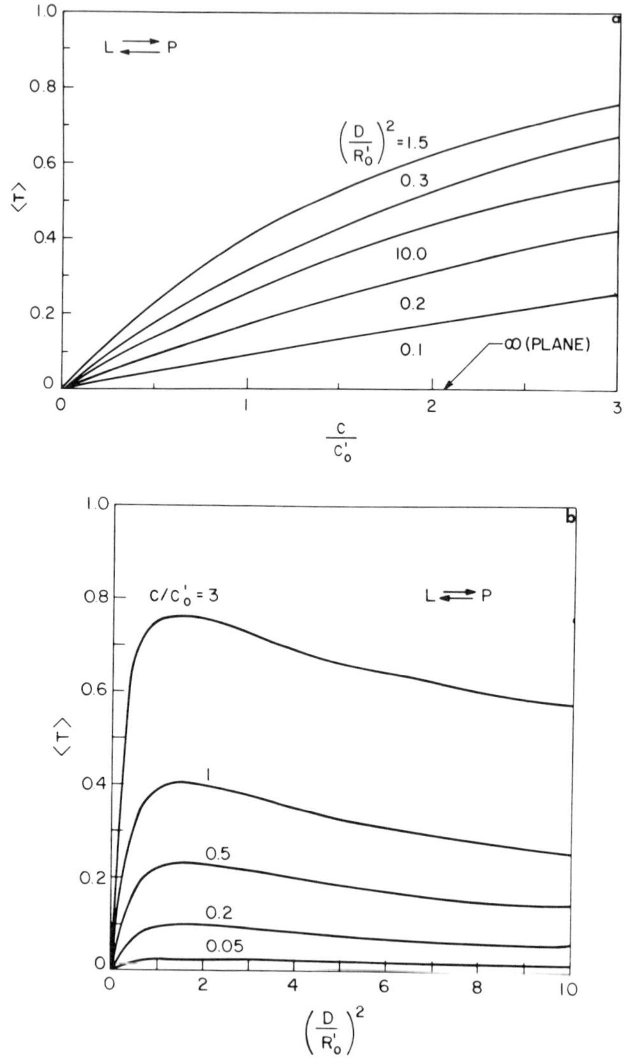

FIGURE 14. $\langle T \rangle$ as a function of (a) acceptor concentration and (b) sphere size, as in FIGURE 9, but for transfer of the type $L \rightleftarrows P$ (cf. TABLE 1). Note that for this case *no* transfer occurs in the planar limit due to the orthogonality of the ***D*** and ***A*** transition moments. For sphere diameters of the order of the Förster distance, quite efficient EET is expected to take place.

ACKNOWLEDGMENTS

We wish to thank Bea Chambers and Adrian Simons for their assistance with the numerical integrations reported in this work.

REFERENCES

1. STRYER, L. 1978. Ann. Rev. Biochem. **47:** 819–846.
2. FAIRCLOUGH, R. H. & C. R. CANTOR. 1978. *In* Methods in Enzymology. C. H. W. Hirs & S. N. Timasheff, Eds. Vol. **48:** 347–379. Academic Press, New York, N. Y.
3. BADEA, M. G. & L. BRAND. 1979. *In* Methods in Enzymology. C. H. W. Hirs & S. N. Timasheff, Eds. Vol. **61:** 378–425. Academic Press. New York, N. Y.
4. KINOSITA, K., JR., S. KAWATO & A. IKEGAMI. 1977. Biophys. J. **20:** 289–305.
5. DALE, R. E., J. EISINGER & W. E. BLUMBERG. 1979. Biophys. J. **26:** 161–194.
6. KOPPEL, D. E., P. J. FLEMING & P. STRITTMATTER. 1979. Biochemistry **18:** 5450–5457.
7. FÖRSTER, T. 1948. Ann. Physik **2:** 55–75.
8. FÖRSTER, T. 1951. Fluoreszenz Organischer Verbindungen. Vandenhoeck & Ruprecht, Göttingen.
9. PERRIN, F. 1926. J. Phys. (Paris) **7:** 390–401.
10. SOLEILLET, P. 1929. Ann. Phys. (Paris) **12:** 23–97.
11. HAAS, E., E. KATCHALSKI-KATZIR & I. Z. STEINBERG. 1978. Biochemistry **17:** 5064–5070.
12. TU, S.-C., C.-W. WU & J. W. HASTINGS. 1978. Biochemistry **17:** 987–993.
13. ANGELIDES, K. J. & G. G. HAMMES. 1979. Biochemistry **18:** 5531–5537.
14. FUNG, B.K.-K. & L. STRYER. 1978. Biochemistry **17:** 5241–5248.
15. ESTEP, T. N. & T. E. THOMPSON. 1979. Biophys. J. **26:** 195–207.
16. FÖRSTER, T. 1949. Z. Naturforsch. **4a:** 321–327.
17. TWEET, A. G., W. D. BELLAMY & G. L. GAINES, JR. 1964. J. Chem. Phys. **41:** 2068–2077.
18. WOLBER, P. K. & B. S. HUDSON. 1979. Biophys. J. **28:** 197–210.
19. JABLONSKI, A. 1958. Bull. Acad. Pol. Sci. Ser. Sci. Math. Astron. Phys. **6:** 663–669.
20. LAKOWICZ, J. R., F. G. PRENDERGAST & D. HOGEN. 1979. Biochemistry **18:** 508–519.
21. KAWATO, S., K. KINOSITA, JR. & A. IKEGAMI. 1977. Biochemistry **16:** 2319–2324.
22. ANDRICH, M. P. & J. M. VANDERKOOI. 1976. Biochemistry **15:** 1257–1261.
23. AXELROD, D. 1979. Biophys. J. **26:** 557–573.

ROTATIONAL DIFFUSION OF BIOLOGICAL MACROMOLECULES BY TIME-RESOLVED DELAYED LUMINESCENCE (PHOSPHORESCENCE, FLUORESCENCE) ANISOTROPY

Thomas M. Jovin, Marty Bartholdi, and Winchil L. C. Vaz

Abteilung Molekulare Biologie
Max Planck Institut f. Biophysikalische Chemie
D-3400 Göttingen, Federal Republic of Germany

Robert H. Austin

Department of Physics
Princeton University
Princeton, New Jersey 08544

The integrity of the biological cell derives from the maintenance of specialized structures and functions. The characteristic motions of complex macromolecules define the specificity and rapidity with which the system can react. The physical processes of diffusion are largely responsible for the (re)distribution of matter and energy within the cell and between the cell and its environment. *Translational* diffusion operates over macroscopic distances whereas *rotational* diffusion reflects the immediate environment of an isolated molecule or of a molecular assembly.

The rotational properties of a macromolecule are determined by the combined effects of its intrinsic structural features (size, shape, flexibility) and interactions with its surrounding (viscous damping, stereochemical constraints, liganding). In fluid (e.g. aqueous) media, the rotations of typical biological macromolecules occur in the nanosecond domain and are, therefore, well suited for the techniques of polarized fluorescence emission spectroscopy (for classical and recent reviews see refs. 1,2, respectively). However, rotational relaxations of larger, more extended molecules (e.g. DNA) or within viscous, anisotropic structures (e.g. membranes) take place in times greater than about 1 microsecond. Thus, other methods based upon electromagnetic absorption, resonance, scattering, and emission phenomena have been developed, particularly for the study of membranes (for reviews see refs. 3,4). In some cases, including the present, the relatively long excited triplet state of certain probes has been exploited both by polarized absorption (flash-induced transient dichroism, reviewed in ref. 3) and polarized emission (phosphorescence, delayed fluorescence).[5–7]

In this report, we describe the basis and utility of the techniques based upon the polarized emission originating directly (phosphorescence) or indirectly (delayed fluorescence) from the triplet state and for which we choose the encompassing designation Delayed Luminescence Anisotropy (DLA). The rotational motions of surface components can be resolved with great inherent sensitivity as required in studies of cellular suspensions as well as of proteins reconstituted into lipid vesicles. The temporal sensitivity appears adequate for the analysis of microaggregation

0077-8923/81/0366-0176$01.75/0 © 1981, NYAS

phenomena thought to be involved in cellular responses triggered, for example, by growth factors (hormones[8]) and allergens.[9]

The first biological applications of DLA using solutions or suspensions in fluid media were reported in 1979.[5–7,30] Traditionally, of course, phosphorescence spectroscopy has been performed at low temperatures in the glassy state so as to maximize quantum yields and excited state lifetimes. Another technique applied at "room temperature" and denoted room temperature phosphorescence (r.t.p.[10]) exploits the emission of organic compounds absorbed on dry substrates for the purpose of quantitative analysis. The requirements for observation of phosphorescence in fluid media are summarized in reference 14.

Photophysical Principles

The luminescence of organic molecules originates in general from the first excited singlet S_1 and triplet T_1 states formed rapidly by internal conversion after the absorption of light. The simplest possible representation of the electronic transitions involved is shown in Figure 1. The various vibrational manifolds as well as other deactivation pathways (resonance energy transfer, solvent relaxation, collisional or binding interactions with other excited or ground state species) are omitted for the sake of simplicity. (Excellent discussions of spectroscopic phenomena associated with the triplet state are found in monograph[11] and review[10,12–14] form.)

The transition from the excited singlet to the less energetic triplet state (intersystem crossing) is formally "forbidden" because of the change in spin multiplicity, but spin orbital coupling can be greatly enhanced as, for example, by introduction of internal or external heavy atoms. The radiative deactivation of the singlet state (fluorescence) is typically efficient (and thus fast) whereas emission from T_1 (phosphorescence) is slower (and less probable) than competing nonradiative processes. The latter include collision with solvent or paramagnetic molecules such as ground state O_2. (It follows

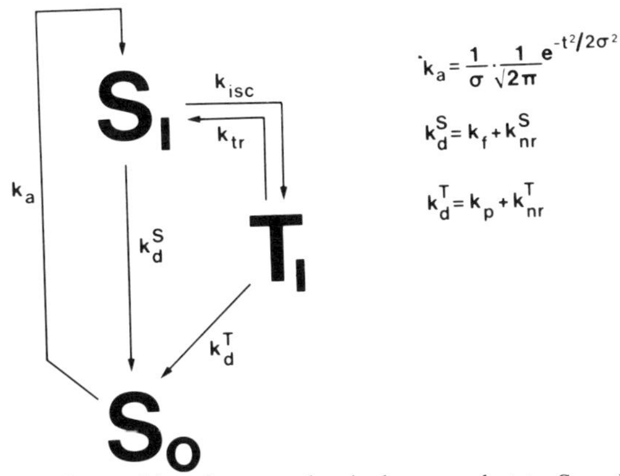

Figure 1. Electronic transitions between the singlet ground state S_o and the first excited singlet S_1 and triplet T_1 states. The rate constant for absorption k_a is ascribed the Gaussian profile of an exciting light pulse. The decay from S_1 (k_d^S) and T_1 (k_d^T) is represented as the sum of radiative (*f*, fluorescence; *p*, phosphorescence) and nonradiative (generally multicomponent) pathways (k_{nr}^S, k_{nr}^T). Intersystem crossing from S_1 to T_1 is denoted by the composite rate constant k_{isc} and the reverse reaction (thermal reactivation of T_1 and intersystem crossing) by k_{tr}.

TABLE 1
DIFFERENT FORMS OF DYE LUMINESCENCE

	Prompt Fluorescence	Delayed Fluorescence	Phosphorescence
Spectrum	←identical→		red-shifted to fluorescence
Lifetime	short <0.1 μsec	←long-identical→ >1 μsec	
Yield	generally high	←generally low→ ↑ ←↑ temperature→ ↓	
Polarization	←identical→ generally high		low

The above properties are representative but by no means universal.

that the observation of delayed luminescence requires in general that the system be purged free of air.) The thermal reactivation of T_1 and triplet-triplet annihilation provide alternative pathways for regenerating S_1 and thus giving rise to delayed fluorescence emission.

A comparison between the various forms of luminescence is given in TABLE 1. From the standpoint of the time-resolved measurements described here, the key feature is the excited state lifetime. Thus, use of the triplet state enables the extension of the traditional fluorescence methods[1,2] into the time domain greater than 10^{-7} seconds.

Decay Laws

The decay processes corresponding to the scheme in FIGURE 1 and excitation with a Gaussian light pulse with standard deviation σ are defined by the following differential equations:

$$\dot{S}_1 + a_{11}S_1 + a_{12}T_1 = 1/\sigma \cdot Z(t/\sigma) \qquad Z(x) = 1/\sqrt{2\pi} \cdot e^{-x^2/2}$$
$$\dot{T}_1 + a_{21}S_1 + a_{22}T_1 = 0$$

$$a_{11} = k_d{}^S + k_{isc} \qquad a_{12} = k_{tr} \tag{1}$$
$$a_{21} = -k_{isc} \qquad a_{22} = k_d{}^T + k_{tr}$$

S_1 and T_1 are given in terms of fractional quantities representing the distribution of excited state species. Time t is referenced to the center of the Gaussian excitation pulse.

At all times,

$$\text{Fluorescence} \propto k_f S_1(t) \tag{2}$$
$$\text{Phosphorescence} \propto k_p T_1(t)$$

where the proportionality factor includes the chemical concentration (ground state), extinction coefficient, and instrumentational parameters (spectral sensitivity, bandwidth, optical collection, electronic amplification) that determine the absolute and relative signal intensities. It is assumed that the latter are proportional to excitation intensity, i.e. that the system is operated well below saturation.

The eigenvalues ($\lambda_1 \gg \lambda_2$) of equations 1 yield the two lifetimes for prompt (τ_1) and delayed (τ_2) emission:

$$\lambda_1 = k_d{}^S + k_{isc} = \tau_1{}^{-1} \qquad \lambda_2 = k_{nr}^T + k_p + k_{tr}(1 - \Phi_{isc}) = \tau_2{}^{-1} \tag{3}$$

The time course of prompt emission (fluorescence and phosphorescence) reflects the convolution of the finite excitation profile with the response function:
for $t \simeq \tau_1, \sigma$

$$\begin{aligned} S_1(t) &\simeq e^{(\lambda_1\sigma)^2/2}P(t/\sigma - \lambda_1\sigma)e^{-\lambda_1 t} + S_d P(t/\sigma) \\ T_1(t) &\simeq T_d[P(t/\sigma) - e^{(\lambda_1\sigma)^2/2}P(t/\sigma - \lambda_1\sigma)e^{-\lambda_1 t}] \\ P(x) &= \int_{-\infty}^{x} Z(\alpha)d\alpha \end{aligned} \tag{4}$$

The ratio T_1/S_1 increases rapidly with the product $\lambda_1\sigma$ and thus is maximal for a long pulse. The delayed emission ($t \gg \tau_1$) is monoexponential according to this simple scheme:

$$S_1(t) \simeq S_d e^{-\lambda_2 t} \qquad T_1(t) \simeq T_d e^{-\lambda_2 t} \tag{5}$$

In addition, for the delayed emission

$$\begin{aligned} \text{fluorescence/phosphorescence} &\propto \Phi_f k_{tr}/k_p \\ \text{time-integrated fluorescence} &\propto \Phi_f \Phi'_{tr}\Phi_{isc}/(1 - \Phi'_{tr}\Phi_{isc}) \\ \text{time-integrated phosphorescence} &\propto \Phi'_p\Phi_{isc}/(1 - \Phi'_{tr}\Phi_{isc}) \end{aligned} \tag{6}$$

The following definitions hold in the above expressions:

$$\begin{aligned} &S_d = k_{tr}/k_{isc}\cdot\Phi_{isc}^2 \qquad T_d = \Phi_{isc} \\ &\Phi_f = k_f/(k_f + k_{nr}^S + k_{isc}) = k_f\tau_1 \\ &\Phi_{isc} = k_{isc}\tau_1 \\ &\Phi'_{tr} = k_{tr}/(k_d{}^T + k_{tr}) \\ &\Phi'_p = k_p/(k_d{}^T + k_{tr}) \end{aligned} \tag{7}$$

The primed quantum yields refer to segments of the deactivation pathways and not, as is more conventional, the overall conversions originating with the absorption process.

From the above it can be appreciated that the time constants, amplitudes, and yields corresponding even to the simplest of decay schemes reflect the combined influence of many intrinsic and extrinsic factors. For example, the heavy atom effects are usually manifested by an increase in k_{isc} (and thus $\downarrow\tau_1$, $\downarrow\Phi_f$, $\uparrow\Phi_{isc}$) and an increase in k_p (and thus generally $\downarrow\tau_2$ and $\uparrow\Phi'_p$, unless k_{nr}^T is enhanced as well[12,14]) (see also TABLE 2).

Actual systems demonstrate complicated behavior due to the superposition of decay processes corresponding to various chemical species possibly in dynamic equilibrium during the excited state lifetime. It is important to emphasize that while the prompt decay is frequently dominated by the radiative pathway, the rate of deactivation of T_1 is generally determined by the nonradiative processes (k_{nr}^T).

Rotational Diffusion and Emission Anisotropy

The photoselection principles operative upon excitation with linearly polarized light lead to the generation of a spatially anisotropic distribution of excited molecules. In the case of a macroscopically isotropic ensemble in the ground state (e.g. homogeneous solutions, suspensions of cells, vesicles, membrane fragments), the linear polarization of luminescence (fluorescence, phosphorescence) is uniquely defined in terms of the emission components polarized parallel ($I_{\|}$) and perpendicular ($I_{\perp}$) to the excitation electric vector. The emission anisotropy (r) is given as a dimensionless function of these components which in the case of time-resolved measurements are determined continuously following the initial excitation.

$$r(t) = [I_{\|}(t) - I_{\perp}(t)]/[I_{\|}(t) + 2I_{\perp}(t)] \quad \textbf{(8)}$$

The denominator can be shown to be a measure of the total polarization-independent emission intensity and is designated $S(t)$ (see, for example, ref. 2). We now generalize to a system of emitting species each characterized by an excited state lifetime τ_i (equations 3,5).

$$S_i(t) = S_{o,i}e^{-t/\tau_i} = [I_{\|}(t) + 2I_{\perp}(t)]_i \quad \textbf{(9)}$$

$$r_i(t) = \sum_j \alpha_{ij}e^{-t/\phi_{ij}} = [I_{\|}(t) - I_{\perp}(t)]_i/S_i(t) \quad \textbf{(10)}$$

$$\langle r(t)\rangle = \sum_i S_i(t)\cdot r_i(t)/\sum_i S_i(t) \quad \textbf{(11)}$$

$$= \sum_i [S_{o,i}e^{-t/\tau_i} \sum_j \alpha_{ij}e^{-t/\phi_{ij}}]/\sum_i S_{o,i}e^{-t/\tau_i} \quad \textbf{(12)}$$

Equation 10 expresses the fact that the anisotropy function for a rigid molecule undergoing rotational relaxation in an isotropic medium is generally representable as a sum of exponential decays. [15] Thus, an irregular body, an ellipsoid of revolution, and a sphere require 5, 3, and 1 component(s), respectively. The time constants ϕ_{ij}, denoted rotational correlation times, are given by different linear combinations of the rotational diffusion coefficients D_i for motions about the various molecular axes. The coefficients α_{ij} are related to the angles made by the absorption and emission transition moments to each other and to the rotation (symmetry) axis. In the absence of Brownian motion, the fundamental anisotropy $r_{o,i}$ obtains which is a function only of the photoselection regime and the angle δ between the absorption and emission transition moments (FIG. 2).

$$r_{o,i} = r_i(0) = \sum_j \alpha_{ij} = \frac{2}{5} P_2 (\cos\delta_i) \qquad (\text{for } i = 1,\ r_{o,i} = r_o) \quad \textbf{(13)}$$

where $P_2(x) = (3x^2 - 1)/2$ is the second Legendre polynomial.[4,16] Experimentally determined *initial* anisotropies r_{in} (= $r(0)$) often differ from values derived from equation 13 due to rapid unresolved motions and/or to other experimental and spectroscopic factors.[17] Equations 11 and 12 are expressions of the addition rule for emission anisotropy and arise directly by superposition from equations 9 and 10.

A more complex temporal behavior for the anisotropy function holds in the event that (a) the molecules possess internal flexibility, i.e. are elastically deformable during the course of the decay,[18] and (b) the surrounding medium is anisotropic, i.e. as in the case of a probe or macromolecule embedded in a membrane. In the latter case, two models have been used extensively: (1) uniaxial rotation of a rigid body about

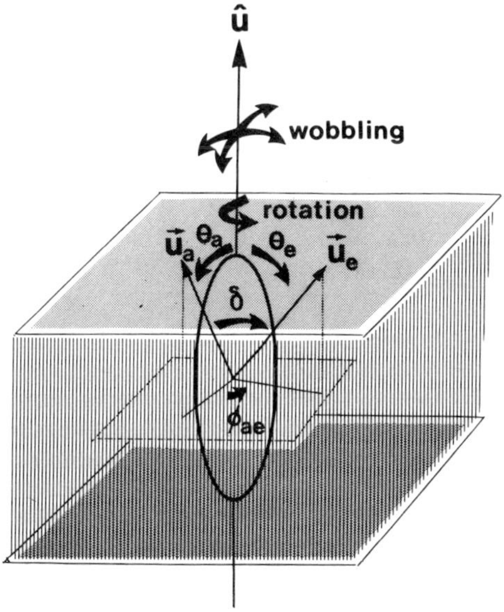

FIGURE 2. Angular parameters defining the rotational motions of a rigid particle embedded in a membrane. The symmetry axis is represented by the vector $\hat{\mu}$ and the absorption and emission transition moments by $\vec{\mu}_a$ and $\vec{\mu}_e$, respectively. These vectors are at angles θ_a and θ_e to the symmetry axis and δ to each other. $\hat{\mu}$ may assume an angle θ to the direction normal to the plane of the membrane (shown is $\theta = 0$). The molecular motions are defined in terms of rotation about and libration (wobbling) of $\hat{\mu}$. The azimuthal projections of $\vec{\mu}_a$ and $\vec{\mu}_e$ define the difference angle ϕ_{ae}.

its symmetry axis, generally assumed to lie normal to the plane of the membrane[3,15]; and (2) free diffusion (wobbling) within a cone about the symmetry axis with semiangle θ_0.[16,19–21] Common to both approaches is the consideration of the constant limiting anisotropy r_∞ at long times ($t \rightarrow \infty$) as a consequence of the hindered rotation. r_∞ has been discussed extensively in relation to the orientational order parameter S employed primarily in magnetic resonance studies.[4,16,21–23] A model-independent general expression for r_∞ has been derived recently by Lipari and Szabo.[16]

$$r_\infty = \frac{2}{5} P_2(\cos\theta_a) P_2(\cos\theta_e) S^2 \qquad \textbf{(14)}$$

where θ_a and θ_e are the angles between the absorption and emission transition moments, respectively, and the rotation (symmetry) axis (FIG. 2), and

$$S = \langle P_2(\cos\theta) \rangle \qquad \textbf{(15)}$$

θ is the instantaneous value of the angle made by the molecular symmetry axis to the membrane normal. $S = 0$, 1, and 0–1 corresponds, respectively, to isotropic, uniaxial (model 1 above), and wobbling (model 2 above) motion.

In general, the rotational and librational (wobbling) excursions of the molecular symmetry axis will be superimposed. Lipari and Szabo (Equation 30 in ref. 16) have provided an approximate expression for the time-dependent anisotropy under these conditions. (Only a single species, $i = 1$, is considered. The specific notation is ours.)

$$r(t) \simeq \sum_{j=1}^{3} \alpha_j e^{-\beta_j{}^r t/\phi_r{}'} [S^2 + (1 - S^2) e^{-\beta_j{}^w t/\phi_w{}'}] \qquad \textbf{(16)}$$

where the subscripts and superscripts r and w refer to the contributions from rotation and wobbling, respectively, and

j	α_j	β_j^r	β_j^w
1	$\frac{3}{10}\sin^2\theta_a\sin^2\theta_e\cos 2\phi_{ae}$	4	2
2	$\frac{3}{10}\sin 2\theta_a\sin 2\theta_e\cos\phi_{ae}$	1	5
3	$\frac{2}{5}P_2(\cos\theta_a)P_2(\cos\theta_e)$	0	6

(17)

Numerous additional observations can be made regarding equation 16 and the terms in 17:

(1) The rotational correlation time ϕ_r' applies to rotation about the symmetry axis and is related directly to the Stokes-Einstein diffusion coefficient from hydrodynamic theory using the "stick" boundary condition. (For models with "slip" see ref. 39.)

$$[\phi_r']^{-1} \equiv D_{\parallel} = kT/\xi_{\parallel} \quad \xi_{\parallel} = 6V\eta \text{ (sphere)}, 4V\eta \text{ (cylinder)} \qquad \mathbf{(18)}$$

Thus, the friction coefficient[24] $\xi_{\parallel}$ is proportional to the molecular volume V and medium viscosity η. (The more complex relationships for ellipsoids are given in refs. 25, 26.) In the case of cylindrical particles embedded in lipid bilayer membranes, the appropriate volume element is defined by the membrane thickness and cross-sectional area of the molecule.[27] In Equation 18, k and T are Boltzmann's constant and absolute temperature. A major problem is the specification of η (see below).

(2) The wobbling correlation time ϕ_w' is inversely related to an effective wobbling diffusion constant D_w but the constant of proportionality is a function of the cone angle θ_0 which in turn can be obtained from r_∞/r_0.[16,20] The various relationships have been determined only for certain orientations of the absorption or emission transition moments to the symmetry axis and are to be used with caution. This applies as well to the monoexponential phenomenological approximation (which results formally from equation 16 when either θ_a or θ_e = 0, thereby canceling the rotational contribution to depolarization).[16,19–23]

$$r(t) \simeq (r_o - r_\infty)e^{-6t/\phi_w'} + r_\infty \text{ and } r_\infty/r_o = S^2 \qquad \mathbf{(19)}$$

(3) The decay of anisotropy is characterized by 5 rotational correlation times and a limiting anisotropy ($\alpha_3 S^2$) given previously (Equation 14). the number of terms decreases for either θ_a or $\theta_e = 0$ ($\alpha_1 = \alpha_2 = 0$), $S = 1$ (absolute order, uniaxial rotation), $S = 0$ (unhindered rotation with $[\phi_w']^{-1} = D_{\perp}$; see ref. 15). The coefficients α_j (equation 17) may be negative for values of the azimuthal angle ϕ_{ae} (FIG. 2) greater than 45° (α_1) or 90° (α_2) or for either of the polar angles θ_a, θ_e greater than 54.7° (α_3). In the case of a macromolecule labeled randomly with a probe lacking a preferred orientation to the symmetry axis

$$\alpha_1 = \alpha_2 = \frac{2}{5}r_o, \alpha_3 = r_o/5, \text{ and } r_\infty/r_o = S^2/5 \qquad \mathbf{(20)}$$

(4) The attachment of the molecule in question to a mobile spherical carrier with rotational correlation time ϕ_M modifies equation 16 by the multiplicative factor e^{-t/ϕ_M}.[15,16]

(5) Molecular heterogeneity can produce effects mimicking the properties of equation 16. For example, the decay of two immobile species having different anisotropies $r_{o,i}$ and lifetimes τ_i will lead to an ascending or descending $r(t)$ function (equation 12). Other complications arise if diffusion within the postulated cone of wobbling is inhomogeneous[16,23] or if successive rotational depolarizations occur in widely separated time domains. In this case, a chain rule applies such that the limiting anisotropy perceived at shorter times defines the initial anisotropy for the next slower process. Order parameters (S) need to be interpreted accordingly.[22] Finally, measurements on oriented anisotropic structures require the determination of many more (>2) polarization components in order to define the system.[4]

Delayed Luminescence Spectrometer

The instrument used for time-resolved measurements of polarized phosphorescence and delayed fluorescence is depicted in Figure 3. The basic features have been described previously.[5] Recent improvements include the use of a commercial tunable dye laser source and versatile data processing for immediate display.[28] The dye laser (Lambdaphysik FL2000 pumped by a M1000 N_2 laser) produces 3–5 nsec wide vertically polarized pulses with an energy of 0.1–0.2 mjoule at 515 nm using Coumarin 307. Cuvettes (pathlength 5, 7, and 10 mm) in a thermostated holder are purged of 0_2 after liquid sample introduction by a combination of argon flow and gentle mechanical agitation. The emission at 90° is collected through a filter combination (generally Schott KV550 + RG645 or RG695 in the case of the halogenated fluorescein family of compounds) or directed into a monochromator equipped with a stepping motor under microprocessor control for obtaining emission spectra. For determinations of emission anisotropy, a rotating Polaroid polarizer controlled by the microprocessor selects between the $I_{\parallel}$ and $I_{\perp}$ components. The latter are digitized in a Biomation 8100 transient recorder after amplification and display in a Tektronix 7603 oscilloscope with either a 7A22 (adjustable bandwidth ≤ 1 MHz) or 7A15A (bandwidth 10 or 75 MHz) vertical plug-in. The EMI 9817 QGB photo-

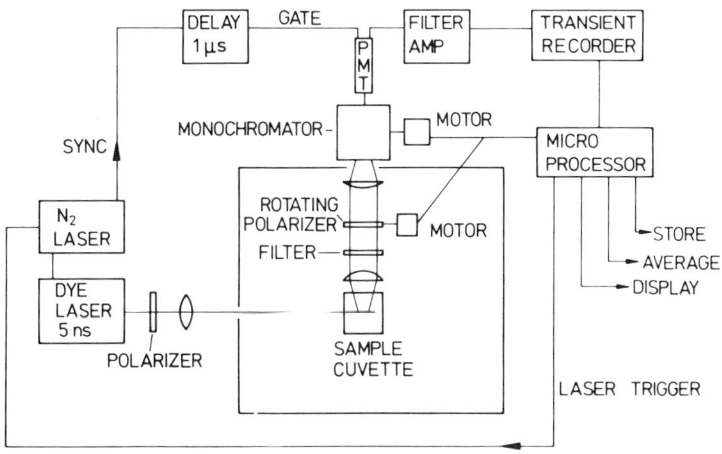

Figure 3. Delayed luminescence spectrometer for time-resolved measurements.

multiplier is generally gated off for 0.2-1 μsec after the laser flash so as suppress the intense prompt emission.

Data acquisition, averaging, and display are under microprocessor control. In general, m records each consisting of n successive determinations of $I_{\parallel}$ and $I_{\perp}$ are superposed. (Typically m = 64 and n = 32.) The generation of blank records for subtraction from the averaged $I_{\parallel}$ and $I_{\perp}$ traces is done in several ways:

(1) A pretrigger segment is recorded and evaluated for each component so as to calculate a straight line extrapolation.

(2) A measurement is made with either the laser source or photomultiplier obstructed.

(3) A measurement is made with a sample lacking the chromophore.

(4) A measurement is made with the original sample supplemented with a triplet quencher (atmospheric O_2 or cyclo-octatetraene, COT[29]). This procedure is demonstrated below.

The acquired data are displayed immediately in various formats according to microprocessor programs implementing hardware floating point arithmetic. Thus $I_{\parallel}$, $I_{\perp}$, S, and r in linear or logarithmic form can be viewed as a function of time. The data are plotted and stored on casette for subsequent analysis in a minicomputer using nonlinear multiexponential regression algorithms. In the case of the emission spectra, the acquisition program integrates the accumulated n traces at a given wavelength and stores that quantity before incrementing the wavelength (generally by about 1 nm).

The delayed emission spectra of two triplet probes used in our studies, eosin and di-iodofluorescein are shown in FIGURE 4. It is seen that the introduction of 2 iodine or 4 bromine atoms into the fluorescein nucleus leads to similar emission spectra characterized by a delayed thermal fluorescence at 540–550 nm (also the position of the prompt fluorescence) and phosphorescence at about 690 nm. (The spectra have not been corrected for wavelength-dependent instrumental sensitivity.) Similar data have been reported for eosin and erythrosin (tetra-idofluorescein) by Garland and Moore.[30]

Time-resolved data for free eosin and erythrosin probes in aqueous solution are shown in FIGURE 5. The anisotropy functions are centered about 0, indicating rapid Brownian depolarization (expected to occur in <1 nsec). The effects of photomultiplier gating and COT blank subtraction are seen in an experiment with erythrosin with higher time resolution (FIG. 6). The spike of intense prompt fluorescence is abolished by the gating pulse (about 0.5 μsec) and only a small residual tail followed by the phosphorescence emission persists after restoring the conductance of the photomultiplier. (Note that most of the prompt fluorescence has been suppressed optically by use of the filter combination specified above.) The high anisotropy of the prompt fluorescence is seen in the lower part of panel A. This results from a combination of a very short fluorescence lifetime (TABLE 2), a somewhat longer rotational correlation time, and the relatively parallel orientation of the absorption and fluorescence transition moments. What is measured is the *steady-state* fluorescence anisotropy $\bar{r}$ $(r_o/\bar{r} = 1 + \tau/\phi \simeq 1)$ since the decay processes (Equations 3 & 4) are faster than the instrumental response. The blank subtraction procedure employing COT (panel B) produces a record free of fluorescence contribution as expected if the triplet quencher does not affect the deactivation of the singlet state (see also glycophorin data below, FIG. 10).

The absolute sensitivity of the apparatus shown in FIGURE 3 is determined by the

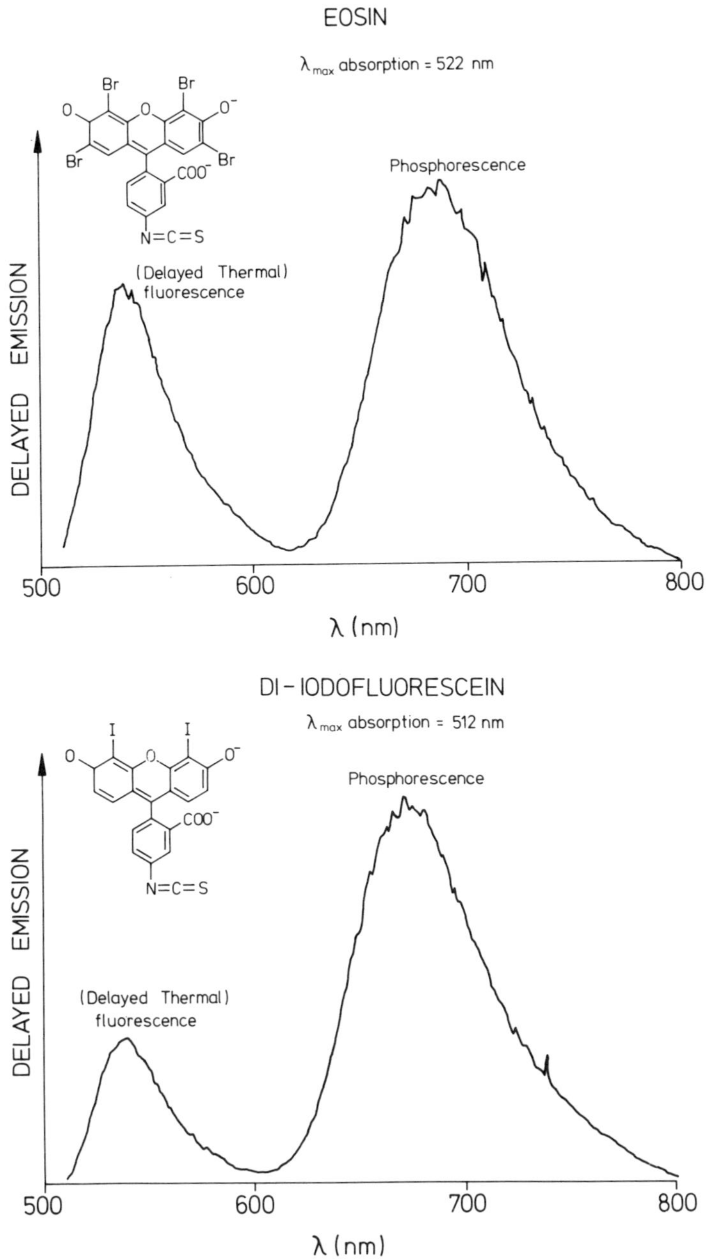

FIGURE 4. Delayed luminescence spectra of eosin and di-iodofluorescein probes. Solutions were 5 μM of the indicated compounds in 50 mM KPO_4 buffer, pH 7.4. Laser excitation was at 505 nm. Each point represents the sum of 32 integrated decay curves measured for 5 and 0.5 msec in the case of eosin and di-iodofluorescein, respectively. The monochromator was set at a bandpass of 6 nm and incremented 1.3 nm per step. The temperature was 20°. Scans of prompt fluorescence obtained by removing the photomultiplier gating and reducing sensitivity overlapped the region ascribed to delayed fluorescence.

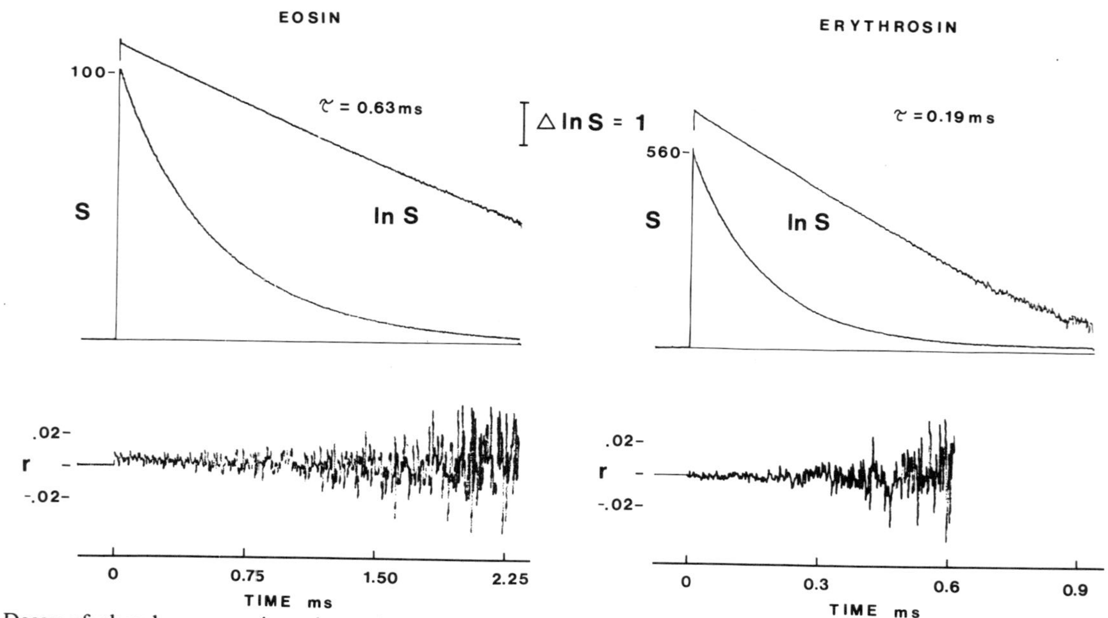

FIGURE 5. Decay of phosphorescence intensity and anisotropy for free eosin and erythrosin probes. Solutions were 1 μM of the respective isothiocyanates in phosphate buffer. 1024 curves were averaged with a resolution of 2 and 5 μsec per point in the case of eosin and erythrosin, respectively.

source energy, optical collection efficiency, photomultiplier response, and sources of systematic noise. However, with averaging usable data can be acquired for erythrosin at 1 nm concentration (below which background emission from solvents becomes appreciable). Anisotropy nulls are routinely below 0.01 (FIGS. 5, 6). It is considered possible to extend the temporal response considerably below the 1 μsec range by faster photomultiplier gating procedures, a matter under current development. In the absorption (linear dichroism) mode, measurements have been obtained at the full 100 MHz rate of the transient recorder.[5]

A summary of currently available spectroscopic decay constants for eosin and erythrosin is given in TABLE 2. These data were not obtained under uniform conditions and need to be redetermined in a systematic manner. Erythrosin is characterized by a high triplet yield and a correspondingly weak fluorescence, features which make it the preferred probe when low concentrations and high temporal resolution are required. Its main disadvantage is the short triplet lifetime relative to eosin.

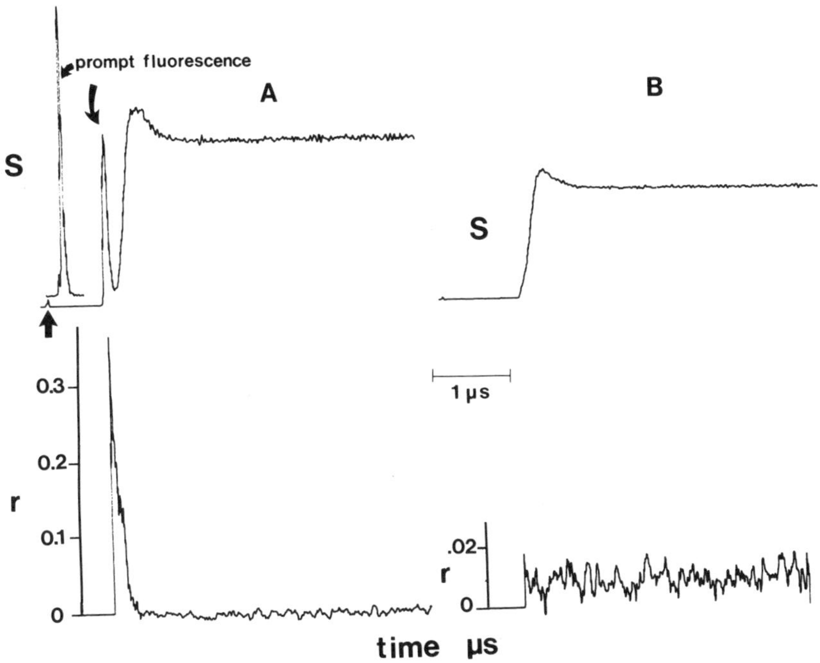

FIGURE 6. Erythrosin decay at high temporal resolution and effect of cyclo-octatetraene COT quenching. A 1 μM solution of erythrosin isothiocyanate was measured with a resolution of 0.01 μsec per point and a 10 MHz electronic bandwidth. (A) data obtained with blank subtraction by interruption of laser beam. The spike of prompt fluorescence shortly following the laser trigger (arrow) coincides with the laser flash and was obtained in a separate experiment with the photomultiplier gating pulse (about 0.5 μsec) turned off. (B) Same as A but the blank subtraction was performed with the original sample to which cyclo-octatetraene (BASF) had been added to about 5 mM. The slight decrease in the phosphorescence level is attributed to incomplete triplet quenching in the sample containing COT.

TABLE 2
SOME PARAMETERS OF FLUORESCEIN, EOSIN, AND ERYTHROSIN EMISSION

	τ_f nsec	ϕ_f	ϕ_{isc}	τ_p msec	ϕ_p	r_o
Fluorescein: free	4	0.92	~0	–		
Eosin: free	0.9	0.19	0.7	1	0.01	
protein-bound	>1.6	0.1	0.1	0.6–2	$\sim 10^{-3}$	0.16
in plastic*			0.98	3.6		0.30
Erythrosin: free	0.1	0.02		0.5		
protein-bound				0.1–0.4	10^{-3}–10^{-2}	0.15
in plastic*				0.65		0.25

The above data are from references 5, 30, 31 and further citations therein. They are typical for fluid aqueous solutions but were not obtained under identical conditions. Neither do they reflect observed heterogeneity in emission kinetics. In particular, values for protein-associated or otherwise conjugated probes may vary considerably from the above.

* Solid solutions in poly(methy methacrylate).[30]

ROTATIONAL MOTIONS OF MEMBRANE-ASSOCIATED PROTEINS

The following systems under investigation in our laboratory illustrate the potential of rotational diffusion measurements using phosphophorescence probes.

Band 3, the Human Erythrocyte Anion Transport Protein

Band 3 comprises about 25% of the human erythrocyte membrane protein and is responsible for the essential function of anion transport (for reviews see 32, 33). It is a glycoprotein of molecular weight about 10^5 which spans the membrane, probably more than once. The rotational behavior of this protein in ghosts derived from cells labeled specifically with eosin probes has been studied extensively by Cherry and associates using flash-induced transient dichroism (see refs. 3, 34 and references therein). Valuable information about the putative associations of Band 3 with itself and other membrane and cytoskeletal components as well as the influence of temperature and membrane lipid manipulations has been obtained. Molecular rotation studies are of particular interest in this case because of the unusually high turnover number ($>10^4$), large activation energies, and evidence for conformational changes associated with the gating function.

The rotational mobility of Band 3 can also be determined by time-resolved phosphorescence anisotropy.[5] The initial measurements performed with samples labeled with eosin isothiocyanate have been extended by use of other probes. FIGURE 7 shows the decay of phosphorescence intensity $S(t)$ and anisotropy $r(t)$ corresponding to Band 3 in ghosts made from erythrocytes labeled with eosin-5′-maleimide (prepared as in ref. 35). At 37°, a multicomponent decay was observed ($\tau = 2.2$, 0.24 msec), a typical finding with as yet unresolved implications regarding possible binding or environmental heterogeneity or the existence of excited state equilibria. The rotational motion as revealed by $r(t)$ is seen to be strongly temperature dependent and highly anisotropic. The rotational correlation times at 37° obtained by a two-component analysis were 1.3 ms and (lower amplitude) 100 μsec. Similar data with identical preparations were obtained by Nigg and Cherry[35] using linear dichroism and interpreted by them in terms of a uniaxial rotation ($S = 1$) and heterogeneity in the state of aggregation or association with other membrane proteins. However, the

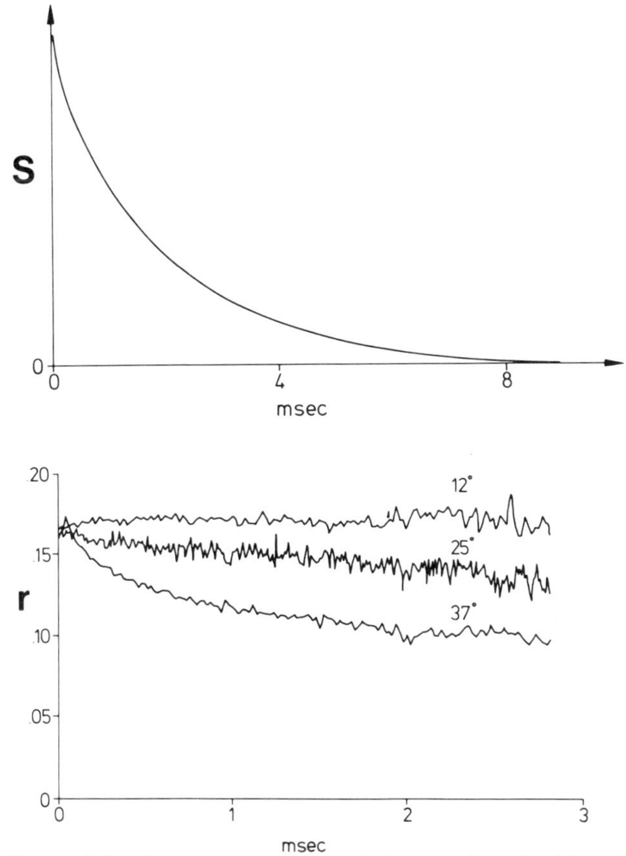

FIGURE 7. Decay of phosphorescence intensity and anisotropy from Band 3 in ghosts labeled with eosin-5′-maleimide. The ghosts were suspended in 5 mM $NaPO_4$, pH 7.4 at a protein concentration of 0.2 mg/ml (about 0.5 μM eosin). Data were obtained at 10 μsec per point at the indicated temperatures and 1024 curves were averaged. In other records (not shown) a limiting anisotropy was measured at 37° extending out to 8 msec. The analysis of the 37° data in this figure yielded two components for both $S(t)$ and $r(t)$: $\tau_1 = 2.2$ msec (87%), $\tau_2 = 0.24$ msec (14%), $\phi_1 = 1.3$ msec (70%), $\phi_2 = 0.10$ msec (30%), $r_\infty = 0.1$.

formalism presented earlier indicates that other molecular models need to be considered. For example, the temperature-dependent decrease in r_∞ seen in FIGURE 7 could result from a change in the distribution of subspecies (Equation 11) and/or in the orientation of the probe to the rotation axis or the hindrance to wobbling excursions (Equation 14). The tantalizing possibility that the shorter correlation times observed in the phosphorescence measurements are associated with *segmental* molecular motions during gating is being explored in our laboratory with other probes which do not appear to abolish Band 3 activity (E. Matayoshi and T. Jovin, unpublished observations). It remains to note that the longer rotational correlation times exceed the predictions of Equation 18 (and ref. 27) for the uniaxial rotation of a Band 3 monomer by at least one order of magnitude, even considering the ambiguities concerned with estimates of membrane microviscosity.[19–23,38]

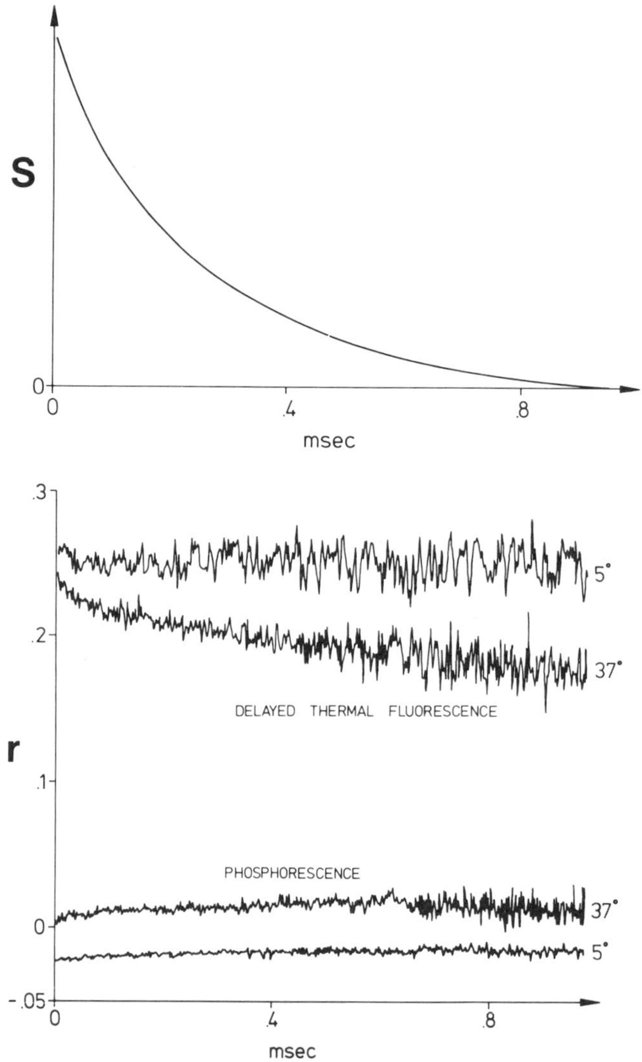

FIGURE 8. Decay of delayed luminescence (phosphorescence, delayed fluorescence) from Band 3 in ghosts labeled with di-iodofluorescein-5′-isothiocyanate. The preparation was made in accordance with the procedures specified for the eosin compound.[35] Suspensions in phosphate buffer at about 50 nM probe concentration were measured at the indicated temperature. Laser excitation was at 512 nm. The phosphorescence was isolated with the filter combination Schott KV550 + RG645, the delayed fluorescence with the Schott bandpass filter SFK 21.

The behavior of di-iodofluorescein (FIG. 4) conjugated to Band 3 provides interesting contrasts to that of eosin. The isothiocyanate of this dye was reacted with erythrocytes and the ghosts examined by both phosphorescence and delayed fluorescence (FIG. 8). The decay of intensity in both cases was the same, in accordance with equation 5, and yielded τ values of 0.3 msec (84%) and 40 μsec (16%). However, the time-resolved anisotropies were quite different. The phosphorescence anisotropy was small in magnitude, negative at 5°, and close to 0 at 37°. Thus, in the case of this molecule containing 2 instead of 4 atoms of iodine as in erythrosin, the absorption and phosphorescence transition moments appear to be more orthogonal ($\delta > 54.7°$ according to equation 13). The kind, number, and position or halogen substitution of aromatic compounds have profound effects on the polarized phosphorescence emission spectra.[12] In contrast to the phosphorescence, the delayed fluorescence emission showed a large positive anisotropy ($r(0) \sim 0.26$) with a temperature-dependent decay similar to that observed with the eosin probe FIG. 7). The transition moments for prompt and delayed fluorescence are necessarily the same (FIG. 1) and for di-iodofluorescein (in common with the other halogenated fluoresceins) lie very nearly parallel to that for absorption (the theoretical maximum for r_0 from Equation 13 is + 0.4). Delayed fluorescence increased while phosphorescence decreased with temperature, an important and general advantage of the former in studies at physiological temperature or of structural transitions.

Glycophorin Reconstituted in Dimyristoyllecithin Vesicles

Glyocophorin is the major sialoglycoprotein of the human erythrocyte and next in abundance to Band 3, with which it may be associated.[34] It has a molecular weight of about 30000 and spans the membrane once with the carbohydrate moieties and N-terminus facing the external medium.[36] The hydrophobic intramembranous portion of the molecule accounts for only about 8% of the total mass.

The lateral[37] and rotational[28] diffusion behavior of this well-characterized integral membrane protein has been investigated after reconstitution into bilayers of dimyristoylphosphatidylcholine at a lipid/protein ratio of 4500. The objective in such studies has been the comparison of the translational and rotational diffusion properties of the same molecule under various chemical and thermodynamic conditions so as to test theoretical concepts of diffusion and lipid-protein interaction.

The rotational diffusion of reconstituted glycophorin measured by DLA (phosphorescence) is presented here in preliminary form and more extensively elsewhere.[28]

Glycophorin was isolated from human erythrocytes, purified, and labeled in aqueous solution with erythrosin-5′-isothiocyanate at a conjugation stoichiometry of about 1 probe per molecule. The rotational behavior of this preparation in 99% glycerol is shown in FIGURE 9. A spectrum of decay processes was observed extending from the 1 μsec to the msec range. The expected rotational correlation time from equation 18 (see also 5) in this highly viscous solvent was about 70 μsec. Thus, the faster relaxations provide evidence for segmental mobility at the random attachment sites of the probe while the finite r_∞ indicates the presence of larger aggregates (a frequent observation in concentrated glycerol).

Upon reconstitution into DMPC vesicles, the labeled glycophorin showed a quite different behavior (FIG. 10). Panel A depicts the system with the highest time resolution. The anisotropy decays rapidly to a limiting anisotropy (Panel B). The inset in Panel A demonstrates the effectiveness of the COT blank subtraction procedure for eliminating artifacts attributed to fluorescence. (However, it should be noted that this reagent under certain conditions promotes the precipitation of other suspended or soluble systems.)

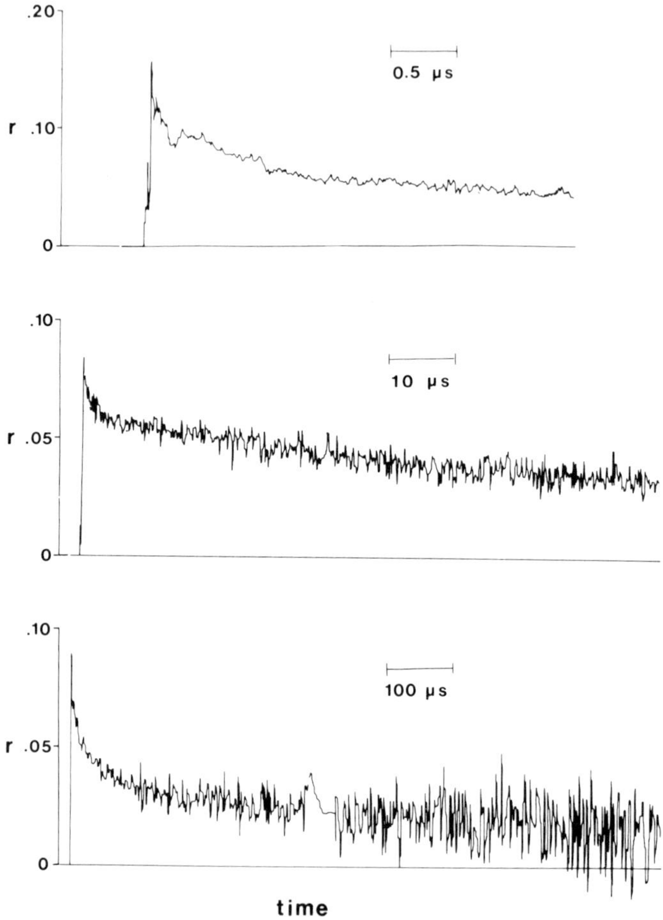

FIGURE 9. Decay of phosphorescence anisotropy of glycophorin labeled with erythrosin-5′-isothiocyanate. Glycophorin was extracted from human erythrocytes and labeled with erythrosin-5′-isothiocyanate to a stoichiometry of about 1.[28,37] The protein was measured in 99% glycerol at a concentration of 50 μg/ml. Records were taken at 7° using resolutions of 0.01, 0.2, and 2 μsec per point and a blank generated by interrupting the laser light. The nominal oscilloscope bandwidth was 75 MHz.

Analysis of data such as those of FIGURE 10 obtained at different temperatures and after various periods of storage following reconstitution has led to the following results[28]:

(a) Two rotational correlation times were seen, one in the range 1–2 μsec and a second of 10–20 μsec.

(b) The longer ϕ appeared only after prolonged storage (50 hr) and was not observed above the lipid phase transition temperature (24°) in either fresh or stored samples.

(c) r_{in} and r_∞ decreased systematically with temperature (FIG. 11).

(d) The lifetime τ at a given temperature was unaffected by changes in the distribution of rotational correlation times.

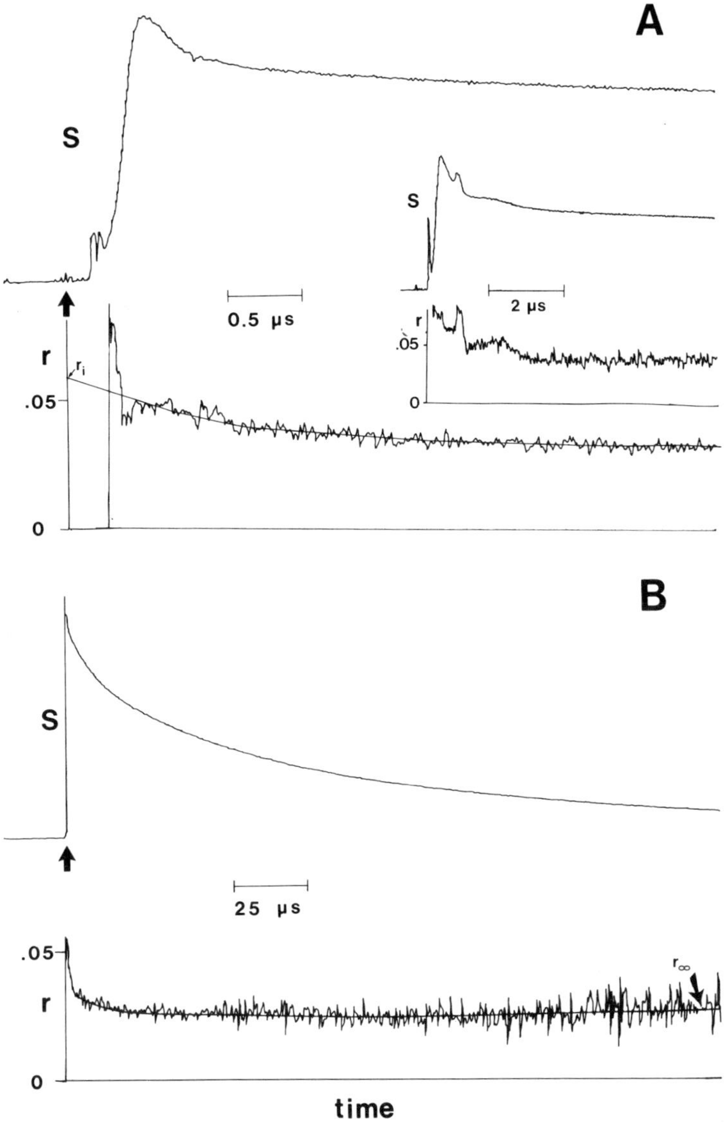

FIGURE 10. Decay of phosphorescence intensity and anisotropy from erythrosin-labeled glycophorin reconstituted in DMPC vesicles. The reconstitution was carried out by mixing the protein and lipid, evaporating the solvent, then rehydrating at 40°.[28] The larger vesicles isolated by low speed centrifugation were measured by DLA at a protein concentration of 10 µg/ml. The data shown are for 22° and a time resolution of 0.02 µsec per point (A) and 1 µsec per point (B). A COT blank was used to suppress fluorescence artifacts in the record. The inset in A shows data obtained without this subtraction procedure.

(e) The observed decays in anisotropy were insensitive to manipulations of solvent viscosity, e.g. addition of glycerol.

These findings are consistent with a model attributing the faster rotational correlation time to monomeric glycophorin and the longer component to an aggregate

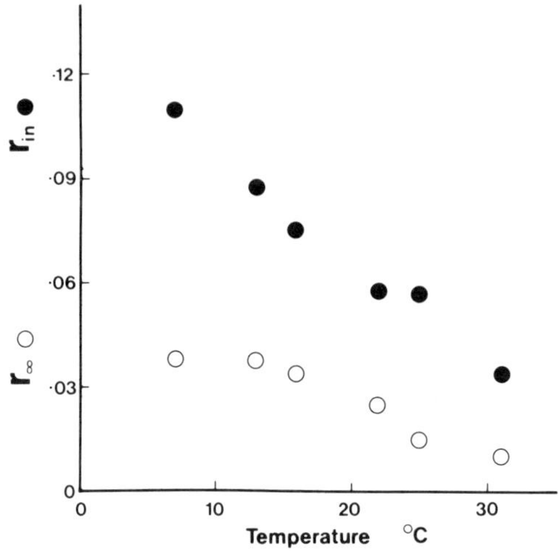

FIGURE 11. Temperature dependence of r_{in} and r_{∞} for the phosphorescence of reconstituted erythrosin-labeled glycophorin. The experiments were performed as in FIGURE 10.

formed within the bilayer membrane but only when the bulk lipid is in the gel phase. Such an aggregation would be consistent with the units of 6–7 monomers thought to exist in the natural membrane environment.[36] Assuming a cylindrical uniaxial rotation model and an effective lipid microviscosity of about 1 poise,[21] $D_{\parallel}$ from Equation 18 would be on the order of 0.1–1 μsec for monomeric glycophorin. However, the systematic decrease in r_{in} with temperature as well as the data of FIGURE 10 suggest that the rigid body assumption may have limited validity in this case. The lateral diffusion measurements on the same system show a highly cooperative transition from a restricted to a relatively free lateral mobility of reconstituted glycophorin at a temperature about 9° below the main lipid phase transition.[37] Thus, it would appear that the protein reduces the order of the lipid in its immediate environment. If this effect is progressive in the region of 15–24°, then conceivably the degree of rotational depolarization due to fast segmental motions of the protein might increase correspondingly, thereby accounting for the drop in r_{in}. If the relatively slow motions resolved by phosphorescence anisotropy reflect global rotational and/or wobbling excursions of the protein, the drop in r_{∞} might be expected to correlate closely with the order of the bulk lipid (e.g. according to equation 20). In fact, a transition is seen in the region of 20–25°. The fact that rotation but not lateral movement is detected below 15° suggests that the shell of boundary lipid is sufficiently large so as to permit a net displacement exceeding a molecular (protein) diameter only above that temperature.

Current efforts are directed at determining the effects of protein concentration and lipid manipulations on the various diffusion processes.

CONCLUDING REMARKS

The theoretical and experimental basis for rotational diffusional measurements employing the emission of triplet probes has been presented. The technique seems particularly well suited for membranes in which rotational dynamics are characteristically slow (μsec-msec). It appears possible to distinguish between disperse and aggregated proteins in the membrane, a result which has prompted related studies of aggregation phenomena on cell surfaces responsible for hormone regulation, synaptic transmission, and antibody-mediated exocytosis (to be reported elsewhere). Adequate models for such systems will have to consider that molecules may appear rigid in one time domain but flexible in another. The combined data from absorption, prompt fluorescence, and delayed luminescence as well as magnetic resonance measurements will be required in order to reduce ambiguity.

Triplet lifetimes in themselves can supply useful information independent of the anisotropy. They have the advantage of great sensitivity to environment (Equation 3) and can be determined with high accuracy. Thus, the susceptibility to diffusible triplet quenchers such as O_2 can reveal the degree of exposure of specific sites. Ligand binding can also have dramatic effects upon the observed lifetimes, as exemplified in studies of acridine binding to DNA (R. Austin, T. Jovin, unpublished results).

ACKNOWLEDGMENTS

We are indebted to Dr. Donna Jovin for di-iodofluorescein-5′-isothiocyanate and to Dr. Peter Garland for erythrosin-5′-isothiocyanate. Drs. Erich Nigg and Richard Cherry kindly supplied samples of human erythrocyte ghosts labeled with eosin-5′-maleimide. We thank colleagues for critical reading of the manuscript.

REFERENCES

1. WEBER, G. 1953. Adv. Prot. Chem. **8**: 415–459.
2. BADEA, M. G. & L. BRAND. 1979. Methods Enzymol. **61**: 379–425.
3. CHERRY, R. J. 1979. Biochim. Biophys. Acta **559**: 289–327.
4. JOHANSSON, L. B. -A. & G. LINDBLOM. 1980. Q. Rev. Biophys. **13**: 63–118.
5. AUSTIN, R. H., S. S. CHAN & T. M. JOVIN. 1979. Proc. Natl. Acad. Sci. USA **76**: 5650–5654.
6. MOORE, C., D. BOXER & P. GARLAND. 1979. FEBS Lett. **108**: 161–166.
7. GREINERT, R., H. STAERK, A. STIER & A. WELLER. 1979. J. Biochem. Biophys. Meth. **1**: 77–83.
8. SCHLESSINGER, J. 1980. Trends Biochem. Sci. **5**: 210–214.
9. METZGER, H. 1978. Immunol. Rev. **41**: 186–199.
10. PARKER, R. T., R. S. FREEDLANDER & R. B. DUNLAP. 1980. Anal. Chim. Acta **119**: 189–205.
11. MCGLYNN, S. P., T. AZUMI & M. KINOSHITA. 1969. Molecular Spectroscopy of the Triplet State. Prentice-Hall, Inc. Englewood Cliffs, NJ.
12. LOWER, S. K. & M. A. EL-SAYED. 1966. Chem. Rev. **66**: 199–241.
13. BIRKS, J. B. 1976. Photochem. Photobiol. **24**: 287–289.
14. TURRO, N. J., K. -C. LIU, M. -F. CHOW & P. LEE. 1978. Photochem. Photobiol. **27**: 523–529.
15. RIGLER, R. & M. EHRENBERG. 1973. Q. Rev. Biophys. **6**: 139–199.
16. LIPARI, G. & A. SZABO. 1980. Biophys. J. **30**: 489–506.
17. STEINBERG, I. 1975. *In* Biochemical Fluorescence: Concepts. R. F. Chen & H. Edelhoch, Eds. Vol. 1: 79–113. Marcel Dekker, Inc. New York.
18. BARKLEY, M. D. & B. H. ZIMM. 1979. J. Chem. Phys. **70**: 2991–3007.
19. KAWATO, S., K. KINOSITA, JR. & A. IKEGAMI. 1977. Biochemistry **16**: 2319–2324.
20. KINOSITA, K., JR., S. KAWATO & A. IKEGAMI. 1977. Biophys. J. **20**: 289–305.
21. LAKOWICZ, J. R., F. G. PRENDERGAST & D. HOGAN. 1979. Biochemistry **18**: 508–519.
22. JÄHNIG, F. 1979. Proc. Natl. Acad. Sci. USA **76**: 6361–6365.
23. HEYN, M. 1979. FEBS Lett. **108**: 359–364.
24. TELLER, D. C., E. SWANSON & C. DE HAËN. 1979. Methods Enzymol. **61**: 103–124.
25. WEGENER, W. A., V. J. KOESTER, & R. M. DOWBEN. 1979. Proc. Natl. Acad. Sci. USA **76**: 6356–6360.

26. WRIGHT, A. K. & J. E. BAXTER. 1976. Biophys. J. **16**: 931–938.
27. SAFFMAN, P. G. & M. DELBRÜCK. 1975. Proc. Natl. Acad. Sci. USA **72**: 3111–3113.
28. BARTHOLDI, M., W. L. C. VAZ & T. M. JOVIN. 1981. Submitted for publication.
29. PAQUETTE, L. A. 1975. Tetrahedron **31**: 2855–2883.
30. GARLAND, P. B. & C. H. MOORE. 1979. Biochem. J. **183**: 561–572.
31. PARKER, C. A. 1968. Photoluminescence of Solutions. Elsevier Publ. Co. Amsterdam, the Netherlands.
32. CABANTCHIK, Z. I., P. A. KNAUF & A. ROTHSTEIN. 1978. Biochim. Biophys. Acta **515**: 239–302.
33. LASSEN, U. V., H. H. USSING, & J. O. WIETH, EDS. 1980. Membrane Transport in Erythrocytes. Munksgaard. Copenhagen, Denmark.
34. NIGG, E. A., C. BRON, M. GIRARDET, & R. J. CHERRY. 1980. Biochemistry **19**: 1887–1893.
35. NIGG, E. A. & R. J. CHERRY. 1979. Biochemistry **18**: 3457–3465.
36. ROMANS, A. Y., D. O. PARRISH & J. P. SEGREST. 1978. *In* Modern Pharmacology-Toxicology. J. R. Smythies & R. J. Bradley, Eds. Vol. 11: 215–256. Marcel Dekker, Inc., New York.
37. VAZ, W. L. C., H. G. KAPITZA, J. STÜMPEL, E. SACKMANN & T. M. JOVIN. 1981. Biochemistry. **20:** 1392–1396.
38. SHINITZKY, M. & Y. BARENHOLZ. 1978. Biochim. Biophys. Acta **515**: 367–394.
39. BAUER, D. R., J. I. BRAUMAN & R. PECORA. 1974. J. Am. Chem. Soc. **96**: 6840–6843.

NANOSECOND FLUOROMETRY AND THE LUMINESCENCE OF BIOLOGICAL MACROMOLECULES*

Ludwig Brand, J. B. Alexander Ross, and William R. Laws

Department of Biology
The Johns Hopkins University
Baltimore, Maryland 21218

Luminescence studies have been exploited for many years in investigations of the structure and function of biological macromolecules. There are many examples where large changes in fluorescence intensity and/or emission energy are observed in association with conformational changes of biopolymers or interactions between large and small molecules.[1-3]

During the last decade there have been significant advances in the instrumentation required for nanosecond time-resolved measurements. Both phase and pulse methods have been used, and either simple or complex fluorescence decay data can be obtained and analyzed in terms of a variety of mechanistic fluorescence decay laws. In our laboratory we have made use of a pulse technique known as the method of single photon counting.[4,5]

In this report we will describe the fluorescence decay of tryptophan in a protein, liver alcohol dehydrogenase and in several peptides. The origin of complex fluorescence decay behavior of the type to be described here may lie in ground state heterogeneity or microheterogeneity or in one of several types of excited-state interactions. We will describe some of the excited-state mechanisms that can give rise to double or multi exponential decay laws.

The monophoton counting procedure for obtaining nanosecond fluorescence decay data is an interval timing procedure. It does not rely on the ability of the photodetector to respond to light changes on the nanosecond time scale in a linear and reproducible manner. Moreover, the monophoton counting method provides a direct measure of the statistics of the excited-state decay process. The nanosecond "clock" utilized by this method is the time to pulse height converter or TAC. This was developed several years ago as a tool to aid in the study of nuclear decay processes. A start pulse obtained by means of an antenna when the nanosecond lamp flashes starts the build up of a linear voltage ramp in the device. An electron pulse initiated by the arrival of a photon at the photocathode stops the charging of the capacitor. The device then puts out a pulse with a voltage height proportional to the time interval between the lamp flash and the arrival time of the single photon. An analog-to-digital converter is used to record the voltage height (proportional to time) and this information is stored in a digital processor or computer memory. The flash rate of typical lamps is 10–40 KHz and a histogram is built up that reflects the statistics of the decay process.

* *Contribution No. 1090 from the McCollum-Pratt Institute. Supported by National Institutes of Health Grant GM-11632.*

0077-8926/81/0366-0197 $01.75/0 ©1981, NYAS

The single photon counting procedure has been discussed in detail in numerous publications.[5-7] A key feature is that nonlinear characteristics of the electronics and detectors on the nanosecond time scale are overcome by the use of a time-interval measurement rather than a time-intensity measurement. Only the time to pulse height conversion must be done on the nanosecond time scale. The other electronic conversions are carried out on the microsecond time scale.

One of the disadvantages often cited in regard to the single photon counting technique is the long time required for data collection. As will be shown below, this is not necessarily a problem.

Clearly the most straightforward fluorescence decay to interpret is that due to a homogeneous ground state population of chromophores with no excited-state interactions taking place during the lifetime of the excited state. In this case a single exponential decay law is predicted.

$$I = I_0 e^{-t/\tau}$$

where τ is the lifetime of the excited state. Results for such a system are shown in FIGURES 1 and 2. FIGURE 1 shows the results of an analysis of the decay of binaphthyl dissolved in cyclohexane. The analysis in terms of a single exponential decay law are shown. The good fit between the experimental and theoretical decay curves, the residuals between them, the autocorrelation of the residuals and the ψ^2 value all indicate that a monoexponential decay law ($t = 2.6$ ns) is an adequate representation of the data. Similar results are obtained with terphenyl dissolved in cyclohexane in this case $\tau = 0.95$. Fluorophores of this type are useful as standards to evaluate systematic errors in the instrumentation.[8]

FIGURE 3 shows an analysis of a fluorescence decay experiment done with a mixture of terphenyl and binaphthyl in dilute solution where there is little or no interaction between the fluorophores. It is clear that good recovery of the decay parameters of this binary mixture can be obtained. FIGURE 4 shows the same experiment but with only a 30 sec counting time for the entire experiment. In spite of the fact that less than 200 counts were collected at the peak of the decay curve there was reasonable recovery of the correct decay times. The difficulty of long collection times in the monophoton counting technique is more apparent than real. In fact the degree of correlation between decay parameters is a more crucial factor in regard to recovery of lifetimes than the total number of counts collected.

The double exponential decay shown in FIGURES 3 and 4 has its origin in a mixture of two quite different chromophores. As will be seen below, biexponential decay observed with proteins or peptides may have its origin in more subtle microheterogeneity such as the ionization state of a charged residue near a fluorophore.

Complex decay behavior and nanosecond time-dependent spectral shifts may also reveal a variety of excited-state interactions taking place on the nanosecond time scale.[5] A two-state, excited-state reaction may lead to *double* exponential decay kinetics. Excited-state proton transfer is an example of a situation where an electronically excited molecule has different reactivity than its ground-state counterpart. The absorption of light raises the energy content of a molecule and in addition significantly alters its physical and chemical properties. The photoexcited species is in fact a new chemical molecule capable of undergoing a variety of chemical and physical interactions during its several nanosecond life span. The excited-state products may

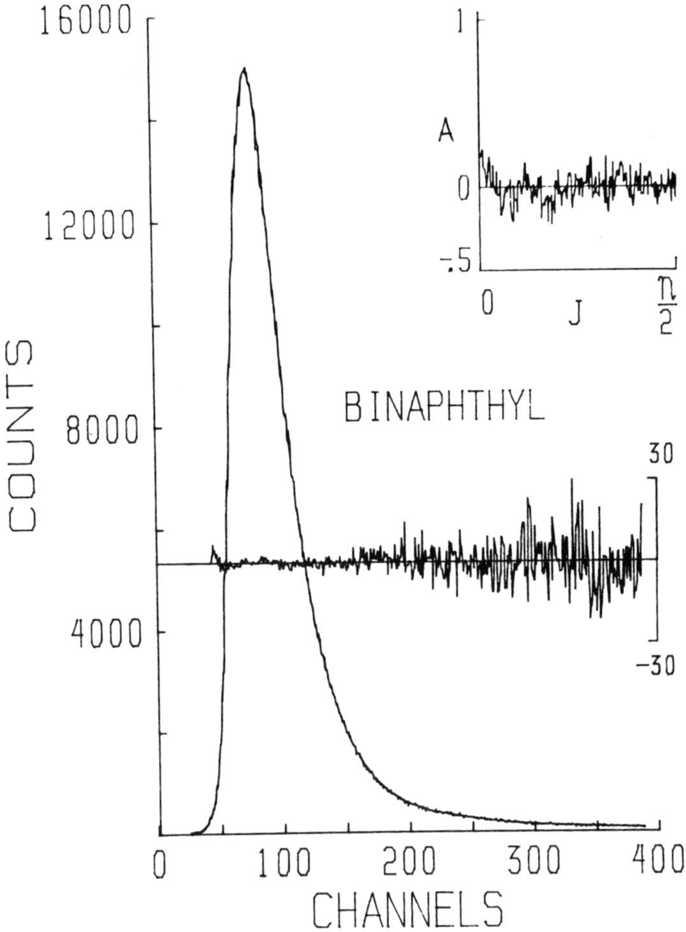

FIGURE 1. Fluorescence decay data of 5×10^{-6} M binaphthyl in cyclohexane at 20°C. Included are the percent residuals and their autocorrelation. The noise free line represents the theoretical parameters convolved with the lamp. The sample was excited with the lamp. The sample was excited with natural light at 296 nm (bandwidth of 10.6 nm) and the emission monitored at 360 nm (bandwidth of 10.6 nm). The time of collection for the decay curve was 60 minutes. The best fit to the data gave single exponential parameters of $\tau = 2.58$ ns, $\alpha = 1.00$ with $\chi^2 = 1.58$.

emit light and so the possibility exists of observing dual or even multiple fluorescence. Even if only one of the excited species emits light complex fluorescence decay kinetics may still be observed provided there is some reversibility in the reaction scheme.

A well studied example of a molecule showing excited proton transfer is represented by 2-naphthol.[9-11] The pK_a in the ground state is 9.5 while the excited state pK_a^* is 2.8. In aqueous solution the rate constants for proton transfer are on the same time scale as the rate constants for light emission. The absorption maximum of naphtholate is well toward long wavelength compared to the absorption spectrum of naphthol,

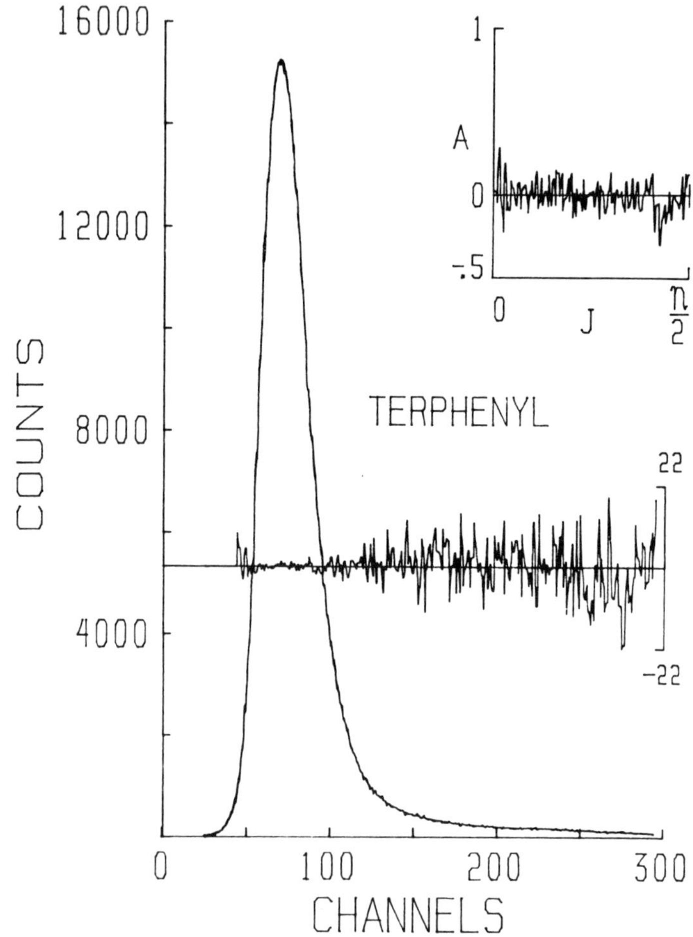

FIGURE 2. Fluorescence decay data of 5×10^{-6} M terphenyl in cyclohexane at 20°C. Included are the percent residuals and their autocorrelation. The noise free line represents the theoretical parameters convolved with the lamp. The sample was excited with natural light at 296 nm (10.6 nm bandwidth) and the emission monitored at 360 nm (10.6 nm bandwidth). The time of collection for the decay curve was 49 minutes. The best fit to the data gave single exponential parameters of $\tau = 0.95$ ns, $\alpha = 1.00$ with $\chi^2 = 1.73$.

and it is thus easy to determine which species exists in the ground state. Between pH 3 and pH 7 little naphtholate is found in the ground state while significant emission due to the ionized species is observed. It is emphasized that the thermodynamics of the system favors only the ionized species in the excited state. Owing to the short decay of naphthol emission the formation of naphtholate is limited by kinetics.

The fluorescence decay kinetics of the excited naphthol-naphtholate system can be understood by reference to FIGURE 5. If A refers to naphthol and B to naphtholate, the differential rate equations for the decay of naphthol and naphtholate are given

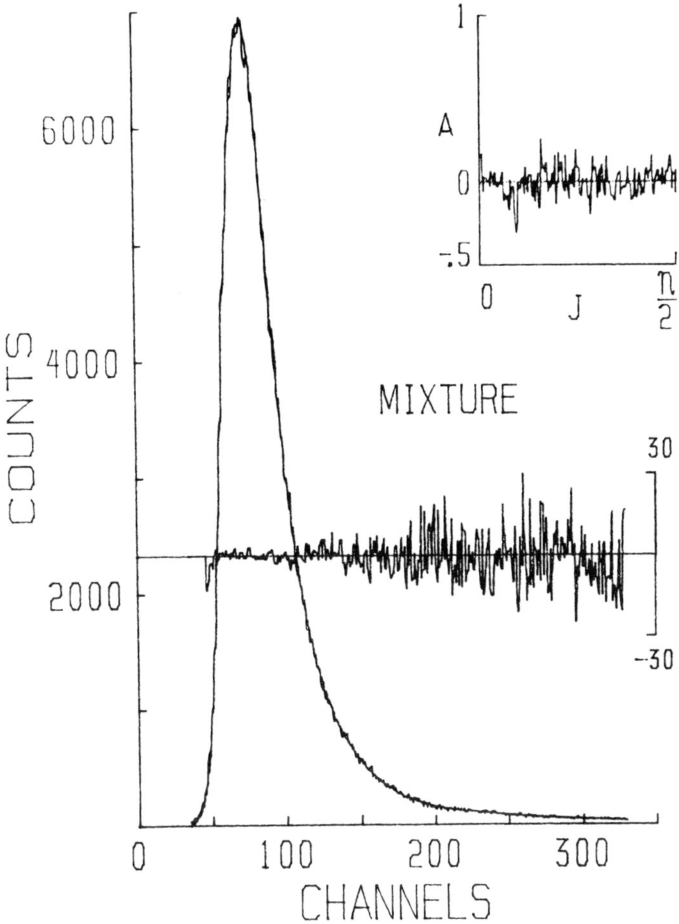

FIGURE 3. Fluorescence decay of 2.5×10^{-6} M binaphthyl and 2.5×10^{-6} M terphenyl in cyclohexane at 20°C. Included are the percent residuals and their autocorrelation for a double exponential analysis. The noise free line represents the theoretical parameters convolved with the lamp. The sample was excited with natural light at 296 nm (10.6 nm bandwidth) and the emission monitored at 360 nm (10.6 nm bandwidth). The time of collection for the decay curve was 18 minutes. The best fit to the data gave $\tau_1 = 0.95$ ns, $\alpha_1 = 0.65$, $\tau_2 = 2.59$ ns, $\alpha_2 = 0.35$ with $\chi^2 = 1.48$.

as below.

$$-d[A^*]/dt = (k_A + k_{BA})[A^*] - k_{AB}\,[H^+][B^*]$$

$$-d[B^*]/dt = (k_B + k_{AB}\,[H^+])[B^*] - k_{BA}[A^*]$$

Integration of these equations with the initial boundary conditions that *only A* is directly excited and that $B^* = 0$ at time $= 0$

$$I_A(t) = \alpha_1\, e^{-t/\tau_1} + \alpha_2\, e^{-t/\tau_2}$$

$$I_B(t) = \beta_1\, e^{-t/\tau_1} - \beta_1\, e^{-t/\tau_1}$$

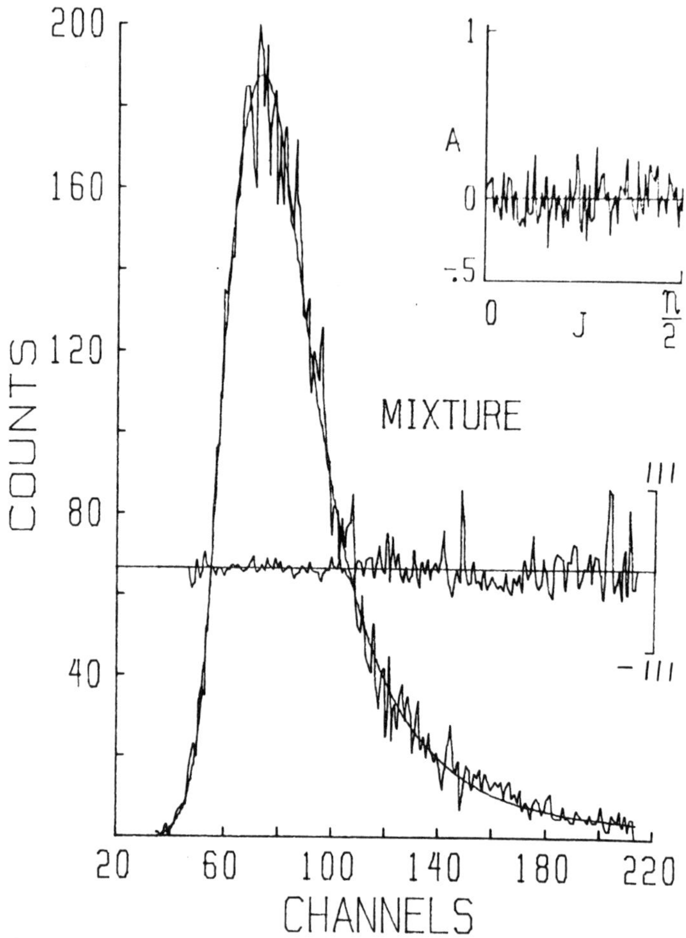

FIGURE 4. Fluorescence decay at 2.5 × 10^{-6} M binaphthyl and 2.5 × 10^{-6} M terphenyl in cyclohexane at 20°C. Included are the percent residuals and their autocorrelation for a double exponential analysis. The noise free line represents the theoretical parameters convolved with the lamp. The sample was excited with natural light at 295 nm (10.6 nm bandwidth) and the emission monitored at 360 nm (10.6 nm bandwidth). The time of collection for the decay curve was 30 seconds. The best fit to the data gave τ_1 = 1.09 ns, α_1 = 0.69, τ_2 = 2.62 ns, α_2 = 0.31 with χ^2 = 0.93.

The amplitudes α_1, α_2 and β_1 as well as the exponential decay constants τ_1 and τ_2 are directly related to the rate constants and concentration terms in the differential rate expressions given above. It is emphasized that the two decay times τ_1 and τ_2 are identical for the decay of A^* and B^*. Thus a two-state reversible excited-state reaction shows double exponential decay kinetics with decay times independent of emission wavelength. In addition, the pre-exponential terms for the decay of B^* are equal in magnitude but opposite in sign. These features provide diagnostics for determining if one is dealing with a reversible two-state excited-state reaction. In the case of naphthol at a pH less than one, the back reaction is so predominant that no proton

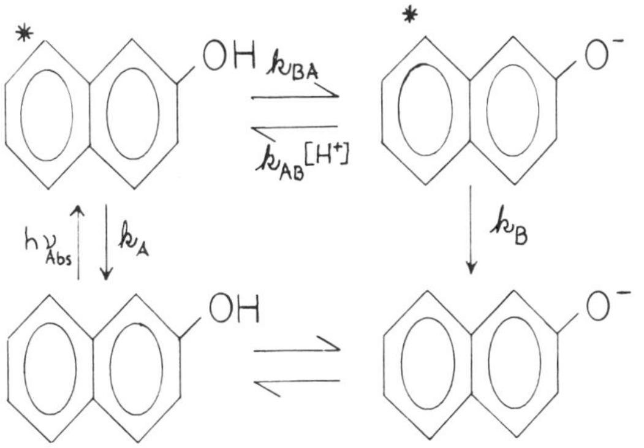

FIGURE 5. The fluorescence of 2-naphthol, a schematic example of a two-state excited-state reaction.

transfer takes place. The decay of naphthol is then a monoexponential. In the pH region 1 to 4 (near the excited state pK_a), the reaction is reversible and the decay of naphthol is described by a double exponential. At a pH near neutrality the hydrogen ion concentration becomes so low that the reaction is not reversible. Under these conditions the decay of naphthol becomes monoexponential again.

Similar kinetic behavior is expected for exciplexes, excimers and other reactions that involve two species in the excited state. These examples are mentioned to emphasize the point that double exponential decay may have its origin *either* in ground state heterogeneity as illustrated by the mixture described with binaphthyl and terphenyl, or in microheterogeneity as will be shown below or in an excited state reaction as indicated with 2-naphthol.

TABLE 1 summarizes the fluorescence decay parameters for several tryptophan models. A comparison of the data shows that different decay laws characterize the indole emission of the different models. Single exponentials fit the fluorescence decay of indole, skatole, *N*-acetyl-tryptophanamide, and acetyl-tryptophan. The emission of these fluorophores in phosphate buffer at ambient temperature occurs on a time scale that is slow compared with solvent interactions during the lifetime of the excited state. For example, *N*-acetyl-tryptophanamide in viscous solvents exhibits an increase in the average fluorescence lifetime from the blue to the red edge of the emission spectrum.[12] This increase is typical of solvent relaxation.

The fluorescence kinetics of tryptophanamide, Gly-Trp, Trp-Gly, and tryptophan are fit by a double exponential decay law. The kinetics of the first three molecules are independent of emission wavelength. With the exception of tryptophan, the α-amino group of these particular model compounds have pKs between 7 and 8.[13] Furthermore, the relative amplitudes of the decay constants change with pH, with the long lifetime persisting when the amino group is unprotonated. Therefore, in this case the double exponential decay has its origin in ground-state heterogeneity since there are two species present at pH 7.

By contrast at pH 7 tryptophan exists as a single species in the ground state. Moreover, its decay kinetics depend upon emission wavelength. The contribution of the subnanosecond component is largest in the blue region of the fluorescence spectrum, and becomes insignificant above 380 nm. In addition, the 0.8 ns decay represents only a small fraction of the emission (10% at 325 nm).

Considering the improvements in methods of data analysis during the past decade,

TABLE 1
FLUORESCENCE DECAY PARAMETERS FOR TRYPTOPHAN AND TRYPTOPHAN MODELS*

Sample	α_1	τ_1	α_2	τ_2
Indole		4.54		
Skatole		8.71		
Acetyl-Trp-amide		3.00		
Tryptophan	0.82	3.26	0.18	0.80
Tryptophan-amide	0.30	6.90	0.70	1.58
Acetyl-tryptophan		4.85		
Trp-Gly	0.12	7.47	0.88	1.86
Gly-Trp	0.56	1.45	0.44	0.51

* Samples (10^{-5}-10^{-4} M) were dissolved in 0.1 M sodium phosphate buffer, pH 7, at 20°C. Excitation was at 295 nm (bandpass of 10.5 nm) and the emission was monitored at 360 nm (bandpass of 8.3 nm).

it is no surprise that with the methods available in 1970, DeLauder and Wahl[14] reported a single exponential for the decay of tryptophan at pH 7. The first report of a double exponential appeared in 1978, and was presented by Rayner and Szabo.[15] They found that the fluorescence decay was more accurately described with time constants of 0.5 and 3.1 ns. This report has provoked considerable controversy. Whereas Robinson and coworkers[16] obtained a double exponential of about 2 and 5 ns, monitoring the red edge of the emission, Alpert *et al.*[17] recovered a single exponential of about 3 ns in agreement with the earlier work of DeLauder and Wahl.[14] Our data agree with that of Rayner and Szabo.[15, 18] We obtain similar time constants, and find the same wavelength dependence.

The double exponential decay of tryptophan at pH 7 has been attributed to rotamers.[18] In this model, a strong interaction with the charged amino group (one possible rotamer), or a concerted interaction involving both the unprotonated carboxyl and the protonated amino groups (second possible rotamer) could be responsible for the subnanosecond component. Since *N*-acetyl-tryptophan decays monoexponentially, it is evident that the carboxyl group alone (third possible rotamer) is not responsible for the quenching interaction. The mechanism responsible for the peculiar decay kinetics of the zwitterion is not known. Favored possibilities include intramolecular proton transfer from the amino group, or electron transfer to this group. In addition, the carbonyl function can quench via charge transfer.

The results for the model compounds, including tryptophan, show that multiexponential decay kinetics are to be expected for tryptophyl residues in polypeptides or proteins. Indeed such kinetics have been observed for a number of single tryptophan-containing proteins.[19] Depending upon the nature of the tryptophan environment, complicated decay kinetics could result from a distribution of ground-state conformations, such as a helix-coil transition or movement of a charged group. Excited-state reactions, as we have already pointed out, can all lead to complex fluorescence decay kinetics.

We have examined the fluorescence decay of tryptophan in three polypeptides, namely ACTH (1–24), glucagon, and a four-peptide fragment of glucagon containing residues 23–26. ACTH (1–24) and glucagon each has only one tryptophan. We have also characterized the fluorescence decay of the two tryptophan residues in the protein horse liver alcohol dehydrogenase (LADH). The results are shown in TABLE

TABLE 2
FLUORESCENCE LIFETIME PARAMETERS FOR TRYPTOPHAN RESIDUES AT NEUTRAL PH*

Residue	α_1	τ_1 (ns)	α_2	τ_2 (ns)
Trp-25:glucagon	0.51 ± 0.01	1.66 ± 0.04	0.49 ± 0.01	5.09 ± 0.04
Trp-25:glucagon (23–26)	0.31 ± 0.01	1.63 ± 0.09	0.69 ± 0.01	4.48 ± 0.01
Trp-9:ACTH (1-24)	0.59 ± 0.02	2.31 ± 0.03	0.41 ± 0.02	5.93 ± 0.09
Trp-15 & Trp 314: ADH	0.61 ± 0.04	3.87 ± 0.16	0.39 ± 0.04	7.21 ± 0.29
Trp-15 & Trp 314: ADH + NAD^+ pyrazole	0.76 ± 0.01	2.39 ± 0.04	0.24 ± 0.01	7.20 ± 0.07

* Sample concentrations were as follows: glucagon and glucagon (23-26), 1.6×10^{-5} M; ACTH (1-24), 8–12 $\times 10^{-5}$ M; ADH, 8×10^{-5} M. A solvent blank was subtracted from the glucagon and glucagon (23–26) data since the background of the blank exceeded 1% of the fluorescence intensity. A 0.1 M sodium phosphate buffer was used except for the ACTH (1-24) solutions where the buffer concentration was 0.01 M. Excitation was at 295 nm with a 10.5 nm bandwidth, and the emission was monitored at 350 nm with a bandpass of 9.9 nm.

2. In all cases double exponential decay laws provide the best fits to the data. After a discussion of the polypeptide data, we will show that the two components in the fluorescence decay of LADH can be associated with each of the two residues. One, Trp-15, is exposed to solvent whereas the other, Trp-314, is buried at the interface between the two subunits.[20]

The decay kinetics of ACTH (1–24) and glucagon (23–26) are similar to those of tryptophanamide and the model dipeptides. That is, the amplitudes and time constants are independent of emission wavelength. According to this similarity, the double exponential decay of these two peptides can be attributed to ground state heterogeneity.

Glucagon exhibits wavelength dependent decay parameters. The change in both α_i and τ_i with emission wavelength is small but significant, and has been previously noted by Grinvald and Steinberg.[19] The wavelength dependence of the fluorescence parameters could be due to a fraction of the polypeptides persisting in a stable conformation such as an α-helix. Deranleau *et al.*[21] have noted that the four peptide fragment, which includes Trp-25, is too short to satisfy the requirements for helix formation. However, there is considerable evidence for helix formation in the region of Trp-25 in glucagon. This evidence has been reviewed by Ross *et al.*[22] If the randomly coiled and helical molecules have similar but not identical tryptophan spectra, and these spectra are characterized by slightly different time constants and pre-exponentials, then the decay kinetics would exhibit a gradual change in both decay parameters (α_i, τ_i) as a function of the emission spectrum. In fact, the steady state fluorescence spectrum of glucagon is about 1.5 ± 0.5 nm broader than that of either glucagon (23–26) or ACTH (1–24) which suggests the possibility of closely overlapping spectra. Therefore, the tryptophan emission of all three molecules can be ascribed to ground state heterogeneity.

The decay results for LADH are also well-described by a double exponential decay law. Both time constants (3.8 and 7.2 ns) are independent of emission wavelength and each appears to be associated with one of the two tryptophans. The pre-exponential terms show a strong wavelength dependence and reflect the relative spectral contribution of each decay. We resolved the spectral contributions of the two decay times using the following relationship:[23]

$$F_i = F_{ss}(\alpha_i\tau_i/\sum\alpha_i\tau_i)$$

where F_i, α_i, and τ_i are the fluorescence, amplitude, and decay constant of the *i*th component, and F_{ss} is the total fluorescence spectrum. With excitation at 295 nm (bandwidth of 10.6 nm) and monitoring the fluorescence decay in 5 nm (bandwidth of 8.3 nm) steps, the time-resolved spectrum of the 3.8 ns component has a maximum at 324 nm and the bandwidth (full-width half height) is 43 nm, while the spectrum of the 7.2 ns component has a maximum at 337 nm and the bandwidth is 54 nm. Under the same optical conditions, the total steady state spectrum has a maximum at 326 nm and the bandwidth is 48 nm. By comparison, the fluorescence maximum of *N*-acetyl-tryptophanamide is at 350 nm and the bandwidth is 60 nm.

Quenching experiments were used to test the hypothesis that each decay time could be assigned to one tryptophan residue. Exposed residues are quenched more than buried residues with KI.[24] We found that at low concentrations of KI the 7.2 ns component was quenched. As the concentration of KI was increased the analysis for a double exponential failed. At about 0.1 M KI the data gave a satisfactory fit for a single exponential with a decay of about 4.5 ns. We therefore concluded that the 7.2 ns decay, which contributes more to the red emission, is due to the exposed tryptophan, Trp-15. Moreover, the spectra assignment from the lifetime data agrees well with the results of Abdallah *et al.*,[25] who used quenching to resolve the steady-state contributions of the "red" and "blue" tryptophan emissions.

In conclusion, nanosecond fluorometry can provide evidence for ground-state heterogeneity or microheterogeneity or decay measurements may indicate excited-state interactions. Useful information about biopolymers is obtained.

References

1. Weber, G. & F. W. J. Teale. 1965. The Proteins. H. Neurath, Ed. Vol 3: 445–529.
2. Brand, L. & B. Witholt. 1967. Methods in Enzymology. S. Colowick & N. O. Kaplan, Eds. Vol. 11: 776–856.
3. Brand, L. & J. R. Gohlke. 1972. Ann. Rev. Biochem. **41:** 843–868.
4. Easter, J. H., R. P. DeToma & L. Brand. 1976. Biophys. J. **16:** 571–583.
5. Badea, M. G. & L. Brand. 1979. *In* Methods in Enzymology. C. H. W. Hirs & S. N. Timasheff, Eds. Vol. 61: 378–425.
6. Yguerabide, J. 1972. *In* Methods in Enzymology. C. H. W. Hirs, & S. N. Timasheff, Eds. Vol. 26, part C: 498–577.
7. Ware, W. R. 1971. *In* Creation and Detection of the Excited State. A. A. Lamola, Ed. Vol. 1, part A: 213–302.
8. Grinvald, A. 1976. Anal Biochem. **75:** 260–268.
9. Weller, A. 1961. Prog. React. Kinet. **1:** 187–222.
10. Laws, W. R. & L. Brand. 1979. J. Phys. Chem. **83:** 795–802.
11. Loken, M. R., J. W. Hayes, J. R. Gohlke & L. Brand. 1972. Biochemistry **11:** 4779–4785.
12. Lakowicz, J. R. & H. Cherek. 1980. J. Biol. Chem. **255:** 831–834.
13. Edelhoch, H., L. Brand & M. Wilchek. 1967. Biochemistry **6:** 547–559.
14. DeLauder, W. B. & P. Wahl. 1970. Biochemistry **9:** 2750–2754.
15. Rayner, D. M. & A. G. Szabo. 1978. Can. J. Biochem. **56:** 743–745.

16. Fleming, G. R., J. M. Morris, R. J. Robbins, G. J. Wolfe, P. J. Thistlethwaite & G. W. Robinson. 1978. Proc. Natl. Acad. Sci. USA **75:** 4652–4656.
17. Alpert. B., D. M. Jameson, R. Lopez-Delgado & R. Schooley. 1979. Photochem. Photobiol. **30:** 479–481.
18. Szabo, A. G. & D. M. Rayner. 1980. J. Am. Chem. Soc. **102:** 554–563.
19. Grinvald, A. & I. Z. Steinberg. 1976. Biochim. Biophys. Acta **427:** 663–678.
20. Branden, C.-I., H. Jörnvall, H. Eklund & B. Furugren. 1975. The Enzymes, 3rd. edit. vol. 11: 103–190.
21. Deranleau, D. A., J. B. A. Ross, K. W. Rousslang & A. L. Kwiram. 1978. J. Am. Chem. Soc. **100:** 1913–1917.
22. Ross, J. B. A., K. W. Rousslang, D. A. Deranleau & A. L. Kwiram. 1977. Biochemistry **16:** 5398–5402.
23. Donzel, B., P. Gauduchow & P. Wahl. 1975. J. Am. Chem. Soc. **96:** 801–808.
24. Lehrer. S. 1971. Biochemistry **10:** 3254–3263.
25. Abdallah, M. A., J. F. Biellman, P. Wiget, R. Joppich-Kuhn & P. L. Luisi. 1978. Eur. J. Biochem. **89:** 397–405.

INSIGHTS INTO THE MODE OF ANTIBODY ACTION FROM INTRINSIC AND EXTRINSIC FLUORESCENT PROBES

Israel Pecht

Department of Chemical Immunology
The Weizmann Institute of Science
Rehovot, Israel

INTRODUCTION

Fluorescence spectroscopy at steady-state as well as its time-resolved forms have been rather useful in the study of the mode of action of antibodies.[1] Both the intrinsic emission due to its tryptophan residues and the fluorescence of extrinsic probes bound noncovalently as haptens at the binding site or covalently at other loci have yielded a wealth of information on the properties and function of this class of proteins. The pioneering work of Velick *et al.*[2] which has shown that excitation energy transfer can be used for quantitative analysis of the binding of haptens to their antibodies is one of the first illustrations of the great potential of the approach. Later this was widely employed in the screening and characterization of a variety of antibodies of different specificities[3] and kinetic analysis of hapten antibody interactions.[4–6] Other properties of the intrinsic emission have also been used yielding very valuable insights. Thus, the circularly polarized component in the fluorescence of antibodies has been studied in an effort to resolve structural changes induced by antigen binding to the antibody.[7,8] The results of these studies have suggested that conformational transitions do occur also in domains of the immunoglobulin molecules which are remote from the combining site. A clear relation between the state of the hinge region and the spectroscopic changes induced by antigen binding has been established.[7] Further examples are numerous so in this paper several illustrations from more recent studies will be described.

Tryptophan residues are distributed throughout all the domains of the immunoglobulins.[9] The crystallographic structure determination of immunoglobulins and their derivatives yielded information about the spatial[10,11] location of these residues and their environment. One invariant tryptophan residue is 35 in V_L and its homologue 36 in V_H and all constant domains. The variable domains of the heavy chain have two further conserved tryptophans (residues 62 and 103).[9] In the 3-D structure, Trp 35 and its homologues are always found in the hydrophobic core of the domain's interior, remarkably close to the intradomain disulfide bond which links the β pleated sheets.[11,12] The second highly conserved residue is located in a region far removed from the inter-domain contact area and is relatively close to the domain surface. Thus, Trp residue 417 in the C_H3 domain and its homologue 313 in C_H2 are in close proximity to phenylalanyl and arginyl residues (P. Phizackerley, personal communication). The second conserved Trp in C_L occupies a similar position in that domain.[11]

This relatively large number of constant tryptophans is contributing a rather low

0077-8923/81/0366-0208 $01.75/0 © 1981, NYAS

quantum yield of the different separate domains[13] and of the immunoglobulins as a whole and as discussed below, extra tryptophan residues appearing in the hypervariable loops of the variable domains are the only exceptions to this rule.[13,14]

Extrinsic Probes

Extrinsic probes have been used primarily to examine the segmental flexibility of immunoglobulins of different classes[15] and for studying the nature of the antigen-combining site.[16,17] Probes for studying the first aspects were either bound noncovalently, as haptens at the combining sites,[18] or covalently, at non-specific loci on the protein surface.[19]

Two types of approaches have been employed for investigating the site itself.

(a) Specific antibodies were raised against the fluorescent dye (e.g., dansyl[16] or fluorescein derivatives[17] and then this fluorophore reports from its binding site. One interesting improvisation on this theme is the use made of the fluorescent nitro compound 4-(α-*N*-L-alanine)-7-nitrobenz-2-oxa-1,3-diazole (NBDA), a structural and functional analogue of the 2,4-dinitrophenyl group (DNP).[20] It was found to bind to all induced anti-DNP antibodies examined and to several monoclonal immunoglobulins with nitro-aromatic specificity. The fluorescence of NBDA is quenched upon binding to these proteins. Similar quenching of NBDA fluorescence was observed in the presence of aromatic amino acid analogues, and also upon binding to serum albumin and apomyoglobin. NBDA does not bind to immunoglob-lins of unrelated specificity. The absorption and fluorescence characteristics of NBDA in different solvents reveal large changes which correlate with medium polarity. A few important exceptions, however, exist suggesting that NBDA is not a simple polarity probe and that its spectral properties are sensitive to specific binding interactions. Thus the observed spectral parameters of NBDA bound to immunoglobulins clearly indicate that binding does not occur through hydrophobic interactions only and suggest the formation of specific interactions such as a charge-transfer complex and hydrogen bonds. The IgA myeloma protein 460 binds NBDA with an association constant of $3.2 \times 10^5\ M^{-1}$ (at 25°C). The bound hapten undergoes full quenching of its fluorescence and marked changes in its absorption spectrum. This property of that fluorophore and its wavelengths of excitation and emission may also make it useful for detection and characterization of anti-nitro-aromatic antibodies, even in crude preparations, and possibly on cell surfaces.[20]

(b) When the hapten itself is nonfluorescent, conjugating it covalently to a fluorescent probe yields an effective reporter group. One illustration of this approach is the characterization of the combining site of the dextran-specific homogeneous IgM MOPC 104-E.[21] For this purpose, several α-(1–3)-linked oligoglucosides were derivatized to their corresponding dansyl hydrazones. The binding of 18 different derivatives to the combining site has been determined by monitoring the quenching of intrinsic emission of the protein and served as the ground for proposing a model for the architecture of the site. This has been further refined on the basis of the observed different extents of quenching by different haptens, the lack of change in the emission band shape and quantum yields of the bound dansylated haptens, the lack of an induced circularly polarized component in their emission and the varying amount of fluorescence anisotropy of the dansyl group. The site is proposed to be cavity-shaped, of 12 Å depth, complementary to a terminal non-reducing nigerosyl group. Further subsites of interaction were also identified.[21]

The method of somatic cell hybridization developed recently by Köhler and Milstein[22] allows raising monoclonal antibodies to a given antigenic determinant of choice. This opened new alleys for studying both the produced combining sites and, perhaps more interesting, the segmental flexibility of single-molecular species of homogeneous antibodies. Thus, for example, known structural differences in hinge regions among subclasses of IgG would be expected to lead to pronounced differences in the motional freedom of the Fab arms carrying the probes. Antibodies of other classes like IgE and IgM specifically raised to bind fluorescent probes of appropriately chosen lifetimes and initial anisotropies would give new insights into the structural properties of these macromolecules. One illustration of the properties of such a homogeneous antibody of the IgG class raised against the well known dimethylamino naphthalene sulfonate (dansyl) hapten is shown in FIGURES 1 and 2. The excitation and emission properties are rather similar to those of the anti-dansyl antibodies normally raised in mice or rabbits.[23] However, the anisotropy spectra in excitation

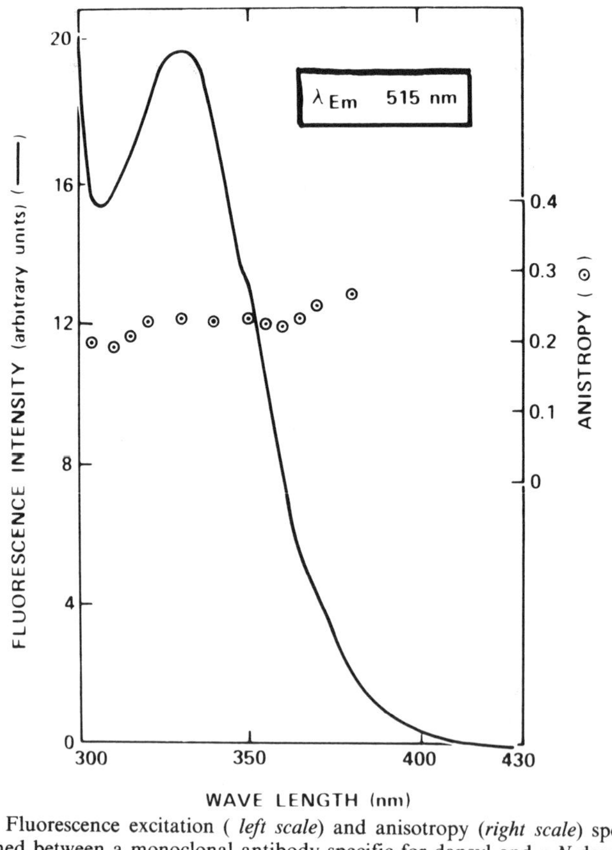

FIGURE 1. Fluorescence excitation (*left scale*) and anisotropy (*right scale*) spectra of the complex formed between a monoclonal antibody specific for dansyl and ε-*N*-dansyl-L-lysine. Protein concentration 5 × 10^{-5} M; 0.5 hapten molecule per site. Temperature 20°C.

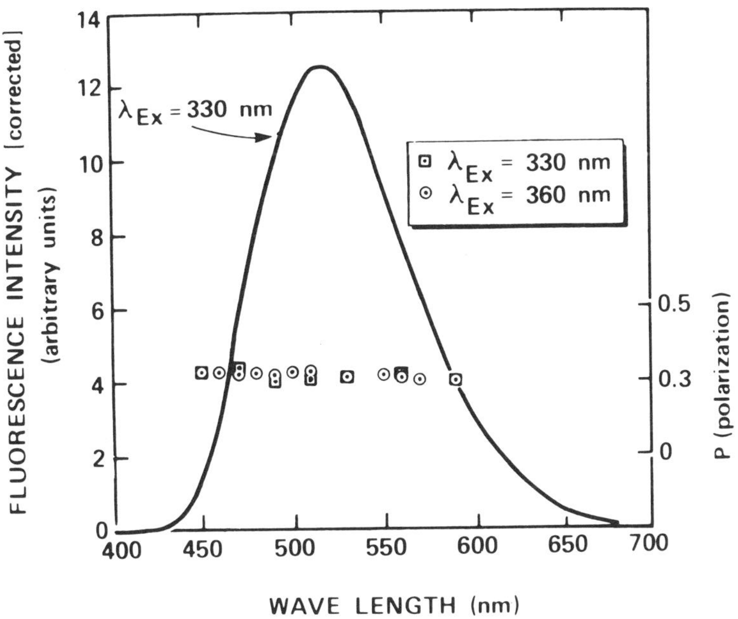

FIGURE 2. Fluorescence emission (*left scale*) and anisotropy (*right scale*) spectra of the above.

and emission which show no variation with wavelength are in sharp contrast to those obtained with the normal, heterogeneous antibody preparations.[23] Other properties of this homogeneous antibody and of others of different classes are currently being investigated.

INTRINSIC EMISSION

The particular value of the intrinsic emission due to the tryptophan residues of the immunoglobulins has already been stated above. Recently, more detailed attention has been focused on the behavior of these residues. This was made feasible by the knowledge of their spatial location in different domain of the immunoglobulins and the availability of light sources by which their relatively short excited-state lifetimes could be better monitored. As already mentioned, quantum yields of the invariant and conserved residues are rather low.[13,14] That this is primarily due to the effective quenching by neighboring residues is clearly borne out by recent studies of the unfolding-refolding process of two different, isolated constant domains of immunoglobulins. Isenman *et al.*[27] have found that already the reduction and alkylation of the C_H3 domain's intramolecular disulfide bridge yields a fourfold enhancement of the emission intensity and ~ 25 nm red shift in its maximum. Essentially the same shift and enhancement were caused by unfolding this protein by 4M guanidinium hydrochloride. Reduction and alkylation in this medium did not cause any significant further changes. As described above, there are only two Trp residues in this domain, one (381) is brought by the protein fold into the close neighborhood of the intradomain disulfide bond and is therefore largely quenched by it[25] as long as it is kept at that position. The second Trp (417) is close to the surface but two neighboring residues (phenylalanyl and arginyl) most probably also quench it effectively in the native state. It is noteworthy that either reduction and alkylation or the denaturant are each fully effective in relieving the quenching: the first by eliminating the disulfide bond and thereby changing the domains structure, the second by unfolding the structure alone. Another recent study of the unfolding-refolding of a constant domain is that of the C_L derived from a human light chain.[26] This domain contains two Trp residues at homologous positions to those in C_H3. It yielded quite similar information about the intrinsic fluorescence. Namely a threefold enhancement in the emission and a red shift of ~ 25 nm occur upon denaturation by 4 M quanidinium hydrochloride. One interesting extra feature reported in this paper is that reduction without alkylation of the disulfide bond relieves most of the quenching yet does not cause the red shift. Alkylation causes the shift but only a very slight further enhancement. This is interpreted to mean that the native fold is maintained as long as the disulfide is not alkylated. The latter modification triggers the unfolding and therefore the exposure of the Trp residues to the solvent. However, the fact that practically the full emission intensity is achieved upon reduction raises the question as to what is the contribution of the second Trp located far from the disulfide? If indeed the reduction per se does not cause a change in the domain fold, then it should cause a lower extent of enhancement. One interpretation could be that the second Trp is quenched (possibly by neighboring charged residues) also in the unfolded state.

The study of tryptophanyl fluorescence of several rabbit antibodies of the IgG class and of their fragments has shown that contributions of the Fc and $F(ab')_2$ fragments to the total yield of the intact IgG are additive.[14] This suggests that the cleavage of the protein has no significant effect on the environment of the Trp residues. The contribution of the Fc compared to that of the $F(ab')_2$ is relatively small. Since the Fc as well as the C_H1 and C_L are common to all rabbit IgG, this leads to the suggestion that the Trp residues in the variable domains are predominantly contributing to the emission and to the variation in quantum yields among different molecules.[13,14] Interestingly marked variations are found among different, homogeneous antibodies specific for the same antigen (TABLE 1). Another interesting feature is the much higher quantum yields of (heterogeneous) antibodies specific to aromatic haptens like dinitrophenyl or dimethylaminonaphthalene derivatives (TABLE 1). The wide variation in quantum yields reflects the diversity of the variable region in these antibodies. The rather high values found in antibodies specific to the hydrophobic haptens are apparently reflecting the increase in the number of tryptophan residues in the variable domains and to the hydrophobicity of these sites.[14] It has been pointed out by Eisen and his associates that antibodies specific to nitroaromatic haptens indeed have a larger number of tryptophan in their variable domain.[27,28]

INTERNAL MOTIONAL DYNAMICS

Synchrotron radiation has recently been shown to be a very useful light source for time-resolved studies of fluorescence.[29,30] The main advantages of this source lie in the sub-nanosecond width of the light pulses, the very high repetition rate and

TABLE 1

Rabbit IgG	Quantum Yield†		Contribution in %*
Homogenous IgG			
anti-peptidoglycan[12]	intact IgG	0.028	100
	red and alk	0.028	100
	$F(ab')_2$	0.034	81
anti-peptidoglycan[864]	intact IgG	0.060	100
	red and alk	0.060	100
	$F(ab')_2$	0.086	93
anti-peptidoglycan[5942]	intact IgG	0.044	100
	red and alk	0.044	100
	$F(ab')_2$	0.058	88
	Fc	0.016	12
Heterogeneous IgG			
anti-polyalanine		0.030	
anti-ovalbumin IgG		0.030	
anti-loop IgG‡		0.050	
anti-DNP IgG		0.120	
anti-DNS IgG		0.120	
normal IgG		0.060	

* Contribution in % of the fragments to the fluorescence of the intact IgG calculated as described in the text, and in reference 14.

† Excitation wavelength −295 nm.

‡ Antibodies specific for the loop region of hen egg white lysozyme.

reproducibility of these pulses and the wavelength range which spans from the vacuum—ultraviolet to the infrared regions. The short lifetime and excitation spectrum of tryptophan residues in proteins lend themselves particularly to investigation using synchrotron radiation.[31] Indeed the problem of internal motional dynamics in proteins has been tackled by using synchrotron radiation to monitor the fluorescence anisotropy decay kinetics of their tryptophan residues.[30] In that study, a group of proteins, all containing only a single tryptophan, were chosen so as to allow a more definitive interpretation of the results. These proteins were: the copper-containing bacterial (*Pseudomonas aeruginosa*) protein, azurin and its apo form (1400 daltons) the bovine myelin basic protein (18000 daltons), *Staphylococcus aureus* nuclease B (20,000 daltons) and human serum albumin (60,000 daltons). Different degrees of rotational freedom of the tryptophan residues were observed for the different proteins. It ranged from essential immobilization within the macromolecular matrix to nearly complete freedom of rotation. The tryptophan of nuclease B exhibits at 20°C a single rotational correlation time of 9.9 nsec which corresponds to the rotation of the whole protein. In human serum albumin 8°C, the tryptophan is also showing fluorescence anisotropy decay kinetics which correspond to its rotation with the whole macromolecule (ϕ = 31.4 nsec). In contrast, the tryptophan in the myelin basic protein which is a component of the myelin membrane and loses its native conformation upon dissolving in water, is having an extremely fast anisotropy decay (ϕ = 0.09 nsec and ϕ = 1.26 nsec), most probably reflecting the free rotation of the indole residue exposed to the solvent. The most interesting behavior has been that of azurin. Its single tryptophan is known (from x-ray structure analysis and its spectroscopic properties) to reside in a rather hydrophobic interior part of the protein. The anisotropy of this internal tryptophan clearly showed (at 8°C) two components of close to equal size in its decay, a fast one ($\phi = 0.51$ nsec) and a slower one ($\phi = 11.8$ nsec). The first most probably reflecting internal flexibility, the slower one, again the motion with the whole protein. Interestingly, the anisotropy decay of human serum albumin is showing, upon raising the temperature to 43°C, a new faster component (0.14 nsec) which may also be due to an internal mobility of the tryptophan.

The direct observation of such fast motional events inside proteins has been predicted earlier by experimental and theoretical studies.[32,33] It raises immediately the question of the relation between the function of the examined protein and its internal flexibility. This obviously is a difficult problem to approach and for the above-mentioned group of proteins has not been attempted yet.

Several aspects make the immunoglobulins an attractive subject for studying the behavior of their tryptophans. Particularly the detailed knowledge described above, of the location and environment of these residues in different domains of these molecules and their unique structure to function relation as expressed in their evolutionarily adapted features. However, the large number of tryptophans in Ig molecules poses a serious problem in the interpretation of the data. In an attempt to overcome this difficulty, a systematic study of the properties of isolated domains of immunoglobulin has been undertaken. The intention is to try to use the information obtained with the fragments for understanding them better as they are and hopefully also reconstruct a self-consistent picture of increasingly larger parts of these complex molecules.

Isolated Ig domains usually dimerize, and therefore we have first looked at the closest model of a single free Ig domain, β_2 microglobulin (11,700 daltons) which is usually found strongly associated with the histocompatibility antigens in mouse and man.[34] In terms of the loci of the tryptophans however, these are different from those in the Ig domains. Both are at different positions in the sequence. As we have indeed found, this difference is markedly affecting the fluorescence properties of that protein compared to all the immunoglobulin domains examined. Thus, the fluorescence observed is dominated by a significantly longer lifetime ($\tau_1 = 0.4$, $\tau_2 = 2.9$ns,$a_2/a_1 \approx 3$) The pFc′ fragment corresponds to the C_H3 domain is dimeric and each subunit contains two tryptophans in the positions described above.[35] The fluorescence lifetime decay fits well to two components of 0.6 and 1.9 nsec with the faster one being 3.6-fold larger. The whole Fc fragment derived from human IgG shows a surprisingly similar lifetime decay (0.37 and 1.8 nsec) yet the shorter one is only about twofold larger than the slower. Though the Fc fragment derived from rabbit IgG has one extra Trp in each of its C_H3 domains, its lifetime decay pattern resembles that of the human Fc.

An unambiguous interpretation of the anisotropy decay kinetics cannot be made with this complex system as, most probably, the heterogeneity in the excited-state decay stems from more than one tryptophan. The data for β_2 microglobulin are dominated by a correlation time which decreases from 5.1 nsec at 5°C to 2.4 nsec at 35°C (with $A_1 = 0.14$). These figures are close to the correlation time of this whole globular protein. A smaller component in the anisotropy decay ($A_2 = 0.08$) is quite fast (0.02 nsec) and cannot be assigned presently to any particular species.

In whole Fc fragments, the assignment is even more difficult but we find it interesting that the slower and dominant ($A_1 = 1.8$) decay component is about 9 nsec (at 5°C) and 6 nsec (at 25°C), by far faster than the expected for the motion of the whole fragment. Naturally, one cannot distinguish between the emitter rotating on its own and with the macromolecule, but judging from the behavior of the pFc and β_2 microglobulin we are inclined to believe that it is due to a segmental motion in the Fc fragment.

A very interesting system is that of the light chain dimer derived from the human IgG-Mcg (or the Bence-Jones protein of this individual). The three-dimensional structures of several crystal modes of this light chain dimer have been studied extensively by A. B. Edmundson and his colleagues.[12,36,37] There are only three tryptophans in each chain, one in the variable domain, and two in the C_L. The protein has been shown to exist in solution in two different conformations which differ very much in their aromatic CD spectra and crystallize in two different crystal modes.[12] One yields trigonal crystals and the other leads to needle-shaped ones. The main structural difference between the two conformers was proposed to involve the region of the more exposed Trp of C_L to which the differences in the CD were assigned.[12] The interconversion between the two conformers was shown to depend on breaking the interchain disulfide. Reduction of the latter bond allows the transition while it is prevented as long as the bond was maintained. Very marked differences were found in the anisotropy decay kinetics between the conformers. As noted above, also drastically different CD spectra were observed for these two conformers in the 260–320 nm range.[12] These results clearly point at changes in the environments of tryptophans residues between the two conformers. Edmundson and his colleagues suggest that these are due to the second Trp residue (No. 189) in the C_L. This residue is present in a helical end of the domain, next to a phenol residue of Tyr 195. On the basis of the atomic model, these authors propose[12] that perturbation of the structure around the neighboring interchain disulfide could readily move the two aromatic side chains and cause the differences in the spectral properties between the two conformations of the protein.

In conclusion, studies using fluorescent probes, intrinsic and extrinsic, yielded a wealth of information on structural and functional properties of immunoglobulins. The recently developed methodologies of raising monoclonal antibodies of well-defind specificities and of different classes along with the refinement of a variety of techniques for studying the excited state properties will result in an even more detailed and quantitative understanding of this family of protein molecules.

Acknowledgments

I wish to thank a large number of colleagues from whose advice and experience I have learned and benefited. Part of the work described here was done together with Drs. L. Stryer and I. Munro while on a sabbatical leave at the Department of Structural Biology of the Stanford University School of Medicine. The monoclonal antibodies specific for fluorescent dyes were raised by Drs. V. T. Oi and L. A. Herzenberg of the Department of Genetics at the above-mentioned university. A sample of β_2 microglobulin was kindly given by Dr. M. D. Poulik.

References

1. Chen, R. F. & H. Edelhoch, Eds. 1975. Biochemical Fluorescence-Concepts. Marcel Dekker, Inc. N.Y.

2. VELICK, S. F., C. W. PARKER & H. N. EISEN, 1960. Proc. Natl. Acad. Sci. USA **46:** 1470–1482.
3. EISEN, H. N., E. S. SIMMS & M. POTTER. 1968. Biochemistry **7:** 4126–4134.
4. DAY, D. R., J. M. STURTEVANT & S. J. SINGER. 1963. Ann. N.Y. Acad. Sci. **103:** 611–625.
5. PECHT, I., D. GIVOL & M. SELA. 1972. J. Mol. Biol. **68:** 241–247.
6. DANDLIKER, W. B. & S. A. LEVISON. 1968. Immunochemistry **5:** 171–183
7. SCHLESSINGER, J., I. Z. STEINBERG, D. GIVOL, J. HOCHMANN & I. PECHT. 1975. Proc. Natl. Acad. Sci. USA **72:** 2775–2779.
8. PECHT, I., B. EHRENBERG, E. CALEF & R. ARNON. 1977. Biochem. Biophys. Res. Commun. **74:** 1302–1310.
9. KABAT, E. A., T. W. WU & H. BILOFSKY. 1977. Variable regions of Immunoglobulin Chains. Medical Computer Systems. Cambridge, MA.
10. POLJAK, R. J. 1978. CRC Crit. Rev. Biochem. **5:** 45–84.
11. DEISENHOFER, J., P. M. COLMAN, O. EPP & R. HUBER. 1976. Hoppe-Seyler's Z. Physiol. Chem. **357:** 1421–1438.
12. FIRCA, J. R., K. R. ELY, P. KREMSER, F. A. WESTHOLM, K. H. DORRINGTON & A. B. EDMUNDSON. 1978. Biochemistry **17:** 148–158.
13. LONGWORTH, J. W., C. L. MCLAUGHLIN & A. SOLOMON. 1976. Biochemistry **15:** 2953–2958.
14. SCHREIBER, A. B., A. D. STROSBERG & I. PECHT. 1978. Immunochemistry **15:** 207–212.
15. CATHOU, R. E. 1979. *In* Advanced Treaties on Immunology. R. A. Good & G. W. Littman, Eds. Plenum Press. N.Y.
16. PARKER, C. W., T. J. YOO, M. C. JOHNSON & S. M. GODT. 1967. Biochemistry **11:** 3408–3416.
17. LOPATIN, D. E. & E. W. VOSS. 1971. Biochemistry **10:** 208–213.
18. YGUERABIDE, J., H. F. EPSTEIN & L. STRYER. 1970. J. Mol. Biol. **51:** 573–590.
19. ZAGIANSKY, Y. A., R. S. NEZLIN & L. A. TUMERMAN. 1969. Immunochemistry **6:** 787–800.
20. LANCET, D. & I. PECHT. 1977. Biochemistry **16:** 5150–5157.
21. SCHEPERS, G., Y. BLATT, K. HIMMELSPACH & I. PECHT. 1978. Biochemistry **17:** 2239–2245.
22. KÖHLER, G. & C. MILSTEIN. 1975. Nature **256:** 495–498.
23. SCHLESSINGER, J., I. Z. STEINBERG & I. PECHT. 1974. J. Mol. Biol. **87:** 724–740.
24. ISENMAN, D. E., D. LANCET & I. PECHT. 1979. Biochemistry **18:** 3327–3336.
25. COWGILL, R. W. 1970. Biochim. Biophys. Acta **207:** 556–559.
26. GOTO, Y. & K. HAMAGUCHI. 1979. J. Biochem. **86:** 1433–1441.
27. MCGUIGAN, J. E. & H. N. EISEN. 1968. Biochemistry **7:** 1919–1927.
28. PARKER, C. W. & C. K. OSTERLAND. 1970. Biochemistry **9:** 1074–1082.
29. LOPES-DELGADO, R., A. TRAMER & I. H. MUNRO. 1974. Chem. Phys. **5:** 72–83.
30. MUNRO, I. H., I. PECHT & L. STRYER. 1979. Proc. Natl. Acad. Sci USA **76:** 56–60.
31. ALPERT, B., O. M. JAMESON, R. LOPES-DELGADO & R. SCHOOLEY. 1979. Photochem. & Photobiol. **30:** 479–481.
32. LAKOWICZ, J. R. & G. WEBER. 1973. Biochemistry **12:** 4171–4179.
33. MCCAMMON, J. A., B. R. BELIN & M. KARPLUS. 1977. Nature **267:** 585–590.
34. PETERSON, P. A., B. A. CUNNINGHAM, I. BERGGARD & G. M. EDELMAN. 1972. Proc. Natl. Acad. Sci. USA **69:** 1697–1671.
35. PHIZACKERLEY, R. P., B. C. WISHNER, S. H. BRYANT, L. M. AMZEL, J. A. LOPES DE CASTRO & R. J. POLJAK. 1979. Immunochemistry **16:** 841–850.
36. ELY, K. R., J. R. FIRCA, K. J. WILLIAMS, E. E. ABOLA, J. M. FENTON, M. SCHIFFER, N. C. PANAGIOTOPOULAS & A. B. EDMUNDSON. 1978. Biochemistry **17:** 158–167.
37. ABOLA, E. E., K. R. ELY & A. B. EDMUNDSON. 1980. Biochemistry **19:** 432–437.

THE MECHANISM OF NEGATIVE COOPERATIVITY IN RABBIT MUSCLE GLYCERALDEHYDE-3-PHOSPHATE DEHYDROGENASE

Yoav I. Henis* and Alexander Levitzki

Department of Biological Chemistry
Institute of Life Sciences
The Hebrew University of Jerusalem
Jerusalem, Israel

INTRODUCTION

The enzyme glyceraldehyde-3-phosphate dehydrogenase (EC 1.2.1.12; GPDH†) from numerous sources is a tetramer composed of subunits exhibiting identical molecular weights[1-3] and almost identical amino-acid sequences.[4-7] The rabbit muscle enzyme exhibits strong negative cooperativity in NAD^+ binding[8,9] and so do the enzymes from lobster muscle[10], sturgeon muscle[11] and *Bacillus stearothermophilus.*[12]

The negatively cooperative binding of NAD^+ to the enzyme was analyzed by different investigators in terms of two models: (a) the sequential model of Koshland *et al.*,[13] which assumes that binding of the coenzyme molecule induces a conformational change which causes a decrease in the affinity of the enzyme towards further ligand binding; this model has been employed to explain the mechanism of the negative cooperativity in the rabbit muscle enzyme,[8,9,15,16] and (b) the pre-existent asymmetry model,[14] which assumes the presence of two or more classes of noninteracting binding sites; this model has been used to explain the behavior of the rabbit muscle enzyme[17,18] and of the sturgeon muscle enzyme.[11,19] The binding curves of NAD^+ and its analogues to the enzyme can be fitted equally well by the two models. Therefore, in order to explore in more detail the mechanisms of negative cooperativity, other experimental approaches are usually needed.

The basic assumption of the pre-existent asymmetry model is that the apoenzyme is constructed such that the subunits are arranged in an asymmetric tetrahedron. X-ray crystallography provides a direct method to test this assumption. However, the data thus far published does not commit itself clearly to whether the enzyme is a symmetric tetramer or not. For example, possible asymmetry in the holoenzyme from lobster muscle was reported,[20,21] but preliminary studies with the apoenzyme indicated a conformational change upon the removal of the last NAD^+ molecule from the enzyme.[21] It is therefore possible that the asymmetry in the holoenzyme could arise.

* Present address: Department of Biochemistry, University of Tel Aviv, Ramat Aviv, Tel Aviv.

† *Abbreviations*: KNF model, the Koshland-Nemethy-Filmer model; PEA model, the pre-existent asymmetry model; MWC model, the Monod-Wyman-Changeux model; HEPES, N-2-hydroxy-ethylpiperazine-N′-2-ethanesulfonic acid; ϵNAD^+, nicotinamide-1-N^6-ethenoadenine dinucleotide; $A_cP_yAD^+$, acetyl-pyridine adenine dinucleotide; ADP-Rib, adenosine diphosphoribose; enzyme, glyceraldehyde-3-phosphate dehydrogenase (EC 1.2.1.12).

0077-8923/81/0366-0217$01.75/0 © 1981, NYAS

from a conformational change induced upon coenzyme binding, especially since differences in the conformation of the adenine-binding pocket and the S-loop were observed between the apo- and the holo-enzyme.[22] Crystallographic studies were performed also on the enzyme from *B. stearothermophilus*, where both the holo- and the apo-enzyme were found to possess a 2:2:2 tetrahedral symmetry.[23,24] These studies also suggest the possibility of a difference between the conformations of the apo- and the holo-enzyme.

Other experimental approaches, such as active site alkylation and acylation.[11,17,25] stabilization of the enzyme against proteolytic degradation,[26] and studies utilizing a variety of spectroscopic methods[15] led to conflicting conclusions. The difficulty in all these approaches is that most of the data can be interpreted by both models. We therefore recognized the necessity to develop a new approach which would be both unambiguous and experimentally feasible and allow a clear decision as to what molecular model should be most appropriate whenever a case of negative cooperativity is encountered. In this communication, we report on a new approach which enables one to delineate the mechanism of negative cooperativity on the basis of ligand binding experiments alone. A full theoretical discussion of this new approach to be described below is presented elsewhere.[27]

Materials and Methods

Materials

Glyceraldehyde-3-phosphate dehydrogenase was prepared from rabbit muscle as previously described,[28] and the apoenzyme was prepared by charcoal treatment.[28] ATP, ADP, AMP, NAD^+, ADP-Rib, and $AcPyAD^+$ were obtained from Sigma (St. Louis, Mo.). $AcPyAD^+$ was further purified from NAD^+ contamination by passing a volume of 24 ml of a solution containing 1×10^{-4} M apoenzyme and 1.32 mM $AcPyAD^+$ on a Sephadex G25 column (5.2 cm × 50 cm), pre-equilibrated with 5 mM NH_4HCO_3 (pH 8.4). Due to the much higher affinity of NAD^+ to the enzyme, essentially all the NAD^+ elutes with the protein peak (right after the void volume), and a peak of $APyAD^+$ with no NAD^+ contamination follows. The complete removal of NAD^+ was verified by including $[^{14}C]$-NAD^+ in the mixture loaded on the column, which resulted in all the radioactivity eluting with the protein peak. After lyophilization, the enzymatically determined purity of the $AcPyAD^+$ preparation was above 99.5%.

ϵNAD^+ was prepared as previously described[29] and purified according to Schlessinger and Levitzki.[15] Nicotinamide-[U-^{14}C]-adenine dinucleotide (302 mCi/mmol) and α-[^{32}P]-ATP (7.7 Ci/mmol) were obtained from the Amersham Radiochemical Center, England.

Binding Measurements of Coenzyme Analogues

Binding and competition measurements were performed in 0.05 M HEPES, 0.01M EDTA, pH 7.5, at 25°C, except where mentioned otherwise. In all measurements, tetramer concentrations above 8 μM were employed, in order to prevent dissociation of the enzyme into dimers, which occurs only at lower enzyme concentrations.[3]

AcPyAD$^+$ binding was measured by the intensity of the difference spectrum formed upon its binding to the enzyme at 350 nm,[30] following the procedure described previously.[28] The intensity of this difference-spectrum band was found to be proportional to AcPyAD$^+$ binding to the enzyme, as is the case with other NAD$^+$ analogues.[15,28,31]. The extinction coefficient for bound AcPyAD$^+$ was ϵ_{350} = 1430 $M^{-1}cm^{-1}$.

ϵNAD$^+$ binding measurements were based on the enhancement in ϵNAD$^+$ fluorescence upon binding to the enzyme. The fluorescence was measured on a Hitachi-Perkin Elmer MPF-4 spectrofluorimeter at 25°C, in a 1.0 mm cuvette positioned at 45° with respect to both the exciting and the emitted light. A barrier filter blocked off emitted light below 350 nm in order to minimize light-scattering. Fluorescence was excited at 325 nm, and emission was measured at 389 nm. For ϵNAD$^+$ concentrations above 1.2×10^{-4} M, a correction for the decrease in fluorescence due to absorption of the exciting light was required. This procedure involved titrating a buffer solution with ϵNAD in the absence of enzyme, employing ϵNAD$^+$ concentrations identical to those used for titrating the enzyme. Up to 1.2×10^{-4} ϵNAD$^+$, a linear plot of free ϵNAD$^+$ fluorescence *versus* its concentration is obtained. At higher ϵNAD$^+$ concentrations, the ratio between the extrapolated straight line and the actual ϵNAD$^+$ fluorescence intensity yields a correction factor for that specific ϵNAD$^+$ concentration. The fluorescence intensity observed for this ϵNAD$^+$ concentration in the presence of the enzyme is then multiplied by that factor to give the corrected fluorescence intensity. The correction did not exceed 5% in all cases.

NAD$^+$ and ATP binding was measured employing radioactively labeled ligands, by the dialysis rate method of Colowick and Womack.[32] To prevent dilution during titration, the titrating solution contained protein and radioactive ligand in addition to unlabeled ligand at concentrations identical to those present in the titrated solution. The dimensions of the dialysis-rate cell were 1 ml and 0.1 ml in the upper and in the lower chambers, respectively. The separating dialysis membrane was Visking cellulose tubing 27/32. The flow rate was 2 ml/min, and 10 samples (1 ml each) were collected subsequent to each addition of ligand to the upper chamber. A steady state of the counts in the effluent was obtained after 5 ml. Bound and free ligand concentrations were computed from the steady-state counts as previously described.[32]

Binding Curve Analysis

Binding data were computer analyzed using a nonlinear regression curve-fitting procedure.[33] The fit was usually done emplloying the general Adair equation for a tetramer.

$$\frac{[X]_B}{[E]_t} = \frac{\dfrac{[X]}{K_1} + \dfrac{2[X]^2}{K_1K_2} + \dfrac{3[X]^3}{K_1K_2K_3} + \dfrac{4[X]^4}{K_1K_2K_3K_4}}{1 + \dfrac{[X]}{K_1} + \dfrac{[X]^2}{K_1K_2} + \dfrac{[X]^3}{K_1K_2K_3} + \dfrac{[X]^4}{K_1K_2K_3K_4}} \quad (1)$$

Where $[E]_t$ is the total tetramer concentration, $[X]_B$ and $[X]$ are the bound and free ligand concentrations, and K_1 through K_4 are the thermodynamic dissociation constants for the binding of the first through the fourth ligand molecules to the tetramer. The intrinsic dissociation constants (the dissociation constants per site) for the binding of the ith ligand molecule to the enzyme, K_i', were computed from the values obtained for the corresponding thermodynamic dissociation constants, K_i, using the formula:[34,35] $K_i' = \frac{(n - i + 1)}{i} K_i$ where n is the number of binding sites on the enzyme (4 in the case of glyceraldehyde-3-phosphate dehydrogenase). When all i-ligand molecules are bound with the same affinity, K_i' represents the exact affinity of the ligand to a binding site in EX_i; if they are bound with different affinities, K_i' represents the average affinity between a binding site and the ligand.

Results

Theory

Negatively cooperative ligand-binding isotherms can usually be fitted by more than one molecular model. Therefore, such curves alone are insufficient to establish a specific mechanism of negative cooperativity. In the present study, we describe the utilization of a new approach[27] to establish the mechanism of negative cooperativity in coenzyme binding to rabbit-muscle glyceraldehyde-3-phosphate dehydrogenase. The approach involves the determination of the effect of one ligand on the mode of binding of a second ligand. Namely, the binding of the primary ligand is examined in the presence and in the absence of a competing ligand and the modulation of the cooperativity in the binding of the primary ligand is examined. It is relatively easy to demonstrate[27] that if the primary ligand binds with negative cooperativity and the competing ligand binds noncooperatively, any change in the extent of negative cooperativity in the binding of the primary ligand in the presence of the noncooperative competitive ligand excludes the pre-existent asymmetry (PEA) mechanism as the mechanism of choice. It should be noted that the same approach can also be used to distinguish between the KNF model and the model of Monod *et al.*[36] in the case of positively cooperative ligand binding.[27] Detailed mathematical analysis shows that the use of a noncooperative competing ligand provides the most clearcut way of distinguishing among the cooperative models, but the use of a cooperative competing ligand—although more complex—is also feasible.[27] We shall now describe the essential features of this approach.

Consider a dimer or a tetramer possessing two classes of noninteracting sites (the simplest case of the PEA model). The Hill coefficient at 50% saturation by a ligand X is given in such a case by:[34,35]

$$n_H(X,Z) = \frac{4}{2 + (K_x'/K_x'')^{1/2} + (K_x''/K_x')^{1/2}} \tag{2}$$

where K_x', K_x'' are the intrinsic association constants for the binding of ligand X to the two classes of sites, respectively. When a competing ligand Z is present in a concentration much higher than the binding-site concentration (so that the free Z concentration remains essentially constant during the titration with X), the Hill coefficient for X binding in the presence of Z, $n_H(X, Z)$, becomes:[27]

$$N_H(X,Z) = \frac{4}{2 + \dfrac{K_x' + K_x'' + (K_x'K_z'' + K_z'K_x'')[Z]}{(K_x'K_x'')^{1/2} + \{1 + (K_z' + K_z'')[Z] + K_z'\,K_z''\,[Z]^2\}^{1/2}}} \tag{3}$$

where K_z' and K_z'' are the intrinsic association constants of Z towards the two classes of sites, respectively. If Z binding is noncooperative, $K_z' = K_z''$, and Equation 3 reduces to Equation 2—namely, a noncooperative competing ligand Z cannot affect the cooperativity of X-binding according to the PEA model. If Z-binding itself is negatively cooperative ($K_z' \neq K_z''$), the presence of Z will have an effect on the cooperativity of X-binding (Ref. 27; see also TABLE 1). In contrast to the PEA model, the KNF model allows for a change in the cooperativity of X-binding in the presence of a noncooperative competing ligand. In the simplest case of the KNF model, where the conformational change is limited to the ligand-binding subunit, one obtains for the binding of X to a dimer or to a multimer of dimers:[34,35]

$$n_H(X) = \frac{2}{1 + (K_{AB}^2/K_{BB})^{1/2}} \tag{4}$$

where A and B are the conformations of unliganded and X-liganded subunits, respectively. K_{AB} and K_{BB} are the subunit interaction constants between an unliganded and an X-binding subunit.[13] In the presence of a high Z concentration, one obtains:[27]

$$n_H(X, Z) = \frac{2}{\left\{\dfrac{1 + K_{AB}^2\left(1 + \dfrac{K_{BC}}{K_{AB}} K_{ZC}K_{t_{AC}}[Z]^2\right)}{K_{BB}\{1 + 2K_{AC}K_{ZC}K_{t_{AC}}[Z] + K_{CC}(K_{Zc}K_{t_{AC}})^2[Z]^2\}}\right\}^{1/2}} \tag{5}$$

where C is the conformation of a Z-liganded subunit. K_{BC} is a subunit interaction parameter describing the interaction between an X-liganded and a Z-liganded subunit, while K_{AC} is the parameter representing the interaction between an unliganded subunit and a Z-liganded subunit. K_{ZC} is the intrinsic association constant of Z to a subunit in the C conformation, and $K_{t_{AC}}$ describes the free energy of the transformation of a subunit from conformation A to C. All these parameters are as originally defined by Koshland *et al.*[13,37]

Comparing Equations 5 and 4, it is apparent that the presence of Z can affect the cooperativity in the binding of X, depending on the relations between all the interaction constants. In the KNF model, a ligand Z binds noncooperatively only if $K^2_{AC}/K_{CC} = 1$.[13] However, even if this is the case, K_{BC}/K_{AB} in Equation 5 can obtain any value. Thus, according to the KNF model, Equation 5 does not necessarily reduce to Equation 4 even when Z-binding is noncooperative, namely, a situation where $n_H(X, Z) \neq n_H(X)$ can still be obtained. Only in the special case where $K_{BC}/K_{AB} = K_{AC}$ *in addition* to the noncooperative binding of Z, will $n_H(X, Z)$ equal $n_H(X)$. In the general KNF model, where conformational changes can be transmitted to neighboring subunits,[37] similar behavior is found, except that more variability in the effect of Z is possible.[27] Qualitatively similar behavior is encountered in the general case of a tetrameric protein, as can be shown using lengthy and more involved algebraic manipulations.[27] Similar to the simpler cases discussed here, it is found that the KNF model can account for a change in the cooperativity of X-binding in the presence of a noncooperative competing ligand, while the PEA model cannot. These conclusions can also be applied to higher oligomers. Analogous treatment of the Monod-Wyman-Changeux model for positive cooperatively[36] shows differences in the predictions of this model and the KNF model for the effect of a competing ligand on the cooperativity of the primary ligand.[27] The qualitative effects of competing ligands on the cooperativity of the displacing ligand according to the different cooperativity models are summarized in TABLE 1.

Ligand Binding Measurements

In order to investigate the mechanism of the negatively cooperative binding of coenzyme to glyceraldehyde-3-phosphate dehydrogenase, one has to find both co-

TABLE 1

THE QUALITATIVE EFFECT OF A COMPETING LIGAND Z ON n_H FOR X BINDING TO AN OLIGOMER*

Model	Effect of a Cooperative Z on n_H for X	Effect of a Non-cooperative Z on n_H for X	
		Type 1	Type 2
PEA	$n_H(X)$ changes but always $n_H(X, Z) \leq 1$.	$n_H(X)$ does not change:	$n_H(X, Z) = n_H(X)$.
MWC† (positive cooperativity)	$n_H(X)$ changes but always $n_H(X, Z) \geq 1$.	$n_H(X)$ changes only to higher values: $n_H(X, Z) > n_H(X)$.	$n_H(X)$ does not change: $n_H(X, Z) = n_H(X)$.
Simplest KNF	Any change in $n_H(X)$ is possible. $n_H(X, Z) > 1$ can be obtained even when $n_H(X) < 1$, and vice versa. $n_H(X, Z) \neq n_H(X)$.	Any change in $n_H(X)$ is possible: $n_H(X, Z) \neq n_H(X)$.	$n_H(X)$ does not change: $n_H(X, Z) = n_H(X)$.
General KNF	Similar to simplest KNF, but more variability possible. $n_H(X, Z) \neq n_H(X)$.	Similar to simplest KNF, But more variability possible. $n_H(X, Z) \neq n_H(X)$.	Any change in $n_H(X, Z) \neq n_H(X)$. Special case: when Z binds with identical affinity to all conformations, $n_H(X, Z) = n_H(X)$.

* In the PEA model, a noncooperative ligand is a ligand which binds with identical affinities to all sites. In the MWC model, a noncooperative ligand of Type 1 is a ligand which binds with a higher affinity to the T conformation in a system where $L \gg 1$ ($L = [T]/[R]$, where R and T are the high affinity and the low affinity forms, respectively[36]). A noncooperative ligand of Type 2 in the MWC model is a ligand which binds with identical affinities to the R and T conformation. In the KNF model, a noncooperative ligand of Type 1 is a ligand whose binding induces a conformational change (from A to C) in the ligand-binding subunits, but this change has no effect on further Z binding to neighboring subunits ($K_{AC}^2/K_{CC} = 1$). A noncooperative ligand of Type 2 is a ligand which does not induce a conformational change upon binding. A noncooperative ligand of Type 1 in the KNF model can behave as a noncooperative ligand of Type 2 in the same model if the conformational change it induces does not affect X binding to the neighboring subunits.[27]

† The MWC model can account for positive cooperativity and not for negative cooperativity. This case is analyzed fully elsewhere.[27]

operative and noncooperative NAD^+ analogues. Of these analogues, the enzymatically active analogue ϵNAD^+[29,38] which binds to the enzyme with negative cooperativity, is very suitable for use as the primary ligand. The enhancement of ϵNAD^+ fluorescence intensity upon binding[15,39] can be used to monitor its binding to the enzyme in the presence of other competing nonfluorescent analogues.

The effect of ϵNAD^+ binding to the enzyme on the ϵNAD^+ fluorescence was studied under conditions where all the ϵNAD^+ present is bound to the enzyme (5.5 μM ϵNAD^+, 50 μM apoenzyme). The blue shift in the ϵNAD^+ fluorescence occurring upon its binding is accompanied by an enhancement in the fluorescence intensity.[15,39] When the fluorescence was excited at 325 nm, a 6-fold enhancement in the fluorescence intensity was obtained at the emission wavelength of 389 nm. A similar enhancement factor was found when the enzyme was saturated with ϵNAD^+, demonstrating that the increase in the quantum yield is identical whether one or four ϵNAD^+ molecules are bound to the enzyme. This is in accord with the finding that the quantum yield per bound ϵNAD^+ is independent of the number of ϵNAD^+ molecules bound to the enzyme.[15,40] $[\epsilon NAD^+]_B/[E]_t$ was computed from the relationship $r = \frac{[\epsilon NAD^+]_B}{[E]_t} = \frac{\Delta F}{\Delta F\infty} \times 4$, where $[\epsilon NAD^+]_B$ is the concentration of bound ϵNAD^+. ϵNAD^+ binding to the enzyme was found to exhibit strong negative cooperativity (FIGURE 1, TABLE 2).

Strong negative cooperativity is also encountered in NAD^+ binding, as was reported earlier.[8,9] The extrapolation of the Scatchard plot of NAD^+ binding to infinite ligand concentration (FIGURE 1) yields $K_4' = (0.54 \pm 0.05) \times 10^{-5}M$, or $K_4 = 2.16 \times 10^{-5}M$, where K_4' and K_4 are the intrinsic and thermodynamic dissociation constants for the binding of the fourth NAD^+ molecule to the enzyme, respectively. Strong negative cooperativity between the third and fourth NAD^+ molecules is observed, and the dissociation constants for the binding of the first three NAD^+ molecules cannot thus be determined from this experiment. Calculation of all four consecutive binding constants can only be a noncooperative competing ligand which considerably diminishes the affinity of NAD^+ binding and makes its determination experimentally feasible.

Unlike the binding of NAD^+ and ϵNAD^+, the binding of $AcPyAD^+$ and ATP to the enzyme was found to be noncooperative (FIGURE 1, TABLE 3). In the case of ATP, the noncooperative binding extrapolates to 4 ATP bound/tetramer as shown in FIGURE 1. ATP binding was measured in the presence of 5 mM P_i, in order to block ATP binding to eight "nonspecific" binding sites which exist on the enzyme.[21,41] This P_i concentration is sufficient to block ATP binding at the concentrations employed (up to 1.1 mM ATP), to the P_i binding sites, since the K_m for P_i is 1.2 mM at pH 7.4.[42] In the absence of P_i, ATP binding extrapolates to approximately 12 ATP/tetramer, where the additional ATP molecules bind with a significantly lower affinity than the first 4 ATP molecules to the 4 P_i sites and presumably to the glyceraldehyde-3-phosphate binding sites (data not shown; see Reference for details).[43] In order to ensure that the 4 ATP molecules which bind in the presence of P_i bind to the NAD^+ sites, ATP binding was also measured in the presence of 5 mM NAD^+, which block ATP binding to the NAD^+ sites. Under these conditions, the four tighter ATP binding sites disappeared, and eight low-affinity sites, which bind ATP with an intrinsic dissociation constant of $(7.0 \pm 1) \times 10^{-4}M$, are observed

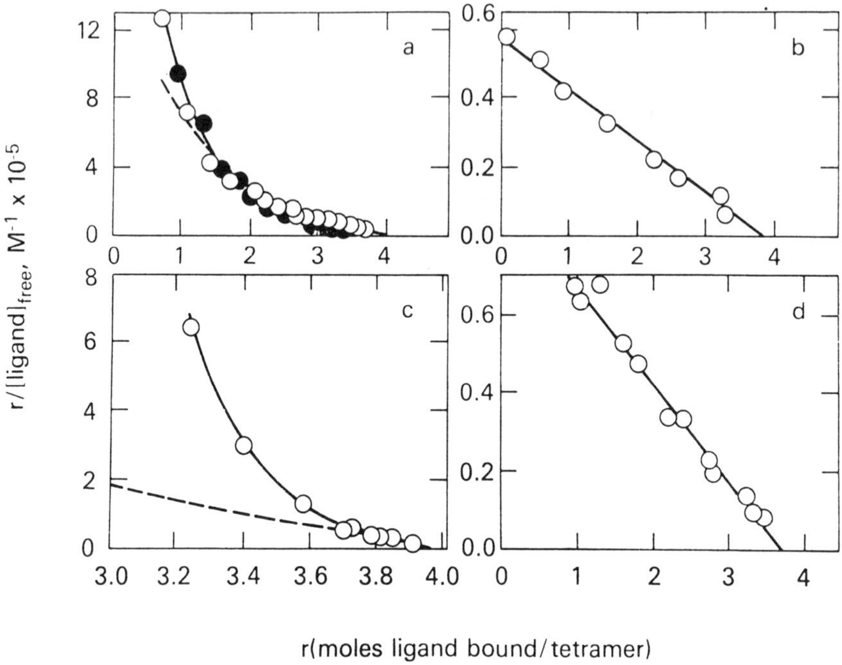

FIGURE 1. Binding of coenzyme analogues to the NAD^+ Binding sites on the enzyme. Measurements were performed as described in MATERIALS AND METHODS. Data are presented as Scatchard plots.[44]

(a) ϵNAD^+ binding, ●,○ represent two separate experiments. Solid lines, best fit to Equation 1. Dashed line, best fit to the two-classes-of-sites model: $[X]_B/[E]_t = 2[X] \{K' + [X]\} + 2(X)/\{K'' + [X]\}$, where K′ and K″ are the intrinsic dissociation constants for the binding of X to the two classes of sites, respectively. The enzyme concentration was 0.84×10^{-5}M. As can be seen from the binding data alone, it is not possible to eliminate the two-classes-of-sites model, although it has only two parameters as compared to four parameters in Equation 1 where all four binding steps are used. The correlation coefficients are 0.9997 for Equation 1 and 0.9995 for the two-classes-of-sites model.

(b) ATP binding, in the presence of 5 mM P_i. The enzyme concentration was 5.1×10^{-5}M, and the [α-^{32}P] - ATP concentration was 2.6 μM (7.7 Ci/mmol). All solutions contained also 5 mM P_i, to prevent binding to the P_i sites (see text). Nonspecific binding of ATP to the dialysis membrane in the flow dialysis method was measured by titrating a solution without enzyme under identical conditions, and subtracted.

(c) NAD^+ binding. The enzyme concentration was 1.0×10^{-5}M, and the [^{14}C]-NAD^+ concentration was 0.74×10^{-5}M (302 mCi/mmol). No nonspecific binding of NAD^+ to the dialysis membrane was observed.

(d) $AcPyAD^+$ binding. The enzyme concentration was 2.75×10^{-5}M. $\epsilon_{350} = 1430\ M^{-1}\ cm^{-1}$ mCi was employed to calculate the concentration of bound $AcPyAD^+$ from the increase in absorption at 350 nm.

(FIGURE 2). This value is in good agreement with the inhibition constant of ATP with respect to P_i in the enzyme-catalyzed reaction, which is $8 \times 10^{-4}M$ at pH 7.4.[42] These findings indicate that the four tighter ATP binding sites correspond to the NAD^+ binding sites on the enzyme, and that ATP also binds to the P_i-binding sites.

TABLE 2
INTRINSIC DISSOCIATION CONSTANTS FOR εNAD+ BINDING TO THE ENZYME*

	Intrinsic dissociation constants, $M \times 10^5$	Constants from Reference 15, $M \times 10^5$	Constants from Reference 39, $M \times 10^5$
K_1' †	0.15 ± 0.03	N.D.‡	N.D.
K_2'	0.63 ± 0.09	N.D.	N.D.
K_3'	1.3 ± 0.2	1.7	3.5
K_4'	1.5 ± 0.2	1.7	3.5

* Curve fitting was performed using Equation 1, and then converting the thermodynamic dissociation constants into the intrinsic dissociation constants. The constants from Reference 15 were obtained by re-analyzing the data employing Equation 1. In Reference 39, an enzyme containing two NAD^+ molecules was used instead of the apoenzyme, and the presence of the bound NAD^+ may be the reason for the higher dissociation constants obtained for K_3' and K_4'.

† K_i' are the intrinsic dissociation constants for the binding of the ith NAD^+ molecule to the enzyme.

‡ N.D. = Not Determined.

TABLE 3
INTRINSIC DISSOCIATION CONSTANTS FOR THE BINDING OF NONCOOPERATIVE NAD+ ANALOGUES*

Ligand	Intrinsic Dissociation Constant, $M \times 10^5$
ATP	6.7 ± 0.7
$AcPyAD^+$	4.0 ± 0.4
ADP-Rib	4.5 ± 0.5

* The constants for ATP and $AcPyAD^+$ were derived from the Scatchard[44] plots shown in FIGURE 1. The effect of ADP − Rib on ϵNAD^+ binding to the enzyme was found to be purely competitive (FIGURE 3; see text), and thus the dissociation constant for ADP-Rib could be derived from its competition with ϵNAD^+ which is given by:[40]

$$[\epsilon NAD^+]_{0.5}^{obs} = [\epsilon NAD^+]_{0.5}\left(1 + \frac{[\text{ADP-Rib}]}{K'}\right)$$

Where $[\epsilon NAD^+]_{0.5}$ and $[\epsilon NAD^+]_{0.5}^{obs}$ are the free ϵNAD^+ concentrations yielding 50% saturation in the absence and in the presence of ADP-Rib, respectively. K' is the intrinsic dissociation constant for the binding of ADP-Rib to the enzyme.

Effect of Noncooperative NAD^+ Analogues on ϵNAD^+ Binding

Competition experiments were carried out primarily with ϵNAD^+ as the displacing ligand. The main reasons for this choice were that, unlike in the case of NAD^+, it was possible to characterize the full binding curve of this analogue to the enzyme in the absence of competing ligands, and that the ϵNAD^+ fluorescence enhancement upon binding provided a quick and accurate measure for its binding. In all experiments, the competing noncooperative NAD^+ analogues were present in a large molar excess over the binding sites and over their respective dissociation constants. This

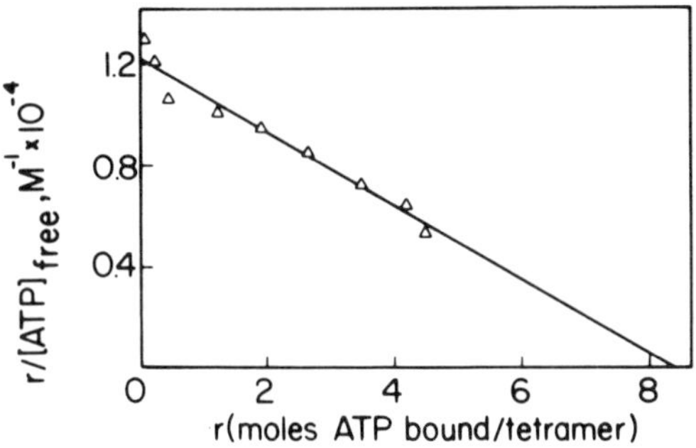

FIGURE 2. ATP binding to the enzyme in the presence of 5 mM NAD^+.
The enzyme concentration was 8.5×10^{-5}M, and the (α^{32}P) - ATP concentration was 2.6 μM (7.7 ci/mmol). All solutions also contained 5 mM NAD^+, to prevent ATP binding to the NAD^+ sites. Nonspecific binding of ATP to the dialysis membrane was measured as described for FIGURE 1b, and substrated.

ensured that the concentration of free-competing ligand does not change during the titration with the displacing ligand, and that the competing ligand saturates the binding sites, thus yielding the maximal possible effect on the cooperativity of the displacing ligand.[27]

The results of the competition experiments with ϵNAD^+ as the displacing ligand are summarized in FIGURE 3 and TABLE 4. In the presence of saturating $AcPyAD^+$ concentrations, the negative cooperativity in the binding of ϵNAD^+ to the enzyme is completely abolished (FIGURE 3, TABLE 4). Similarly, ATP concentrations close to saturation strongly decrease the extent of negative cooperativity observed in ϵNAD^+ binding (FIGURE 3, TABLE 4). Since ATP binds also to the P_i-binding sites, it was necessary to test whether the ATP effect results from competition for the NAD^+ sites, or from noncompetitive interaction with the P_i-binding sites. It was found that the effect of ATP on the cooperativity of ϵNAD^+ binding is identical in the presence and in the absence of 5 mM P_i, and that P_i itself has no effect at all on the cooperativity of ϵNAD^+ binding (FIGURE 3, TABLE 4). Thus it seems that the effect of ATP on ϵNAD^+ binding results from the interaction of the former with the coenzyme binding sites.

Saturating concentrations of ADP and AMP have a similar effect to that of ATP on the cooperativity of ϵNAD^+ binding (FIGURE 3, TABLE 4). The dissociation constants for the binding of AMP, ADP and ATP, as derived from the inhibition of ϵNAD^+ binding, are summarized in TABLE 5.

The effect of $AcPyAD^+$ and ATP on the cooperativity of ϵNAD^+ binding is not due to effects of these ligands on the aggregation state of the tetramer, since ultracentrifugation experiments (sedimentation velocity) under conditions identical with those employed in the competition experiments (0.85×10^{-5}M apoenzyme, 0.6 mM ATP or $AcPyAD^+$) showed identical patterns for apoenzyme alone and in the presence of ATP or $AcPyAD^+$ (data not shown). This is in accord with the finding that at 20°C or above, ATP does not cause dissociation of the tetrameric enzyme.[45]

Unlike the significant effects of $AcPyAD^+$ and ATP on the cooperativity of ϵNAD^+ binding, ADP-Rib had no effect on the shape of the ϵNAD^+ binding curve,

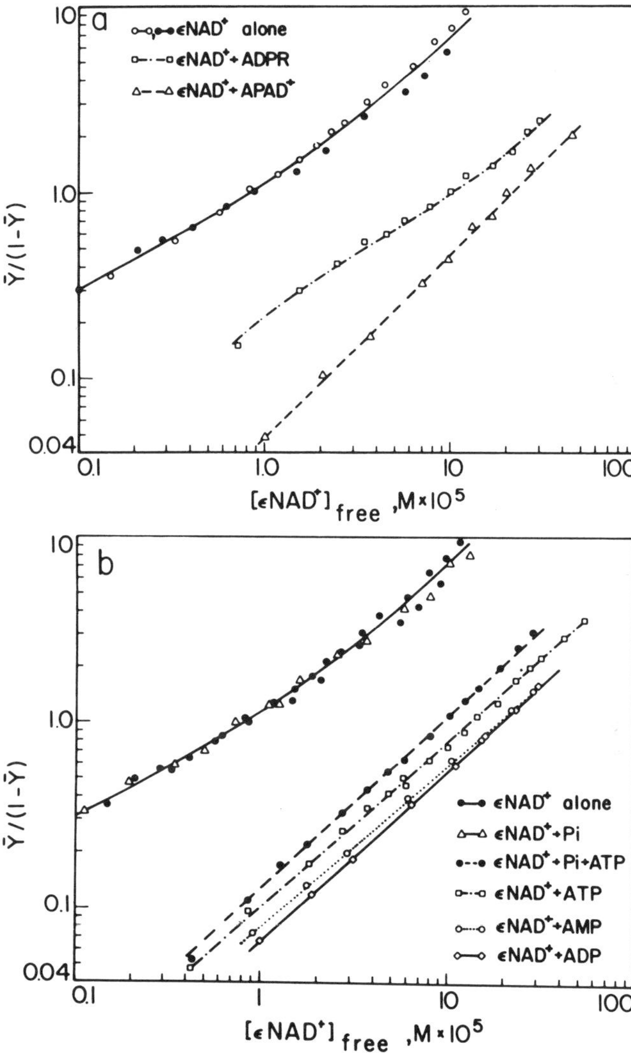

FIGURE 3. Effect of competing ligands on ϵNAD^+ binding to the enzyme. Data are presented in the form of Hill plots[47], $\bar{y}$ is $[NAD^+]^{bound}/4[E]_{total}$, The enzyme concentration was between 8 and 9 μM,

(a) ϵNAD^+ binding alone, and in the presence of 0.41 mM $AcPyAD^+$ ($APAD^+$) (△- -△) or 0.53 mM ADP - Rib (□-·-·□). Two separate experiments for the binding of NAD^+ alone are shown (● and ○), to show the correlation between separate experiments.

(b) ϵNAD^+ binding alone (●-●) and in the presence of 5 mM P_i (△–△), 0.6 mM ATP (□-·-·□), 0.65 mM ATP plus 5 mM P_i (●- -●), 1.63 mM AMP (○···○), and 0.92 mM ADP (◇–◇).

TABLE 4
n_H HILL COEFFICIENTS FOR ϵNAD^+ BINDING TO GPDH IN THE PRESENCE OF SATURATING COMPETING LIGANDS*

Competing Ligand	$n_H(\epsilon NAD^+)$
None	0.64 ± 0.03
ADP-Rib	0.65 ± 0.04
$AcPyAD^+$	1.00 ± 0.04
ATP	0.89 ± 0.02
ATP + P_i	0.90 ± 0.03
ADP	0.93 ± 0.05
AMP	0.88 ± 0.05

* The Hill coefficients were obtained from the slope of the Hill plots at mid-saturation with ϵNAD^+ (FIGURE 3). Each experiment was repeated at least three times, and the standard deviation determined using S.D. $\sqrt{=\sum (\bar{n}_H - n_H/N}$, where $\bar{n}_H$ is the average of the experimental n_H values, and N is the number of individual n_H measurements.

TABLE 5
THE APPROXIMATE INTRINSIC DEVIATION CONSTANTS FOR THE BINDING OF AMP, ADP, AND ATP TO GPDH*

Ligand	Intrinsic Dissociation Constant, $M \times 10^5$
AMP	8.1
ADP	4.4
ATP	3.8

* The intrinsic dissociation constants were derived from the competition experiments with ϵNAD^+ (FIGURE 3), employing the same equation used in TABLE 3 to derive the dissociation constant for ADP-Rib binding from its competition with ϵNAD^+. The constants obtained are only approximate, as this equation is exact only for cases where the competing ligand does not affect the cooperativity of the displacing ligand. Thus, for ADP-Rib this equation is exact, whereas in the cases of AMP, ADP, and ATP, it yields only approximate values. The deviation from the true values is not substantial, as can be seen by the finding that the value obtained for ATP is close to the value obtained by direct ATP binding measurements ($6.7 \times 10^{-5} M$).

and only shifted it to higher ϵNAD^+ concentrations (FIGURE 3, TABLE 4). Such a behavior can be encountered only if the competing ligand binds noncooperatively and, in addition, its binding is not affected by (and does not affect) the binding of ϵNAD^+ to neighboring subunits.[27] Thus, the effect of ADP-Rib on ϵNAD^+ binding is purely competitive. This is in accord with the earlier finding by Eby and Kirtley that ADP-Rib binds noncooperatively to the enzyme.[46]

Effect of Noncooperative NAD^+ Analogues on NAD^+ Binding

The binding of the first three NAD^+ molecules to the enzyme is too tight to permit accurate determination of n_H for the binding of NAD^+ alone (FIGURE 1). Therefore, we were restricted to the comparison between the n_H values for NAD^+ binding in the presence of different noncooperative competing analogues. The results are depicted in FIGURE 4 and TABLE 6. The effects of ATP and ADP-Rib on the negative cooperativity of NAD^+ binding are very different from each other. These results immediately rule out the PEA model also for NAD^+ binding, since this model predicts similarly shaped binding curves (no effect on the cooperativity of NAD^+ binding) for *all* noncooperative competing ligands. The picture obtained for ϵNAD^+ binding is also encountered in the binding of the natural coenzyme, NAD^+, and the conclusions drawn from the experiments with ϵNAD^+ seem to be valid also for the mechanism of NAD^+ binding.

Assuming that the ADP-Rib binding has no effect on the cooperativity of NAD^+ binding in the same manner that it does not affect the cooperativity of ϵNAD^+ binding (as indicated by the similar n_H values obtained for ϵNAD^+ and NAD^+ binding in the presence of ATP or ADP-Rib), all four intrinsic dissociation constants for NAD^+ binding could be calculated from the binding curve of NAD^+ in the presence of 10.46 mM ADP-Rib (TABLE 7). It should be noted that this is the first reported accurate determination of the dissociation constants of the first two NAD^+ molecules from the enzyme, as the binding of these is too tight to permit an accurate *direct* measurement at enzyme concentrations high enough to keep it as a tetramer.‡

DISCUSSION

Role of Coenzyme-induced Conformational Changes in the Negative Cooperativity of the Enzyme

As shown in the theoretical section (see RESULTS), the different models for cooperativity make different predictions as to the effects of competing ligands on the cooperativity of the primary ligand (TABLE 1). Specifically, it is shown that the PEA model does not allow for any modulation in the cooperativity of the primary ligand due to competition by a noncooperative ligand, whereas the KNF model allows for such a modulation (TABLE 1). Thus, the use of competition experiments provides an experimental tool for testing the validity of the various cooperativity models in specific cases.[27]

In the case of rabbit muscle glyceraldehyde-3-phosphate dehydrogenase with which this study is concerned, strong negative cooperativity is observed in the binding of ϵNAD^+ and NAD^+, whereas $AcPyAD^+$ and ATP bind to the NAD^+ sites noncooperatively. The use of the ϵNAD^+ fluorescence enhancement enabled us to measure the full binding curve of ϵNAD^+ to the enzyme, and derive all four intrinsic dissociation constants (TABLE 2).

The negative cooperativity in ϵNAD^+ binding to the enzyme is completely abolished in the presence of saturating $AcPyAD^+$ concentrations, and strongly weakened

‡ This assumption is corroborated by the fact that calculating the NAD^+ binding curve in the presence of ADP-Rib from the data on the binding of NAD^+ alone (FIGURE 1), employing the competition-derived ADP-Rib dissociations constant from TABLE 3, yields a very good fit to the actual experimental data on the binding of NAD^+ in the presence of the same ADP-Rib concentration (see FIGURE 4a). The similarity in the changes of the n_H values for ϵNAD^+ and NAD^+ binding upon changing the competing ligand from ADP-Rib to ATP (TABLES 4 and 6) also support this assumption.

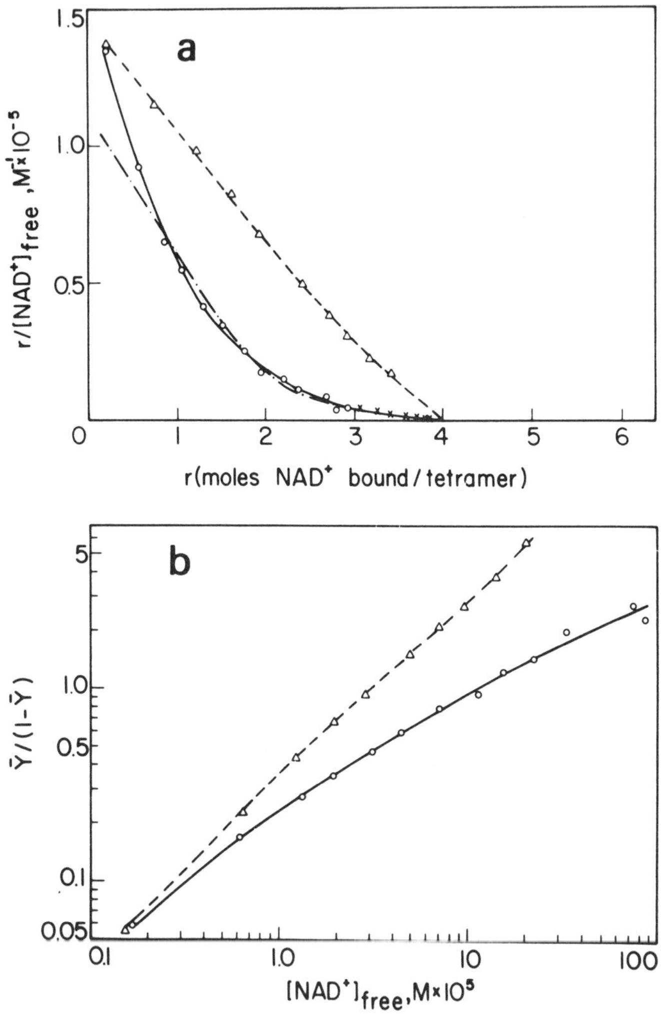

FIGURE 4. NAD$^+$ binding to the enzyme in the presence of ADP - Rib and ATP. The enzyme concentration was 2.7×10^{-5}M, and the [^{14}C]-NAD$^+$ was 0.74×10^{-5}M(302 mCi/mmol). O– –O, titration in the presence of 10.45 mM ADP - Rib, best fit according to Equation 1; O·–·O, same titration, best fit according to the two-classes-of-sites model (see legend of FIGURE 1a for details); Δ– –Δ, titration in the presence of 11.85 mM ATP, best fit according to Equation 1, As for NAD$^+$ binding, NAD$^+$ binding can be rather well-fitted also to the two-classes-of-sites model.

(a) Scatchard plots. X—X are points taken from NAD$^+$ binding in the absence of competing ligand (FIGURE 1C), and transferred to the conditions of 10,47 mM ADP - Rib on the assumption of a purely competitive inhibition; namely, by multiplying $[NAD^+]_{free}$ by [ADP - Rib]/$K'_{ADP\text{-}Rib}$, which is 1045/4.5 under the conditions employed. As can be seen, the fit is very satisfactory.

(b) Hill plot·$\bar{y}$ is $[NAD^+]_{bound}/4[E]_{total}$

in the presence of saturating concentrations of ATP, ADP or AMP (FIGURE 3, TABLE 4). All these experiments were performed under conditions where the enzyme remains in its tetrameric form [3; see also RESULTS].

As ATP and AcPyAD$^+$ have been shown to bind noncooperatively to the NAD$^+$ binding sites, the results of the competition experiments eliminate a pure PEA model, where intersubunit interactions do not exist, as a possible explanation for the negative cooperativity in coenzyme binding to this enzyme. In fact, these findings demonstrate the existence of intersubunit interactions within the tetrameric enzyme, since the binding of a noncooperative competing ligand affects the cooperativity of the primary ligand. Such an effect can occur only if there is a conformational communication within the tetrameric enzyme, and easily accounted for by the KNF sequential model. Clearly, a mechanism which involves site heterogeneity, but which also allows

TABLE 6
n_H VALUES FOR NAD$^+$ BINDING TO GPDH*

Competing Ligand	n_H(NAD$^+$)
ADP-Rib	0.58 ± 0.03
ATP	0.88 ± 0.04

* The Hill coefficients were obtained from the slope of the Hill plots at mid-saturation with NAD$^+$ (FIGURE 4). n_H for NAD$^+$ alone could not be determined due to the tight binding of the first three NAD$^+$ molecules to the enzyme. The standard deviation was determined as described in TABLE 4.

TABLE 7
INTRINSIC DISSOCIATION CONSTANTS FOR NAD$^+$ BINDING TO GPDH*

	Intrinsic Dissociation Constants, $M \times 10^6$	Constants from Reference 8, $M \times 10^6$	Constants from Reference 9, $M \times 10^6$
K_1'	0.11 ± 0.03	$<4 \times 10^5$	<0.1
K_2'	0.35 ± 0.09	$<1.5 \times 10^{-3}$	<0.1
K_3'	0.59 ± 0.13	0.3	2.7
K_4'	5.0 ± 1.0	5.0	8.2

* K_1' through K_4' are the intrinsic dissociation constants for the binding of the first through the fourth NAD$^+$ molecules, obtained from the respective best-fit intrinsic dissociation constants for the binding of NAD$^+$ in the presence of 10.47 mM ADP-Rib (FIGURE 4). This was done employing the formula:

$$K_{X_i}^{\prime obs} = K'_{X_i}\left(1 + \frac{[\text{ADP-Rib}]}{K'_{\text{ADP-Rib}}}\right)$$

where $K_{X_i}^{\prime obs}$ and K'_{X_i} are the intrinsic dissociation constants for the binding of the ith molecule of the primary ligand X in the presence and absence of ADP-Rib, respectively. This formula was obtained with the assumption that all four intrinsic dissociation constants for NAD$^+$ binding are influenced in an identical manner by ADP-Rib competition, as is the case with ϵNAD$^+$.

It should be noted that the values reported earlier for K_1' and K_2'[8, 9] are only rough estimates, due to the high tetramer concentration employed and the tight coenzyme binding.

for ligand-induced conformational changes, is also compatible with these results.[27]

Unlike AcPyAD$^+$ and ATP, the noncooperative NAD$^+$ analogue ADP-Rib has no effect at all on the cooperativity of εNAD$^+$ binding (FIGURE 3, TABLE 4). Such behavior can be accounted for by either the PEA or the KNF models. These findings indicate the advantage of employing more than one noncooperative competing ligand in competition experiments, since different competing ligands may have (or lack) different effects on the cooperativity of the displacing ligand, depending on their specific interactions with the enzyme.[27]

The first three NAD$^+$ molecules bind very tightly to the enzyme, therefore it was not possible to derive the n_H value for the binding of NAD$^+$ to the enzyme in the absence of competing ligands. It was, however, possible to determine the n_H values for NAD$^+$ binding in the presence of saturating concentrations of ADP-Rib and of ATP. The n_H values obtained are similar to those measured for εNAD$^+$ binding in the presence of saturating concentrations of these two ligands. It seems therefore, that the modes of NAD$^+$ and εNAD$^+$ binding are similar, and the conclusions drawn from the competition experiments with εNAD$^+$ are also valid for the natural coenzyme, NAD$^+$.

The advantage of the competition method described is that, unlike other methods, it employs exclusively equilibrium-binding measurements.

The Mode of Coenzyme Binding

Previous evidence, obtained by spectroscopic and chemical methods, indicated that the adenine subsite of the coenzyme binding site mediates the interactions giving rise to negative cooperativity in coenzyme binding.[15, 16] Similar methods did not detect any conformational changes in the nicotinamide subsites.[28] However, modifications of the pyridine ring (which binds to the nicotinamide subsite) have a significant effect on the extent of negative cooperativity exhibited by the enzyme towards the coenzyme analogue. AcPyAD$^+$, which is an enzymatically active NAD$^+$ analogue,[48] binds noncooperatively to the enzyme (Reference 30; FIGURE 1), although it differs from NAD$^+$ only in the pyridine moiety. Similarly, ligands which lack the pyridine ring altogether, such as ADP-Rib and ATP, also bind to the enzyme in a noncooperative manner, although they have an intact adeonsine moiety. These findings strengthen the notion that the nicotinamide subsites participate indirectly in the intersubunit interactions, although they do not show themselves any ligand-induced conformational changes.[28]

The pyridine moiety and the mode of binding apparently play an important role also in the ability of a noncooperative analogue to affect the cooperativity of the primary ligand. This is indicated by the finding that AcPyAD$^+$ and ATP alter the negative cooperativity of NAD$^+$ binding, whereas ADP-Rib has no such effect—although all three analogues bind noncooperatively to the NAD$^+$ sites. The lack of effect of ADP-Rib may be due to the fact that this analogue lacks a pyridine ring altogether, unlike AcPyAD$^+$, and thus binds to the enzyme in a manner that its binding does not affect (and is not affected by) the binding of εNAD$^+$ to neighboring sites. Since the difference between ADP-Rib and AcPyAD$^+$ is in the pyridine ring only, and since it has been shown that conformational changes are transmitted through the adenine subsites,[15,22,28] it is reasonable to assume that the acetyl-pyridine derivative orients the adenine portion of the $AcPyAD^+$ vis-à-vis the enzyme differently from the orientation of that moeity in the ADP-Rib enzyme complex. Hence, bound $AcPyAD^+$ affects the behavior of a neighboring site whereas ADP-riboside does not.

ATP affects the mode of ϵNAD^+ and NAD^+ binding to the enzyme (FIGURES 3 & 4; TABLES 4 & 6), although it lacks a pyridine moiety. However, this may well be due to the mode of binding of ATP to the enzyme which is very different from that of other NAD^+ analogues, most likely due to the strong ionic character of ATP binding. Unlike ϵNAD^+, NAD^+ $AcPyAD^+$, ADP-Rib and other NAD^+ analogues, ATP also binds to the eight P_i-binding sites of the enzyme (FIGURE 2), and its affinity to the NAD^+ sites on the enzyme is significantly decreased (approximately 8-fold) by the presence of 0.1 M KCl (data not shown). The unique mode of ATP binding is also indicated by the fact that its effect on the circular polarized luminescence of a fluorescent probe covalently attached to the active-site SH groups of the enzyme is different from the effects induced by NAD^+, ADP-Rib and $AcPyAD^+$.[49]

The evidence presented here for intersubunit interactions affecting ligand binding and for the role of the pyridine moiety in the negative cooperativity shown by the enzyme toward coenzyme analogues leads us to the following model (FIGURE 5): Coenzyme binding induces a conformational change, which is transmitted to neighboring subunits via the adenine subsites. The mode of binding of the pyridine moiety determines the mode of binding of the adenine moiety and thus the nature of the conformational effects transmitted to neighboring subunits. Therefore, different coenzyme analogues have different effects on ϵNAD^+ and on NAD^+ binding.

This model is also supported by our previous physio-chemical studies which showed, using circular polarized fluorescence techniques, that the binding of a coenzyme molecule to one subunit induces conformational changes in the *neighboring* subunits.[49] The recent finding that conformational changes between the apo- and holo-enzyme from lobster muscle exist only in the adenine-binding pocket and the "S loop",[22] also agrees with this model.

Possibility for Physiological Relevance of the ATP Effect

ADP and AMP have an effect similar to that of ATP on the cooperativity of ϵNAD^+ binding, and all three ligands bind to the NAD^+ sites on the enzyme with similar affinities (FIGURE 3; TABLES 4 & 5). In agreement with these findings, Oguchi *et al.*[42] reported that ADP inhibits the enzyme to the same extent as ATP, and AMP shows only a slightly weaker inhibitory effect. This suggests that the mode of ADP and AMP binding is similar to that of ATP. Furthermore, this finding indicates the NAD^+ binding to the enzyme is independent of the fluctuations in the ATP to ADP (and AMP) ratio which may occur *in vivo*.

Under physiological conditions, the ionic strength resembles that of 0.1 M KCl, conditions under which the inhibition of NAD^+ binding to the enzyme by ATP is not significant. Furthermore, the cytoplasmic Mg^{2+} and ATP concentrations are in the millimolar range,[50,51] ensuring that most of the intracellular ATP is Mg^{2+}-bound. This further lowers the affinity of ATP to the enzyme, as the Mg-ATP complex

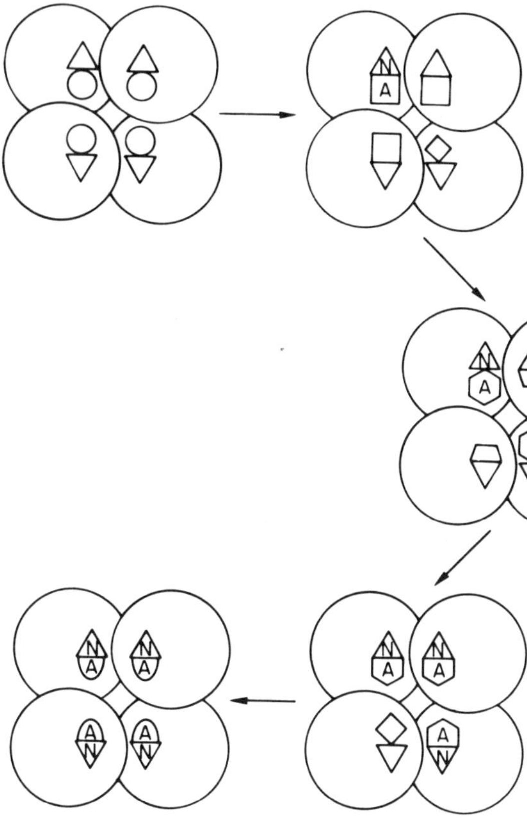

FIGURE 5. Schematic representation of the conformational changes induced upon coenzyme binding. The adenine subsite in the apoenzyme is represented by a circle, and the nicotinamide subsite by a triangle (upper left). The coenzyme is represented as NA, where N is the nicotinamide portion, and A is the adenine portion of the coenzyme. Coenzyme binding induces a conformational change, which propagates to the neighboring adenine subsites. Successive coenzyme binding induces further conformational changes, which in turn induce a sequential decrease in the affinity towards the coenzyme. Conformational changes are not induced from one nicotinamide subsite to another, but they affect the transfer of conformational changes through the adenine subsites by defining the orientation of the adenine portion in the adenine subsite.

exhibits a significantly lower affinity to the enzyme than free ATP.[42,52] These observations eliminate ATP as a possible regulatory ligand which controls directly the extend and the cooperativity of NAD^+ binding to the enzyme *in vivo*.

It should be pointed out, however, that in spite of the weaker binding of ATP to the enzyme under physiological conditions, feedback inhibition by ATP is still feasible at the level of the acyl-enzyme intermediate, reported to be the predominant enzyme species *in vivo*.[53] This is due to the fact this species has a significantly reduced affinity for NAD^+,[54] whereas its affinity towards ATP is similar to that of the non-acylated enzyme. If ATP binds to the acyl-enzyme with an affinity higher than that of ADP (or AMP), a direct effect of the ATP to ADP (or AMP) ratio on the extent and cooperativity of NAD^+ binding to the enzyme would be obtained.

References

1. Harrington, W. F. & G. M. Karr. 1965. J. Mol. Biol. **13:** 885–893.
2. Jaenicke, R., O. Schmid & S. Knof. 1968. Biochemistry **7:** 919–926.
3. Hoagland, V. D. & D. C. Teller. 1969. Biochemistry **8:** 594–602.
4. Harris, J. I., B. P. Merriwether & J. H. Park. 1963. Nature **198:** 154–157.
5. Davidson, B. E., M. Sajgo, H. F. Noller, & J. I. Harris. 1967. Nature **216:** 1181–1185.
6. Harris, J. I. & R. N. Perham. 1968. Nature **219:** 1025–1028.
7. Jones, G. M. T. & J. I. Harris. 1972. FEBS Lett. **22:** 185–189.
8. Conway, A. & D. E. Koshland, Jr. 1968. Biochemistry **7:** 4011–4022.
9. DeVijlder, J. J. M. & E. C. Slater. 1968. Biochem. Biophys. Acta **167:** 23–34.
10. DeVijlder, J. J. M., W. Boers & E. C. Slater. 1969. Biochem. Biophys. Acta **191:** 214–220.
11. Seydoux, F., S. A. Bernhard, O. Pfenninger, M. Payne & O. P. Malhotra. 1973. Biochemistry **12:** 4290–4300.
12. Allen, G. & J. I. Harris. 1975. Biochem. J. **151:** 747–749.
13. Koshland, D. E., Jr., G. Nèmethy & D. Filmer. 1966. Biochemistry **5:** 365–385.
14. Seydoux, F., O. P. Malhotra & S. A. Bernhard. 1974. CRC Crit. Rev. Biochem. **2:** 227–257.
15. Schlessinger, J. & A. Levitzki. 1974. J. Mol. Biol. **82:** 547–569.
16. Levitzki, A. 1974. J. Mol. Biol. **90:** 451–458.
17. Malhotra, O. P. & S. A. Bernhard. 1968. J. Biol. Chem. **243:** 1243–1252.
18. MacQuarrie, R. A. & S. A. Bernhard. 1971. Biochemistry **10:** 2456–2466.
19. Seydoux, F. & S. A. Bernhard. 1974. Biophys. Chem. **1:** 161–174.
20. Moras, D., K. W. Olsen, M. N. Sabesan, M. Buehner, G. C. Ford & M. G. Rossmann. 1975. J. Biol. Chem. **250:** 9137–9162.
21. Olsen, K. W., R. M. Garavito, M. N. Sabesan & M. G. Rossmann. 1976. J. Mol. Biol. **107:** 577–584.
22. Murphy, M. R. N., R. M. Garavito, J. E. Johnson & M. G. Rossmann. 1980. J. Mol. Biol. **138:** 859–872.
23. Biesecker, G. & A. J. Wonacott. 1977. Biochem. Soc. Trans. **5:** 647–652.
24. Biesecker, G., J. I. Harris, J. C. Thierry, J. E. Walker & A. J. Wonacott. 1977. Nature **266:** 328–333 London.
25. MacQuarrie, R. A. & S. A. Bernhard. 1971. J. Mol. Biol. **55:** 181–192.
26. Feneslau, A. 1972. J. Biol. Chem. **247:** 1074–1079.
27. Henis, Y. I. & A. Levitzki. 1979. Eur. J. Biochem. **102:** 449–465.
28. Henis, Y. I. & A. Levitzki. 1977. J. Mol. Biol. **117:** 699–716.
29. Barrio, J., J. A. Secrist & N. J. Leonard. 1972. Proc. Natl. Acad. Sci., USA **69:** 2039–2042.
30. Furfine, S. F. & C. S. Velick. 1965. J. Biol. Chem. **240:** 844–855.
31. Bloch, W. A. 1970. Ph.D. Thesis, University of Oregon, Eugene, Oregon. University Microfilms, Order No. 71-1296, Ann Arbor, Michigan.
32. Colowick, S. P. & F. C. Womack. 1969. J. Biol. Chem. **244:** 774–777.
33. Booth, R., G. W. Box, M. E. Muller & T. I. Peterson. 1959. IBM Corp. Mimes (IBM share program No. 687, WL-NL-1).
34. Levitzki, A. & D. E. Koshland, Jr. 1976. Curr. Top. Cell. Regul. **10:** 1–40.
35. Levitzki, A. 1978. Quantitative aspects of allostery *In* Molecular Biology, Biochemistry and Biophysics. Springer-Verlag, Heidelberg. Vol. **28.**
36. Monod, J., J. Wyman & J. P. Changeux. 1965. J. Mol. Biol. **12:** 88–118.
37. Koshland, D. E., Jr. 1970. *In* The Enzymes P. D. Boyer, Ed., 3rd edition, Vol. **1:** 342–396. Academic Press. New York.
38. Greenfield, J. C., N. J. Leonard & R. I. Gumport. 1975. Biochemistry **14:** 698–706.

39. Luisi, P. L., A. Baici, F. J. Bonner & A. A. Aboderin. 1975. Biochemistry **14:** 362–368.
40. Henis, Y. I. & A. Levitzki. 1980. Eur. J. Biochem. **112:** 59–73.
41. Buehner, M., G. C. Ford, D. Moras, K. W. Olsen, & M. G. Rossmann. 1974. J. Mol. Biol. **99:** 25–49.
42. Oguchi, M., B. P. Merriweather & J. Harting Park. 1973. J. Biol. Chem. **248:** 5562–5570.
43. Henis, Y. I. 1980. Ph.D. Thesis, The Hebrew University of Jerusalem, Israel.
44. Scatchard, G. 1949. Ann. N. Y. Acad. Sci. **51:** 660–666.
45. Constantinides, S. M. & W. C. Deal, Jr. 1969. J. Biol. Chem. **244:** 5695–5702.
46. Eby, D. & M. E. Kirtley. 1976. Biochemistry **15:** 2168–2171.
47. Hill, A. V. 1913. Biochem. J. **7:** 471–480.
48. Kaplan, N. O. & M. M. Ciotti. 1956. J. Biol. Chem. **221:** 823–832.
49. Henis, Y. I., A. Levitzki & A. Gafni. 1979. Eur. J. Biochem. **97:** 519–528.
50. Dickerson, J. W. T. & E. M. Widdowson. 1960. Biochem. J. **74:** 247–257.
51. Hohorst, H. J., M. Reim & H. Bartels. 1962. Biochem. Biophys. Res. Commun. **7:** 137–141.
52. Schwendimann, B., D. M. Ingbar & S. A. Bernhard. 1976. J. Mol. Biol. **108:** 123–138.
53. Bloch, W. A., R. A. MacQuarrie & S. A. Bernhard. 1971. J. Biol. Chem. **246:** 780–790.
54. Malhotra, O. P. & S. A. Bernhard. 1973. Proc. Natl. Acad. Sci., USA **70:** 2077–2081.

A NEW COMPONENT IN PROTEIN FLUORESCENCE*

J. W. Longworth

Biology Division
Oak Ridge National Laboratory
Oak Ridge, Tennessee 37830

Introduction

Because they have a large tyrosyl content and a small number of tryptophyl residues, mammalian serum albumins have been useful globular proteins for studies of the fluorescence of protein tyrosyl residues. Human serum albumin has 1 tryptophan and 18 tyrosine residues, and the protein from cattle has 2 tryptophans and 19 tyrosines.[1] Comparison between the amino acid sequences of human and cattle serum albumins indicates there is a conserved tryptophan (position 212 in cattle), with the cattle protein having a second tryptophan located at 134. Only a single tryptophan is found in the protein from man, horse, rabbit, and dog, whereas there are two tryptophans in cattle, pig, and sheep proteins.[2] Repeating patterns within the amino acid sequence are observed.[1,3] It is possible that the two tryptophan locations of cattle serum albumin are homologous though situated in different domains with a similar folding pattern. There is nearly complete sequence homology surrounding the majority of tyrosyl residues of human and cattle sequences, but the large number of tyrosines in a given albumin molecule precludes any simple analysis.

Radiation at 280 nm excites both tyrosyl and tryptophyl residue fluorescence, and a shoulder is found on the short-wavelength side of the fluorescence from serum albumins which have a single tryptophan.[4] This shoulder is in the spectral region where tyrosine compounds fluoresce. Albumins with two tryptophyl residues lack a discernible shoulder,[4] but appreciable tyrosine contribution to the fluorescence is nevertheless apparent if not always recognized.

Because tryptophan absorbs at longer wavelengths than tyrosine, selective excitation of tryptophyl residue fluorescence in proteins is possible through a suitable choice of exciting radiation wavelength. Tryptophan fluorescence has a greater Stokes' shift than does tyrosine, and the spectral maximum is at longer wavelengths. These two spectral features make it possible to determine the contribution from tryptophan when both tyrosine and tryptophan fluoresce. Such a selective spectroscopic procedure was first applied to the fluorescence of acid-denatured serum albumins,[5,6,7] as there is a marked increase in tyrosyl fluorescence quantum yield accompanying the denaturation. After removal of the tryptophyl fluorescence, the

* Research sponsored by the Office of Health and Environmental Research, United States Department of Energy, under contract W-7405-eng-26 with the Union Carbide Corporation.

0077-8923/81/0366-0237$01.75/0 © 1981, NYAS

appreciable remainder of the fluorescence of denatured albumin was shown to be similar to that of tyrosine compounds in both spectral maximum and width.

There is a low quantum yield for tyrosyl fluorescence in native serum albumins which has made detailed analysis of its fluorescence difficult.[8] In a preliminary study,[9] note was made of the presence of a long-wavelength component in addition to the expected fluorescence of tyrosine in the fluorescence of native serum albumin. The existence of a second tyrosyl component can also be seen in published figures by other workers,[10] and this second component is illustrated and discussed in my review article.[11]

Single photon-counting spectrofluorimeters permit the attainment of a high spectral resolution and still achieve good signal-to-noise ratios. This allows scattering of exciting radiation to be minimized, which is essential in separation of tyrosyl and tryptophyl fluorescences. Collecting the spectra as computer files by stepping a wavelength drive makes it simple to compare spectra and to then decompose them into constituent components. These technical advances have allowed a more convincing study of the tyrosyl fluorescence of mammalian serum albumins.

In this report, evidence is given for a second tyrosyl fluorescence component in serum albumin fluorescence with a maximum at 345 nm. This component is found in the fluorescence of native serum albumins, but is markedly reduced in denatured conformers of albumins. The excitation spectrum of the 345-nm component is similar to that of the 304-nm component, and both are similar to that of *N*-acetyl-L-tyrosinamide. A possible source for the 345-nm emission component is an excited-state protolysis in one or more of the buried tyrosyl residues in the interior of mammalian serum albumins.

Materials and Methods

Human serum albumin used was rechromatographed quality obtained from Worthington Biochemical Corp. *N*-Acetyl-L-tyrosinamide was obtained from Cyclo Chemical Corp. Other reagents used were of the highest quality obtainable, and all materials were used without further purification.

A Spex spectrofluorimeter, interfaced to a PDP11/system time-shared minicomputer, was used to collect data. Single photon-counting detection was used, and there was a 5-Hz background counting rate. Fluorescence spectra were collected in 1-nm wavelength steps at a spectral resolution of 5 nm for excitation and 2.5 nm for observation with a 1-sec integration. Excitation spectra were collected with 5-nm resolution for excitation and 10 nm for observation with 5-sec integrations. Difference spectra were smoothed with a 5-point Chebyshev polynomial.

Results

Though the model compounds for tyrosyl and tryptophyl residues, the *N*-acetyl carboxamides, absorb with a similar wavelength for the maximal absorbance (280 nm), the tryptophyl compound has appreciable absorption at longer wavelengths than the tyrosyl compound. *N*-Acetyl-L-tyrosinamide has a molar absorption coefficient of 3 $m^2\ mol^{-1}$ at 295, whereas for *N*-acetyl-tryptophanamide the coefficient is 220 $m^2\ mol^{-1}$. Human serum albumin, with its 18 tyrosyl residues and 1 tryptophyl residue, is calculated to have 0.20 of 295-nm radiation absorbed by its tyrosyl residues; the corresponding value for the cattle protein is 0.11. Thus, radiation at 295 nm will selectively excite the tryptophyl residues in both proteins.

There is a major difference in the Stokes' shifts: tyrosyl residues have a shift of 350 mm^{-1}; for tryptophyl residues, the shift is 630 mm^{-1} for human and 650 mm^{-1} for cattle proteins.[8] Thus, the longer wavelength spectral region of the protein fluorescence will be dominated by its tryptophyl residue contribution. To calculate the relative proportion at a given wavelength in the long-wavelength region of protein fluorescence, I have used as an estimate for the tyrosyl fluorescence quantum yield 0.1 of the tryptophyl fluorescence yield.[8] Tatischeff and Klein[12] have reported fluorescence spectral quantum yields for tyrosine and tryptophan zwitterions. These fluoresce with spectral maxima comparable to those of the albumin fluorescence components. The zwitterions have nearly equal quantum yields and a ratio of spectral quantum yields at 370 nm of 1:5. The fraction of light absorbed by tyrosyl residues at 280 nm in human serum albumin is calculated to be 0.80.† Hence, approximately 0.02 of the radiation at 370 nm from human serum albumin fluorescence excited at 280 nm is calculated to be emitted by its tyrosyl residues. It is reasonable to consider all fluorescence radiated at wavelengths longer than 370 nm to have been emitted by tryptophyl residues when this protein is excited at either 280 or 295 nm.

For determination of the contribution of tyrosyl residues to the fluorescence of human serum albumin, two fluorescence spectra are collected. One is excited at 280 nm, where both tyrosyl and tryptophyl residues contribute to the protein fluorescence. A second spectrum is excited with 295 nm radiation, and this is considered to be entirely from tryptophyl residues. The average of the ratios between a 280- and a 295-nm excited fluorescence spectrum between 440 and 480 nm in 1-nm intervals is determined and then is used as a multiplier to equalize the two spectra in this interval.

The difference between a 280- and a normalized 295-nm excited fluorescence spectrum was determined. These spectra, with their difference, are given in FIGURE 1. The remainder of the fluorescence has a maximum at 320 nm, a longer wavelength than that at which tyrosyl components have their maximal fluorescence. Nevertheless, it is instructive to consider the extent of the tyrosyl contribution expected. Tatischeff and Klein[12] report the ratio of spectral quantum yields of tyrosine and tryptophan zwitterions as 7:1 for 305-nm radiation. So the calculated fractional tyrosyl contribution at 305 nm, upon 295-nm excitation, is 0.15 for human serum albumin. Thus, it is reasonable to consider the difference spectrum given in FIGURE 1 as an estimate for the tyrosyl fluorescence of human serum albumin.

The fluorescence of phenol has a Stokes' shift which is not markedly dependent upon the solvent dielectric constant.[13] Hence, the fluorescence of *N*-acetyl-L-tyrosinamide can be taken as a model for the fluorescence of tyrosyl residues in proteins. The fluorescence of this model tyrosine compound is normalized between 305 and 310 nm to the tyrosyl difference spectrum obtained from human serum albumin, and these two spectra are given in FIGURE 2. Both the shift in spectral maximum and the

† At 280 nm, the molar absorption coefficients used are: tyrosyl residues, 140 $m^2\ mol^{-1}$ when exposed, 180 $m^2\ mol^{-1}$ when buried; tryptophyl, 550 and 600 $m^2\ mol^{-1}$; cystinyl links, 13 $m^2\ mol^{-1}$. The composition of human serum albumin used is 1 W, 18 Y, 17 C-C, with 1 W and 6 Y exposed; for the cattle protein, 2 W, 19 Y, 17 C-C, with 2 W and 8 Y exposed. The fractional tyrosyl absorbance at 280 nm is 0.8 for the human protein and 0.7 for cattle protein.

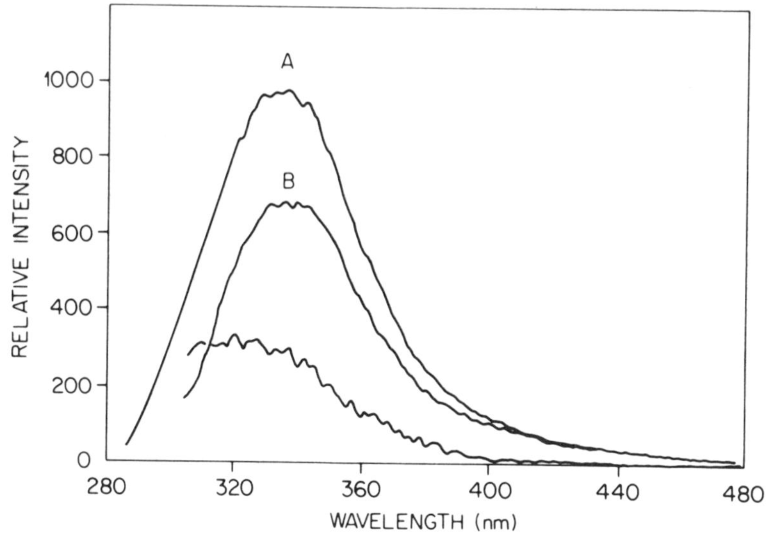

FIGURE 1. Fluorescence of human serum albumin. 0.5 mg/ml human serum albumin (pH 7.0), 10 mM phosphate buffer, 0.1 M NaCl. (A) Excited 280 nm, 5 nm; observed 2.5 nm. (B) Excited 295 nm.

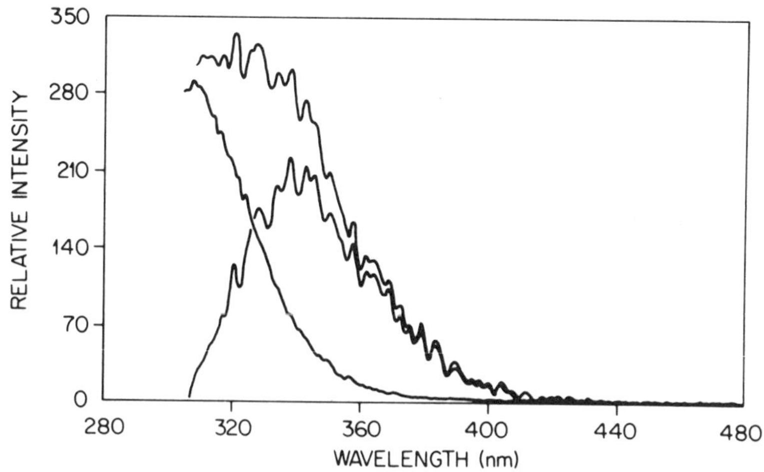

FIGURE 2. Tyrosine fluorescence of human serum albumin. 1 mg/ml *N*-acetyl-L-tyrosinamide.

width suggest that there is an additional component in the long-wavelength region of the protein tyrosyl fluorescence spectrum. When the model tyrosyl fluorescence spectrum is subtracted, a second tyrosyl fluorescence component with a 345-nm maximum is derived from the emission of human serum albumin excited by 280 nm radiation.

Excitation spectra of fluorescence radiation at 310, 340, and 380 nm were determined between 260 and 305 nm in 1-nm steps for human serum albumin. The excitation spectrum for 380-nm fluorescence is considered to be entirely that of the protein tryptophyl residues. The fluorescence at 310 nm is regarded as a mixture of tryptophyl and un-ionized tyrosyl fluorescence. Since the radiation absorbed in the long-wavelength region of the protein absorption is predominantly caused by tryptophyl residues, the two fluorescence excitation spectra were equalized over an interval by the average ratio between 292 and 296 nm obtained in 1-nm steps. The difference between these normalized excitation spectra is given in FIGURE 3. It is apparent in a comparison (FIG. 3) that the excitation spectrum of a tyrosyl model compound, *N*-acetyl-L-tyrosinamide, is similar to the excitation spectrum of protein tyrosyl fluorescence.

A similar difference excitation spectrum was determined for 340-nm fluorescence radiation; this is compared in FIGURE 4 with the 310-nm excitation difference spectrum, and reasonable agreement is shown. As the tyrosyl fluorescence at 340 nm is largely contributed by the second tyrosyl fluorescence component, this result suggests that the long-wavelength tyrosyl component has an excitation similar to that of the short-wavelength tyrosyl fluorescence and of a model tyrosyl compound.

Human serum albumin is denatured in acid solution, and at pH 2 all the tyrosyl residues are exposed to the solvent. The tyrosyl difference spectrum and the fluorescence of *N*-acetyl-L-tyrosinamide are shown in FIGURE 5. There is a marked reduction in the contribution of tyrosinate to the tyrosyl fluorescence. Similar results are found when serum albumin is denatured in 10 M guanidinium hydrochloride solution at pH 7.0.

With cattle serum albumin in native and denatured conformers, results are similar to those for human serum albumin.

DISCUSSION

The phenolic hydroxyl of tyrosine ionizes in alkaline water, and this causes a pronounced long-wavelength shift of both absorption[14] and fluorescence.[15] Although

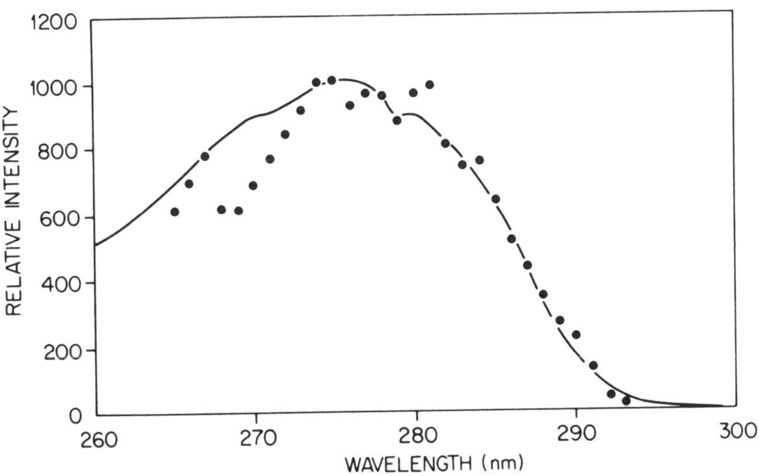

FIGURE 3. Tyrosine fluorescence excitation spectrum of human serum albumin. 0.1 mg/ml human serum albumin (pH 7.0), 10 mM phosphate buffer, 0.1 M NaCl. Observed 310 nm, 10 nm; excited 5 nm.

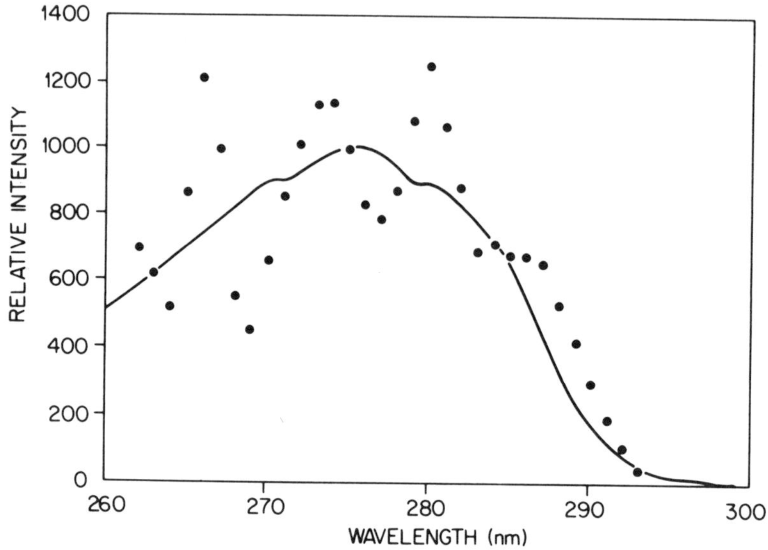

FIGURE 4. Tyrosine fluorescence excitation spectra of human serum albumin. Observed 310 and 340 nm.

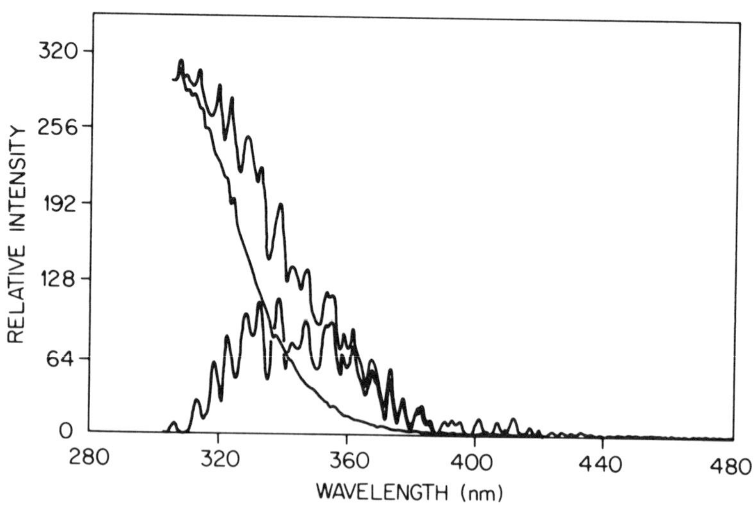

FIGURE 5. Tyrosine fluorescence of acid-denatured human serum albumin. 0.5 mg/ml human serum albumin (pH 2), 0.1 M NaCl.

in water the fluorescence of tyrosinate has a very small quantum yield of fluorescence, in neutral organic solvents much higher yields are found.[16] Thermodynamic cycle analyses of the enthalpies derived from the Stokes' shift[17,18] suggest a 10^5 increase in acidity for the first excited singlet state only by assuming equal entropies of ionization in ground and excited states. The rate of proton loss is dependent upon the basicity of acceptor and follows the Brönsted linear Gibbs energy relationship.[19] Strong bases such as phosphate[20] and acetate[21] can react with the excited singlet state within the lifetime of this state, whereas water, being a much weaker base, does not react to any significant extent.[16,19,22] A kinetic analysis of the reaction of the excited singlet of tyrosine zwitterion with acetate, which assumed that an equilibrium was attained, gave estimates for the increase in acidity which were comparable to those from the thermodynamic cycle analyses.[21] Thus, like other hydroxyarenes,[23] phenols will photoionize from their excited singlet state when given a suitable acceptor. This photoionization causes a pronounced alteration in the fluorescence spectrum. Water is an unsuitable acceptor for tyrosine because of a short fluorescence lifetime and a slow reaction rate. Since water has been the most frequently used acceptor, even if inadvertently so, the phenomenon of an excited-state protolysis from tyrosine has not received wide attention.

The presence of a long-wavelength tyrosyl fluorescence component with a spectrum in the same spectral interval as tyrosinate in the fluorescence of native serum albumin is suggestive of an excited-state protolysis. The excitation spectrum of this emission is similar to that of the un-ionized tyrosyl fluorescence. This is expected for an excited-state chemical reaction.[24]

Exposed tyrosyl residues of serum albumins are not likely sources of the tyrosinate fluorescence of native albumins, as they are solvated by water, a poor acceptor base. Serum albumins are denatured by acid, and a major unfolding of the structure takes place.[25,26] In native serum albumin, two-thirds of the tyrosyl residues are buried,[27–29] whereas in acid all of the tyrosyl residues are exposed.[29] The marked reduction in the tyrosinate component in acid-denatured serum albumin fluorescence suggests that the tyrosinate fluorescence originates from the buried tyrosyl residues. The buried tyrosyl groups which photoionize are probably associated with suitable acceptor bases, and as there are a large number of buried tyrosyl residues in mammalian albumins, many possible complexes are probably present.

Buried tyrosyl residues in proteins have been found hydrogen-bonded with seryl or threonyl hydroxyl groups, peptide-chain amide groups, asparaginyl and glutaminyl amide groups and water.[30] All of these groups are sufficiently weak bases that no excited-state protolysis is expected. Other tyrosyl residues have been found hydrogen-bonded to the carboxyl group of glutamyl or aspartyl residues. The ionized form of this grouping would act as a suitable acceptor; it is the neutral form which is expected to be buried, and this would be an unsuitable proton acceptor. Tyrosyl residues have been found hydrogen-bonded to histidyl residues in thermolysin of *Bacillus thermoproteolyticus* and cattle ribonuclease A and to a lysyl residue in lobster glyceraldehyde-3-phosphate dehydrogenase. Both are suitable proton-accepting groups, but the fluorescence of ribonuclease has no evidence of a tyrosinate component. Thus, only the complex with lysine remains. The lobster enzyme has three tryptophan residues, but the selective excitation studies developed here have not been applied to this enzyme.

Proteins which lack tryptophyl residues but have tyrosyl residues are known that fluoresce with a long-wavelength tyrosyl component. Cattle adrenodoxin has a single tyrosyl residue and fluoresces with a 330-nm peak from native conformer[31,32] but has a 304-nm tyrosyl fluorescence when denatured.[33] An anomalous tyrosyl fluorescence is found from parsley plastocyanin,[34,35] a protein with three tyrosyl groups. Here the native enzyme has a fluorescence peak at 320 nm which could arise by a mixture of

304- and 345-nm components. Normal tyrosyl fluorescence is found from acid-denatured plastocyanin.[34] Two cytotoxins from the venom of the Indian cobra, each with four tyrosyl groups, have a 345-nm fluorescence component from a native conformer and a normal 304-nm component from a denatured conformer.[36,37]

Though the atomic structure of poplar plastocyanin, which is homologous in sequence to parsley, has been reported, the detailed analysis of the side chains is not yet available. A sea snake short neurotoxin and Siamese cobra long neurotoxin also have published atomic structures. These proteins are homologous in sequence to Indian cobra cytotoxin, but again the detailed side-chain conformation data are not yet available. Inspection of the published diagrams of the chain folding patterns of the three proteins suggests that a single buried tyrosyl residue is probable in both plastocyanin and cytotoxin, although it is not readily apparent what the hydrogen-bonded partner could be. Lysyl residues are located in the vicinity of the probable buried tyrosyls, therefore emission from a buried tyrosyl complexed with a lysine amino group may possibly account for the anomalous fluorescence of these proteins.

SUMMARY

Tryptophyl residues in proteins absorb at longer wavelengths than tyrosyl residues, thus the tryptophyl fluorescence can be selectively excited. In addition, tryptophyl residues have a fluorescence maximum at much longer wavelengths than tyrosyl residues[20] and are the predominant source of fluorescence in the long-wavelength region. The contribution of tyrosyl fluorescence to protein fluorescence can be determined by exploiting these spectral properties. The fluorescence of the tyrosyl residues in native human serum albumin is different from the fluorescence of *N*-acetyl-L-tyrosinamide: the spectral maximum is at a longer wavelength, and the spectral width is greater. This difference is caused by a second component with a maximum at 345 nm. The excitation spectrum of the 345-nm component is similar to that of the normal 304-nm tyrosyl component. The 345-nm component is largely absent from denatured serum albumin. An excited singlet state protolysis from the buried tyrosyl residues explains the appearance of the 345-nm component. A possible acceptor base is an amino group of buried lysyl residue.

REFERENCES

1. BROWN, J. R. 1978. *In* Albumin: Structure, Biosynthesis, Function. T. Peters & I. Sjoholm, Eds.:1–10, Pergamon Press. Oxford.
2. FELDHOFF, R. C. & T. PETERS. 1976. Biochem. J. **159:** 529.
3. MACLACHLAN, A. D. & J. E. WALKER. 1977. J. Mol. Biol. **112:** 543.
4. CHEN, R. F. 1967. *In* Fluorescence: Theory, Instrumentation and Practice. G. G. Guildault, Ed.:483–485. Marcel Dekker. New York.
5. WEBER, G. 1961. Biochem. J. **79:** 29P.
6. TEALE, F. W. J. 1961. Biochem. J. **80:** 14P.
7. VLADIMIROV, Y. A. 1961. Dokl. Akad. Nauk SSSR **136:** 960.
8. TEALE, F. W. J. 1960. Biochem. J. **76:** 381.
9. LONGWORTH, J. W. 1962. Ph.D. Thesis. University of Sheffield.
10. VLADIMIROV, Y. A. & G. M. ZIMIMA. 1966. Biokhimiya **30:** 1105.
11. LONGWORTH, J. W. 1971. *In* Excited States of Proteins and Nucleic Acids. R. F. Steiner & I. Weinryb, Eds.:407. Plenum Press. New York.

12. TATISCHEFF, I. & R. KLEIN. 1976. *In* Excited States of Biological Molecules. J. B. Birks, Ed.:375–387. John Wiley. London.
13. WEBER, G. Quoted in LOMBARDI, J. A. 1969. J. Chem. Phys. **50:** 3780.
14. BEAVEN, G. H. & E. R. HOLIDAY. 1952. Adv. Protein Chem. **7:** 319.
15. CORNOG, J. L. & W. R. ADAMS. 1963. Biochim. Biophys. Acta **66:** 356.
16. LONGWORTH, J. W. & R. O. RAHN. 1967. Biochim. Biophys. Acta **247:** 526.
17. VLADIMIROV, Y. A. & C. K. LI. 1962. Biofizika **7:** 270.
18. FEITELSON, J. 1964. J. Phys. Chem. **68:** 391.
19. AVIGAL, I., J. FEITELSON & M. OTTOLENGHI. 1969. J. Phys. Chem. **50:** 2614.
20. SHIMIZU, O. & K. IMAKUBO. 1977. Photochem. Photobiol. **26:** 541.
21. RAYNER, D. M., D. T. KRAJCORSKI & A. G. SZABO. 1978. Can. J. Chem. **56:** 1238.
22. FAYET, M. & PH. WAHL. 1971. Biochim. Biophys. Acta **229:** 102.
23. WELLER, A. 1961. Prog. React. Kinet. **1:** 189.
24. WEBER, G. 1961. Nature **190:** 2723.
25. PETERS, T. 1975. *In* Plasma Proteins. 2nd edit. F. W. Putman, Ed. Vol. **1:** 133–181. Academic Press. New York.
26. FOSTER, J. F. 1977. *In* Albumin: Structure, Function and Uses. V. M. Rosenoer, M. Oratz & M. A. Rothschild, Eds.:53–84. Pergamon Press. Oxford.
27. TANFORD, C. & G. L. ROBERTS. 1952. J. Am. Chem. Soc. **74:** 2509.
28. TANFORD, C., S. A. SWANSON & W. S. SHORE. 1955. J. Am. Chem. Soc. **77:** 6409.
29. HERSKOVITS, T. T. & M. LASKOWSKI. 1960. J. Biol. Chem. **235:** PC56; ibid. 1962. **237:** 2481.
30. FELDMAN, R. J. 1976. Atlas of Macromolecular Structure on Microfiche, Tracor Jitco, Inc. Rockville, MD.
31. KIMURA, T. & J. J. TING. 1971. Biochem. Biophys. Res. Commun. **45:** 1227.
32. KIMURA, T., J. J. TING & J. J. HUANG. 1972. J. Biol. Chem. **247:** 4476.
33. LIM, B. T. & T. KIMURA. 1980. J. Biol. Chem. **255:** 2440.
34. GRAZIANI, M. T., A. FINAZZI AGRO, G. ROTILIO, D. BARRA & B. MONDOVI. 1974. Biochemistry **13:** 804.
35. MCMILLAN, D. R., R. C. ROSENBERG & H. B. GRAY. 1974. Proc. Natl. Acad. Sci. USA **71:** 4760.
36. SZABO, A. G., K. R. LYNN, D. T. KRAJCORSKI & D. M. RAYNER. 1978. FEBS Lett. **94:** 249.
37. SZABO, A. G., K. R. LYNN, D. T. KRAJCORSKI & D. M. RAYNER. 1979. J. Lumin. **18:** 582.
38. BUSHUEVA, T. L., E. P. BUSEL & E. A. BURSTEIN. 1978. Biochem. Biophys. Acta. **534:** 141.

[**Note Added in Proof:** There is no tyrosinate fluorescence from the cytotoxin of Middle Asian Cobra venom.[38] This cytotoxin has one less tyrosinate residue than the homologous cytotoxin of Indian cobra, assigning tyrosinate emission from the additional site.]

PART III. LUMINESCENCE STUDIES OF CELLS AND CELL COMPONENTS

FLUORESCENCE STUDIES ON CHROMATIN STRUCTURE*

Charles R. Cantor, Ann E. Dieterich, and Christopher P. Prior

Departments of Chemistry and Biological Sciences
Columbia University
New York, New York 10027

INTRODUCTION

Fluorescence techniques have proven to be powerful methods for analyzing the structure of macromolecular assemblies and the dynamics of interaction of these assemblies with different types of ligands.[1] One assembly system of great current interest is the complex of DNA, histones, and numerous other proteins known as chromatin.[2] This system lends itself very well to several fluorescence techniques, and a number of laboratories have begun to use fluorescence to study various aspects of chromatin and its constituents. Here we will review some of the basic properties of chromatin and outline some of the potential advantages and difficulties in using fluorescence to study this system. Then a number of the recent results, which we have obtained in collaboration with other groups, will be summarized.

Intact chromatin consists of DNA, in single chromosome-sized pieces, condensed with a relatively constant amount of histones, and quite a variable amount of non-histone chromosomal proteins. In a typical preparation, there will be eight core histones (two each of histones H2a, H2b, H3, and H4) and one histone H1 or its equivalent, for every 150 to 230 base pairs of DNA. The amount and type of nonhistone proteins depends on many things, including whether or not the particular segment of DNA is from a transcriptionally active region. The overall system is much too complicated to permit definitive structural studies without some kind of fractionation.

Most structural studies to date have focused on the nucleosome core particle. This can be isolated by a mild nuclease digestion of nuclei or chromatin, a salt treatment to remove loosely bound proteins, and several chromatographic steps, followed by sucrose gradient centrifugation. The result is a fairly homogeneous preparation consisting of the eight core histones and 146 base pairs of DNA. The core particle can be disassembled and reassembled easily *in vitro* which greatly facilitates the incorporation of fluorescent probes or other labels. However, it has no assayable functional activity, which severely restricts an assessment of the significance of many of the properties which have been observed for this particle. Studies on larger and more complex chromatin fractions have been reported using a number of techniques.

* Supported by National Science Foundation Grant PCM 77-09254 and United States Public Health Service Grant GM 14825.

0077-8923/81/0366-0246$01.75/0 © 1981, NYAS

However, no reliable assembly system exists yet for such systems, and indeed it is not yet known which nonhistone proteins should be present to mimic particular *in vivo* functional states. This would appear to greatly limit possibilities for fluorescence studies. However, we will show that the difficulties may be bypassed by a new *in vivo* labeling procedure.

The variety of ways in which fluorescence can be used to look at biological molecules are exemplified well by other contributions in this volume. A requirement for most types of studies is one or more specifically placed fluorescent chromophores. Nucleosome core particles have as intrinsic components only the relatively weak fluorescence from a number of tyrosines—about thirty per particle distributed among the four types of core histones.[3] None of these proteins contain tryptophan, and H1 species contain little or no tyrosine or tryptophan. Thus, to have particles with specific or strong fluorophores, it is necessary to introduce extrinsic probes. Given the complexity of even the core particle, it is a virtual necessity that such probes be introduced covalently.

The only sulfhydryl in the nucleosomes from most species is the highly conserved cysteine 110 of histone H3. This residue is not accessible to reaction with most bulky reagents in the intact nucleosome. However, specific modification of this position can be obtained by dissociating nucleosomes with salt and urea, reacting the entire mixture of components with fluorescent iodoacetyl or maleimide derivatives, and then reconstituting nucleosomes by stepwise removal of urea and salt.[4] This is the procedure which has been used in most studies to date. Occasionally, other sulfhydryls are found among the core histones. For example, the sea urchin *A. lixula*, contains a cysteine in position 73 of histone H4 in addition to the universal H3 cysteine.[5] By labeling purified histone, it is possible to make a specific fluorescent conjugate in this position which is still capable of *in vitro* reconstitution.[6] The third position in the nucleosome which has been explored to date, is the ends of the DNA which can be labeled by adding unusual nucleotides to the ends enzymatically, and then reacting these specifically with fluorescent probes.[7] In the future, it will be possible to make other specific fluorescent derivatives of DNA by extending such procedures to internal bases, and other derivatives of histones will be accessible by using proteins from other species. Nonhistone chromosomal proteins appear generally to be richer in cysteines than histones, and fluorescent conjugates of these materials should be made easily from individual purified components. However, the problem of reconstituting these proteins into functional chromatin remains.

Nucleosome Core Particles

Nucleosome core particles reconstituted with histone H3 containing various fluorescent dyes covalently attached to cysteine 110 are indistinguishable from native core particles by a wide variety of physical criteria and also by the detailed pattern of sites which are susceptible to DNAse I cleavage.[4,6,8] This is reasonable evidence that the presence of one or two bulky dyes does not cause a significant perturbation in the structure of the particle. When subjected to environmental perturbations, the fluorescent particles display a strikingly complex series of structural changes. Often these appear as small but significant alterations in the emission properties of the dyes; however when examined by more informative techniques, most of these changes appear to reflect fairly substantial structural alterations.[8] We have used the dyes 1,5-IAEDANS, Iodoacetylaminofluorescein (IAAF) and Iodoacetoxypyrene (IAP) in our studies; others have seen similar results with pyrenemaleimide.[9]

Salt, temperature, pH, urea, and divalent ions are among the variables that have been used by us and by others to probe the range of structures accessible to nucleosome cores. We have focused on a single variable, ionic strength, which has the advantage that some of its effects should be relatively easy to rationalize. For example, lower ionic strengths should favor extension of the DNA, strengthen many histone-DNA contacts, and weaken histone-histone contacts, because of the large net positive charge of the histones and the larger net negative charge on the DNA. Counterion effects will tend to screen all charge–charge interactions, and new methods allow these to be analyzed quantitatively.[10]

As the ionic strength is varied from 0.1 mM to 1 M, core particles show a series of four fairly distinct conformations.[8] The low salt form, stable at ionic strengths below about 1 mM, is fairly flexible and expanded structure. Fluorescence polarization measurements on H3-labeled particles show a rotational correlation time about one-fourth that expected for a rigid compact disk-shaped nucleosome as seen by scattering techniques at higher ionic strengths. Energy transfer measurements show that the cysteines of the two copies of H3 are about 45 Å apart[8]; the H3 cysteines are located about 50 Å from the DNA termini, and H4 cysteines are located less than 30 Å from the DNA termini.[6] The two H3 cysteines are relatively accessible to the solvent as judged by acrylamide quenching.

Throughout a broad range of salt, from about 5 mM to 100 mM, the core particle behaves as a rigid structure. The rotational correlation time is somewhat longer than that expected from current models of the compact nucleosome. Pyrene labels on both copies of H3 cysteine 110 result in almost exclusively pyrene excimer emission,[9,11] and H3-disulfide-linked dimer is capable of reconstitution.[12] This indicates that the two cysteines must be very close. The distances between H3 cysteine and the DNA terminus, or H4 cysteine and the DNA terminus are the same in the compact form of the core particle as they are in the low salt form.[6] The two H3 cysteines are virtually inaccessible to solvent as judged by acrylamide quenching. All of these results indicate that the transition from the compact form to the low salt form involves a dramatic expansion of the particle which appears to arise mainly by breaking protein–protein contacts with little or no alteration in protein–DNA contacts. Other studies indicate that the relative stability of the low salt form is enhanced when the core contains DNA longer than 146 base pairs.[13] Thus, intramolecular repulsions within the coiled DNA molecule must be a major determinant in causing the low salt expansion.

In the ionic strength range between 0.1 and 0.3 molar, another structural change in the core particle occurs. This leads to no observable alteration in size, shape, or the specific distances measured by energy transfer. The change in structure is evidenced only by a distinct increase in the accessibility of H3-cysteine labels to solvent quenchers, and a decrease in the quantum yield of IAEDANS-labeled H3.[8] Presumably, this structural change is rather localized but it may be biologically significant since it occurs at physiological ionic strength.

At an ionic strength around 0.5, there is an abrupt and dramatic change in the properties of the core particle. The structure becomes very flexible, as judged by fluorescence polarization. Fluorescent dyes on H3 cysteine are totally exposed to the solvent. The two H3 cysteines now move apart to a distance greater than 70 Å.[8] We know that these changes occur without physical removal of histones from the DNA; however, it is likely that at such high ionic strengths it is relatively easy for histones to move on DNA or to transfer from one DNA to another. In the specific case of H3, we can show that the distance from H3 to the DNA termini remains unaltered even at this high ionic strength.[6] Thus, H3, at least, does not change its position on the DNA during the transition to the high salt form.

The properties of the four states of the core particle revealed by the fluorescence studies discussed above are summarized schematically in FIGURE 1. This figure should not be taken literally since there is far too little hard structural information available about any but the compact forms to support specific features of the drawing. However the figure conveys, at least as a mnemonic, the rather dramatic nature of these structural changes. We know that FIGURE 1 is an oversimplification of the true

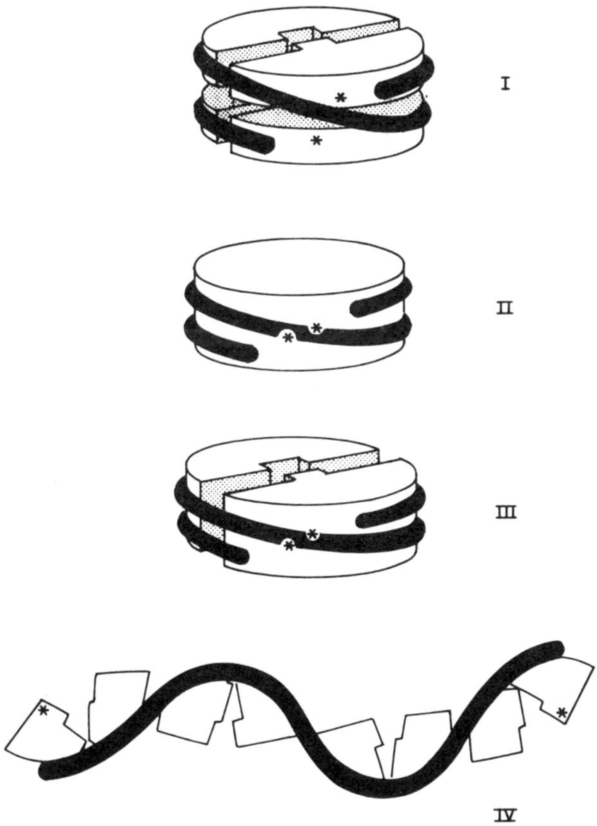

FIGURE 1. Schematic illustration of the structure of the nucleosome core particle at different ionic strengths. C_2 symmetry is assumed for all states. The DNA is shown as a rod; proteins as slabs. The position of cysteine 110 of histone H3 is shown by an asterisk. In all states except IV, this asterisk is meant to be near the center of the particle. State I, at low ionic strength; State II, the compact form; State III, near physiological ionic strength; State IV, above an ionic strength of 0.5 molar.

situation, since kinetic studies reveal that there are additional structural states which can be seen as kinetic intermediates as the salt is varied.[14]

Nucleosomes must participate in a variety of functional events *in vivo*. They must be able to participate in DNA replication, transcription and DNA repair. They must be able to accommodate a variety of levels of higher order chromatin packaging. Little is known about how nucleosome structure responds to these various states. It is clear, for example, that nucleosomes on DNA regions capable of active transcription are more sensitive to DNAse digestion,[15,16] but the detailed mechanism for this sensitivity is not yet revealed. Similarly, we know that the organization of nucleosomes covering newly replicated regions of DNA is different,[17,18] but the structural consequences of this at the level of individual nucleosomes is not known. The fluorescence studies have uncovered an amazing range of nucleosome structures *in vitro*. Presumably, this range somehow reflects a variety of nucleosome states *in vivo*. What is needed is a way to look at nucleosome structure in specific functional states of chromatin. One possible approach which may ultimately permit this is described in the next section.

Uptake and Utilization of Fluorescent Histones

A nagging question in all studies with fluorescent probes is whether introduction of a covalent label has perturbed the structure or function so much as to invalidate any conclusions drawn from spectroscopic measurements. This problem was especially severe with fluorescent histone labels because we had no *in vitro* functional assays and because physical studies indicated that the labeled particles had a range of accessible structures. These concerns led us to start to examine whether histones containing covalently attached dyes could be inserted into organisms, find their way to the nucleus, and incorporate into nucleosomes. Microinjection of histones was one obvious possibility, but since we wanted large quantities of labeled material, the possibility that cells might be able to take up histone directly from their medium was examined. It has been known for quite a while that histones can enter cells,[19] but little is known about the fate of these histones.

Most cells show a toxic response to high levels of histone in their medium. However, the acellular slime mold *Physarum polycephalum*, shows no ill effects of growth on 40 μg/ml total Physarum histone.[11] The growth curve of the organism is unaltered even when the histone contains IAP-labeled H3, as shown in FIGURE 2. Under these conditions, there is a rapid uptake of histone into the cytoplasm of the organism as judged by the appearance of blue pyrene monomer fluorescence. After a few-hour lag, blue emission can also be seen in nuclei, and shortly thereafter green excimer fluorescence is seen.[11] This comes almost exclusively from nuclei; after several generations of growth (18-hour doubling time under our conditions), the nuclei appear as brilliant green dots in the fluorescence microscope.

Appearance of the pyrene excimer is indirect evidence that the histone has incorporated into nucleosomes. Under physiological conditions, there is virtually no tendency for free IAP-labeled H3 to associate into DNA-free structures which display excimer emission.[11] To confirm that the excimers were in nucleosomes, nuclei were isolated from *Physarum* grown in the presence of IAP-labeled H3 and core particles were isolated from these nuclei. Such particles contained almost exclusively excimer

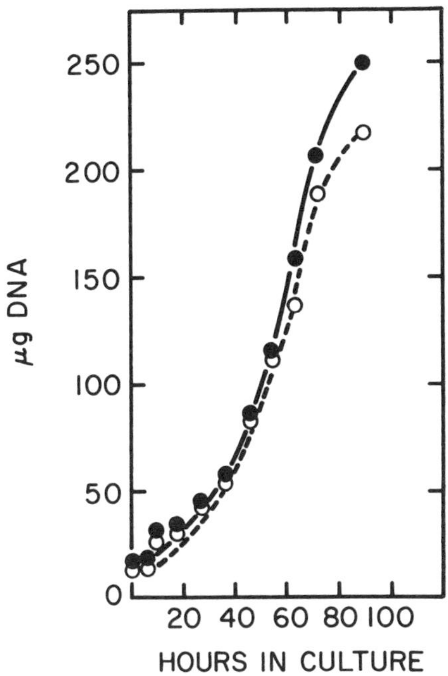

FIGURE 2. The effect of histone on the growth of *Physarum Polycephalum* growth. Shown is the DNA content of constant aliquots of a culture of microplasmodia, taken at the time intervals indicated. The solid points represent growth for 36 hours in the presence of IAP-labeled histones followed by 54 hours of growth in the presence of unlabeled histones, all at 40 μg/ml. The hollow points represent growth in a medium identical except for the absence of histones.

fluorescence. The amount of pyrene-labeled H3 incorporated after several generations was more than 40% of the total H3 present.[11] This shows that the uptake and utilization of the exogenously fed histone is strikingly efficient. The finding of only excimer fluorescence in nucleosomes means that, after uptake, the histones are used in pairs. This is consistent with the general finding that the pool of free histone is very small,[20] and that the replication of core histones is conservative.[21,22]

By using *Physarum*, it is possible to isolate nucleosomes containing a fluorescent-labeled histone that have been assembled *in vivo*. Since the organism seems to function well even with massive amounts of pyrene-labeled nucleosomes, the presence of the label must not be causing much functionally significant structural perturbation. The optical properties of nucleosomes isolated from these organisms are identical with those prepared by *in vitro* reconstitution,[11] which validates many of the fluorescence studies done on such particles. More importantly, it is now possible to isolate more complex states of chromatin containing a specifically placed fluorescent label. In this way, we should be able to prepare samples like active chromatin for physical studies. At present, there is no way to prepare such samples by *in vitro* reconstitution. We are optimistic that similar procedures should allow the *in vivo* incorporation of other fluorescent-labeled proteins.

ACKNOWLEDGMENTS

The authors are indebted to Vincent Allfrey, Richard Axel, Donald Crothers, Hilbert Eshaghpour and Edward Johnson who collaborated with us in various aspects of the work summarized in this manuscript.

References

1. Fairclough, R. & C. R. Cantor. 1978. *In* Methods in Enzymology. C. H. W. Hirs & S. N. Timasheff, Eds. Vol. **48:** 347. Academic Press. New York.
2. Garrett, R. A. 1979. Int. Rev. Biochem. **25:** 179–203.
3. Isenberg, I. 1979. Ann. Rev. Biochem. **48:** 159–191.
4. Dieterich, A. E., R. Axel & C. R. Cantor. 1978. Cold Spring Harbor Symp. Quant. Biol. **42:** 199–205.
5. Strickland, M., W. N. Strickland, W. F. Brandt & C. Von Holt. 1974. FEBS Lett. **40:** 346–348.
6. Eshaghpour, H., A. E. Dieterich, C. R. Cantor & D. M. Crothers. 1980. Biochemistry **19:** 1797–1805.
7. Eshaghpour, H. & D. M. Crothers. 1979. Nucleic Acids Res. **7:** 1485–1495.
8. Dieterich, A. E., R. Axel & C. R. Cantor. 1979. J. Mol. Biol. **129:** 587–602.
9. Zama, M., P. N. Bryan, R. E. Harrington, A. L. Olins & D. E. Olins. 1978. Cold Spring Harbor Symp. Quant. Biol. **42:** 32–41.
10. Record, M. T., T. M. Lohman & P. De Haseth. 1976. J. Mol. Biol. **107:** 146–158.
11. Prior, C. P., C. R. Cantor, E. M. Johnson & V. G. Allfrey. 1980. Cell. **20:** 597–608.
12. Camerini-Otero, R. D. & G. Felsenfeld. 1977. Proc. Natl. Acad. Sci. USA **74:** 5519–5523.
13. Dieterich, A. E., H. Eshaghpour, D. M. Crothers & C. R. Cantor. 1980. Nucleic Acids Res. **8:** 2475–2487.
14. Dieterich, A. E. & C. R. Cantor. 1980. Biopolymers. In press.
15. Weintraub, H. & M. Groudine. 1976. Science **193:** 848–856.
16. Garel, A. & R. Axel. 1976. Proc. Natl. Acad. Sci. USA **73:** 3966–3970.
17. Murphy, R. F., R. B. Wallace & J. Bonner. 1978. Proc. Natl. Acad. Sci. USA **75:** 5903–5907.
18. Worcel, A., S. Han & M. C. Wong. 1978. Cell **15:** 969–977.
19. Ryser, H. J. P. & R. Hancock. 1965. Science **150:** 501–503.
20. Stein, G. S., S. Hochhauser & J. L. Stein. 1979. *In* The Cell Nucleus. H. Busch, Ed. Vol. **7:** 259–307. Academic Press. New York.
21. Leffack, J. M., R. Grainger & H. Weintraub. 1977. Cell **12:** 837–845.
22. Seidman, M. M., A. J. Levine & H. Weintraub. 1979. Cell **18:** 439–449.

SPECTROSCOPIC PROPERTIES OF RHODAMINE B-LABELED THYROID HORMONE

Harold Edelhoch, Sheve-Yann Cheng, and Gaetano Irace*

Clinical Endocrinology Branch
National Institute of Arthritis, Metabolism and Digestive Diseases
National Institutes of Health
Bethesda, Maryland 20205

The fluorescence of Rhodamine has been used recently to follow the localization of labeled hormones and proteins on membranes and their endocytosis by cells.[1,2] We have undertaken a study of the molecular properties of rhodamine derivatives of the thyroid hormones so that the environmental factors controlling their structure, and consequently their fluorescence, can be more readily understood.

Results

In order to understand the behavior of T_3-Rhodamine†, we have compared its properties with those of two simpler analogs, i.e., Ala-Rhodamine† and, in some cases, with the chromophore itself, Rhodamine.† Of these two compounds, Ala-Rhodamine† is the more appropriate analog since it also contains the thiourea link present in T_3-Rhodamine.

Ethanol

Absorption

Increasing concentrations of ethanol produce rather large changes in the oscillator strength of the rhodamine chromophore in T_3-Rhodamine between 490 and 610 nm. The marked effects of 20, 40, and 100% ethanol (v/v) on the spectra of T_3-Rhodamine are shown in Figure 1. The extinction coefficient of the major peak at 556 nm increases almost twofold without a significant shift in wavelength maximum between 0 and 20% ethanol. In contrast, Ala-Rhodamine and Rhodamine show only slight increases in extinction coefficient at 565 nm, their wavelength maximum (Figure 2). This change in absorption of T_3-Rhodamine is not due to dissociation of a dimer, since the concentration-dependent changes in absorption that are observed in water

* On leave from the Institute of Biological Chemistry, 1st Faculty of Medicine, University of Naples, Italy.

† The following abbreviations are used:
T_3-Rhodamine—
Ala-Rhodamine—
Rhodamine—This is the B compound.

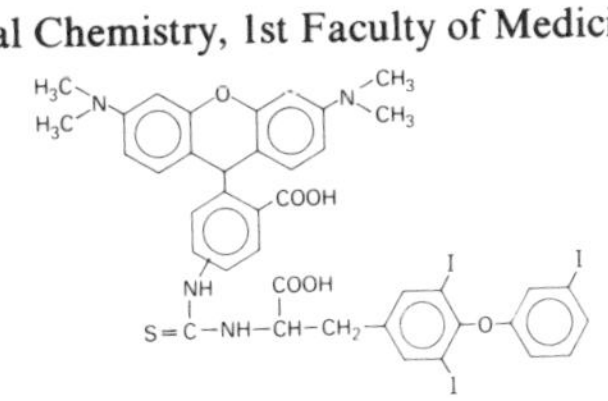

The above three compounds are capitalized in the text. When rhodamine is used without capitalization it refers to the chromophoric moiety in any of the above three compounds

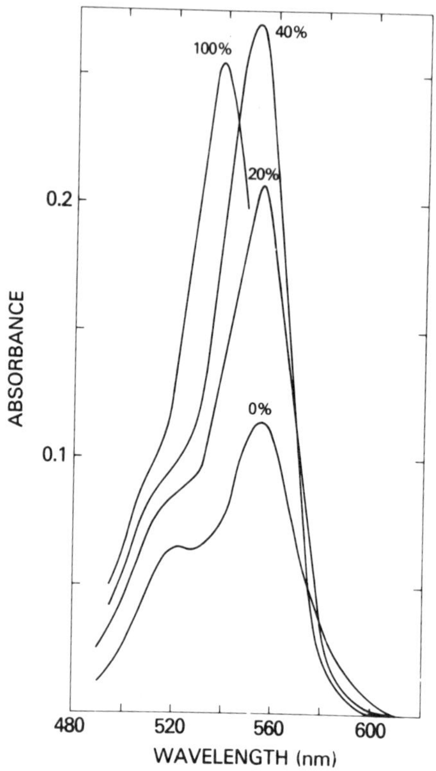

FIGURE 1. Effect of ethanol concentration on the absorption spectrum of T_3-Rhodamine. Measurements were made at 4°C with mixtures of 20% (v/v), and 40% (v/v) ethanol/water solutions. All solutions contained 0.01M phosphate and were pH 7.2 in water. The concentration of T_3-Rhodamine was 2.7 10^{-6}M.

or 20% ethanol occur only at higher concentrations of T_3-Rhodamine than we have employed. Consequently, the rather large change in absorption spectra observed with T_3-Rhodamine in ethanol is attributed to an intramolecular unfolding in which the two chromophores no longer interact with each other.

When the alcohol concentration was increased from 20 to 40%, a further, significant increase in intensity of the major and minor bands of T_3-Rhodamine occurred with a small blue shift (FIGURE 1). The absorption spectra of Ala-Rhodamine and Rhodamine, however, increased only slightly in 40% alcohol while their maxima were also only slightly blue-shifted (not shown). In 60% alcohol, somewhat larger blue shifts were observed for all three rhodamine compounds without much further change in their oscillator strengths. In 100% ethanol, the absorption maxima of T_3-Rhodamine, Ala-Rhodamine and Rhodamine are further blue-shifted and their oscillator strengths decreased slightly. The greater blue-shift in absorption peak and much larger increase in oscillator strength of T_3-Rhodamine when compared with Ala-Rhodamine between water and alcohol suggest that the aromatic chromophores in T_3-Rhodamine which interact to form an intramolecular complex in water are no longer in contact by 40% ethanol.

Fluorescence

The emission (i.e. quantum yield) of the rhodamine chromophore is strongly quenched in T_3-Rhodamine when compared with Ala-Rhodamine in aqueous solution. The quenching observed with T_3-Rhodamine is largely eliminated in 40% ethanol (FIGURES 3 and 4). The strong quenching observed in water which is relieved in alcohol is due to a transition from a folded to an unfolded molecule. Relatively little enhancement in Ala-Rhodamine fluorescence occurs in alcohol solutions.

The quantum yield of Rhodamine changes very differently with ethanol concentration when compared with Ala-Rhodamine although the wavelength maxima are essentially the same (FIGURE 5). The quantum yield of T_3-Rhodamine approaches that of Ala-Rhodamine at 60% ethanol and becomes indistinguishable at 100% ethanol. It is clear, therefore, that the properties of T_3-Rhodamine should be compared with those of Ala-Rhodamine rather than Rhodamine.

Circular Dichroism

Chromophoric groups generally show much greater induced optical activity when they form parts of a rigid structure than when they possess rotational freedom. We have observed a significant CD spectrum between 500 and 600 nm for the rhodamine chromophore in T_3-Rhodamine but not in Ala-Rhodamine or Rhodamine (FIGURE

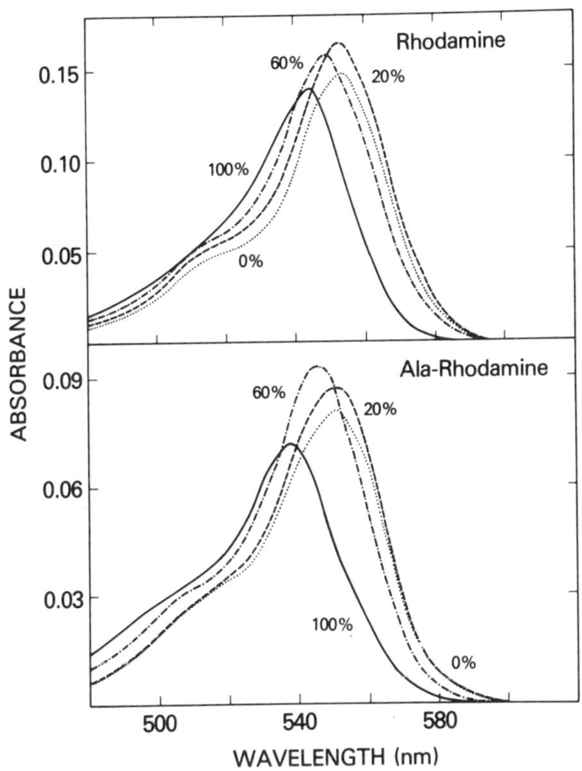

FIGURE 2. Effect of ethanol concentration on the absorption spectra of Ala-Rhodamine and Rhodamine. Measurements were made at 4°C in 20% (v/v) and 60% (v/v) ethanol/water mixture (in 0.01M phosphate buffer, pH 7.2). The concentrations of Rhodamine and Ala-Rhodamine are 1.75 10^{-6}M and 0.90 10^{-6}M, respectively.

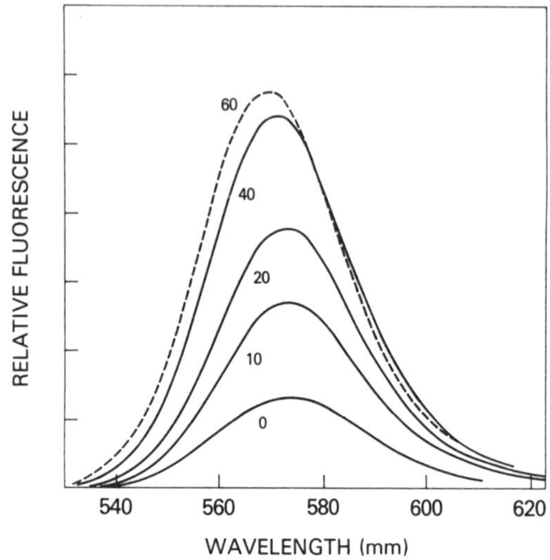

FIGURE 3. The effect of ethanol concentration on the emission spectrum of T_3-Rhodamine. Excitation was at 515 nm. Solutions were at 4°C and contained 0.01 M phosphate, pH 7.2.

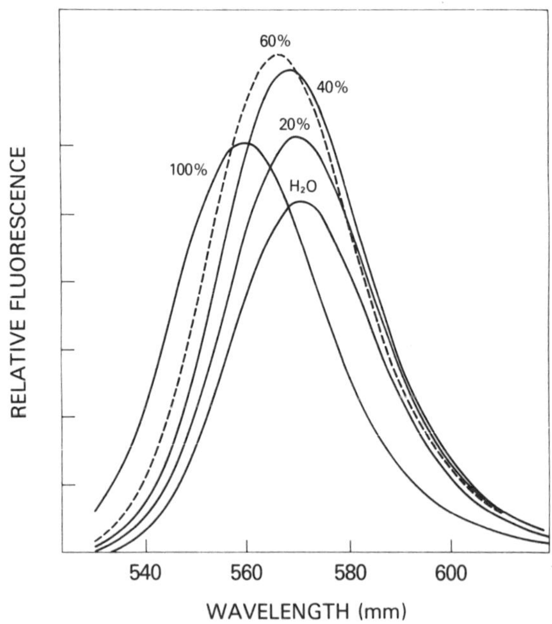

FIGURE 4. The effect of ethanol concentration on the emission spectrum of Ala-Rhodamine. Same conditions as FIGURE 3 except temperature was 25°C.

6). The spectrum exhibits at 4°C a strong, positive peak centered at 583 nm and two weaker, overlapping negative bands with extrema near ~556 and 528 nm. The latter two bands correspond approximately to the two major absorption peaks of T_3-Rhodamine at 556 and 523 nm (FIGURE 1) while the positive CD peak corresponds to the negative difference absorbance peak at 583 nm generated with T_3-Rhodamine by increasing the temperature (see below—FIGURE 7). A similar negative difference absorption band at 583 nm is evident in the absorption spectra of T_3-Rhodamine in 20% (or 40%) alcohol when compared with aqueous solution (FIGURE 1). This band is absent in 20% alcohol solutions of either Ala-Rhodamine or Rhodamine (FIGURE 2).

Temperature

Absorption

The loss of the interaction between T_3 and rhodamine observed in ethanol can also be seen with temperature. The difference spectra produced by temperature with

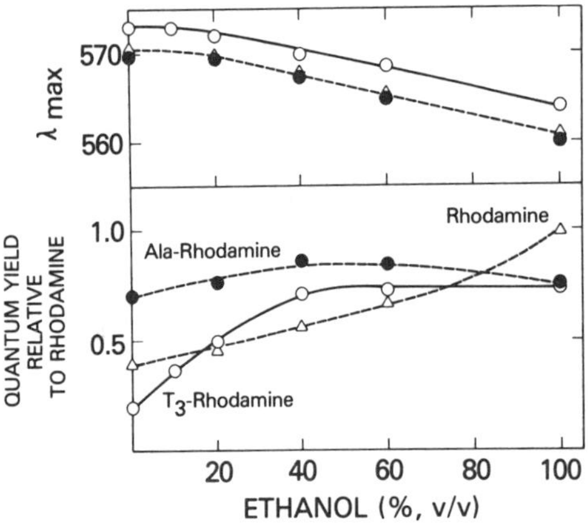

FIGURE 5. The dependence of the emission wavelength maximum and fluorescence quantum yield of Rhodamine (△), Ala-Rhodamine (●), and T_3-Rhodamine (○) on ethanol concentration. Same conditions as in FIGURES 3 and 4.

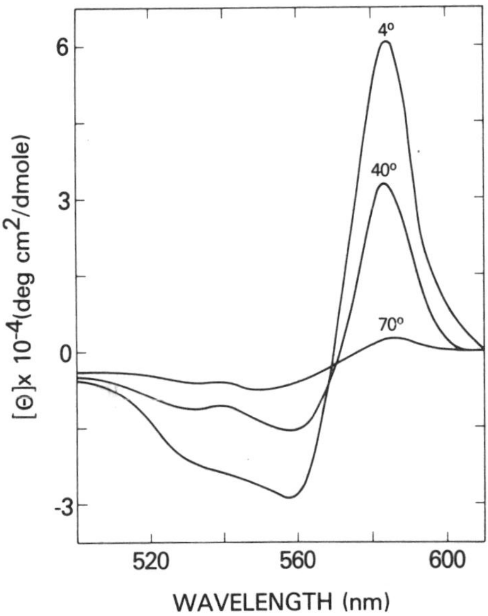

FIGURE 6. The CD spectra of T_3-Rhodamine at 4°C, 40°C and 70°C. Solution contained 13.4 10^{-6}M T_3-Rhodamine dissolved in 0.01M phosphate, 0.1, NaCl, pH 7.2.

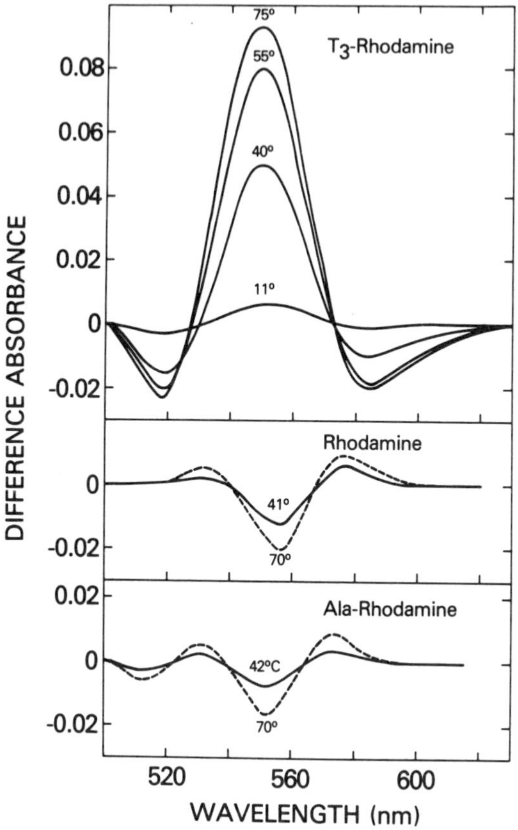

FIGURE 7. The temperature dependence of the difference absorbance spectra of T_3-Rhodamine, Rhodamine and Ala-Rhodamine. Reference solutions were T_3-Rhodamine (4.3 10^{-6}M), Ala-Rhodamine (1.8 10^{-6}M), and Rhodamine (1.8 10^{-6}M) dissolved in 0.01M phosphate, 0.1M NaCl, pH 7.2 at 4°C.

the three rhodamine compounds are shown in FIGURE 7. The thermal difference spectra of T_3-Rhodamine are clearly very different from those of Ala-Rhodamine and Rhodamine between 4 and 75°C. The signs of the three difference bands in T_3-Rhodamine are opposite from those in Ala-Rhodamine (or Rhodamine) (FIGURE 7). Consequently, the magnitude of the three difference peaks in T_3-Rhodamine should be even greater than observed if the effects of temperature on the free chromophoric group were accounted for.

There is reciprocal dependence on temperature between the two long wavelength difference absorption peaks of T_3-Rhodamine at 550 nm and 583 nm. The transition is very broad, being essentially complete at high temperatures, but still incomplete at low temperatures, i.e., 4°C. The broad transition observed with temperature resembles that found with purine and pyrimidine dinucleotides[4,5] and nicotinamide adenine dinucleotide (NAD)[6] which also form folded structures at low temperatures.

Fluorescence

The emission maxima of aqueous solutions of T_3-Rhodamine, Ala-Rhodamine and Rhodamine are very similar, i.e., 573, 570, and 571, respectively. The quantum yield of T_3-Rhodamine, however, was only 29% of Ala-Rhodamine at pH 7.2, 0.01M phosphate, 0.1M NaCl, 4°C. Increasing temperature produced a monotonic decrease in the fluorescence spectra of Ala-Rhodamine (FIGURE 8). This is the expected result for a freely rotating chromophore. The behavior of T_3-Rhodamine, however, was very different since with increasing temperature, the fluorescence intensity first decreased slightly between 4 and 11°C, then increased somewhat between 11 and 35°C, and finally fell again at temperatures above 45°C (FIGURE 9). The fluorescence, i.e., quantum yields, of T_3-Rhodamine and Ala-Rhodamine were similar by 70°C.

Circular Dichroism

The strong rhodamine CD spectrum observed with T_3-Rhodamine at 4°C decreased with increasing temperature and was almost eliminated by 70°C (FIGURE 6). When the temperature dependence of the major CD peak at 583 nm and the difference spectra peak at the same wavelength is compared, the two sets of points fall on the same curve and clearly represent the same transition.

Guanidinium Hydrochloride

The interaction between the two chromophoric groups of T_3-Rhodamine were also evaluated in guanidinium hydrochloride (GdmCl). The effect of 6.0M GdmCl on the absorption spectra of T_3-Rhodamine and Ala-Rhodamine is shown in FIGURE 10. With the latter compound, there is a red-shift of 7 nm, a positive difference peak at 565 nm and little change in oscillator strength. In the case of T_3-Rhodamine, there is a large increase in oscillator strength which is accompanied by a blue shift of 4 nm. The difference spectrum of T_3-Rhodamine in 6M GdmCl resembles the difference

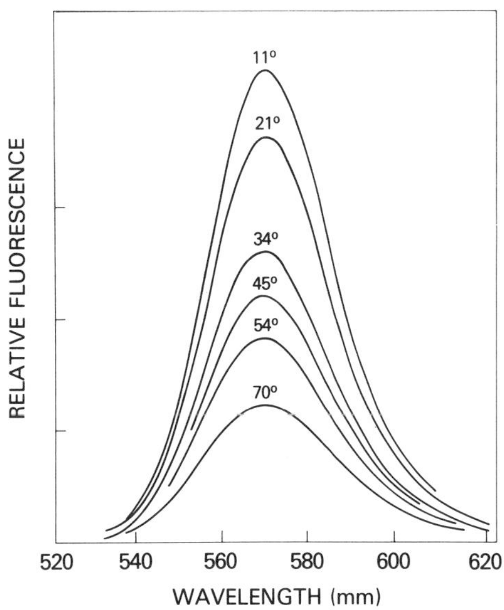

FIGURE 8. The temperature dependence of the fluorescence intensity of Ala-Rhodamine (1.9×10^{-6} M). Conditions the same as in FIGURE 9.

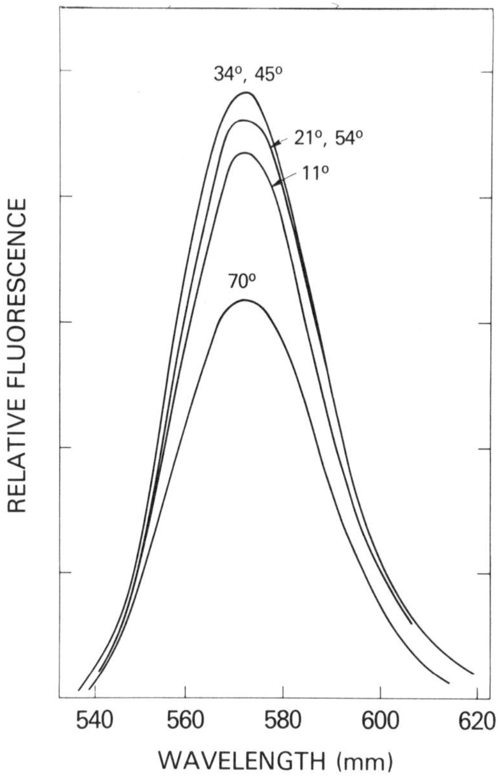

FIGURE 9. The temperature dependence of the fluorescence intensity of T_3-Rhodamine (1.0×10^6 M). Solutions contained 0.01M phosphate, 0.1M NaCl, pH 7.2. Excitation at 495 nm.

spectrum produced at elevated temperatures but is quite different from the difference spectrum of Ala-Rhodamine in 6M GdmCl.

The effect of charge interactions in stabilizing the folded form of T_3-Rhodamine was evaluated with NaCl. Increasing NaCl resulted in a decrease in oscillator strength of T_3-Rhodamine (FIGURE 11). A negative difference spectrum was observed in 1.0M or 3.0M NaCl which is opposite from that observed in concentrated solutions of GdmCl. It appears, therefore, that increasing ionic strength, i.e. NaCl, favors the low-temperature folded form of T_3-Rhodamine since the transition between the two forms is still incomplete at 4°C in 0.1M NaCl and 0.01 phosphate.

The increase in oscillator strength of T_3-Rhodamine begins at about 1M GdmCl and ends near 5M GdmCl (FIGURE 10—inset). The transition with GdmCl is much broader than is found with proteins and presumably reflects a noncooperative transition. The difference spectrum of T_3-Rhodamine observed in 1.2M GdmCl resembles that in 1M NaCl. The major effect of 1.2M GdmCl therefore appears to be that of ionic strength in displacing the equilibrium to the folded form. At higher GdmCl concentrations, the equilibrium is displaced towards the unfolded form.

DISCUSSION

All the spectral properties of the rhodamine chromophore that we have evaluated are significantly altered when T_3 is covalently conjugated to it by a thiourea link. In contrast, the spectral properties of the same chromophore with the alanyl group, i.e. Ala-Rhodamine, are not significantly modified.

The transitions in molecular properties are most readily explained in terms of a structural transition between a folded or sandwich structure of T_3-Rhodamine in water at low temperature and an unfolded, freely rotating structure in alcohol or 6M GdmCl solutions or at high temperatures since at the concentration studied T_3-Rhodamine is almost completely dissociated.

The optical changes of rhodamine produced by T_3 interaction in T_3-Rhodamine include a major hypochromism, strong quenching of rhodamine fluorescence and induced optical activity. The magnitude of these changes preclude simple solvent effects and require a specific interaction between the two chromophores. Unfortunately, with chromophores as complex as rhodamine and T_3, it is not easy to distinguish between different types of interactions.

The three conditions used to unfold T_3-Rhodamine are also those used to denature proteins. Unfortunately, the beginning of the thermal transition of T_3-Rhodamine is below 4°C and consequently the complete transition curve is not available. Therefore, the equilibrium constants for the transition as a function of temperature could not be

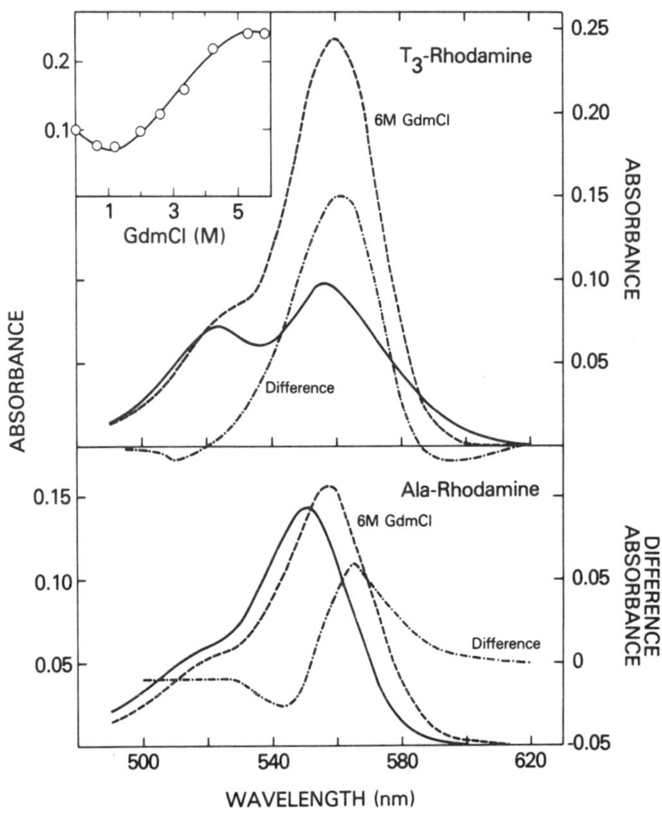

FIGURE 10. Absorption of T_3-Rhodamine (2.6×10^{-6}M) and Ala-Rhodamine (1.8 10^{-6}M) in water (−) at 4°C and in 6M GdmCl. The difference spectra are also shown. *Insert:* Effect of GdmCL on absorbance at 560 nm.

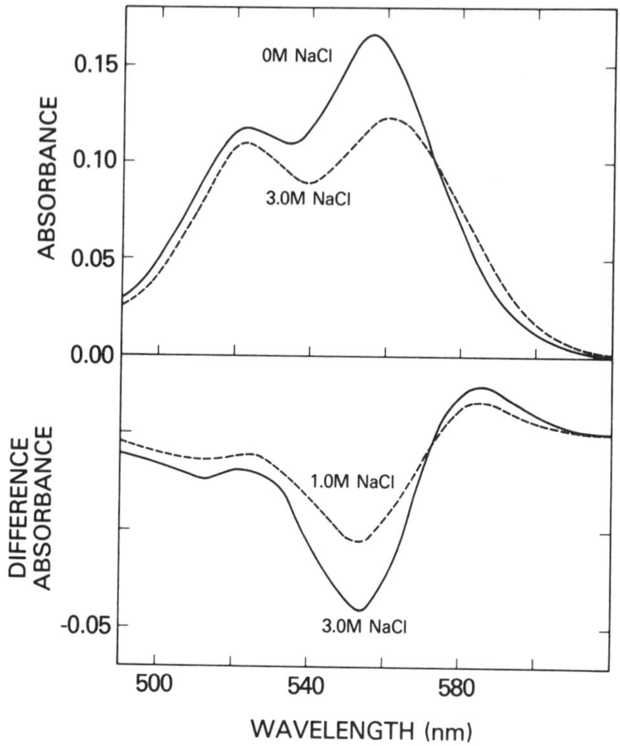

FIGURE 11. Absorbance spectra (*upper half*) at 4°C of T_3-Rhodamine (4.4 10^{-6}M) in water and in 3.0M NaCl. Difference absorbance spectra between water and 1M NaCl or 3.0M NaCl at 4°C and neutral pH.

calculated. Calculation of free energy changes would indicate whether there is a heat capacity change and would allow confirmation of a hydrophobic basis to the folding.[7] The unfolding produced by GdmCl, however, supports a hydrophobic contribution to the interaction. This type of interaction does not preclude other types of bonding acting in concert. There is a small but important electrostatic contribution since addition of neutral salt (NaCl) drives the molecule further into the folded state.

METHODS

Absorption

Absorption spectra were obtained on a Cary 14 equipped with a temperature-controlled cell holder. Most of the measurements were made at 4°C since the absorption spectrum changes with temperature (see above). Difference spectra were obtained using tandem cell technique. The oscillator strengths were compared by measuring the areas under the absorption curves at the same concentration of rhodamine.

Fluorescence

Fluorescence intensities and spectra were obtained at 25°C with a Perkin-Elmer MPF III fluorometer equipped with a temperature-controlled cell holder. Fluorescence measurements were made in the range where emission was linear with chromophore concentrations. The absorbance of all solutions was less than 0.1 OD at the excitation wavelength. The relative quantum yields of T_3-Rhodamine, Ala-Rhodamine and Rhodamine were obtained by comparing the areas (A) of the emission spectra with the absorbance (OD) at the excitation wavelength.

$$Q = \frac{A}{A \text{ standard}} \times \frac{\text{OD standard}}{\text{OD}} \times Q \text{ standard}$$

Circular Dichroism

CD measurements were carried out on a Cary model 60 spectropolarimeter, equipped with a temperature-controlled cell holder. The molar ellipticities were calculated by:

$$[\theta] = \frac{100 \times \theta_{obs}}{\ell \times M}$$

where $[\theta]\lambda$ is the molar ellipticity at a given wavelength λ; θ_{obs}, the observed ellipticity; M, the molarity of the chromophore, and ℓ, the optical path length in cm.

Concentration

A value of 72,443 at 546 nm was used for the molar extinction coefficient of Rhodamine and Ala-Rhodamine in ethanol. The molar extinction coefficient at the absorption peak (538 nm) in methanol was determined for T_3-Rhodamine and found to be 94,000.

The concentrations of the three rhodamine compounds were determined from their absorption in ethanol at the peak values listed above.

Summary

The spectroscopic properties (absorption, fluorescence and circular dichroism) of two amino acid derivatives of rhodamine have been compared with those of rhodamine. The oscillator strength and quantum yield of rhodamine when coupled to triiodothyronine (through a thiourea link) are strongly reduced in aqueous solution when compared with that of rhodamine coupled to alanine. Moreover, a significant circular dichroic band of rhodamine is observed with the triiodothyronine but not the alanyl derivative. All the aberrant properties of rhodamine when coupled to triiodothyronine are normalized by ethanol, quanidinium chloride or high temperatures. It is concluded that the aberrant behavior of the rhodamine chromophore is due to an intramolecular interaction with the triiodothyronyl chromophore.

References

1. Cheng, S.-Y., N. L. Eberhardt, J. Robbins, J. D. Baxter, & I. Pasten. 1979. FEBS Lett. **100:** 113.
2. Cheng, S.-Y., F. Maxfield, J. Robbins, M. Willingham, & I. Pasten. 1980. Proc. Natl. Acad Sci. USA **77:** 3425.
3. Selwyn, J. E., & J. I. Steinfeld. 1972. J. Phys. Chem. **76:** 762.
4. Brahms, J., J. C. Maurizot, & A. M. Michelson. 1967. J. Mol. Biol. **25:** 481.
5. Davis, R. C., & I. Tinoco. 1968. Biopolymers **6:** 223.
6. Oppenheimer, N. J., L. J. Arnold, Jr., & N. O. Kaplan. 1978. Biochemistry **17:** 2613.
7. Formisano, S., M. L. Johnson, & H. Edelhoch. 1977. Proc. Natl. Acad. Sci. USA **74:** 3340.

ROTATIONAL DIFFUSION OF RHODOPSIN IN MEMBRANES AS DETERMINED BY STEADY-STATE POLARIZED PHOTOCHROMISM

C. Pasternak, D. Corda, H. Stieve, and M. Shinitzky*

Department of Membrane Research
The Weizmann Institute of Science
Rehovot, Israel

Dynamic characteristics of membrane proteins, which relate to diffusion, are now believed to be directly implicated in physiological processes.[1] Proteins in membrane undergo simultaneous lateral and rotational diffusions, which are of two distinct operational categories. Lateral diffusion of membrane proteins is related to a membrane region that is much greater than the diameter of the protein and is, therefore, affected by both the lipid viscosity and protein density.[2,3] Rotational diffusion of the membrane proteins is mostly affected by the immediate lipid environment, which is characterized by the average viscosity of 1-10 poise.[4] The expected average rotational diffusion constant of most membrane proteins is in the range of 10^3–10^6 radian^2/sec, which is about 2–3 orders of magnitude smaller than in aqueous solutions.[5]

Spectral probes, with transient lifetimes of 1 μsec–1 msec, could be employed for monitoring rotational diffusion of membrane proteins by optical depolarization. The two spectral processes that can match this range of lifetime, and thus provide suitable techniques for biological membranes, are phosphorescence[6] and photochromism.[7,8] In the latter process, the excited probe molecule is converted into a physical (e.g. triplet state) or chemical (e.g. *cis* or *trans* isomer) transient which spontaneously isomerizes back to the parent compound. This reversible process is analogous to fluorescence, where the absorption spectrum of the phototransient corresponds to the fluorescence spectrum and its lifetime corresponds to the excited-state lifetime in fluorescence. Similarly, photochromism can be measured with polarized excitation and polarized monitoring light in either decay after pulse excitation[7,9] or at steady-state.[8,10] The technique of steady-state polarized photochromism is now being employed in our laboratory for monitoring rotational diffusion of membrane proteins. The analysis of the depolarization data can be carried out with the aid of expression for fluorescence depolarization, as described in the following.

In unrestricted rotation, as in an isotropic fluid, the steady-state depolarization of emission or photochromism is described by the well-known Perrin equation[5]:

$$\frac{r_0}{r} = 1 + \frac{3\tau}{\rho} \qquad (1)$$

where τ is the excited-state lifetime, in fluorescence or phosphorescence, or the lifetime of the phototransient in reversible photochromism, and ρ is the rotational

* To whom correspondence should be addressed.

0077-8923/81/0366-0265$01.75/0©1981, NYAS

relaxation time which includes all the hydrodynamic parameters of the system. The measured steady-state anisotropy, r, approaches the limiting value of r_0 when no rotation takes place during τ. The limiting anisotropy r_0 is an intrinsic molecular parameter which relates to the angle α between the excitation and the responding oscillators[5]:

$$r_0 = \frac{3\cos^2\alpha - 1}{5} \tag{2}$$

For most globular structures, ρ is proportional to the ambient viscosity, η, and inversely proportional to the absolute temperature, T. Equation 1 can thus be converted to

$$\frac{r_0}{r} = 1 + C\frac{T\cdot\tau}{\eta} \tag{3}$$

where C is a constant which includes the hydrodynamic structural features of the rotating body. For a rotating sphere, $C = \kappa/v$, where v is the effective volume and κ is Boltzmann constant.

In the case of $\tau \gg \rho$ the corresponding r (termed as r_∞) approaches 0 (see Equation 1). Isotropic unrestricted rotation is thus characterized by complete randomization of the well-defined order in the excited molecules which is induced by photo-selective excitation. At the stage of complete randomization, the average angular displacement θ_∞ equals *arc* $\cos^2 1/3$, namely $\theta_\infty = 54°\ 44'$, corresponding to $r_\infty = 0$ (see also Equation 4). Restricted rotation is characterized by residual order in the photo-selected molecules after extensive rotational randomization and is expressed in $\theta_\infty <$ $54°\ 44'$ or $|r_\infty| > 0$. The dependence between these parameters is given in Equation 4.[11]

$$\frac{r_\infty}{r_0} = \frac{3\cos^2\theta_\infty - 1}{2} \tag{4}$$

For the common experimental case of $r_0 = 0.4$, Equation 4 turns to

$$r_\infty = \frac{3\cos^2\theta_\infty - 1}{5} \tag{5}$$

which indicates an interesting analogy between the angle separating the excitation oscillator from the responding oscillator in the frozen state and the average angle of rotational displacement (θ_∞) in the dynamic state (see Equations 2 and 5).

For restricted rotation, the steady-state Perrin equation (Equations 1 and 3) is converted to the form[12]:

$$r = r_\infty + \frac{(r_0 - r)\cdot\rho}{3\tau} = r_\infty + \frac{1}{C}\frac{(r_0 - r)\cdot\eta}{T\tau} \tag{6}$$

which can be employed experimentally.

The actual geometrical boundaries of the restricted rotation are not disclosed by the experimentally evaluated r_∞, and one is bound to use specific models for interpretation of r_∞. For restricted rotation of a membrane protein, two pertinent models may be considered: the wobbling motion[13,14] and the unidimensional rotation,[15] which are illustrated in FIGURE 1.

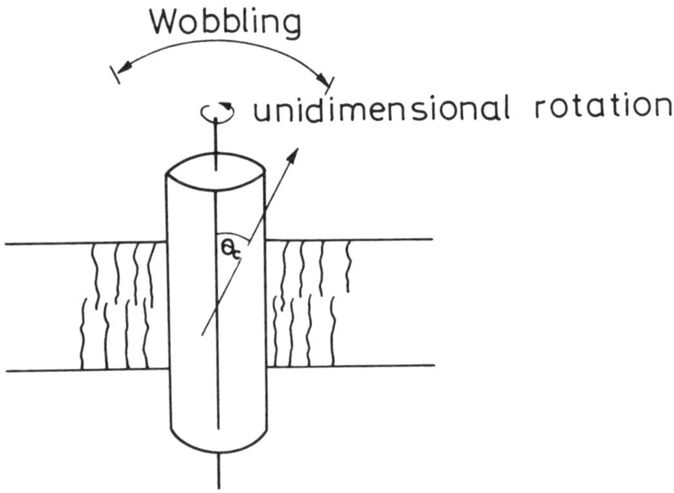

FIGURE 1. Two realistic models for rotation of proteins in biological membranes monitored by a spectral transition-dipole of a well-defined orientation.

In the wobbling motion, the angular displacement is restricted to a cone angle of $\theta_c < 90°$, and is presented by[13,14]:

$$\frac{r_\infty}{r_0} = \left(\frac{\cos\theta_c(1 + \cos\theta_c)}{2}\right)^2 \tag{7}$$

In this model r_∞ declines with increase of θ_c and at unrestricted rotation $\theta_c = 90°$ and $r_\infty = 0$ (corresponding to $\theta_\infty = 54°44'$ in Equation 5).

The unidimensional rotation is characterized by a fixed angle (θ_c) between the responding oscillator and the axis of rotation (see FIGURE 4). The dependence of r_∞ on θ_c in this model can be explicitly evaluated with the aid of Euler angles.[4] The procedure is, however, complex and we shall present only the final equation:

$$\frac{r_\infty}{r_0} = \left(\frac{3\cos^2\theta_c - 1}{2}\right)^2 \tag{8}$$

This function reaches a minimum of $r_\infty = 0$ at $\theta_c = 54°\ 44'$, while at $\theta_c = 0°$, where rotational depolarization is eliminated, r_∞ approaches r_0 and at $\theta_c = 90°$ $r_\infty = r_0/4$. Equation 8, for the specific case of $r_0 = 0.4$, could be deduced from relations obtained for decay of polarized emission.[15]

Evaluation of ρ, by steady-state depolarization methods, can be accomplished by altering τ at a constant T and η (see Equations 1 and 6) and monitoring the corresponding r. This approach is feasible in fluorescence, phosphorescence or triplet absorption, where τ can be shortened with the aid of an external quencher (e.g. oxygen). In some photochromic processes (e.g. *cis–trans* isomerization) the lifetime of the phototransient, presented by τ, can be shortened upon excitation and is, therefore, dependent on the intensity of the monitoring light. However, in the cases where τ cannot be changed appreciably, one is bound to change η or T and monitor the consequent changes in the product $T \cdot \tau/\eta$ and in r (see Equations 1 and 6). When this procedure is applied to rotational diffusion of membrane proteins, one faces the problem of evaluating the effective membrane viscosity, which could be of formidable complexity. The methods available for this purpose generally result in the apparent average η value of the lipid hydrocarbon core[4] and can serve only as an approximation for η in rotational depolarization measurements.

The instrumentation and technique for measuring steady-state polarized photochromism[10] have similar features to those employed in the phase-modulation of polarized fluorescence.[16] Polarized excitation of relatively low intensity is modulated mechanically, and the absorption of the phototransient is monitored by a frequency-matched polarized beam. Frequency scanning provides a simple means for resolution of intermediates with different lifetimes.[10,16]

A few membrane proteins bear natural photochromic substances, which can be utilized for spectral monitoring of diffusion characteristics. Rhodopsin is undoubtedly the most prominent in this group. Vision in vertebrates initiates a series of photoisomerization of the rhodopsin parent chromophore, *11-cis* retinal, which ends up in chemical dissociation of all-*trans* retinal. This series is characterized by irreversible bleaching upon light absorption, which does not permit the use of the reversible photochromism techniques. The visual membrane of invertebrates, however, is characterized by a complete photochromic cycle without significant bleaching.[17]

The photosensory membranes of cephalopods (e.g. cuttlefish) below ~15°C, were confirmed by us to display a reversible photochromism upon light absorption (λ_{max} = 492 nm). The shape of the absorption band of the major phototransient in the cycle, *meta* rhodopsin, almost completely overlaps the initial absorption spectrum, although with significant changes in extinction coefficient, ϵ. The change in ϵ at the absorption maxima during the reversible photochromism of cuttlefish rhodopsin is sufficient for monitoring the rotational diffusion of rhodopsin in a dilute suspension of visual membranes. Alternatively, excitation can be carried out at the red edge of the rhodopsin absorption band (578 nm Hg band) with simultaneous monitoring of *meta* rhodopsin at the blue edge of its absorption spectrum (430 nm). At these wavelengths, the increments of the ϵ values of rhodopsin and *meta* rhodopsin are not equal. Changing of the product $T \cdot \tau/\eta$ in these membranes could be carried out in the limited range of $(2–5) \times 10^{-3}$ °k·sec·poise^{-1}, by changing the temperature in the narrow range of 4–12°C. In this range, the lifetime of *meta* rhodopsin, τ, was found to change in the range of 0.1–0.2 msec. The value of η was related to the lipid core and was determined by fluorescence depolarization of 1,6-diphenyl-1,3,5-hexatriene (DPH), as described.[4] The range of $T \cdot \tau/\eta$ could be extended by incorporation of different amounts of stearic acid to increase η.

The dependence of the photochromic anisotropy, r, on the product $T \cdot \tau/\eta$, for cuttlefish visual membranes, is presented according to Equation 6 in FIGURE 2. The effective volume of an equivalent sphere of 1.6×10^5 Å^3 is obtained from the slope. This volume corresponds to a diameter of 54 Å, which is about the thickness of the lipid bilayer itself, and to a molecular weight of about 120,000. The molecular weight of monomeric rhodopsin of invertebrates was estimated to be about one-third of this figure,[18] which suggests that the average rotating rhodopsin structure, following photoactivation, is oligomeric. It could be that this oligomer is a transient structure, which is formed only upon light induction and can thus acquire some physiological function.[17] The apparent ρ value of this system is approximately 200 μsec at 25°, which is about 10-fold greater than that estimated for ρ of mammalian membrane rhodopsin.[19] More subtle information could be extracted from the intercept of r_∞ = 0.05 (see FIGURE 2). According to the unidimensional rotating cylinder (FIGURE 1), which fits well with accepted models for mammalian membrane rhodopsin,[19] this intercept corresponds to $\theta_c = 40°$ or 73° (see Equation 8). The latter value is very close to that estimated for the transition dipole orientation, which was derived from dichroic measurements.[20]

A classical case of a protein with reversible photochromic properties is bacteriorhodopsin from the purple membrane of *Halobacterium halobium*. The parent pigment of this protein is all-*trans* retinal, which displays an absorption maximum at 570 nm. Excitation at this absorption band initiates a photocycle with several resolvable intermediates, which is completed without any detectable bleaching.[21] One of the major transients in the cycle (λ_{max} = 410 nm) displays a relatively long lifetime of about 10 msec with r_0 = 0.335, and is especially suitable for monitoring rotational depolarization.[8]

In the intact membrane, the recorded photochromic anisotropy, r, upon excitation at 578 nm (Hg band) and monitoring at 410 nm, practically matches with the r_0 value. Furthermore, changes in temperature in the range of 4–40° do not reduce this high r value, which indicates that even during the relatively long lifetime of 10 msec the bacteriorhodopsin molecules are immobile and maintain their orientation in the membrane. This interpretation is consistent with the well-documented crystalline structure of bacteriorhodopsin *in situ*.[22]

Isolated bacteriorhodopsin can be incorporated into artificial lipid membranes with reconstituted proton-pump activity.[23] In reconstituted membranes with dipalmitoyl lecithin or dimyristoyl lecithin, and protein/lipid of 3 (w/w) as in the natural

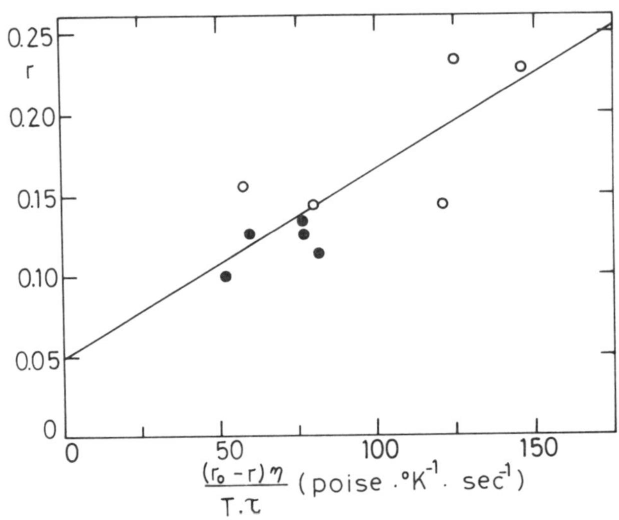

FIGURE 2. Rotational depolarization of the photochromic transition in a suspension of cuttlefish photosensory membranes (~50 μg protein/ml sea water), presented according to Equation 6. Measurements were carried out with untreated membranes (●), or membranes enriched with stearic acid (○) in the range of 4–12°C. Polarized modulated excitation was at 578 nm (Hg band) and the phototransient was monitored at right angle with polarized 430 nm beam (r_0 = 0.360).

membrane, r is high below the lipid-phase transition and is reduced sharply above the melting temperature. The transient lifetime, τ, in these systems, is severalfold greater than in the natural membrane and is affected by the lipid phase transition, as shown in FIGURE 3, for dipalmitoyl lecithin reconstituted membranes. Around the phase transition of the membrane ($T_c = 36°$), the transient lifetime is minimal and at both the liquid crystalline state (above 45°C) and the gel state (below 25°C), it approaches the values displayed by the natural membrane. The photochromic anisotropy, r, and the transient lifetime, τ, in the bacteriorhodopsin-dipalmitoyl lecithin membrane, were measured simultaneously at different temperatures in the range of 10–50°C. The corresponding η values were taken from available data on lipid microviscosity of dipalmitoyl lecithin liposomes.[4] For the arbitrary value of $r_0 = 0.4$, the obtained dependence of r on $T \cdot \tau/\eta$ is shown in FIGURE 4.

The apparent curvature obtained in FIGURE 4 indicates that the correct r_0 value of the monitored transition is significantly lower than 0.4. As clearly shown, the determined r levels off slightly below 0.3 at the gel-state phase. This value is somewhat lower than $r_0 = 0.335$ of the natural membrane, and it indicates a fast wobbling motion of at least 5° of the retinal in the cleft of protein in the reconstituted membrane (see Equation 7). Employing $r_0 = 0.3$ and replotting the same experimental data as in FIGURE 4 indeed eliminate the curvature (see FIGURE 5). The apparent rotating-sphere volume obtained from the slope of FIGURE 5 is very large ($\sim 4 \times 10^9$ Å^3) and corresponds to a radius of about 1000 Å. These dimensions are obviously much greater than monomeric or any reasonable oligomeric assembly of bacteriorhodopsin. When the experiments were carried out in viscous solutions of sucrose, no increase in r was apparent, which excluded the possibility that the monitored rotating body is the membrane itself. A reasonable explanation of this discrepancy could be that the effective η in the system is actually much greater than that of the lipid core

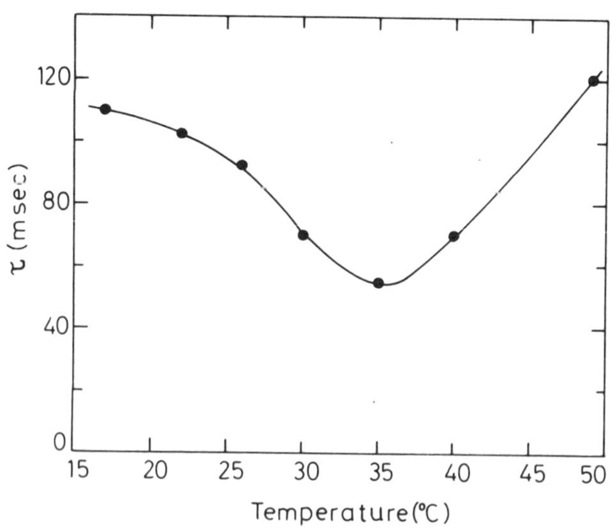

FIGURE 3. Lifetime, τ, of the phototransient ($\lambda_{max} = 410$ nm) of bacteriorhodopsin reconstituted in artificial membranes of dipalmitoyl lecithin at different temperatures.

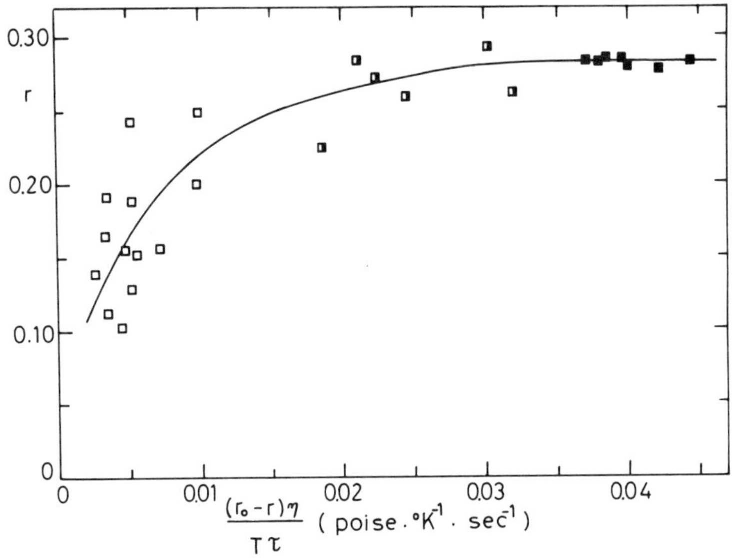

FIGURE 4. Rotational depolarization of bacteriorhodopsin in dipalmitoyl lecithin membranes, using $r_0 = 0.4$. Measurements were taken below (■), above (□) and around (◨) the phase transition of reconstituted system (~36°).

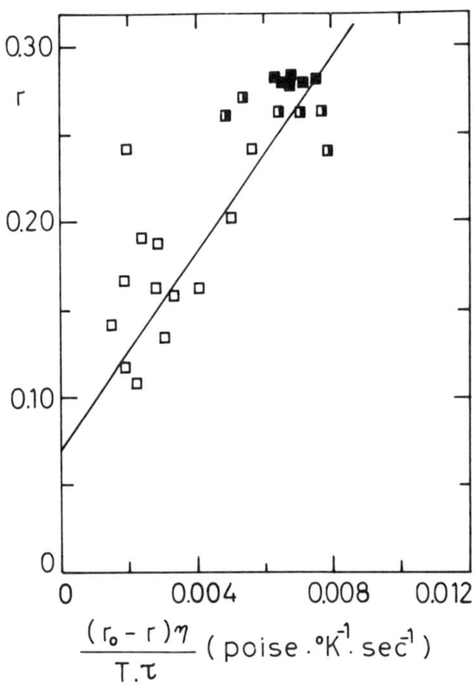

FIGURE 5. Replot of the data given in FIGURE 4, using $r_0 = 0.3$.

and results mostly from protein-protein interaction at the high local concentration of 3 mg protein/mg lipid in the reconstituted membrane. The apparent dependence of r on lipid viscosity could be mediated via modulations of the protein-protein interaction. The extrapolated value of $r_\infty = 0.065$ in FIGURE 5 further supports monomeric-type rotation. This value corresponds to $\theta_c = 43°$ or $76°$ which are very similar to those obtained for the cuttlefish visual membranes (FIGURE 2). A unidimensional rotating cone (FIGURE 1), with an angle of about 75°, was found by others to fit bacteriorhodopsin in the purple membrane.[24]

It is interesting to compare structural and dynamic features of rhodopsin in the bacterial membrane, in the photosensory membrane of cuttlefish and in the vertebrate rod outer segment, which belong to three distinct evolutionary categories. In the three membranes, the orientation of the retinal chromophore is similar and is about 75° with respect to the rotation axis perpendicular to the membrane surface. The rotational freedom of rhodopsin in these membranes is, however, markedly different. In the vertebrate membrane rhodopsin rotates virtually freely ($\rho_{25°} \sim 20$ μsec, Reference 19), in the cephalopod membrane its rotation is hindered markedly ($\rho_{25°} \sim 20$ μsec), whereas in the bacterial membrane the rhodopsin is practically immobile. This difference in dynamics could be implicated in the special patterns of ion fluxes that follow light absorption in these membranes.[25]

REFERENCES

1. SHINITZKY, M. & P. HENKART. 1979. Fluidity of cell membranes: current concepts and trends. Int. Rev. Cytol. **60:** 121–147.
2. VAZ, W. L. C., K. JACOBSON, E. S. WU & Z. ZERKO. 1979. Lateral mobility of an amphipathic apolipoprotein, Apo C-III, bound to phosphatidyl choline bilayers with and without cholesterol. Proc. Natl. Acad. Sci. USA **76:** 5645–5649.
3. CHERRY, R. J., U. MULLER & G. SCHNEIDER. 1977. Rotational diffusion of bacteriorhodopsin in lipid membranes. FEBS Lett. **80:** 465–469.
4. SHINITZKY, M. & Y. BARENHOLZ. 1978. Fluidity parameters of lipid regions determined by fluorescence polarization. Biochim. Biophys. Acta **515:** 367–394.
5. WEBER, G. 1953. Rotational brownian motion and polarization of the fluorescence of solutions. Adv. Prot. Chem. **8:** 415–459.
6. AUSTIN, R. H., S. SUILING CHAN & T. M. JOVIN. 1979. Rotational diffusion of cell surface components by time-resolved phosphorescence anisotropy. Proc. Natl. Acad. Sci. USA **76:** 5650–5654.
7. CHERRY, R. J. 1979. Rotational and lateral diffusion of membrane proteins. Biochim. Biophys. Acta **559:** 289–327.
8. PASTERNAK, C. & M. SHINITZKY. 1978. Polarized photochromism in the purple membrane from *halobacterium halobium. In* Energetics and Structure of Halophilic Microorganisms. S. R. Caplan & M. Ginzburg, Eds. Elsevier/North-Holland, pp. 309–314.
9. CHERRY, R. J. 1978. Measurements of protein rotational diffusion in membranes by flash photolysis. Methods in Ezymol. **34:** 47–61.
10. PASTERNAK, C., M. A. SLIFKIN & M. SHINITZKY. 1978. Polarized photochromism in solid solutions. J. Chem. Phys. **68:** 2669–2673.
11. WEBER, G. 1978. Limited rotational motion: Recognition by differential phase fluorometry. Acta Phys. Pol. **A54:** 859–865.
12. LAKOWICZ, J. R., F. G. PREDERGAST & D. HOGAN. 1979. Fluorescence anisotropy measurements under oxygen quenching conditions as a method to quantify the depolarizing rotations of fluorophores: Application to diphenylhexatriene in isotropic solvents and in lipid bilayers. Biochemistry **18:** 520–527.
13. KINOSITA, K., S. KAWATO & A. IKEGAMI. 1977. A theory of fluorescence polarization decay in membranes. Biophys. J. **20:** 289–305.

14. KAWATO, S., K. KINOSITA & A. IKEGAMI. 1977. Dynamic structure of lipid bilayer studies by nanosecond fluorescence techniques. Biochemistry **16:** 2319–2324.
15. RIGLER, R. & M. EHRENBERG. 1976. Fluorescence relaxation spectroscopy in the analysis of macromolecular structure and motion. Quart. Rev. Biophys. **9:** 1–19.
16. SPENCER, R. D. & G. WEBER. 1970. Influence of brownian rotations and energy transfer upon the measurements of fluorescence lifetime. J. Chem. Phys. **52:** 1654–1663.
17. MONTAL, M. & J. I. KORENBROT. 1976. Rhodopsin in cell membranes and the process of phototransduction. *In* The Enzymes of Biological Membranes. A. Martonosi, Ed. Plenum Press. New-York. Vol. **4**, pp. 365–405.
18. HAGINS, F. M. 1973. Purification and partial characterization of the protein component of squid rhodopsin. J. Biol. Chem. **248:** 3298–3304.
19. CONE, R. A. 1972. Rotational diffusion of rhodopsin in the visual receptor membrane. Nature (London) New Biol. **236:** 39–45.
20. GOLDSMITH, T. H. & R. WEHNER. 1977. Restriction on rotational and translational diffusion of pigment in the membranes of a rhabdomeric photoreceptor. J. Gen. Physiol. **70:** 453–490.
21. LOZIER, R., R. BOGOMOLNI & W. STOECKENIUS. 1975. Bacteriorhodopsin: A light-driven proton pump in *halobacterium halobium*. Biophys. J. **15:** 955–962.
22. HENDERSON, R. 1975. The structure of the purple membrane from *halobacterium halobium*: Analysis of the X-ray diffraction pattern. J. Mol. Biol. **93:** 123–138.
23. BAKKER, E. P. & S. R. CAPLAN. 1978. Phospholipid substitution of the purple membrane: The stoichiometry of light-induced proton release by phospholipid-substituted purple membranes. Biochim. Biophys. Acta **503:** 362–379.
24. HEYN, M. P., R. J. CHERRY & U. MULLER, 1977. Transient and linear dichroism studies on bacteriorhodopsin: Determination of the orientation of the 568-nm all-trans retinal chromophore. J. Mol. Biol. **117:** 607–620.
25. STIEVE, H. 1977. Thoughts on the comparative biology of photosensory functions. Verh. Dtsh. Zool. Ges. 1–25.

DYNAMICS OF HORMONE RECEPTORS ON CELL MEMBRANE*

Joseph Schlessinger

Department of Chemical Immunology
The Weizmann Institute of Science
Rehovot, Israel

Insulin and epidermal growth factor (EGF) are two small polypeptide hormones that regulate various important physiological processes.[1-5] Many investigators believe that these hormones act directly on the plasma membrane of their target cells, though internalization has been demonstrated[6-10] and postulated to yield an active species or fragment.[11] Many substances have been proposed as second messengers of insulin.[3] At present, there is insufficient evidence for any of the proposed second messengers as mediators of the diverse responses of the hormone.

In this paper, we demonstrate that clustering (cross linking) of hormone receptors plays an important role in the molecular events which lead to biological responses. We propose that hormone-mediated receptor clustering (microaggregation) produces or activates a molecular species which can trigger the various membrane responses of the hormone and eventually lead to delayed bioactivity. Our model is an extension of previous models which emphasize the significance of receptor mobility in the mechanism of action of hormones which bind to membrane receptors.[12-15]

Our model is based on recent findings which concern the dynamic aspects of hormone receptor interactions. We have studied the lateral motion[16,17] and distribution[18] of fluorescent analogues of insulin and EGF on cultured fibroblasts. We have shown that insulin and EGF are initially mobile[17] on the plane of the membrane with $D = 5 \times 10^{-10}$ cm^2/sec and homogeneously distributed over the cell surface.[18] Within approximately ten minutes at 37°C, the hormone receptor complexes aggregate into visible immobile[17] patches on the cell surface (see FIG. 1). Within 15 minutes at 37°C, much of the aggregated hormone (patches) become internalized into endocytic vesicles.[18] The aggregation process does not require metabolic energy, but the subsequent internalization of hormones does.[18] Using fluorescein-labeled α_2-macroglobulin and rhodamine-labeled insulin or EGF, we have shown that all three polypeptides accumulate at the same patches on the cell surface and are internalized within the same vesicles.[19] Since α_2-macroglobulin was shown to patch over coated regions and to be internalized by coated vesicles,[20] we concluded that insulin, EGF and α_2-macroglobulin indeed enter the cell via clustered receptors at coated regions of the plasma membrane. Recently Gorden *et al.* have demonstrated by electron microscopic autoradiography that a fraction of ^{125}I-EGF becomes internalized by coated vesicles.[21]

Kahn and his coworkers have shown that sera of some patients with insulin-resistant diabetes contain autoantibodies to insulin receptors. These antibodies bind

* This study was supported by grants from the National Institutes of Health (CN-25820) and from the U.S.-Israel National Science Foundation.

0077-8923/81/0366-0274 $01.75/0

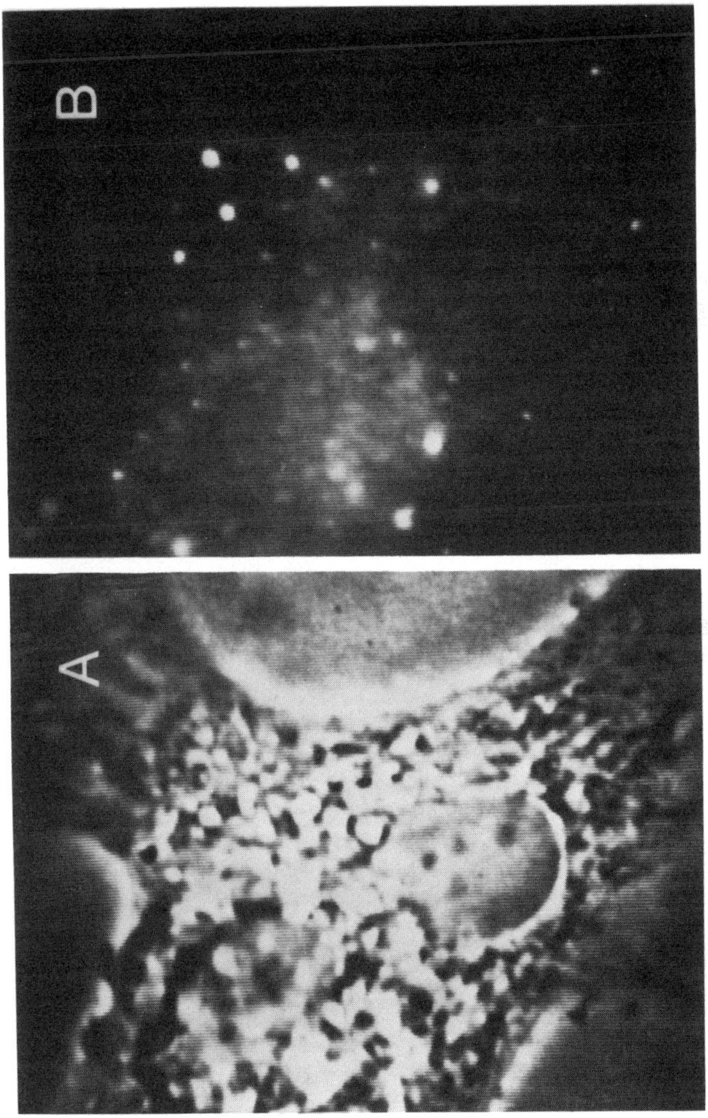

FIGURE 1. Distribution of rhodamine EGF on the cell membrane of 3T3 cells. The cells were labeled with rhodamine EGF for 20 min at 37°C (20 ng/ml). The cells were washed with phosphate-buffered saline, fixed with formaldehyde and observed with an image-intensified video flurescent microscope as previously described.[17] (A) phase micrograph and (B) fluorescent micrograph of the same field. Magnification × 1755.

to insulin receptors,[22] block insulin binding[23] and initiate many of the biological effects of insulin.[24] Both monovalent Fab′ and divalent (Fab′)$_2$ fragments of the anti-receptor antibody are fully active in inhibiting the binding of ^{125}I-insulin and are approximately as potent as insulin on a molar basis. However, bivalent (Fab′)$_2$ *retains* its insulin-like biological potency while the monovalent Fab′ *loses* almost all ability to stimulate glucose oxidation. The Fab′ fragment acts as a competitive antagonist of insulin-stimulated glucose oxidation,[25] demonstrating that the difference between the divalent and monovalent antibodies is not one of affinity towards the receptor but is a consequence of receptor clustering. Addition of anti-Fab′ antibodies which cause the cross linking of the Fab′ receptor complexes results in a restoration of the insulin-like activity of the antibody. Furthermore, the exposure of cells to submaximal doses of insulin followed by addition of low amounts of anti-insulin antibodies enhances the biological activity of insulin.[25]

It was reported that antibodies raised against purified insulin receptor can also initiate some of the biological effects of insulin, though those antibodies do not compete with insulin for the same sites.[26] Similar results were obtained when fat cells were sensitized with anti-idiotype directed against anti-insulin antibodies.[27]

We have recently reported on the preparation of a cyanogen bromide-cleaved EGF (CNBr-EGF) derivative. This molecule retains 10% of the binding capacity of the native hormone and 0.3% of its ability to stimulate thymidine incorporation into 3T3 fibroblasts.[28] Rhodamine conjugates of CNBr-EGF appeared to be diffusely distributed on the cell membrane of 3T3 fibroblasts even after incubations as long as 90 min at 37°C. This is in contrast to rhodamine EGF[14] which aggregates on the cell membrane (see FIG. 1) within 15 min at 37°C. When anti-EGF antibodies were added to the cells which were initially labeled with rhodamine-CNBr-EGF visible patches were readily seen on the cell surface.[28] Strikingly, the bioactivity of the inactive CNBr-EGF analogue was restored by cross linking with specific bivalent antibodies to EGF. Addition of monovalent Fab′ fragments does not cause restoration of bioactivity. The concentration of bivalent antibodies required for full restoration of bioactivity is far below the concentration at which visual patches can be seen. Thus, *microaggregation* of perhaps only a few receptor molecules rather than the formation of visible patches may be the relevant mechanism for bioactivity.

Two additional experiments demonstrate the significance of receptor clustering. First, concentrations of native EGF far below those that normally show activity can elicit dramatic responses upon addition of bivalent anti-EGF antibodies. Monovalent Fab′ fragments are not capable of enhancing bioactivity. The affinity of EGF to its surface receptors is not affected by the addition of anti-EGF antibodies.[29] Thus, the enhanced bioactivity in the presence of antibodies is the consequence of receptor clustering and is not due to increased affinity towards the membrane receptors. Second, cells that are relatively refractory to EGF, despite unimpaired binding, can be markedly sensitized and respond to the hormone after cross linking with anti-EGF antibodies. Clearly, receptor occupation alone is a biologically unproductive process without a subsequent microaggregation of cross linking of receptors.

MICROAGGREGATION OF RECEPTORS AS A MECHANISM FOR TRANSMEMBRANE SIGNALING

We have summarized the results which show that aggregation of receptors for insulin and EGF can induce bioactivity and that hormones binding to their membrane receptors trigger receptor clustering. We propose that hormone-mediated receptor clustering is an important physiological step which activates the various responses to the hormone and eventually leads to bioactivity. In the case of insulin, for example, hormone-induced receptor aggregation is coupled to the induced membrane hyperpolarization and other membrane electrical changes.[30] We postulate that the aggregated hormone receptor complex forms an ion channel or triggers some active transport of ions across the membrane. This in turn could activate some cytoplasmic or membrane enzymes and thereby trigger the various biological responses to insulin.

The hormone-induced aggregation of receptors can be divided into two processes: *local* aggregation and *global* aggregation.

Local aggregation of receptors is defined as a process in which occupied receptors encounter each other after moving a distance in the range of 50–200 nm. This aggregation involves only a few receptor molecules and the time scale of such a process will be in the range of 50 msec up to 1 sec.

Global aggregation of occupied hormone-receptor complexes is defined as the process in which the microaggregates of hormone-receptor complexes move a distance of several microns to form visible patches which contain hundreds of receptor molecules. The time scale of such a process lies between a few seconds to 10 minutes.

Our main assumptions are that local receptor aggregation is related to the rapid membrane responses to insulin and initiates the process of thymidine incorporation activated by EGF. The global aggregation is related to receptor down regulation and precedes the internalization and degradation of insulin and EGF.

We have divided the molecular events in our model into two chemical reactions. Reaction I describes a fast reversible binding of hormone (H) to its membrane receptor (R). Reaction II describes the interaction between two occupied receptors (HR) which can diffuse in the plane of the membrane with a two-dimensional diffusion coefficient D to form a product (C) through an intermediate encounter complex (HRRH). Finally, we assume that the biological response is proportional to (C) ("unit signal"†). For simplicity, we assume a minimum model and our "unit signal" is composed of two occupied receptors. The mechanism by which the bound hormone triggers receptor aggregation is unknown. For insulin, one simple mechanism would be hormone self-aggregation on the cell membrane, as it is known that insulin dimerizes and aggregates in solution at concentrations above 10^{-6} M[31] and its local concenttration on the cell membrane is about 10^{-6} M. On the other hand, it is argued that if surface aggregation is a biologically important process it must differ from aggregation in solution because there are several insulins which do not dimerize but are biologically active.[31] A more plausible mechanism is that the bound hormone induces a conformational transition in the receptor molecules that increases the binding affinity between mobile receptors. At present, the molecular mechanism of the aggregation process is not known. Insulin binding to its membrane receptors exhibits negative cooperativity.[32] The basis for the negatively cooperative binding has been interpreted to be due to site-site interactions among multivalent receptors. Our observations are consistent with the idea that the *occupied hormone* receptors are either multivalent towards their hormones or are oliogomers of monovalent receptors.

† "Unit signal" is defined as the minimal number of clustered receptors that produces the biological response. More precise models could be formalized after the valence of the receptor toward the hormone and the "unit signal" are elucidated.

Currently, it is impossible to measure the rate of the local aggregation process and thus correlate the kinetics of receptor clustering with the time scale of insulin action. Studies with video intensification microscopy[18,19] have shown that visible patches of hormone-receptor complexes appear after approximately ten minutes of incubation at 37°C. Although this method is not accurate enough to determine the exact kinetics of the global aggregation process, it is still possible to conclude that the appearance of visible patches and internalized hormone is too slow to account for the rapid events stimulated by insulin such as the uptake of 2-deoxy-glucose,[33,34] α-aminoisobutyrate[35] or uridine[36,37] or similar processes which are stimulated by EGF binding.[35]

A rough estimate of the time scale of a random encounter between two diffusing receptors can be calculated by assuming that the cell possesses 50,000 receptors per (μm).[2] The mean distance between two molecules in two dimensions is given by $r = 0.5\sqrt{1/a}$,[38] where a is the concentration of the occupied receptors. For $a = 50$ receptors/$(\mu m)^2$, the mean distance r equals 0.07 μm. For $D = 5 \times 10^{-10}$ cm^2/sec the encounter time $r^2/4D = (0.07 \times 10^{-4})^2/(4 \times 5 \times 10^{-10}) = 0.03$ sec. This corresponds to 30 collisions per second.

Based on the classical theory of Smoluchowsky[39] with some modifications by Eigen[40] and Bell,[41] our solution for this set of equations has been previously published[34]:

$$\text{I.}\quad R + H \underset{k_{-1}}{\overset{k_1}{\rightleftharpoons}} RH$$

$$\text{II.}\quad RH + RH \underset{k_{-2}}{\overset{k_2}{\rightleftharpoons}} HRRH \underset{k_{-3}}{\overset{k_3}{\rightleftharpoons}} C$$

$$\text{III.}\quad C \Rightarrow \text{Biological Response}$$

We assume that the system is in equilibrium therefore the overall equilibrium constant of Reactions I and II, $K = k_2 \cdot k_3/k_{-2} \cdot k_{-3}$ does not depend on the diffusion coefficient of the receptor. Therefore, the amplitude of the biological response which is proportional to (C), does not depend on the rate of diffusion of the receptors.

The complete solution of the rate equation is presented elsewhere.[42] If we assume that $K_{-3} \ll k_3$ which means that the second step of Reaction II is irreversible, then a complicated expression can be reduced to the following simple solution.[42]

$$(C) = \frac{4\Pi a^2 Dt}{1 + 8\Pi aDt}$$

At $t \ll 1/8\Pi aD$, (C) is approximately equal to $4\Pi a^2 Dt$.

At $t \gg 1/8\Pi aD$, (C) reaches plateau which equals a/2. The time required to reach the plateau equals $1/8\Pi aD$.

The conclusion from this solution is that receptor clustering is very rapid even at very low levels of receptor occupancy. Even at the level of a single occupied receptor $(\mu m)^2$ (which corrresponds to 2500 receptors per cell) the time required for this process is 0.8 sec. Thus, formation of dimers through diffusion of receptors does not seem to be the limiting step in the cascade of events which leads to bioactivity.

Receptor Patching and Endocytosis as a Model for Hormone-mediated Receptor Down Regulation.

Cells of obese insulin-resistant diabetic patients or obese rodents that have increased levels of circulating insulin have abnormally few insulin receptors.[43] In both, the number of binding sites increases toward normal when serum insulin is lowered by prolonged restriction of food intake[48] suggesting that insulin induces a specific loss of insulin-binding sites from the cell surface. A similar phenomenon can be demonstrated *in vitro*. Gavin *et al.* found a time-dependent loss[44] of insulin receptors when human IM-9 lymphocytes were cultured in media containing insulin. With each insulin concentration, a particular rate of loss and steady-state level was achieved. The ability of insulin to "down regulate" its receptors is dependent on hormone concentration, time and temperature. When the cells are returned to a medium free of insulin, there is a return of receptors to normal levels within 16 hours. A similar phenomenon was also observed for EGF.[45,46]

The biological significance of down regulation of receptors is not clear. Many investigators believe that the function of receptor down regulation is to adjust the cell's sensitivity to the hormone. Raff proposed[47] that the function of receptor down regulation is to regulate the level of the cell surface receptors to achieve maximal biological response at fractional ligand occupancy. It is probable that the aggregation and endocytosis of insulin and EGF through coated vesicles provided the mechanism of receptor down regulation and precedes the degradation of insulin and EGF.[10,18,19,42–48]

Goldstein and Brown[48] have shown that coated regions of the plasma membrane of human fibroblasts are involved in the process of clustering and endocytosis of low density lipoprotein (LDL). On the basis of these results, Anderson *et al.*[49] proposed a mechanism for the regulation of the receptors for low density lipoprotein on fibroblasts involving binding of LDL to preexisting clusters of receptor on coated areas of the plasma membrane. They raised the possibility that the same mechanism might exist for the internalization of EGF on fibroblasts.[49] We find that insulin and EGF receptors are initially diffuse before the hormones are bound and that patching and endocytosis are triggered by hormone binding.[18] LDL-receptors may cluster without the binding of LDL, however, the binding of ligands to clustered receptors is a less efficient process than binding to diffuse mobile receptors which cluster after becoming occupied. Berg and Purcell[50] have developed a theory which allows us to compare these two (and other) possibilities. The following equation can be used for our purpose:

$$J = J_{max}\, N\dot{S}/(NS + A)$$

where J is the diffusional flux of a ligand which can bind to membrane receptors; N is the total number of receptors per cell; S and A are the radius of the receptor and the cell respectively. We have calculated J for three different geometries. In all three examples, the total number of the receptors is 40,000, the radius of the receptor is 20 Å, and the radius of the cell is 10 μm. In the first example, the receptors are diffuse over the cell membrane. In the second example, the receptors are clustered within 100 randomly distributed patches each containing 400 receptors. In the third example, all the receptors are clustered in a single cap. The results are that diffusional flux to clustered receptors (patches) is 25% of the flux to diffuse receptors and the flux to a single cap is only 3% of the flux to the diffusely distributed receptors.

This calculation demonstrates that simple topological considerations can drastically affect the kinetic properties of the binding of a ligand to the cell membrane. However, the equilibrium properties of the binding are not affected by these considerations. This simple calculation demonstrates that diffuse distribution of mobile receptors followed by patching is more efficient for selectively adsorbing solutes than binding to preclustered receptors. Since the level of LDL in the serum is high (1–3 mg/ml), it is possible that the cell may use less efficient processes when the solute is largely available. This does not seem to be the case for insulin, EGF and other peptide hormones which are available at very low concentrations (1 ng/ml).

A Possible Relationship Between Receptor Aggregation and Hormone Bioactivity

The results summarized in this paper provide experimental evidence for the notion that hormone-mediated receptor clustering provides a "unit signal" for the generation of some of the responses activated by insulin or EGF binding to their membrane receptors. We propose that hormone-mediated clustering activates the various transport systems (glucose transport, amino acid transport, etc.) and also provides the mechanism for the hormone-induced changes in the electrical properties of the plasma membrane (Stage III in Fig. 2). The following mechanisms could cause the electrical changes:

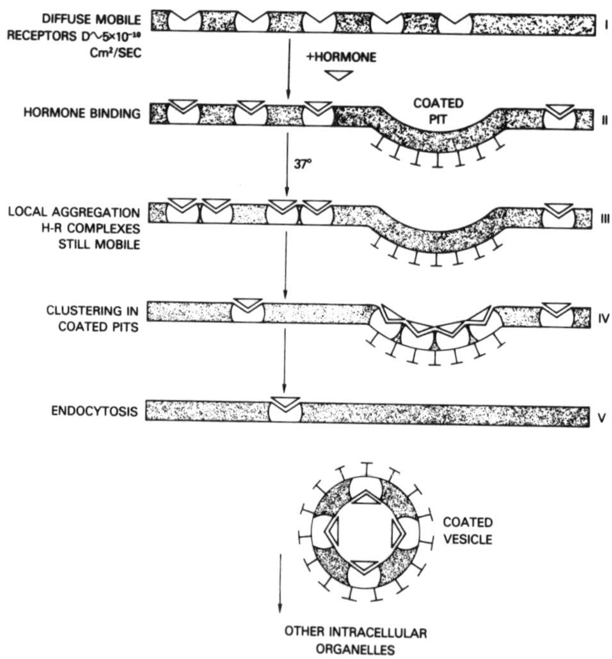

Figure 2. Steps in insulin and EGF endocytosis. (I) The unoccupied receptors are mobile and homogenously distributed over the cell membrane. (II) The hormone binds to diffuse receptors. (III) Hormone binding induces local aggregation of hormone receptor complexes. We postulate that this process provides the "unit signal" for triggering the various transport systems which are activated by hormone binding. (IV) The mobile hormone receptor complexes clusters at coated regions of the plasma membrane and become *immobile*. We call this step as the global aggregation process. (V) The clustered hormone receptor complexes and the coated pits pinch off and form a coated vesicle. Later the coated vesicle interacts with other intracellular organelles.

(a) The molecular species (unit signal) C which is formed from two HR moieties has the properties of an ion channel.

(b) The molecular species C activates a preexisting ion channel.

(c) The molecular product C activates an active transport system.

Each of these possibilities could cause the influx of Ca^{+2} ions (following efflux of potassium ions) and thereby cause the observed hyperpolarization of the plasma membrane. The increase in the intracellular Ca^{+2} would cause the activation of various cytoplasmic enzymes which are known to be activated by insulin, EGF or other growth factors. It was recently shown by Kyte[51] that a (Na^{+} + K^{+}) ATPase which is responsible for the active transport of sodium and potassium across the plasma membrane of animal cells is indeed a dimer of two identical subunits. Kyte provided experimental evidence that the ion channel passes between the subunits through the symmetry axis of the dimer.[51]

Several interesting points emerge from our *local* aggregation model. According to this model, the absolute magnitude of the biological response is not limited by the diffusion of the receptor while the kinetics of the biological response is proportional to the diffusion coefficients of the occupied receptors. Therefore, changing the diffusion coefficient of the membrane receptor should in principle cause a decrease in the rate of the biological response but should not affect the absolute value of the hormone effect. Addition of cholesterol, which is known to decrease membrane "fluidity," should decrease D (receptor) thereby attenuating the rate of the aggregation process and decreasing the kinetics of the hormone-induced hyperpolarization if the latter is dependent on the former. Thus, the kinetics of the hyperpolarization should be affected by addition of cholesterol to the membrane while the amplitude of the hyperpolarization should remain unchanged. Another conclusion which emerges from the aggregation model is that the formation of a "unit signal" is a very rapid process even at receptor occupancy of 1% or less. This means that the aggregation process is not a rate-determining step in the cascade of events which lead to the metabolic effects of the hormone.[33,52]

The next stage in the aggregation process is what we call global aggregation (State IV in FIG. 2). The mobile hormone receptor complexes (microaggregates) cluster at coated regions of the plasma membrane and become *immobile.* According to our hypothesis, the formation of microaggregates causes elevation of intracellular Ca^{++} which could activate, among other things, an enzyme that anchors the hormone receptors to clathrin, which is a major constituent of coated vesicles.[53] Other possible structures that could be involved in the process of clustering and immobilization of receptors over coated pits are contractile proteins of the cytoskeleton of the cell. It was previously postulated by Edelman[54] and Singer[55] that cytoskeletal elements participate in the process of patching and capping of cell surface receptors.

Finally, the clustered hormone receptor complexes and the coated pits pinch off and form coated vesicles (State V in FIG. 2). This step is presumably the mechanism of hormone-mediated receptor loss (down regulation) of insulin, EGF and α_2-macroglobulin. Recently, we have shown that when IM-9 lymphocytes were incubated with rhodamine-labled insulin, the hormone receptor complexes form a single cap on one pole of the cell.[56] Since the fraction of the capped cells correlates with the extent of receptor down regulation and since both phenomena depend on time, temperature, and hormone dose, it is possible that the observed insulin-mediated receptor loss on these cells is caused by insulin-induced capping of hormone receptors.

There are a few other biological systems in which ligand-mediated receptor clustering seems to play an important role. The most convincing biological system in which a relationship between receptor motion, receptor aggregation, and biological response was demonstrated is the process of degranulation of mast cells and basophils. In a set of elegant experiments performed primarily by Metzger and coworkers, it was shown that the Fc receptors are monovalent toward the IgE molecule[57,58] and that the IgE:Fc receptor complex can move laterally in the plane of the membrane with $D = 3 \times 10^{-10}$ cm^2/sec.[58] The degranulation of mast cells and basophils can be accomplished by each of the following treatments:

(a) The binding of IgE to the membrane receptor followed by cross linking with anti-IgE antibodies.[59]

(b) Binding of chemically cross-linked IgE dimers.[59]

(c) Binding of antibodies against IgE receptors.[60] Fab′ fragments bind to the IgE receptor but do not cause degranulation; $(Fab')_2$ binds to the membrane receptors and induces degranulation.[60] The presence of Ca^{+2} ions is necessary in order to obtain degranulation in each of these treatments.

Thus, clustering of two Fc receptors provides the unit signal for triggering degranulation.[60] The fact that antibodies against Fc receptors can trigger histamine release [60] is similar to the results with antibodies against insulin receptor which were shown to have insulin-like activity. It seems that in both systems receptor occupation by ligand alone without an additional step of microaggregation is *not* sufficient for the generation of biological response. Furthermore, Fc receptors and insulin receptors (possibly also EGF receptors) probably contain the biological information necessary to initiate histamine release and hormone response, respectively.

There are at least two more biological systems in which aggregation of receptors (and their lateral motion) seem to play a physiological role. It was recently demonstrated that intact (divalent) antibodies or $(Fab)_2$ fragments of antibodies against acetylcholine receptor (isolated from patients with myasthenia gravis) induce receptor degradation,[61] while their Fab fragments bind to the receptor but do not trigger degradation of receptors.[62] It was also shown that antibodies against TSH-receptors (isolated from patients with Grave's disease) can mimic TSH and activate the adenylate cyclase of thyroid cells.[63,64]

We predict that our simple aggregation model will be applicable to other biological systems. However, the generality of these proposals for other hormones and cell types should be established by further studies as more membrane receptors are purified and antibodies against them prepared.

References

1. Krahl, M. E. 1961. *In* The Action of Insulin on Cells. Academic Press. New York.
2. Fain, J. N. 1974. *In* MTP International Review of Science, Biochemistry Series 1. H. V. Rickenberg, Ed. Vol. 8. Park Press. Baltimore, MD. pp. 1–23.
3. Czech, M. P. 1977. Ann. Rev. Biochem. **46:** 359–384.
4. Cohen, S. & J. M. Taylor. 1974. Recent Prog. Horm. Res. **30:** 533–550.
5. Cohen, S. & C. R. Savage. 1974. Recent Prog. Horm. Res. **30:** 551–574.
6. Terris, S., D. F. Steiner, J. Biol. Chem. **250:** 8389–8398.
7. Steiner, D. F. 1977. Diabetes **26:** 322–340.
8. Goldfine, I. D., G. J. Smith, K. Y. Wong & A. L. Jones. 1977. Proc. Natl. Acad. Sci. USA **74:** 1368–1372.
9. Carpenter, G., K. J. Lembach, N. M. Morrison & S. Cohen. 1975. J. Biol. Chem. **250:** 4297–4304.

10. CARPENTER, G., & S. COHEN. 1976. J. Cell Biol. **71:** 159–171.
11. DAS, M., & C. F. FOX. 1978. Proc. Natl. Acad. Sci. USA **75:** 2644–2648.
12. CUATRECASAS, P. 1974. Ann. Rev. Biochem. **43:** 169–214.
13. DEHAEN, C. 1976. J. Theoret. Biol. **58:** 383–400.
14. BOEYNAEMS, J. M. & J. E. DUMONT. 1977. Mol. Cell. Endocrinol. **7:** 33–47.
15. TOLKOVSKY, A. M. & A. LEVITZKY. 1978. Biochemistry **17:** 3795–3810.
16. SHECHTER, Y., J. SCHLESSINGER, S. JACOBS, K. J. CHANG & P. CUATRECASAS. 1978. Proc. Natl. Acad. Sci. USA **75:** 2135.
17. SCHLESSINGER, J., Y. SHECHTER, P. CUATRECASAS, M. C. WILLINGHAM & I. Pastan. 1978. Proc. Natl. Acad. Sci. USA **75:** 5353–5357.
18. SCHLESSINGER, J., Y. SHECHTER, M. C. WILLINGHAM & I. PASTAN. 1978. Proc. Natl. Acad. USA **75:** 2659–2663.
19. MAXFIELD, F. R., J. SCHLESSINGER, Y. SHECHTER, I. PASTAN & M. C. WILLINGHAM. 1978. Cell **14:** 805–811.
20. WILLINGHAM, M. C., F. R. MAXFIELD & I. PASTAN. 1979. J. Cell. Biol. **82:** 614–625.
21. GORDEN, P., J. L. CARPENTIER, S. COHEN & L. ORCKI. 1978. Proc. Natl. Acad. Sci. USA. **75:** 5025–5024.
22. JARRETT, D. B., J. ROTH, C . R. KAHN & J. S. FLIER. 1976. Proc. Natl. Acad. Sci. USA **73:** 4115–4119.
23. FLIER, J. S., C. R. KAHN, D. B. JARRETT & J. ROTH. 1976. J. Clin. Invest. **58:** 1442–1449.
24. KAHN, C. R., K. L. BAIRD, J. S. FLIER & D. B. JARRETT. 1977. J. Clin. Invest. **60:** 1094–1106.
25. KAHN, C. R., K. L. BAIRD, D. B. JARRETT & J. S. FLIER. 1978. Proc. Natl. Acad. Sci. USA **75:** 4209–4013.
26. JACOBS, S., K. J. CHANG & P. CUATRECASAS. 1978. Science **200:** 1283–1284.
27. SEGE, K. & P. A. PETERSON. 1978. Proc. Natl. Acad. Sci. USA **75:** 2443–2448.
28. SHECHTER, Y., L. HERNAEZ, J. SCHLESSINGER & P. CUATRECASAS. 1979. Nature **278:** 835–838.
29. SHECHTER, Y., K. J. CHANG, S. JACOBS & D. CUATRECASAS. 1979. Proc. Natl. Acad. Sci. USA In press.
30. ZIERLER, K. L. 1977. Handb. Physiol. Sec. Endocrinol. **7:** 347–368.
31. BLUNDELL, T., G. DODSON, D. HODGKIN & D. MERCOLA 1972. Adv. Prot. Chem. **26:** 279–402.
32. DE MEYTS, P., A. R. BIANCO & J. ROTH. 1976. J. Biol. Chem. **251:** 1877–1888.
33. OLEFSKY, J. M. 1978. Biochem. J. **172:** 137–145.
34. CIARALDI, T. P. & J. M. OLEFSKY. 1979. Arch. Biochem. Biophys. **193:** 221–231.
35. HOLLENBERG, M. D. & P. CUATRECASAS. 1975. J. Biol. Chem. **250:** 3845–3853.
36. ROZENGURT, E. & W. D. STEIN. 1977. Biochem. Biophys. Acta **464:** 417–432.
37. KORN, R., E. SHOHAMI, P. BIBI & W. P. STEIN. 1978. FEBS Lett. **86:** 71–75.
38. CHANDRASELECHAR, S. 1943. Rev. Mod. Phys. **15:** 1–143.
39. SMOLUCHOWSKY, M. 1976. Phys. Z. Chem. **92:** 557–585.
40. EIGEN, M. 1978. *In* Quantum Statistical Mechanics in the Natural Sciences. S. L. Mintz & S. M. Wiedermayer, Eds. Plenum. New York.
41. BELL, G. I. 1978. Science **200:** 618–627.
42. SCHLESSINGER, J. 1979. *In* Physical Chemical Aspects of Cell Surface Events in Cellular Regulation. C. Delisi & R. Blumenthal, Eds. Elsevier. N.Y. pp. 89–111.
43. KAHN C. R. 1976. J. Cell Biol. **70:** 261–286.
44. GAVIN, J. R., J. ROTH, D. M. NEVILLE, P. DE Meyts & D. N. NUELL. 1974. Proc. Natl. Acad. Sci. USA. **71:** 84–88.
45. CARPENTER, G. & S. COHEN. 1976. J. Cell. Physiol. **71:** 159–171.
46. AHARONOV, A., R. M. PRUSS & H. R. HERSCHMAN. 1978. J. Biol. Chem. **253:** 3970–3977.
47. RAFF, M. 1976. Nature **259:** 265–266.
48. GOLDSTEIN, J. L. & M. S. BROWN. 1977. Ann. Rev. Biochem. **46:** 897–930.
49. ANDERSON, R. G. W., J. L. GOLDSTEIN & M. S. BROWN. 1977. Nature **279:** 695–699.
50. BERG, H. C. & E. M. PURCELL. 1977. Biophys. J. **20:** 193–219.
51. KYTE J. E. 1970. Biophys. J. **24:** 153a.
52. CIARADLI, T. & J. OLEFSKY. 1978. Proc. of 38th Ann. Meeting of the American Diabetes Assoc. Diabetes Abstr. **87:** 452.
53. PEARSE, B. M. F. 1976. Proc. Natl. Acad. Sci. USA. **73:** 1255–1259.

54. Edelman, G. M. 1976. Science **192:** 218–226.
55. Ash, J. F., D. Louvard & S. J. Singer. 1977. Proc. Natl. Acad. Sci. USA **74:** 5585–5588.
56. Schlessinger, J., E. Van Obberghen & C. R. Kahn. 1980. Nature **286:** 729–731.
57. Mendoza, G. & H. Metzger. 1976. Nature **264:** 548–549.
58. Schlessinger, J., W. W. Webb, E. L. Elson & H. Metzger. 1976. Nature **264:** 550–551.
59. Segal, D. M., S. T. Taurog & H. Metzger. 1977. Proc. Natl. Acad. Sci. USA **74:** 2293–2997.
60. Isersky, C., J. D. Taurog, G. Poy & H. Metzger. 1978. J. Immunol. **121:** 549–558.
61. Stanley, E. F. & D. B. Drachman. 1978. Science **200:** 1285–1287.
62. Drachman, D. B., C. W. Angus, R. N. Adams, J. O. Michaelson & G. J. Hoffman. 1978. N. Engl. J. Med. **298:** 1116–1122.
63. Smith, B. R. & R. Hall. 1974. Lancet **ii:** 427–430.
64. Mehdi, S. A. & S. S. Nussey. 1975. Biochem. J. **145:** 105–111.

CONFORMATIONAL DYNAMICS OF TROPOMYOSIN IN SOLUTION: EVIDENCE FOR TWO CONFORMATIONAL STATES*

Sherwin S. Lehrer,† Philip Graceffa, and David Betteridge

Department of Muscle Research
Boston Biomedical Research Institute
Boston, Massachusetts 02114 and
Department of Neurology
Harvard Medical School
Boston, Massachusetts 02115

Introduction

Tropomyosin, a coiled-coil of two parrallel α-helical polypeptide chains that interact in register,[1-3] functions as a component in the Ca^{2+}-dependent regulatory system of skeletal and cardiac muscle.[4-6] Rabbit skeletal tropomyosin preparations contain two types of chains, α and β, which differ slightly in amino acid sequence;[7-9] they are combined into two predominant molecular species, αα and αβ.[10,11,3] Rabbit cardiac tropomyosin contains only α chains and hence only αα species. An interesting difference is that the α chain contains a single Cys at position 190, whereas the β chain has two Cys at residues 190 and 36.[9]

In the native coiled-coil arrangement, the two Cys 190 residues of tropomyosin are in close proximity as shown by the ability to form a disulfide bond by thiol-disulfide exchange with 5-5′-dithiobis(2-nitrobenzoate) (Nbs_2)[3] and by Cu^{2+}-catalyzed air oxidation.[1,2,12] We verified their proximity with the observation of excimer fluorescence from pyrene maleimide covalently attached to the Cys 190's of αα tropomyosin.[13] Recently, we showed that pyrene-labeled tropomyosin in solution exists in an equilibrium between two states, an excimer-forming state and a non-excimer-forming state.[14] In this paper, we will present additional evidence that suggests that the two states correspond to localized "chain open" and "chain closed" states. We will also show that the ability to form a disulfide bond between the Cys 190's on adjacent chains[3] and the destabilizing effect of this disulfide bond on the conformation of tropomyosin in the physiological temperature region[12] can be explained by the presence of these two states. Finally, we will suggest that the two states of tropomyosin on the thin filament associated with the presence and absence of Ca^{2+}, are related to the two conformational states of tropomyosin in solution.

Pyrene-labeled Tropomyosin

Our initial studies, which involved reacting Cys 190 of αα-tropomyosin with pyrene maleimide at pH 6.0 (to minimize possible reaction with Lys), gave a product

* Supported by grants from the National Institutes of Health (HL 22461) and the Muscular Dystrophy Association of America.

† Author to whom correspondence should be addressed at the BBRI address.

0077-8923/81/0366-0285 $01.75/0 © 1981, NYAS

(Type I) which showed only a small degree of excimer formation at 480 nm, despite the relatively high degree of labeling, as well as a highly structured monomer fluorescence with peaks at 374 nm and 394 nm and a shoulder at 416 nm (FIG. 1). At pH values greater than 7.5, a different product (Type II) was formed as shown by a shift of the major monomer peak to 384 nm with broadened shoulders at 404 nm and 426 nm, and a greater degree of excimer formation. A time-dependent fluorescence change from the Type I to Type II spectrum was also observed when the pH of a solution containing the Type I product was adjusted to values above 7.5. The fluorescence spectrum of each product was not affected by pH adjustment *per se*. These data indicated that a reaction of the type described by Wu *et al.*[15] occurred where the bound succinimido ring opens due to a secondary reaction with a neighboring amino group (Type IIA, FIG. 2). From studies of model compounds, we have noted that a similar shift in spectrum occurs due to ring hydrolysis (Type IIB, FIG. 2). In both cases of ring opening, either by aminolysis or hydrolysis (Type II),

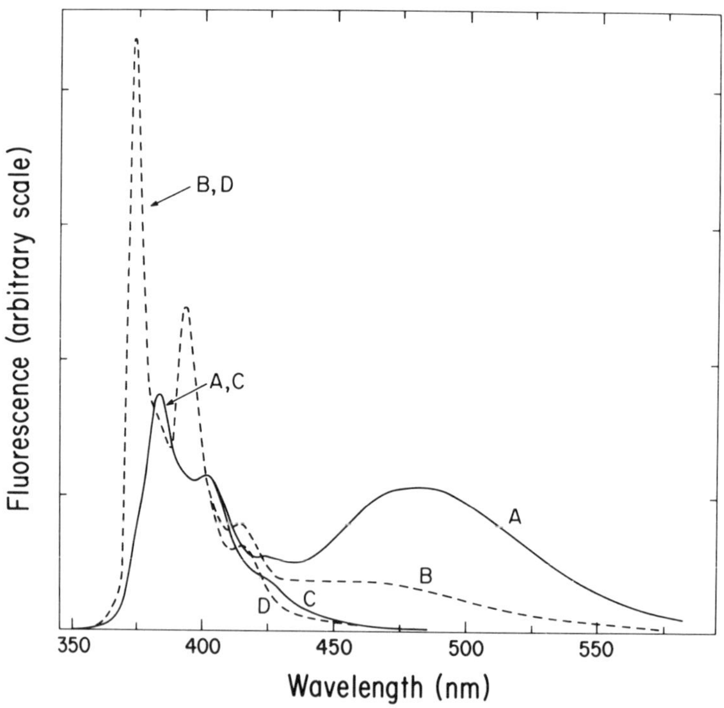

FIGURE 1. Fluorescence emission spectra of the two products of pyrene maleimide-labeled $\alpha\alpha$ tropomyosin in the native and denatured states. (– – –), Type I product labeled at pH 6; (——), Type II product labeled at pH 7.5; A and B, native; C and D, denatured. Conditions of measurement: pyrene-tropomyosin at 0.1 mg/ml in: A, 5 mM Hepes buffer, pH 7.5, 1M NaCl, 1 mM EDTA; B, 20 mM phosphate buffer, pH 6.0, 1M NaCl, 1 mM EDTA; C and D included 4M GdmCl in the A and B solutions. Conditions of labeling: reduced tropomyosin was labeled at 40°C with excess pyrene maleimide in 20 mM phosphate buffer, 1M NaCl, 1mM EDTA at pH 6.0 for 4 ½ h (Type I) or pH 7.5 for 22 h (Type II) which gave degrees of labeling of 0.6 and 0.9 at pH 6.0 and 7.5, respectively. See Reference 12 for details.

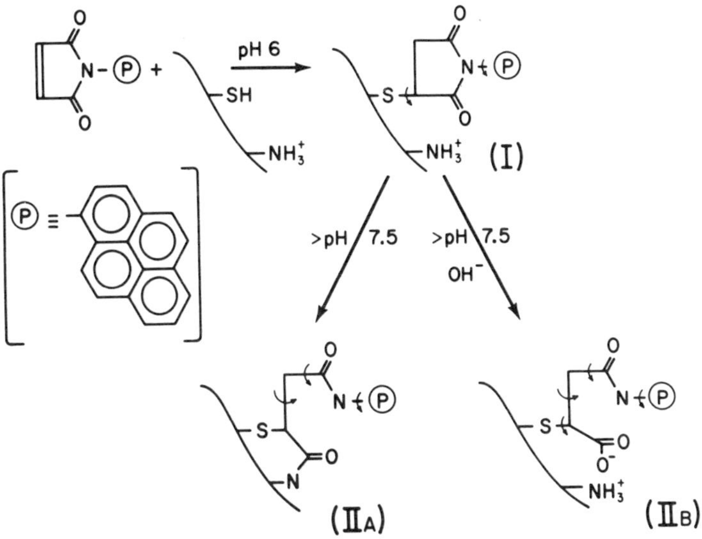

FIGURE 2. Reaction scheme of pyrene maleimide with tropomyosin. Two possible products for the ring opening either by aminolysis, Type IIA,[15] or hydrolysis, Type IIB, are indicated. Possible rotation around single bonds is indicated by curved arrows.

the pyrene moiety would have greater rotational freedom around the bonds indicated in FIGURE 2 than in the case of the product of the primary reaction (Type I). This would increase the chances for the pyrene to find an appropriate non-polar environment. The increased excimer fluorescence, seen after the secondary reaction, which is due to the ability of the pyrenes to interact more closely, is consistent with the expected increased rotational flexibility. Because of the increased excimer yield of the Type II product, we have routinely carried out the conversion.

In this highly labeled $\alpha\alpha$-tropomyosin system (FIG. 1), a larger monomer contribution was obtained than expected from an estimation of the fraction of singly labeled molecules present, assuming that the quantum yields of excimer and monomer are similar.[16] Thus, it appeared that a fraction of the doubly labeled tropomyosin molecules could not form excimer. This latter possibility was supported by studies of the spectrum of a sample of rabbit skeletal tropomyosin that was fully labeled with pyrene maleimide.[14] In general, the monomer fluorescence is expected to be quite heterogeneous with contributions from pyrenes at Cys 190 which do not form excimer, as well as from pyrenes at Cys 36 on the β chain, and also from singly labeled pyrenes at Cys 190. Since monomer fluorescence can thus be obtained from pyrenes that cannot form excimer, we do not report values of excimer/monomer intensity ratios, but rather relative excimer yield.

Temperature dependence studies of the fluorescence of the fully labeled skeletal sample provided information regarding the mechanism of excimer formation (FIG. 3). In the 5–30°C range, it appears that the excimer fluorescence competes with monomer fluorescence and also that dissociation of excimer into monomer does not occur, since there is an isoemissive point in this temperature range.[16–20] At T > 30°C, the excimer intensity decreased and the isoemissive point was lost, indicating that

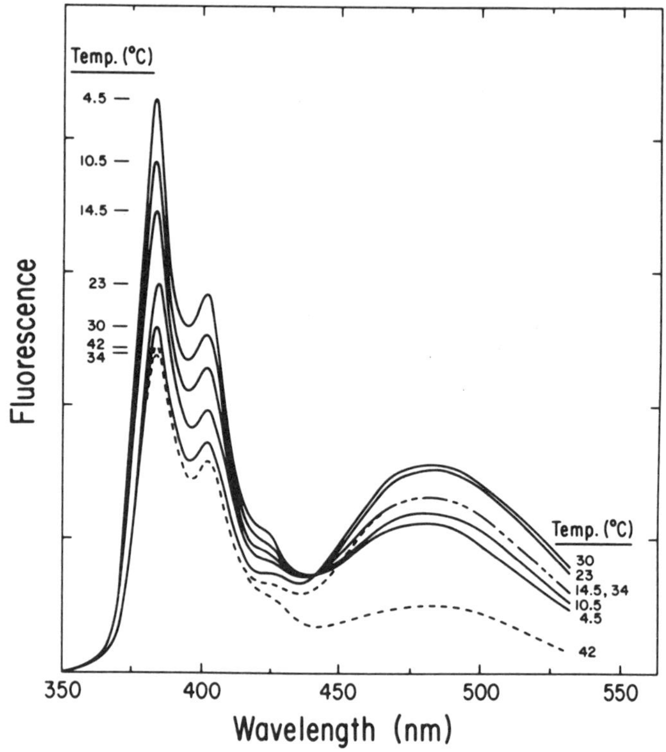

FIGURE 3. Temperature dependence of the fluorescence spectrum of pyrene maleimide-labeled rabbit skeletal tropomyosin. (——), 4.5°C to 30°C; (- - -), 34°C and 42°C. 1.3 mg/ml in 5 mM Mops, buffer, pH 7.5, 1M NaCl, 1 mM EDTA. λ_{exc} = 342 nm, degree of labeling = 1.

excimer dissociation occurred, most probably due to complete chain separation, by analogy with denaturant-induced chain separation.[21,22] Even though there was almost a 50% increase in excimer fluorescence prior to chain dissociation (FIG. 4), lifetime studies indicated that the decay time of the excimer only slightly decreased from 60 nsec to 54 nsec between 9°C and 25°C, probably due to a small temperature dependence on the excimer radiationless decay rate[16] (FIG. 5). There was no appreciable time-lag associated with excimer build-up due to rotational diffusion of the pyrenes, and no indication of a fast monomer fluorescence decay associated with the excimer rise at short times.[18,19] Thus, there did not appear to be any competition *in the excited state* between excimer and monomer fluorescence when T < 30°C. Evidence for competition through a *ground-state* equilibrium, however, was obtained from a close look at excitation spectra. The excitation spectra for the monomer and excimer fluorescence were very similar but not identical, as seen by a slight red shift and broadening of the vibrational structure for the excimer compared to the monomer (FIG. 6). Thus, pyrene moieties that lead to excimer fluorescence are in a different environment from pyrenes that produce monomer fluorescence. The spectral shift suggests that pyrenes that form excimer interact hydrophobically in a close interac-

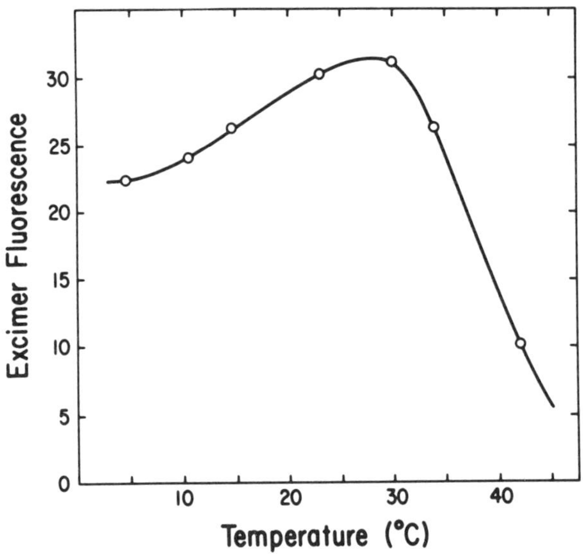

FIGURE 4. Temperature dependence of the excimer fluorescence of rabbit skeletal pyrene-tropomyosin. Same conditions as FIGURE 3.

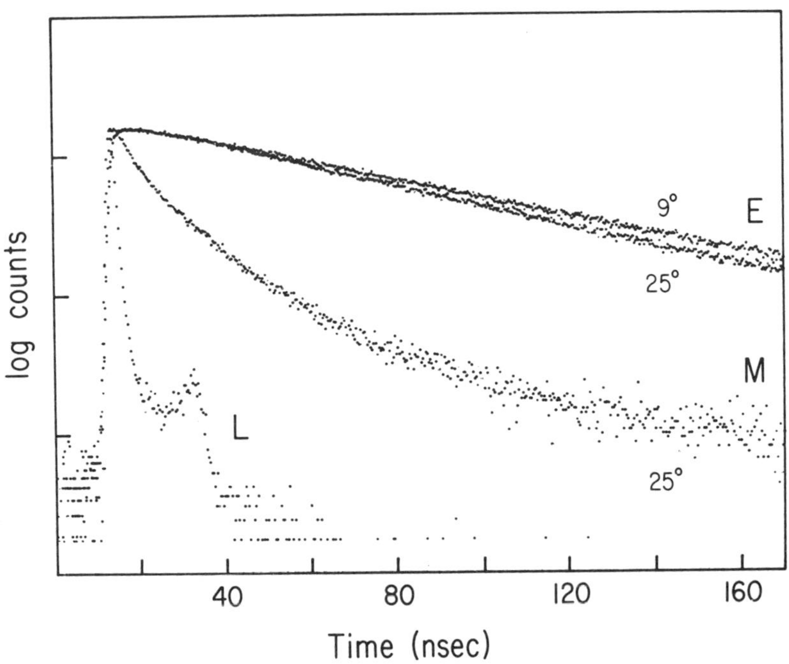

FIGURE 5. Temperature and salt dependence of the decay of excimer (E) and monomer (M) fluorescence of rabbit skeletal pyrene-tropomyosin. Conditions as for FIGURE 3. Identical decay curves were obtained at 0.01M and 1M NaCl at 25°C. L, lamp pulse.

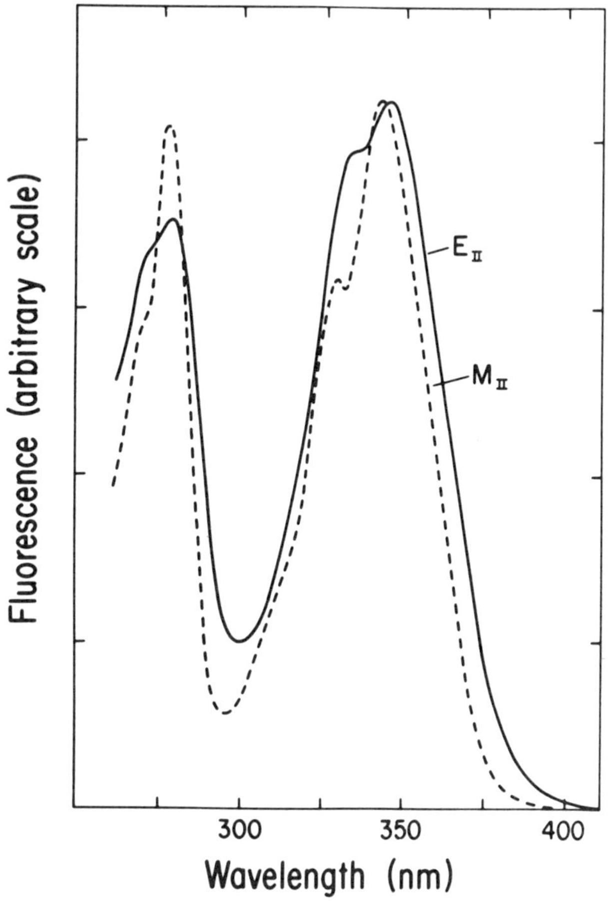

FIGURE 6. Normalized fluorescence excitation spectra of the excimer (E_{II}) and monomer (M_{II}) of the Type II product of pyrene-labeled $\alpha\alpha$-tropomyosin (see FIGURE 2). See Reference 12 for details.

tion, since similar spectral shifts were observed due to hydrophobic interaction among pyrene moieties bound to polymers in aqueous solution.[23]

Considerable information regarding the relationship between the conformation of pyrene-tropomyosin and the ability to form excimer was obtained from the salt dependence, since salt would not be expected to affect the spectroscopic properties of pyrene *per se*. The salt dependence of the fluorescence of pyrene-labeled tropomyosin indicated a dramatic increase in the excimer fluorescence as the salt concentration was raised from zero to 0.2M (FIG. 7). The excimer rise and decay times were identical (FIGURE 5), despite this large increase in excimer yield with salt concentration. These data indicate that the fluorescence properties of the species that form excimer have not been altered by salt so that the intensity changes must be due to a shift in equilibrium between two ground states of the pyrene moiety, a state that can

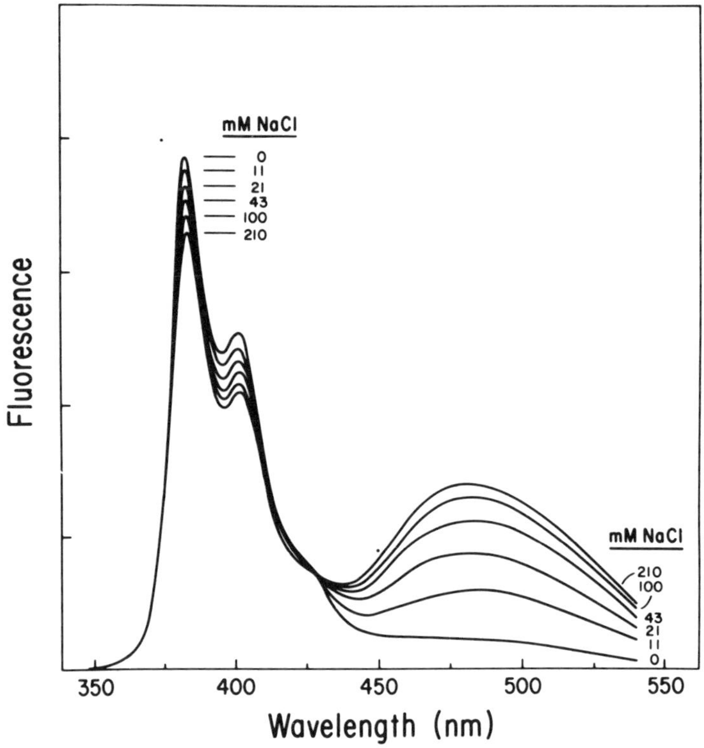

FIGURE 7. Salt dependence of the fluorescence spectrum of rabbit skeletal pyrene-tropomyosin at 25°C. 0.04 mg/ml in 2 mM Mops buffer, pH 7.5, 0.3 mM EDTA. λ_{exc} = 342 nm.

form excimer and a state that cannot form excimer, rather than changes in excited-state association and dissociation.

A possible model to explain these data (Model I, FIG. 8) can be found in the original proposal of Birks to explain intermolecular excimer formation appropriately modified to the case of intramolecular excimer formation.[16–20] In this model, all pyrenes are in equivalent environments and excitation spectra of excimer and monomer should therefore be identical. Thermal effects are mainly due to the temperature dependences of the association and dissociation constants, k_a and k_d, respectively. For the reasons discussed above, this model does not explain the data. In Model II, we consider that tropomyosin fluctuates between two states, an excimer-forming state (B) and a monomer-forming state (A). In this model, we tentatively associate State A with a "chain closed" state and State B with a "chain open" state which allows the pyrenes to interact hydrophobically with each other, since space-filling models showed that the pyrene moieties cannot stack to form excimer in the state which allows maximum interaction between chains unless the chains are locally separated (FIG. 9). All aspects of the above fluorescence data can be explained by Model II. The difference in excitation spectra is due to the different environments of

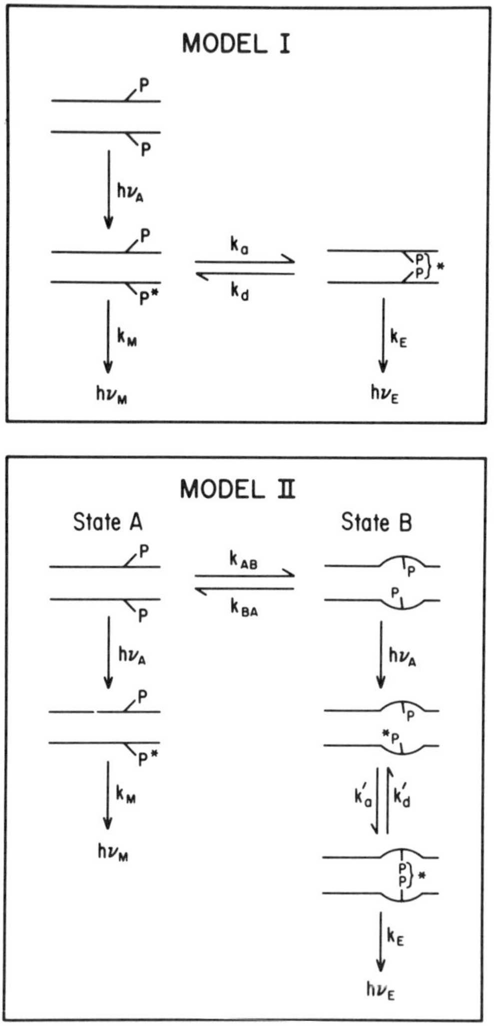

FIGURE 8. Two possible models to explain the relationship between the monomer (M) and excimer (E) fluorescence of pyrene-labeled $\alpha\alpha$-tropomyosin. Model I does not consider any contribution from tropomyosin conformational fluctuations. Model II includes a ground-state localized conformational fluctuation that brings adjacent pyrenes close together.

the pyrenes in the two states. The fast excimer rise is due to the close interaction of the pyrenes, resulting in a high association rate, k_a'. The temperature and salt dependence can be explained by effects on the equilibrium constant, k_{BA}/k_{AB}. The loss of excimer above 30°C can be explained by the beginning of denaturation which results in complete chain separation.

DANSYL-LABELED TROPOMYOSIN

Fluorescence studies on rabbit cardiac tropomyosin ($\alpha\alpha$)[7,8] specifically labeled at Cys 190 with a dansyl group also provided evidence for the presence of two states in solution. The labeling reaction involved sulfhydryl-disulfide exchange between the

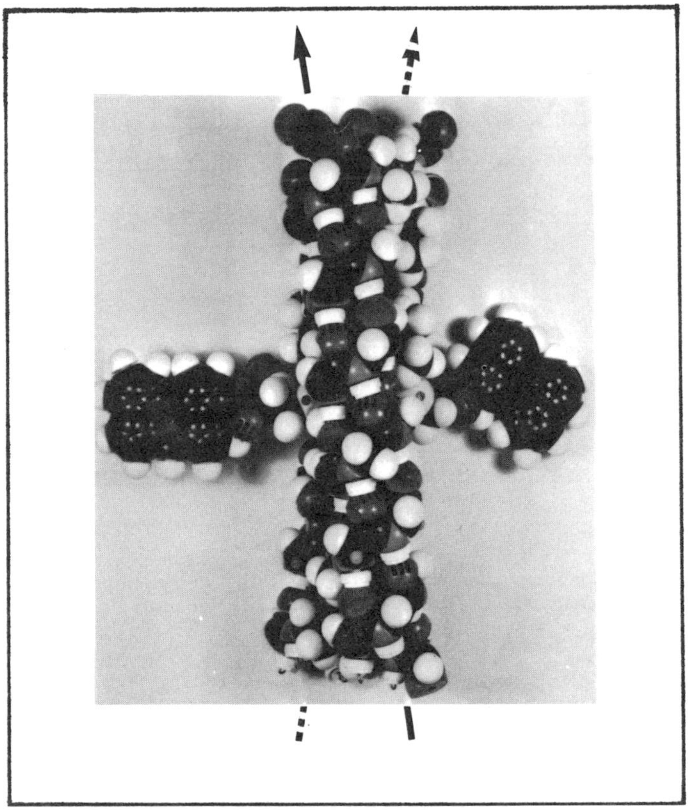

FIGURE 9. Photographs of a space-filling CPK model of the region of $\alpha\alpha$-tropomyosin near Cys 190 (residues 180–200) labeled with pyrene maleimide converted to the Type II product (see FIGURE 2). Only the hydrophobic side chains located between the α-helices have been included in order to see the S atoms (marked with a dot). The amino ends are at the bottom. From this top view most of one chain is seen. (→); the second chain lies below, (- -→). In this chain closed conformation (State A) the pyrenes are located on opposite sides of the molecule and are not sterically able to interact with each other.

reduced protein and didansylcystine in GdmCl. This resulted in tropomyosin whose Cys 190's were labeled to a high degree with dansylcystine via a disulfide bridge. Analyses of lifetime data showed that to a first approximation, two components were sufficient to fit the decay (FIG. 10); a major component having a relatively short lifetime (7 nsec) and a minor component having a long lifetime (17 nsec). In GdmCl, essentially one component was observed, since the residual long-lived component was less than 1%. Decay data obtained as a function of temperature indicated that the contribution of the long-lived component increased to 10% and showed a maximum, just prior to the main unfolding transition, before decreasing toward zero (FIG. 11). A small spectral blue shift (~3 nm) was also observed, which reached a maximum over the same temperature range. The similarity of the temperature dependence of the contribution of the long-lived dansyl component with the temperature dependence of the excimer fluorescence suggests that the dansyl long-lived

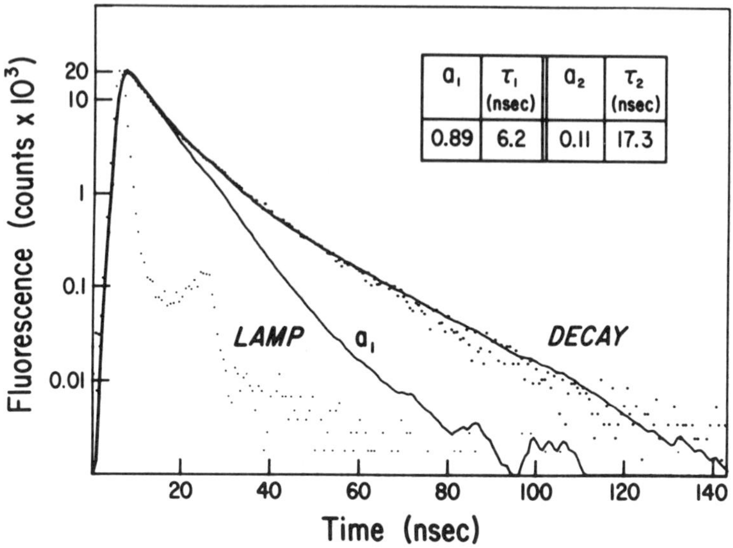

FIGURE 10. Fluorescence decay of dansylcystine-labeled cardiac tropomyosin. The decay data (points) were analyzed with a two-component method of moments procedure. Two lifetime components, τ_1 and τ_2, with indicated weights, a_1 and a_2, satisfactorily fit the decay (line). 0.1 mg/ml in 2 mM Hepes buffer, pH 7.5, 1M NaCl, 1 mM EDTA, 40°C. Degree of labeling = 0.8; 20,000 counts at peak.

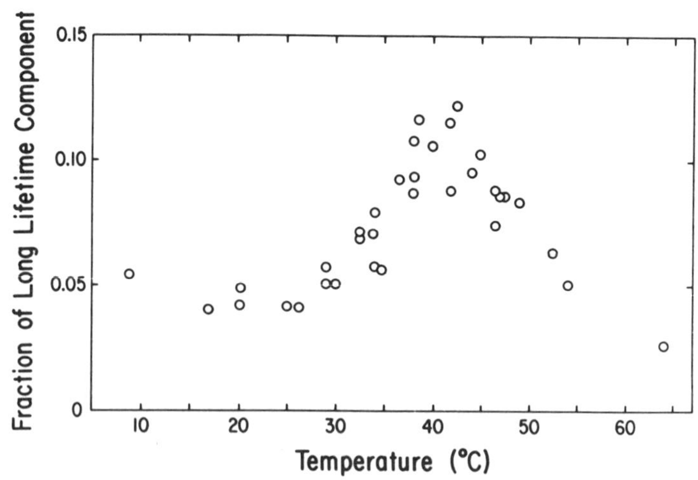

FIGURE 11. Temperature dependence of the fractional contribution of the long-lived component, a_2, in the fluorescence decay of dansylcystine-cardiac tropomyosin. Conditions as for FIGURE 10.

fluorescence is probing the contribution of State B to the overall conformation in the temperature region before complete chain separation occurs. This is consistent with the longer lifetimes and blue-shifted dansyl spectra expected[24] when the dansyl group interacts with the hydrophobic interior between the two chains in the "chain open" state. This appears to be an example of the case where multiple fluorescence lifetimes are the result of multiple conformations since there is no possibility that the probe was linked to groups other than Cys 190. If we can assume that the dansyl is efficiently probing each state, we can calculate an approximate equilibrium constant at 25°C, $K = a_2/a_1 = .05$ where a_2 is the fraction of the long-lived component and a_1 is the fraction of the short-lived component. Since it is possible that the dansyl group is not probing the "open" state with high efficiency, either for thermodynamic or kinetic reasons, the value of $K = 0.05$ is a lower limit.

DISULFIDE-CROSSLINKED TROPOMYOSIN

Several years ago, we showed that Nbs_2 quantitatively crosslinks the two chains of tropomyosin via interchain disulfide bond formation between Cys 190's.[3] Further insight into the mechanism of this crosslinking can be obtained with the knowledge of the two-state equilibrium. That a disulfide crosslink was formed at high Nbs_2 concentration rather than the usual Nbs-blocked species was somewhat surprising, particularly since space-filling models indicated that the SH groups are about 7 Å apart separated by the peptide backbone (FIG. 12). A disulfide bond could be rapidly formed, however, in the "chain open" state (State B), via the intermediate which has one SH group exchanged (brackets, FIG. 13), since in that state the two sulfur atoms can get close together. Thus, the ability to crosslink means that the internal reaction rate, k_3, is greater than the external reaction rate, k_1, under conditions of rapid equilibrium. Studies are currently being carried out to obtain the details of the reaction with Nbs_2.

The differences between disulfide crosslinked and uncrosslinked tropomyosin observed in both the thermal- and denaturant-induced unfolding profiles[12] can now also be qualitatively understood in the light of the two-state equilibrium. The unfolding profile of uncrosslinked tropomyosin indicated that some helix is lost in a monotonic manner, prior to the main unfolding transition.[12] This is most clearly shown in the thermal unfolding profiles, as a non-zero value of the derivative for temperatures below 40°C (FIG. 14). Similar data for tropomyosin that appeared to be uncrosslinked have been published by Woods.[25] There are two effects of the crosslink on the unfolding transitions: a shift of the main transition by about 8°C to higher temperatures and a decreased stability in the 25°–40°C region associated with a well developed pre-transition. The increased stability at high temperatures can be understood as being due to a decrease in entropy of the unfolded state caused by the crosslink.[26] It was suggested[12] that this pre-transition was due to the introduction of strain by the presence of the crosslink. In view of the two-state theory, the source of this strain can be explained as being due to the inability of the crosslinked tropomyosin produced in State B to form the proper closed state (State A). Space-filling models indeed indicate that proper chain closure is not quite complete when the S-S crosslink is present (FIG. 12, *bottom*), compared to the uncrosslinked model (FIG. 12, *middle*).

Woods has interpreted unfolding data of various tropomyosins in terms of a unique partially unfolded intermediate, X, between the native and denatured states.[25] For rabbit skeletal tropomyosin, the estimated temperature and GdmCl concentration, at the midpoint of the N=X transition, are 31°C and 1.3M GdmCl (at 25°C),

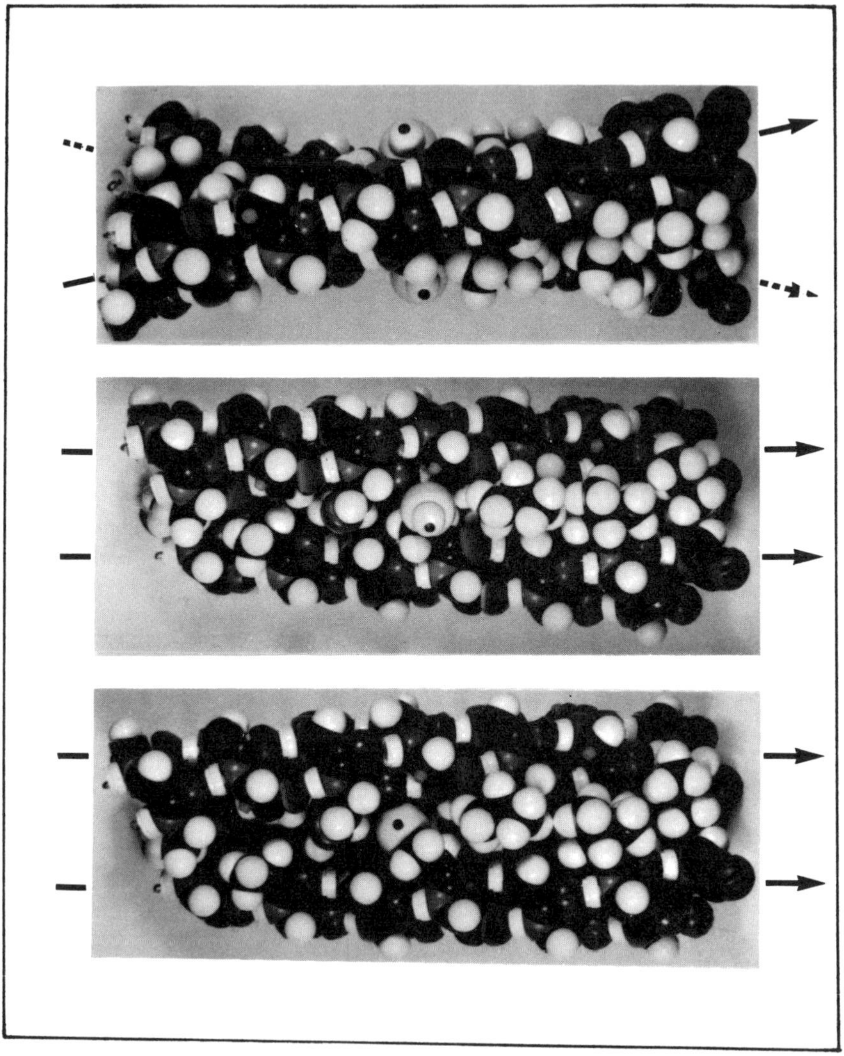

FIGURE 12. Photographs of space-filling CPK models of uncrosslinked and disulfide-crosslinked $\alpha\alpha$-tropomyosin near the Cys 190 region (residues 180–200) in the "chain closed" state (State A). Only the hydrophobic sidechains that are located between the α-helices have been included in order to see the Cys 190 side chains (dot on SH or S). *Top*, uncrosslinked tropomyosin, top view with one chain above (→), and one chain below (--→). *Middle*, uncrosslinked tropomyosin, side view. *Bottom*, disulfide crosslinked tropomyosin side view. The SH groups are located too far apart (*top*) and are sterically blocked (*middle*) from possibly forming a disulfide bond in this "chain closed" configuration. A disulfide bond can be introduced if the chains are locally separated resulting in a crosslinked molecule (*bottom*) in which the chains cannot be brought as close together as for the uncrosslinked structure (*middle*).

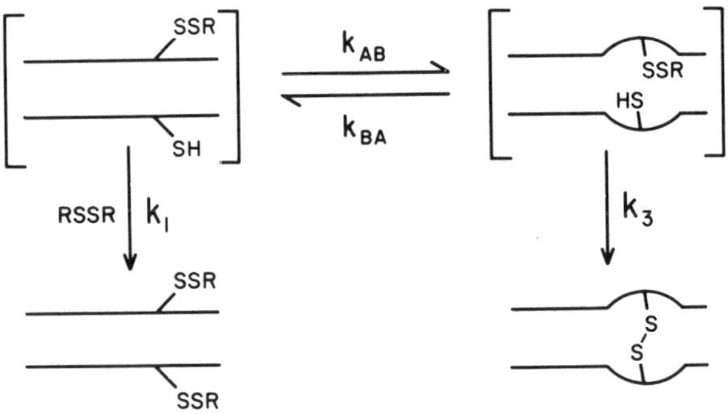

FIGURE 13. Two possible products of the reaction of reduced $\alpha\alpha$-tropomyosin with Nbs_2 (RSSR) arising from the intermediate with one SH group exchanged (brackets). It is postulated that internal exchange leading to the crosslink can only occur in the chain open state (State B).

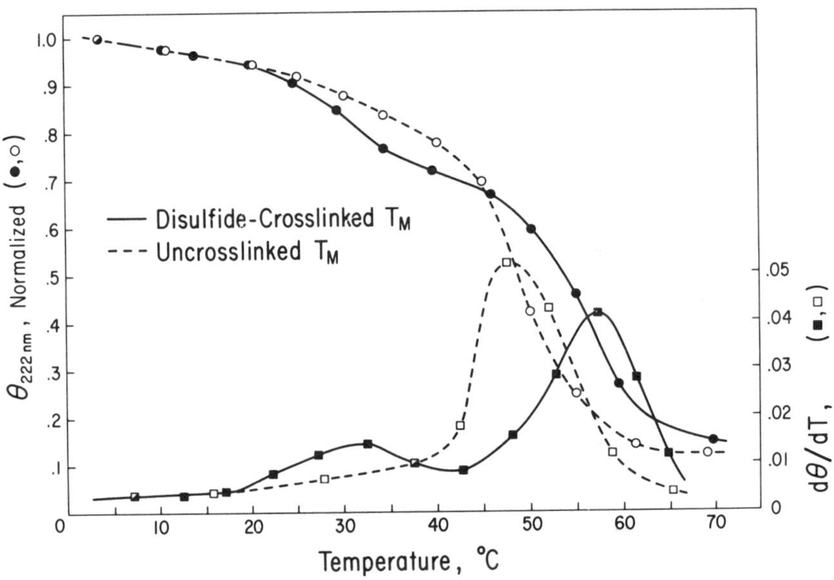

FIGURE 14. Unfolding of uncrosslinked (– – –) and disulfide-crosslinked (——) rabbit skeletal tropomyosin (●, ○), $\theta_{222\,nm}$; (■, □), $d\theta/dT$. Disulfide-crosslinked tropomyosin was prepared by Nbs_2 reaction as outlined in Reference 12. 0.1 mg/ml in 2 mM Hepes buffer, pH 7.5, 1M NaCl, 1 mM EDTA.

respectively.[25] These values are very close to the midpoint for the pre-transition of the thermal (FIG. 14) and denaturant-induced[12] unfolding of the crosslinked sample, the GdmCl-induced loss of ability to form S-S crosslinks,[12] the GdmCl-induced loss of excimer fluorescence of pyrene-labeled tropomyosin[13] and the maxima in the thermal dependence of the excimer intensity and dansyl long-lived components, shown above. In view of these similar pre-transition values, it seems reasonable to associate the open chain conformation (State B) with the intermediate, X.

CONCLUDING REMARKS

The data presented above on pyrene and dansylcystine-labeled tropomyosin, provide strong evidence for a model of tropomyosin in which two states are present in solution in equilibrium with each other. It was shown that the ability of Nbs_2 to disulfide crosslink tropomyosin and the unfolding properties of crosslinked tropomyosin can now be understood in terms of the presence of two states. The more stable state (State A) appears to be a "chain closed" state and the somewhat less stable state (State B), which can be associated with an intermediate in the unfolding, is one in which the chains are separated near Cys 190. It is not clear how extensive this unfolding is but it appears that 10–20% of the helix is lost before the main transition. Thus, although we have been probing a very localized region near Cys 190, it is possible that this unstable region extends as far as 80 Å. Evidence for preferential instability over a longer range was indicated by a comparison of the loss of polarization of Tyr fluorescence with changes in ellipticity at 222 nm caused by GdmCl.[12] This showed that the polarization was lost prior to the helix loss, suggesting that the carboxyl terminal half of the molecule was more unstable since five out of six Tyr are located in that half.[8] Recent X-ray diffraction studies on crystalline tropomyosin also indicate that the carboxyl terminal arm is more unstable than the amino terminal arm.[27] NMR[28] and ESR[29] studies have also suggested multiple regions of differing stability.

Is there any relationship between the conformational states of tropomyosin in solution and the structural changes within the actin-tropomyosin-troponin thin filament associated with Ca^{2+} binding? X-ray studies on muscle[30, 31] and electron microscopy-optical diffraction techniques on thin filament paracrystals[32, 33] have shown that tropomyosin occupies two different azimuthal positions in the groove of the actin thin filament helix as a result of the binding and removal of Ca^{2+} from troponin. Early experimental studies suggested that troponin simply controls the position of tropomyosin by anchoring and releasing it from actin response to Ca^{2+}.[34, 35] The possibility that the conformational state or flexibility of tropomyosin can have a role in determining the position of tropomyosin on the thin filament and its resulting activity has been considered.[12, 27] From the studies discussed above, it is known that the region of tropomyosin near Cys 190, where the bulk of troponin is bound[36, 37] is in the region where the chain separartion occurs associated with state B. Perturbation of this region by chemical modification of Cys 190 influences the conformation of the molecule far from Cys 190. For example, increased chain unfolding at physiological temperatures results from the introduction of a disulfide crosslink between the Cys 190 side chains of adjacent subunits.[12] Most recently, the modification of Cys 190 with pyrene maleimide was shown to cause the inhibition of end-to-end polymerization of tropomyosin.[38] This observation is particularly relevant since end-to-end interactions of tropomyosin have been shown to greatly contribute to the strength and manner of binding of tropomyosin to actin.[39] This suggests that the mode of binding of tropomyosin to actin can be altered by troponin via a perturbation in the region of Cys 190. Although in this model troponin exerts its effect on tropomyosin locally resulting in a long-range affect to its ends, a direct interaction between a component of troponin and the ends of tropomyosin is possible. Further experiments will clarify these points and lead to an understanding of the role of the conformational dynamics of tropomyosin in the regulation of muscle contraction.

ACKNOWLEDGMENTS

We thank Mr. Sunny Wong and Ms. Justine Strungis for excellent technical assistance and Dr. Terence Tao for help in lifetime measurements.

REFERENCES

1. JOHNSON, P. & L. B. SMILLIE. 1975. Biochem. Biophys. Res. Commun. **64:** 1316–1322.
2. STEWART, M. 1975. FEBS Lett. **53:** 5–7.
3. LEHRER, S. S. 1975. Proc. Natl. Acad. Sci. USA **72:** 3377–3381.
4. EBASHI, S. & M. ENDO. 1968. Prog. Biophys. Mol. Biol. **18:** 123–183.
5. COHEN, C., D. CASPAR, J. P. JOHNSON, K. NAUSS, S. MARGOSSIAN & D. PARRY. 1973. Cold Spring Harbor Symp. Quant. Biol. **37:** 287–297.
6. WEBER, A. & J. M. MURRAY. 1973. Physiol. Rev. **53:** 612–673.
7. STONE, D. & L. B. SMILLIE. 1978. J. Biol. Chem. **253:** 1137–1148.
8. SODEK, J., R. S. HODGES & L. B. SMILLIE. 1979. J. Biol. Chem. **253:** 1129–1136.
9. MAK, A. S., W. G. LEWIS & L. B. SMILLIE. 1979. FEBS Lett. **105:** 232–234.
10. EISENBERG, E. & W. W. KIELLEY. 1974. J. Biol. Chem. **249:** 4742–4748.
11. YAMAGUCHI, M., M. GREASER & R. G. CASSENS. 1974. J. Ultrastruct. Res. **48:** 33–58.
12. LEHRER, S. S. 1978. J. Mol. Biol. **118:** 209–226.
13. BETCHER-LANGE, S. & S. S. LEHRER. 1978. J. Biol. Chem. **253:** 3757–3760.
14. LEHRER, S. S. & P. GRACEFFA. 1980. Fed. Proc. **39:** 1622.
15. WU, C.-W., L. R. YARBROUGH & F. Y.-H. WU. 1976. Biochemistry **15:** 2863–2868.
16. BIRKS, J. B. 1970. *In* Photophysics of Aromatic Molecules. Wiley-Interscience. London. pp. 301–371.
17. CHANDROSS, E. A. & C. J. DEMPSTER. 1970. J. Am. Chem. Soc. **92:** 3586–3593.
18. JOHNSON, G. E. 1974. J. Chem. Phys. **61:** 3002–3008.
19. AVOURIS, P., J. KORDAS & M. A. EL-BAYOUMI. 1974. Chem. Phys. Lett. **26:** 373–376.
20. GOLDENBERG, M., J. EMERT & H. MORAWETZ. 1978. J. Am. Chem. Soc. **100:** 7171–7177.
21. LEHRER. S. S. & S. BETCHER-LANGE. 1979. *In* Motility in Cell Function. Academic Press. New York. pp. 381–385.
22. WOODS, E. F. 1969. Int. J. Protein Res. **1:** 29–45.
23. PRANIS, R. A. & I. M. KLOTZ. 1977. Biopolymers **16:** 299–316.
24. CHEN, R. F. 1967. Arch. Biochem. Biophys. **120:** 609.
25. WOODS, E. F. 1976. Aust. J. Biol. Sci. **29:** 405–418.
26. KAUZMANN, W. 1959. Adv. Protein. Chem. **14:** 1.
27. PHILLIPS, G. N., E. E. LATTMAN, P. CUMMINS, K. Y. LEE & C. COHEN. 1979. Nature **278:** 413–417.
28. EDWARDS, B. F. P. & B. D. SYKES. 1978. Biochemistry **17:** 684–689.
29. CHAD, Y-Y. H & A. HOLTZER. 1975. Biochemistry **14:** 2164–2170.
30. HUXLEY, H. E. 1972. Cold Spring Harbor Symp. Quant. Biol. **37:** 361–376.
31. HAZELGROVE, J. C. 1972. Cold Spring Harbor Symp. Quant. Biol. **37:** 341–352.
32. WAKABAYASHI, T., H. E. HUXLEY, L. A. AMOS & A. KLUG. 1975. J. Mol. Biol. **93:** 477–497.
33. GILLIS, J. M. & E. J. O'BRIEN. 1975. J. Mol. Biol. **99:** 445–459.
34. POTTER, J. & J. GERGELY. 1974. Biochemistry **13:** 2967–2703.
35. HITCHCOCK, S. 1975. Eur. J. Biochem. **52:** 255–263.
36. NONOMURA, Y., W. DRABIKOWSKI & S. EBASHI. 1968. J. Biol. Chem. **64:** 419–422.
37. STEWART, M. 1975. Proc. Roy. Soc. London **B190:** 257–266.
38. GRACEFFA, P & S. S. LEHRER. 1980. J. Biol. Chem. **255:** 11296–11300.
39. WEGNER, A. 1979. J. Mol. Biol. **131:** 839–853.

LUMINESCENCE MEASUREMENTS OF MACROMOLECULAR MOBILITY*

W. W. Webb

Applied Physics
Cornell University
Ithaca, New York 14853

Fluorescence radiation provides optical markers for measurements of macromolecular mobility. High quantum efficiency fluorophores that can be stimulated by visible light and attached to macromolecules serve as indicators of the distribution and concentration of specifically labeled macromolecules on the plasma membranes of cells in tissue culture and on model lipid membranes. Specificity of labeling is frequently provided by attaching the fluorescent label to toxins or monovalent antibody fragments. Surface concentrations of quantum efficient fluorophores as sparse as 200 per square micron are frequently detectable. Molecular mobility measurements on cells exploit a feature of fluorophores that is usually undesirable—their propensity for irreversible photolysis in bright light. Nonuniform distributions of fluorophore are generated by photobleaching small areas or patterns with a bright light pulse. The time course of recovery of fluorescence after photobleaching an area provides a measure of the dynamics of the process by which fresh fluorophore enters the area; often this process is lateral diffusion.

Application of these fluorescence photobleaching recovery (FPR) techniques during the 1970s has provided a substantial body of knowledge about the diffusion of macromolecules and lipids on cell surfaces and on lipid model membranes. As this research has progressed, it has required more powerful technology to deal with complex nondiffusive processes of molecular motion, smaller fluorophore populations, difficult cell types, and heterogeneous cellular structures. This paper describes some of the new fluorescence technology that is evolving now to meet the challenge posed by some fundamental problems in cell biology. To illustrate these developments, I have drawn on current research in my laboratory at Cornell. Some provocative new biophysical data are included to demonstrate the potential of the available technology but more complete results will be found in our research papers.

LATERAL DIFFUSION IN CELL MEMBRANES—BACKGROUND

Systematic procedures to measure lateral diffusion on cell membranes have evolved from early experiments of Peters *et al.*[1] and Edidin and Fambrough.[2] Our version of the procedure is called fluorescence photobleaching recovery[3,4] (FPR), or alternatively, fluorescence recovery after bleaching (FRAP).[5] Several variations of the

* Research supported by the National Institutes of Health through Grants GM 21661 and CA 14454 and the National Science Foundation through PCM 76-83068

0077-8923/81/0366-0300$01.75/0 © 1981, NYAS

experimental geometry and analysis have been developed recently.[6,7] All share the same basic operations:

(1) Fluorescent markers are dissolved in the membrane or coupled to the membrane macromolecules to be studied;

(2) Some of the fluorophore is photobleached in a well-defined area of the cell membrane by a brief pulse of light focused into a small spot or pattern of well-defined geometry on the cell surface to create an inhomogeneous non-equilibrium distribution; and

(3) The fluorescence intensity in a well-defined area, usually just the same bleached spot or pattern, is measured as fresh fluorophore diffuses along the membrane into the bleached spot allowing some or all of the initial fluorescence to recovery.

Analysis of the time course of this fluorescence recovery process yields specific details of the transport mechanism responsible for the recovery.[3] Most of the research to date has concentrated on the lateral diffusion of macromolecules on cell surfaces and lipid-analog molecules in model membranes. Diffusion consists of the quasi-random Brownian motion of individual molecules. It produces a net transfer of molecules only in the presence of a chemical potential gradient, usually due to a simple concentration gradient. This diffusive flux gradually smooths out the induced inhomogeneity of concentration in accord with the diffusion equation $\partial c/\partial t = D\nabla^2 c$ where D is the diffusion coefficient and c is the fluorophore concentration.

Before summarizing the accumulated results of diffusion experiments on vertebrate tissue cells in culture, which have led us to the study of more complex processes, I want to illustrate the current state-of-the-art for diffusion measurements, by describing the latest model of our apparatus, reporting some new measurements of lipid-analog diffusion in intact erythrocytes and describing some morsels of fluorescence technology that may be of general interest.

In order to improve the precision of diffusion measurements on biological systems, the technology of FPR continues to be developed in many laboratories. We count now about 20 laboratories around the world that are capable of FPR experiments. The latest version of our fluorescence photobleaching recovery apparatus at Cornell is illustrated schematically in FIGURE 1. Mr. John Bloom, Ms. Marilyn Schneider, Mr. Larry Barak and Dr. Winston Chan are responsible for recent developments. This figure and its caption provides details for the expert in FPR but also indicates the general scheme of FPR experimental apparatus. New features include:

(1) The use of channel plate image intensifier (Ni-Tec) to allow high quantum efficiency observations of visible light images and detection and sharp focusing of invisible near-infrared fluorescence,

(2) Variable reflectivity beam splitters; and

(3) Microscope stage thermostats controllable to 0.01°C even while using water immersion objective lenses.

The quality of data now obtainable in a measurement of lipid-analog diffusion on an intact erythrocyte is illustrated in FIGURE 2. Here many photobleaching recovery curves are combined by signal averaging for high precision. This procedure is particularly convenient if all of the chromophore is diffusable.

To illustrate the technology involved in recent diffusion research, some of Mr. Bloom's measurements of lipid probe diffusion on intact erythrocytes are presented. Intact erythrocytes provide a challenge for FPR because the hemoglobin strongly absorbs visible light which, at high intensity, heats the cells. However, exposure to

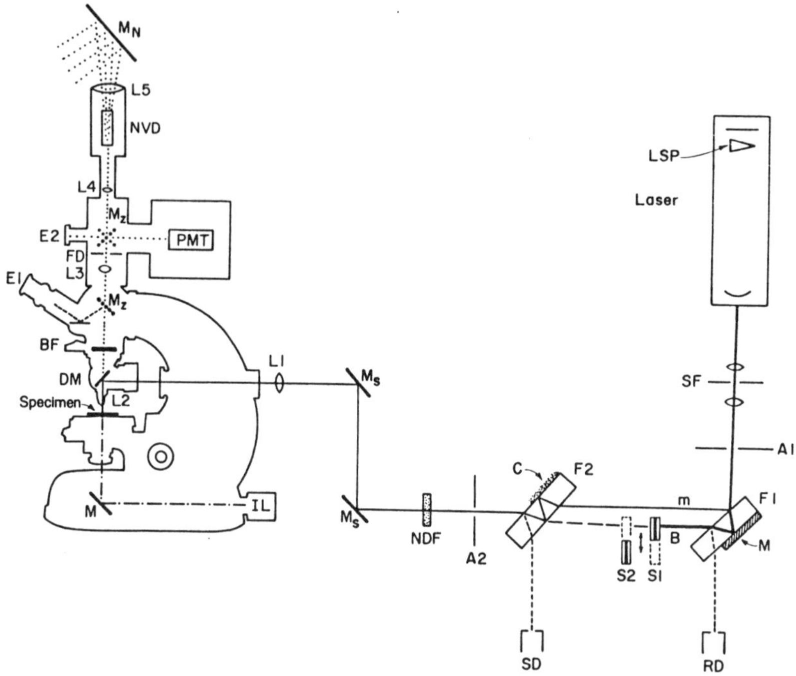

FIGURE 1. One state-of-the-art apparatus for fluorescence photobleaching recovery experiments. Light emerging from a Krypton ion laser (model 52K; Coherent Radiation, Palo Alto, CA) is spatially filtered (SF) to give a reasonably gaussian beam profile. The beam is then split into separate monitor and bleaching beams by reflecting it from a mirrored optical flat (F1). The bleaching beam is normally blocked by shutter (S1); a bleach is initiated by opening this first shutter and ended by blocking the bleach beam with a second, independent shutter (S2) after a selectable time interval. The monitor and bleach beams are then recombined by a matching optical flat (F2) so that they follow the same path to the specimen. Stationary neutral density filters (NDF) may be inserted into the beam path after recombination to adjust the final laser intensity. The beam is then directed to the vertical illuminator of a Zeiss Universal fluorescence microscope where an adjustable focusing lens (L1) brings the beam into focus at the object plane of the microscope objective. A dichroic mirror (DM) directs about 90% of the beam to the specimen via the Zeiss 100X, 0.8 numerical aperture oil-immersion dark field objective lens (L2), and transmits greater than 80% of the sample's fluorescence up through the barrier interference filter (BF) to a sensitive photomultiplier (PMT). The viewing area of the PMT is controlled by an adjustable diaphragm (FD). The sample is aligned through convenient eyepieces (E1, E2) and movable mirrors (Mz). In order to focus the beam properly for the DiI[5] experiments, a night vision device (NVD; Ni-Tec, Niles, IL) was fitted to the microscope to enhance the visibility of the near-infrared (676 nm) probe fluorescence. (J. Bloom)

bright light can lyse erythrocytes even with negligible heating. Fortunately, this effect can be suppressed by deoxygenation of the cells. Reductions of photosensitivity by a factor of 10^4 can be achieved. Various common chemical protectants work, but the most effective in our hands are small additions of submitochondrial particles (SMP) which provide "nature's own" deoxygenator. Their effectiveness in inhibiting red cell lysis on prolonged exposure to bright focused laser light is demonstrated by the data shown in FIGURE 3. This simple method for deoxygenation may be quite

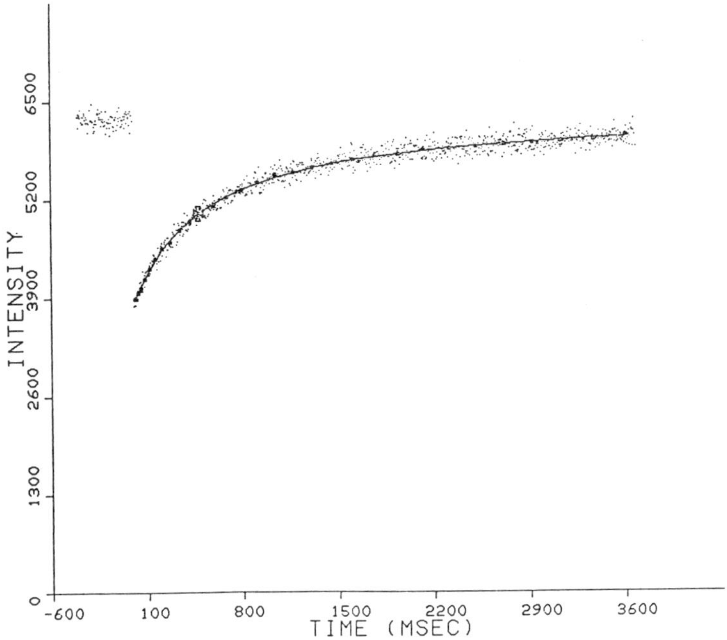

FIGURE 2. Typical fluorescence photobleaching recovery data for lipid probe in intact erythrocyte. Photobleaching light pulse is applied at time zero, individual intensity measurements, small points, are locally averaged to obtain large points. Nonlinear least squares fit yields characteristic recovery time t. Successive experiments can be combined for high percision in this system with complete fluorescence recovery.

generally useful. It was originally suggested to us by Prof. Peter Hinkle. Another helpful application in fluorescence experiments is presented later.

Measurements on erythrocytes are further simplified by use of fluorophores that can be excited with deep red light from the Krypton laser at 647 nm, and which fluoresce in the infrared around 680 nm. A useful lipid probe molecule is 3,3′-dioctadecylindodicarbocyanine. This molecule has been generously provided by Dr. Alan Waggoner who is responsible for the development of many helpful fluorophores and innovations in fluorescence technology that have proven useful in cellular biophysics.

The first data on lipid probe diffusion in intact erythrocytes from mouse, man and frog are shown in FIGURE 4. The temperature dependence of the diffusion coefficient in erythrocytes is much stronger than in vertebrate tissue cells which would appear as a nearly horizontal line at about 100×10^{-10} on this logarithmic plot. Note also the small difference between mouse and man and even cold-blooded frogs acclimated at normal room temperature.

Now we return to a summary of the larger body of diffusion measurements on tissue culture cells. The diffusion coefficients of most membrane proteins that have been studied on vertebrate tissue cells share some common properties:

(1) Small diffusion coefficients $D \lesssim 10^{-10} cm^2/sec$;

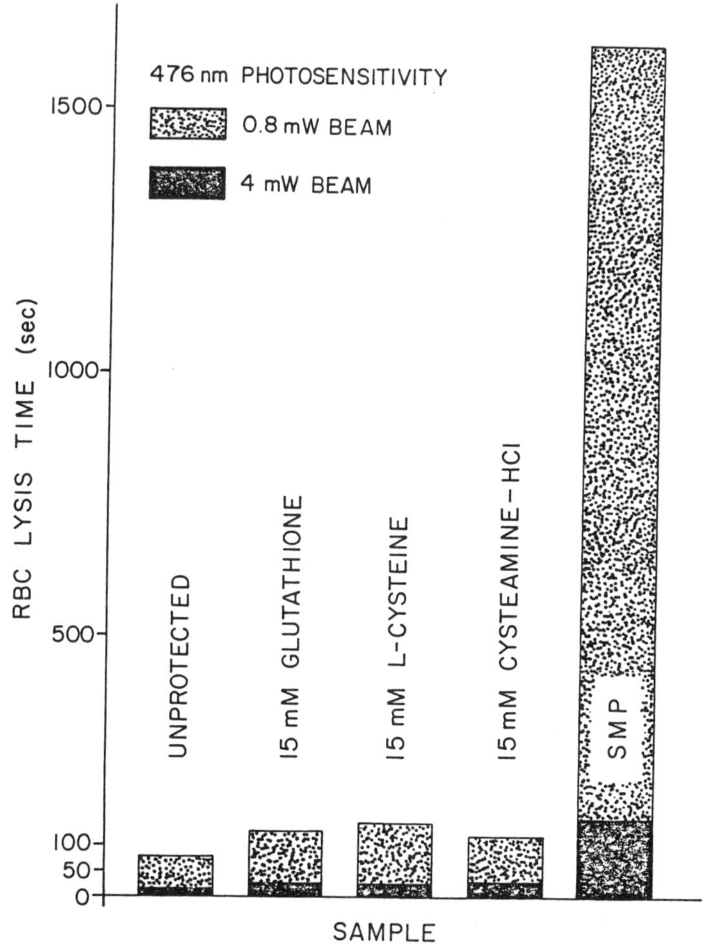

FIGURE 3. Lysis time for intact erythrocytes exposed to laser light. Deoxygenation with submitochondrial particles (SMP) provides strong protection. (Controls showed that lipid composition of erythrocytes is not altered by SMPs under these conditions.)

(2) A substantial fraction of each molecular species, 25–100%, appears essentially nondiffusable; and

(3) Weak temperature dependence, increases by about 3 times from, say, 10°C to 40°C.

Some important membrane macromolecules display special properties. For example, the exoskeletal glycoprotein fibronectin is virtually immobile on fibroblasts,[8] and the acetylcholine receptors on primary myotubes show two populations—an immobile population in patches of receptor and a low-density population that diffuses slowly $D \sim 10^{-11} cm^2/sec$.[9]

The effective fluidity of cell membrane has been appraised by measuring the lateral diffusion of fluorescent-labeled lipids and lipid-analog molecules in the

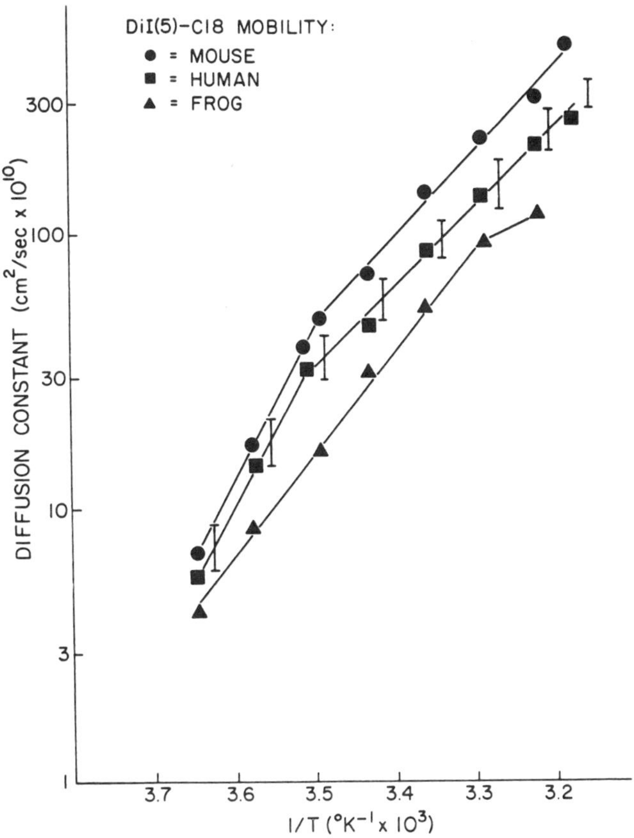

FIGURE 4. Lateral diffusion coefficients of lipid probe in intact erythrocytes show strong temperature dependence.

plasma membrane where they diffuse with a diffusion coefficient $D \sim 10^{-8} cm^2/sec$. Temperature dependence of the lipid probe diffusion, like protein diffusion, is rather weak, varying by only a factor of 3 from 4°C to 37°C.[10,11] In vertebrate tissue cells, essentially all of the lipid analog appears to be diffusable, indicating that it is not trapped in "lakes" embedded in the protein network or permanently bound to protein. In solvent-free phospholipid model membranes above the phospholipid phase transitions, lipids diffuse with about the same diffusion coefficients as in tissue cells. Varying the hydrocarbon chain length, saturation, and head groups cause only small effects on lipid and lipid-analog diffusion above the phase transitions. However, inclusion of hydrocarbon solvents as in Mueller-Rudin "black lipid" membranes increases lipid probe diffusion by as much as a factor of ten.

The dramatic difference between the diffusion coefficients of membrane proteins and lipids presents a puzzle because a reasonable theory of the fluid mechanics of diffusion of macromolecules in a homogeneous viscous membrane shows clearly that the dependence of the diffusion coefficient on molecular size should be very weak.[12,13]

Thus the diffusion coefficients of lipids and proteins in a membrane should differ very little if only viscous drag in the membrane limits both. This discrepancy with experimental data focuses on alternative factors that may regulate diffusion.

Interactions that might be implicated in restriction of membrane protein diffusion are the binding or viscous coupling of proteins to the cytoskeleton or to exoskeletal structures, or coupling among membrane proteins. The existence of an immobile fraction in all populations of membrane receptors that have been studied on mammalian tissue cells demonstrates the ubiquitous importance of such interactions. Measurements of the effects of cytotoxic drugs confirm the existence of interactions with cytoskeletal macromolecules that are not intrinsic to the lipid bilayer.

Observations of protein diffusion in model lipid membranes that are free of cytoskeleton and exoskeleton show that proteins can indeed diffuse more rapidly if free of these constraints. Recently, for example, experiments by other methods have shown that macromolecules in mitochondrial membranes known to be free of gross cytoskeleton do diffuse substantially faster, say, $10^{-9}cm^2/sec$.[14] Early experiments of Poo and Cone showed that rhodopsin diffused rapidly, $D \sim 10^{-9}$ in the cytoskeleton-free visual disc membranes of rod outer segments.[15] Very recently, E.-S. Wu, P. S. Loew and I (unpublished) have reconstituted the integral membrane protein glycophorin into multilamellar lipid membranes. They find that the diffusion coefficient of glycophorin in egg lecithin and in dimyristoylphosphotidyl choline (DMPC) is slightly larger than $10^{-8}cm^2/sec$ with a temperature variation of about a factor of 2 in 10°C. Below a sharp phase transition around 23.2 to 23.7°C in DMPC, the diffusion coefficient drops by a factor of greater than 100. In 50% cholesterol–50% DMPC the temperature dependence is stronger and smoothly increasing from 4°C to 40°C. David Tank, E.-S. Wu, and I have found that gramicidin C dimer in the form of a cylindrical rod threading the membrane behaves similarly. Details will be published elsewhere. Koppel, Sheetz and Schindler have shown that degradation of the spectrin network in erythrocyte ghosts releases the Band III proteins for rapid diffusion.[16] With this exception, there has been little success in identifying the specific interactions controlling diffusability of macromolecules.

A Current Problem: The Search for Membrane–Cytoskeleton Interactions

Our current approach to this problem includes a search for specific interactions between membrane macromolecules and the cytoskeleton. Immunofluorescence methods have proven extremely powerful for studies of cytoskeleton in *fixed* cells. We have sought specific fluorescent labels for cytoskeleton that can be introduced into *living* cells to trace the dynamics of cooperative motions; we hope to identify the coupling with cell surface receptors.

In our laboratory, Mr. Larry S. Barak has conjugated the mushroom toxin phallacidin with a small chromophore, nitrobenzoxadiazole (NBD) to produce a highly specific F-actin stain, NBD-phallacidin (NBD-Ph).[17] We have applied this cytoskeletal marker to a variety of problems, several of which are illustrated. Figure 5 shows a typical example of NBD-Ph staining of actin cytoskeleton in fixed cells in this case, a kangaroo rat fibroblast.

A particularly interesting application of cytoskeletal and membrane receptor stains has been worked out by Mr. David Tank. Here two specific fluorescent markers are

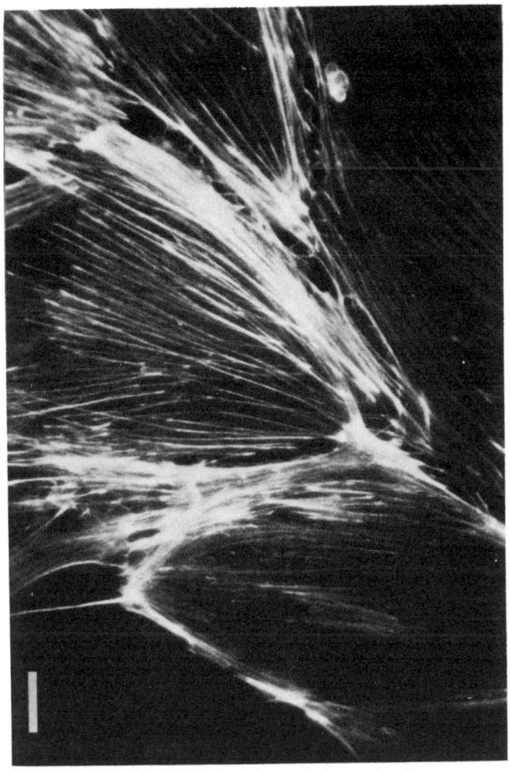

FIGURE 5. Actin filamentary cytoskeleton in cultured kangaroo rat cells fixed and stained with NBD-Phallacidin. Note distinctive tufted ends of actin fibers characteristic of these cells. (L. Barak, unpublished)

used on a rat myotube. NBD-Ph marks the actin cytoskeleton and an α-bungarotoxin labeled with rhodamine marks the acetylcholine receptor in the patches of the receptor commonly seen on the plasma membranes of these cells. The patch appears to occupy a cavity in the actin network when located on the bottom surface of the cell. However, the receptor and the actin network are not directly connected.

Nondiffusive Movement on the Cell Surface

The existence of nondiffusive molecular movement on the cell surface through what we call concerted or driven motion is evident in many cellular processes. Cellular motility clearly involves directed movement of the entire cell with systematic, cooperative movement on the cell membrane necessarily involved in the formation of ruffles at the leading edge of a motile cell, etc. The mechanism of mass transport that delivers the lipid to these forming ruffles is not known, nor is the motive power propelling the advance of the cell edge. The actin cytoskeleton is clearly involved but a self-consistent mechanism remains to be found.

Some Special Methods for Observing Cell Dynamics with Fluorescence

To study the mechanisms of processes involving these concerted molecular movements, it is useful to observe the molecular movements topographically as they occur all over the cell surface. Fluorescent markers can be used for this purpose in time-lapse fluorescence microscopy. However, the problem is that fluorescent markers of membrane components on living cells must have low concentrations. Therefore, the fluorescence is weak and the recording of a single photomicrograph usually involves photobleaching nearly all of the fluorophore. To circumvent this problem, we use high quantum efficiency channel plate image intensifiers and silicon image intensifier video cameras to increase the quantum efficiency of fluorescence detection by several orders of magnitude. Their effectiveness is illustrated in Figure 6 which shows fluorescence photomicrographs of actin cables in the giant alga Chara stained with

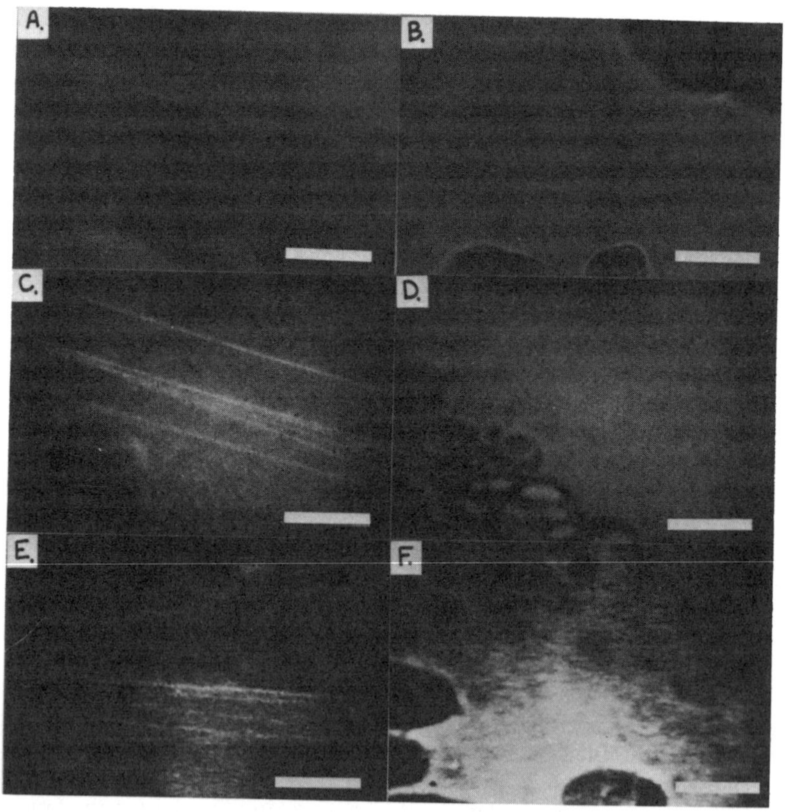

Figure 6. Fluorescence micrography techniques applied to actin cables in *Chara.* Left hand column: compares conventional photography (A), Ni-Tec night vision device (B), and SIT image intensifier video (C). The fluorescence excitation for B and C has been reduced by a factor of 200. Right-hand columns shows corresponding brightfield microscopy images. Scale bars are 10 μm. (E. Nothnagel, unpublished)

NBD-Ph. These cables are thought to pump rapid cytoplasmic streaming in these cells and Mr. Eugene Nothnagel has been able to perfuse these cells with NBD-Ph and observe the actin cables during streaming. FIGURE 6 shows three similar scenes within these cells through a perforated layer of chloroplasts by brightfield microscopy and by NBD-Ph fluorescence; three different methods of detection and recording of the fluorescence are illustrated. The first, a photograph on ASA 400 film pushed to about 1600 exhausted nearly all of the fluorescence by photobleaching. The second two were recorded with a factor of 200 reduction in excitation intensity and modest exposure times. There remains in these two scenes enough fluorescence for at least one hundred more images. The recording method for the second image uses a Ni-Tec image intensifier recorded on film and the third is a silicon image intensifier (SIT) video camera displayed and photographed from the video monitor on film.

A simple procedure for substantial reduction of photobleaching of the dye during photography has also been developed. It increases the total amount of fluorescence light available by about a factor of three. The procedure is simply to deoxygenate the cells using submitochondrial particles (SMP) to remove oxygen. This procedure works quite well in practice for cells that can tolerate low oxygen. The effectiveness of deoxygenation with SMPs for reduction of the photobleaching of NBD-Ph is illustrated in FIGURE 7. The total available fluorescence energy is proportional to the

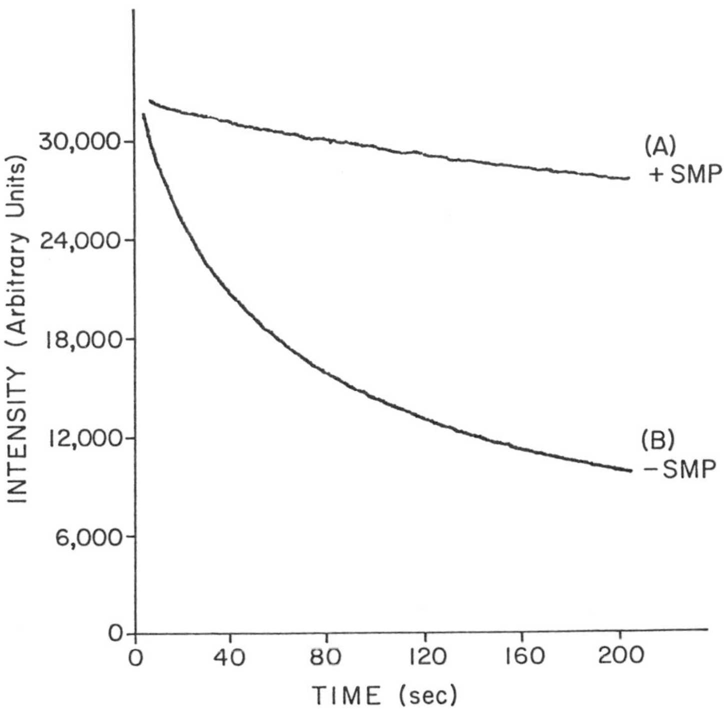

FIGURE 7. Reduction of the photobleaching of fluorescent stains on cells by deoxygenation with submitochondrial particles. Improvement by about a factor of three increases sensitivity of fluorescence microscopy. (L. S. Barak and J. Bloom)

ratio of the quantum efficiency of fluorescence to quantum efficiency for photobleaching. The photobleaching is reduced by about a factor of 3.

Capping

The formation of caps in the active response of cells, particularly lymphocytes, to cross-linking by antibodies or multivalent lectins has been studied relentlessly for many years. Directed flow of the lipid membrane, contractile aggregation of the cytoskeleton and differential motility in the presence of inhomogeneous anchorage of membrane receptors are suggested processes. However, a new facet of this problem has been introduced by the observations of Dragsten *et al.* that diffusion of certain cappable receptors on lymphocytes is promptly arrested by cross-linking with antibodies.[18] Therefore, diffusion cannot provide the mechanism of movement by which the cappable receptors aggregate. Wolf, Henkart and Webb[19] have found that an extrinsic antigenic receptor that is capped by an active process on fibroblasts is also prevented from diffusing by antibody cross-linking during the cap formation process. This result is even more suggestive since the extrinsic receptor cannot be expected to interact with the cytoskeleton.

Some current studies of capping on RDM4 virus-transformed T lymphocytes in our laboratory by E.-S. Wu illustrate the power of the technology now available to study cell membrane dynamics with specific fluorescent labels. The expected coincidence of caps formed by tetravalent Concanavalin A (Con A) and the actin cytoskeleton aggregated underneath these caps is shown in FIGURE 8. The Con A is rhodamine labeled and the actin is stained with NBD-Ph. In contrast, FIGURE 9

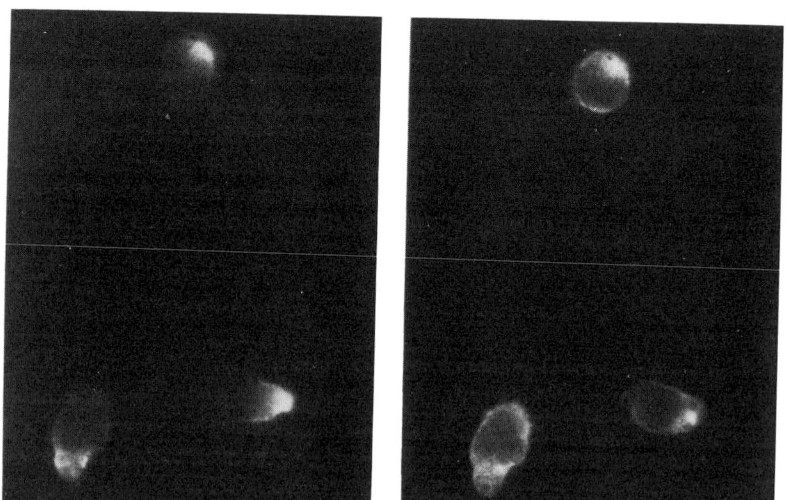

FIGURE 8. Formation of caps by Con A receptor on transformed lymphocytes. (a) Fluorescence labeled Con A caps, (b) NBD-Ph stained cytoskeleton aggregates under caps. Prior to capping, most of the actin cytoskeleton appears in fluorescence microscopy as a diffuse layer uniformly spread under the plasma membrane. (E.-S. Wu, unpublished)

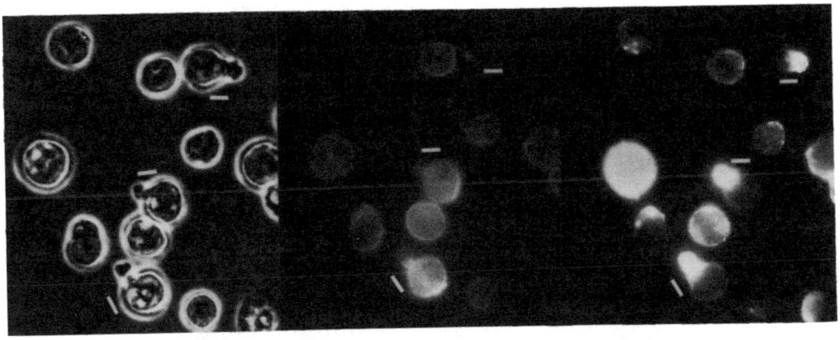

FIGURE 9. Formation of "anti-caps" in actin cytoskeleton after capping of H-2 antigen with monoclonal antibodies. (a) Phase contrast, (b) Cytoskeletal actin aggregate stained with NBD-Ph, (c) Caps of H-2 antigen stained by indirect immunofluorescence. Bars indicate examples of "anti-capping" (E.-S. Wu).

shows the capping of the H-2 antigen by monoclonal antibodies subsequently stained with rhodamine—conjugated antimouse IgG from rabbit. The interesting feature here is the occurrence of "anti-caps" that is, cells in which the H-2 antigen and the actin aggregate in different locations. This striking phenomenon has interesting implications for mechanisms of cap formation and is being studied further.

RECEPTOR DYNAMICS

In a previous paper, Dr. J. Schlessinger has demonstrated powerful fluorescence procedures in studying the mobility and aggregation of hormone receptors at very low concentrations on cell surfaces. A helpful procedure is the attachment of several fluorophores to each hormone and thus to each receptor to increase the sensitivity of fluorescence microscopy and FPR experiments. We have pushed this strategy to the limit in the study of the interaction of low density lipoproteins (LDL) with its receptor cell surfaces and are able to see individual LDL molecules on the cell surface as recounted below.

LDL is a remarkably large object, having a molecular weight about 6 megadaltons and a diameter of about 220Å. It ordinarily sequesters up to 1500 cholesterol esters and carries ~1000 phospholipids in the shell. By substituting fluorophores following variations of the method of Krieger *et al.*[20] Mr. L. S. Barak has succeeded in producing sufficiently bright fluorescence to detect individual LDL molecules bound to its receptors on the surfaces of certain mutant cells called JD cells that are apparently unable to internalize LDL because the receptor does not bind to the coated pits.[21] FIGURE 10 shows such a distribution. These cells have been fixed promptly after treatment with LDL to minimize possible aggregation. The fluorescent intensities of the individual spots agree with expectations for individual LDL particles within our experimental uncertainty. Controls do indicate the appropriate specificity of the modified LDL and the scarcity of aggregation. Cross-linking with antibodies to LDL on live cells gradually aggregates these receptors into small patches. Warming

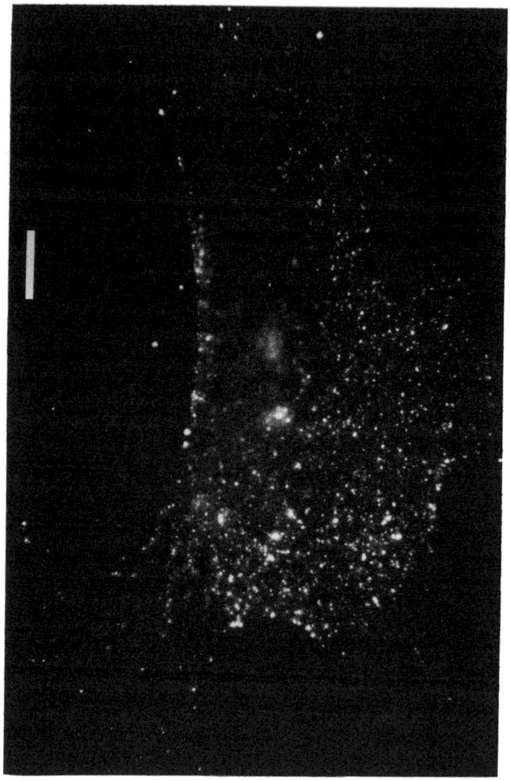

FIGURE 10. Fluorescent LDL on JD cells. Individual LDL molecules and very small clusters yield punctate stain patterns visible due to low receptor concentration and brilliant stain of individual LDL molecules. Stained at 4°C, focused on cell surface near periphery at right, surface out of focus at left. (L. S. Barak, unpublished)

to 37°C without cross-linking also induces aggregation as shown in FIGURE 11. The occurrence of the aggregation process implies lateral diffusion of the individual LDL molecules on the cell surface. However, the slow aggregation implied slow diffusion. Restricted mobility was seen by recording a time-lapse video motion picture at one frame about every 10 seconds and replaying it at normal speed to observe molecular Brownian motion visually. Detailed studies of this restricted, thermally excited molecular motion are in progress. They show evidence that these receptors are not able to diffuse freely but are constrained to local Brownian motion.

Direct observation of fluorophore concentration fluctuations is the basis of a method for diffusion coefficient measurement that we have previously developed and call fluorescence correlation spectroscopy (FCS)[22–24] This new application of the concept to individually resolvable molecules introduces new experimental possibilities that we are currently exploring. The ability to detect individual molecules of LDL should help to analyze the interaction of LDL receptors with coated pits at which LDL is supposed to be internalized.

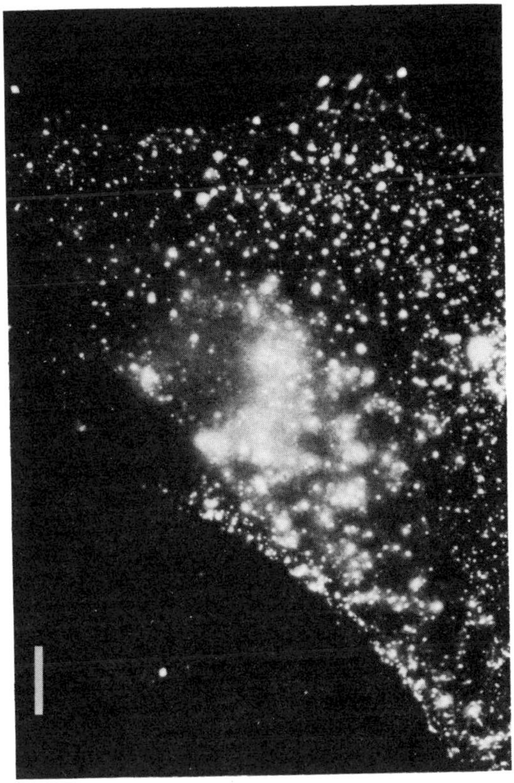

FIGURE 11. Fluorescent LDL on JD cells. After incubation at 37°C, some aggregation and possibly some internalization has occurred during 30 minutes at 37°C. Focused on surface near cell periphery. (L. S. Barak, unpublished)

CONCLUSIONS

We may anticipate a busy future for fluorescence technology in studying molecular mobility and other aspects of cellular dynamics. Technology will certainly continue to advance providing capabilities that will deliver more new information about the complex molecular dynamics of biological systems.

ACKNOWLEDGMENTS

I am grateful to my colleagues Jim Wu, John Bloom, Larry Barak, Eugene Nothnagel, and David Tank for permission to use unpublished illustrations and to present data prior to detailed publication.

REFERENCES

1. PETERS, R., J. PETERS, L. TEWS & W. BAKER. 1974. Biochim. Biophys. Acta **367:** 282.
2. EDIDIN, M. & D. FAMBROUGH. 1973. J. Cell Biol. **57:** 27.
3. AXELROD, D., D. E. KOPPEL, J. SCHLESSINGER, E. L. ELSON & W. W. WEBB. 1976. Biophys. J. **16:** 1055.
4. KOPPELL, D. E., D. AXELROD, J. SCHLESSINGER, E. L. ELSON & W. W. WEBB. 1976. Biophys. J. **16:** 1315.
5. JACOBSON, K., Z. DERZKO, E.-S. WU, Y. HOU & G. POSTE. 1976. J. Supramol. Struc. **5:** 565.
6. SMITH, B. A. & H. M. MCCONNELL. 1978. Proc. Nat. Acad. Sci. USA **75:** 2759.
7. KOPPEL, D. E. 1979. Biophys. J. **28:** 281.
8. SCHLESSINGER, J., L. S. BARAK, G. G. HAMMES, K. M. YAMADA, I. PASTAN, W. W. WEBB & E. L. ELSON. 1977. Proc. Natl. Acad. Sci. USA **74:** 2909.
9. AXELROD, R., P. RAVDIN, D. E. KOPPEL, J. SCHLESSINGER, W. W. WEBB, E. L. ELSON & T. R. PODLESKI. 1976. Proc. Natl. Acad. Sci. USA **73:** 4594.
10. AXELROD, D., A. WIGHT, W. WEBB & A. HORWITZ. 1978. Biochemistry **17:** 3604.
11. VAZ, W. L., K. JACOBSON, E.-S. WU & Z. DERZKO. 1979. Proc. Natl. Acad. Sci. USA **76:** 1645.
12. SAFFMAN, P. G. & M. DELBRÜCK. 1976. Proc. Natl. Acad. Sci. USA **72:** 3111.
13. SAFFMAN, P. G. 1976. J. Fluid Mech. **73:** 593.
14. HÖCHLI, M. & HACHENBROCK. 1979. Proc. Natl. Acad. Sci. USA **76:** 1236.
15. POO, M. M. & R. CONE. 1974. Nature **247:** 438.
16. SCHINDLER, M., D. E. KOPPEL & M. P. SHEETZ. 1980. Proc. Natl. Acad. Sci. USA **77:** 1457.
17. BARAK, L. S., R. R. YOCUM, E. A. NOTHNAGEL & W. W. WEBB. 1980. Proc. Natl. Acad. Sci. USA **77:** 980.
18. DRAGSTEN, P. R., P. HENKART, R. BLUMENTHAL, J. WEINSTEIN & J. SCHLESSINGER. 1979. Proc. Natl. Acad. Sci. USA **76:** 5163.
19. WOLF, D. E., P. HENKART & W. W. WEBB. 1980. Biochemistry **19:** 3893.
20. KRIEGER, M., L. C. SMITH, R. G. W. ANDERSON, J. L. GOLDSTEIN, Y. J. KAO, H. J. POWNALL, A. M. GOTTO, JR. & M. S. BROWN. 1979. J. Supramol. Struc. **10:** 467.
21. BROWN, M. S. & J. L. GOLDSTEIN. 1976. Cell **9:** 663.
22. MAGDE, D., E. L. ELSON & W. W. WEBB. 1972. Phys. Rev. Lett. **29:** 705.
23. ELSON, E. L. & D. MAGDE. 1974. Biopolymers **13:** 1.
24. MAGDE, D., E. L. ELSON & W. W. WEBB. 1974. Biopolymers **13:** 29.

BIOLUMINESCENCE OF BACTERIAL LUCIFERASE*

J. Woodland Hastings
The Biological Laboratories
Harvard University
Cambridge, Massachusetts 02138

Shiao-Chun Tu
Department of Biophysical Sciences
University of Houston
Houston, Texas 77004

INTRODUCTION

All enzymes that catalyze reactions in which a singlet excited intermediate is formed in high quantum yield, thus emitting (visible) light, are termed luciferases. Biochemically, they may be classed together as oxygenases, since they utilize molecular oxygen and incorporate it into products. Chemically, the luciferases may be very different in different species of organisms, ranging in molecular weight (Mr) from as low as about 20,000 up to 400,000 or more (TABLE 1).

Luciferin, like luciferase, is also a generic term which, irrespective of the exact chemical structure of the molecule, is used to refer to the substrate that is oxidized in the light-emitting reaction. Thus, in systems from different organisms, the luciferase, luciferin and the reaction scheme may differ; the cellular disposition of the luminescence and the control of the reaction (e.g., flashing or continuous) may differ; and the color of the light may differ (TABLE 1). A review of the comparative biology and biochemistry of light emission in different luminous species[1–4] is outside the scope of this presentation, which will focus on luciferase of luminous bacteria.

BACTERIAL BIOLUMINESCENCE

Bacteria which are capable of emitting light occur ubiquitously in the oceans; they also occur in terrestrial habitats.[5] Such bacteria can be isolated by enrichment and plating, and selected by observation in the dark. The luminescent system is inducible in most strains examined;[6] the inducer is produced and excreted into the medium by the cells themselves. Its structure in one species has recently been reported.[7] In fully

* This work was supported by National Science Foundation Grant PCM 77-19917 (to J. W. H.) and Robert A. Welch Foundation Grant E-738 and National Institute of General Medical Science Grant GM 25953 (to S.-C. T.). J. W. H. was an awardee of the Alexander Von-Humboldt Foundation.

† The lysates of *B. harveyi* cells contain at least three forms of flavin reductatase.[16,17,18] In a recent experiment, purified NADH-flavin reductase was tested for its ability to complex with immobilized luciferase, but little binding was observed. Since a rather crude flavin reductase sample was used in the studies described above, the identity of the reductase species complexed with the immobilized luciferase is not known. It is also not certain whether the binding of reductase and immobilized luciferase involves other protein species.

0077-8923/81/0366-0315$01.75/0 © 1981, NYAS

TABLE 1
FEATURES OF BIOLUMINESCENT SYSTEMS OF DIFFERENT ORGANISMS

Organism	Luciferase Mr	Luciferin	Emission Maximum (nm)	Display
Bacteria	80,000	$FMNH_2$ + RCHO	495	continuous
Dinoflagellate	420,000	polypyrrole	465	flash
Coelenterate	21,000	coelenterazine	460	flash
Firefly	100,000	benzthiazole	560	flash

induced cells, the content of luciferase is very high—5% or more of the soluble protein.

The light-emitting reaction may be viewed as a branch of the electron transport pathway in which electrons from reduced substrates are shunted via pyridine nucleotides to oxygen at the level of flavin, catalyzed by a flavin reductase.

$$NAD(P)H + H^+ + FMN \rightarrow NAD(P)^+ + FMNH_2$$

The light-emitting reaction itself involves the concomitant oxidation of $FMNH_2$ and a long-chain aliphatic aldehyde (RCHO) (thus, there are two "luciferins") in a reaction that is catalyzed by luciferase.

$$FMNH_2 + RCHO + O_2 \rightarrow h\nu + FMN + RCOOH + H_2O$$

IMMOBILIZED LUCIFERASE

In the isolated system, the reduced flavin produced by the first (reductase) step is free and subject to rapid autoxidation.[8–10] But it has always seemed to us that in the living cell, the supply of reduced pyridine nucleotide would not be subject to an uncontrolled dissipation via a shunt leading to freely autoxidizable flavin and H_2O_2 production. One would imagine that this might be circumvented if *in vivo* there was some sort of molecular complex between the reductase and the luciferase so that control could be exerted over this pathway.[11] In fact, one report did suggest the existence of such a complex.[10] Some evidence has also been published to indicate that luciferase functions *in vivo* in association with membrane proteins.[12] If so, then the association between luciferase and reductase might be stronger if one of the two is immobilized. As a model for such a system, we covalently attached *Beneckea harveyi* luciferase to Sepharose 6B and examined its interaction with soluble flavin reductase.[13] The results indicated that both binding and electron transfer between the two proteins were enhanced with the immobilized luciferase.

The properties of immobilized luciferase are shown in TABLE 2. Its apparent specific activity was lower than that of the soluble enzyme by a factor of about 20, but its catalytic constants (k_{cat}), as judged by the decay rate in the nonturnover assay, were not greatly altered. Thus, the decreased light-emitting activity cannot be attributed to decreased catalytic efficiency. The lower apparent activity might have been due to inactivation during the immobilization; higher activities have been obtained more recently. The immobilized luciferase was different from the soluble enzyme with regard to its thermal stability; in 0.05 M phosphate, pH 7, its heat

TABLE 2
PROPERTIES OF SOLUBLE AND IMMOBILIZED LUCIFERASES

Parameter*	Soluble	Immobilized
Relative specific activity (%)	100	4†
k_{cat} (min^{-1})		
Octanal	2.4	2.3
Decanal	17.8	14.1
Dodecanal	2.4	2.2
Thermal inactivation		
Denaturation temperature‡ (°C)	44	48
E_a ($kcal \cdot mol^{-1}$)	64	28

* Examined in 0.05 M phosphate, pH 7, at 23°C for relative specific activity and k_{cat}, and over a range of 42–55°C for the thermal inactivation.

† The amount of bound luciferase was estimated as the difference of total quantity of luciferase used for the immobilization less that of soluble luciferase recovered in supernatant washes after the attachment reaction.

‡ Temperature at which half the activity was lost in 5 min.

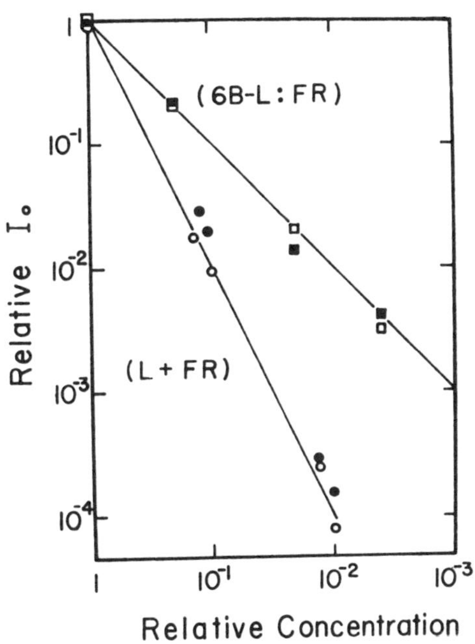

FIGURE 1. Effect of dilution on the initial velocity of the coupled reductase-luciferase reaction, comparing bound and soluble forms of the enzymes. The initial maximal intensity (I_o) of the coupled bioluminescence was determined at 23°C in 1 ml of 0.02 M phosphate (pH 7) containing 0.2% bovine serum albumin, 0.001% decanal, 50 μM FMN, 0 .25 mM (● and ■) or 25 μM (○ and □) NADH, and soluble luciferase plus flavin reductase (○ and ●) or the reductase-luciferase complex (□ and ■). The total amount of luciferase plus reductase was varied by dilution, so that the relative amounts of the two enzymes were constant. For samples of soluble luciferase plus reductase, an arbitrary unit of 1 is equivalent to 0.12 mg of luciferase plus 0.24 mg of flavin reductase per ml in total enzyme concentration and 4.5 × 10^{12} (●) or 3.4 × 10^{12} (○) q/sec in bioluminescence intensity. For samples of the reductase-luciferase complex an arbitary unit of 1 is equivalent to 3.1 × 10^{11} (■) or 1.2 × 10^{11} (□) q/sec in bioluminescence and 50 μg of immobilized luciferase plus the adsorbed flavin reductase per ml of assay solution in total enzyme concentration. (Reproduced with permission from S.-C. Tu and J. W. Hastings. 1980. Pro. Natl. Acad. Sci. USA **77:** 249).

inactivation required a higher temperature and exhibited a much lower energy of activation (Ea).

When partially purified flavin reductase was mixed for 5 min at 0°C in 0.05 M phosphate (pH 7) with the immobilized luciferase, about 75% of the reductase was bound, as determined by the reductase activity remaining in the five washes. This reductase-immobilized luciferase complex was active in the NADH-initiated coupled bioluminescence reaction. A key feature of this protein-protein complex is that $FMNH_2$, the product of the first enzyme, is evidently passed on to the luciferase with an enhanced efficiency. With mixtures of the two enzymes in the soluble state, the activity in the coupled assay decreased 1:100 when the mixture was diluted 1:10, whereas with the immobilized complex a similar dilution resulted in only a 1:10 decrease in activity (FIG. 1).

The enchanced efficiency of electron transfer between the reductase-immobilized luciferase pair is also evident from kinetic analysis of the coupled bioluminescence (FIG. 2). In the soluble system, slower kinetics of the coupled bioluminescence were observed with more diluted samples. As is shown, the kinetics of the immobilized system were independent of dilution and the time needed to reach the steady state was relatively short.

Prior studies involving the immobilization of multi-enzyme systems have been carried out by immobilizing the different enzymes on the same particle, thereby

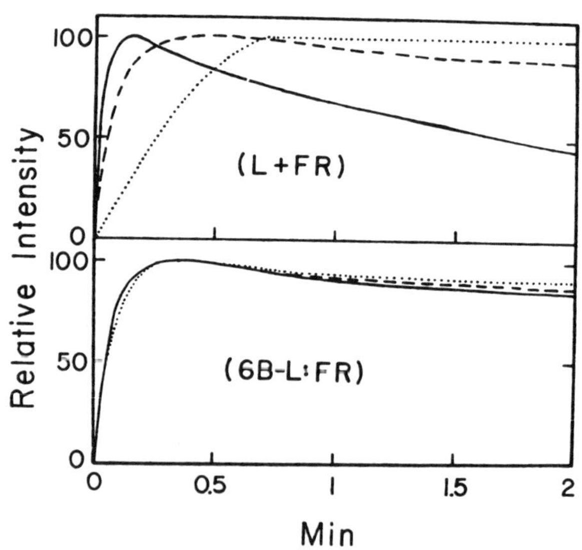

FIGURE 2. Effect of dilution on the kinetics of the coupled bioluminescence with soluble luciferase plus flavin reductase (*upper*) or the Sepharose 6B-luciferase-flavin reductase complex (*lower*). *Upper*: Bioluminescence kinetics of a sample containing 50μg of luciferase and 100 μg of of flavin reductase (——) are compared with that observed with samples diluted 1:5 (– – –) and 1:50 (·····). *Lower*: Similar kinetic comparisons are shown for the complex containing 50 μg of immobilized luciferase and the adsorbed flavin reductase (——), and the 1:5 (– – –) and 1:50 (·····) diluted samples. The coupled bioluminescence was initiated with 25 μM NADH, as described in FIGURE 1. The light intensities of different samples are normalized to the same value at the peaks. (Reproduced with permission from S.-C. Tu and J. W. Hastings. 1980. Proc. Natl. Acad. Sci. USA 77: 249.)

creating an "artificial" multistep enzyme system.[14,15] In such systems, there will be higher local concentrations of intermediates within the microenvironment of the enzymes, feeding the next enzyme in the sequence.[14] In our experiments, we presume that similar considerations apply, but possibly even more so; the product of the first enzyme may be fed directly to the second by virtue of binding.†

SUBUNIT EXCHANGE

Although there are significant differences, all bacterial luciferases studied possess a heterodimeric structure,[19] with Mr values in the range of 76,000 ± 4,000. The α (Mr 42,000) and β (Mr 37,000) subunits of *B. harveyi* luciferase have been isolated in quantity by DEAE Sephadex chromatography in 5 M urea.[20] Individual subunits are inactive, but can be recombined along with removal or dilution of the urea to yield reconstituted luciferase with activity and properties like those of the native enzyme. From chemical modification studies[21,22] and the analysis of mutant luciferases,[23,24] it is known that the catalytic properties are associated largely with the α subunit. The β subunit is also required for activity and appears to be involved in the binding of $FMNH_2$.[25,26]

There is evidently a high affinity between the subunits under nondenaturing conditions. This conclusion is based on the absence of detectable quantities of free subunits in chromatographic, ultracentrifugal, and subunit titration analyses,[27,28] and on the fact that the activity of luciferase is directly proportional to dilution over many orders of magnitude.[29,30]

It is nevertheless expected that some subunit exchange should occur under non-denaturing conditions. In order to demonstrate this,[31] we used enzymes from two dim mutants, altered-kinetics (AK)-6 and flavin binding (FB)-1. Both mutant luciferases have lower specific activities than that of the wild type enzyme; [24,25] the AK mutant has a defect in the α subunit of luciferase and a slower turnover rate than the wild type, whereas the FB mutant possesses a lesion in the β subunit and has a lower affinity than wild type luciferase for $FMNH_2$.

The activity of a mixture of purified AK and FB luciferases increased promptly and substantially upon incubation in 0.05 or 0.35 M phosphate buffer, pH 7, at 25°C (FIG. 3). At the higher phosphate concentration, the increase in activity was considerably slower, but the total activity ultimately obtained was about the same. High phosphate concentration has been previously shown to stabilize wild type luciferase against proteolysis[32] and heat denaturation (unpublished results); it may impede subunit exchange by stabilizing the dimeric form of mutant luciferases. At 0°C, the increase in activity was much slower, but the total activity obtained was about the same.

The kinetics of the *in vitro* luminescence reaction for each of the purified luciferases are shown in FIGURE 4. At 23°C, the half-times for decay of luminescence were about 35, 3, and 2 seconds for AK, FB, and wild type luciferases, respectively. The activity obtained after purified mutant luciferases were incubated together exhibited decay kinetics and a Mr corresponding to those of wild type luciferase. The phenomenon is attributed to subunit exchange between mutant luciferases and the consequent formation of wild type luciferase. It does not seem surprising that this would occur. The mutant enzymes appear less stable than the wild type,[25] and a subunit equilibrium would be strongly pulled towards the formation of wild type enzyme if, as expected, the association constant for its formation is much greater than for the mutant enzymes.

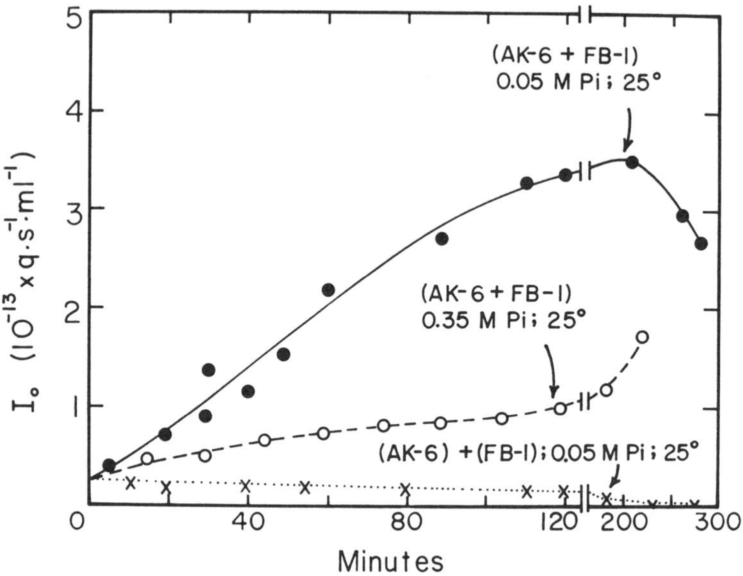

FIGURE 3. Increases in activity upon mixing low activity mutant luciferases. Equal volumes (about 4 mg in 0.5 ml) of AK-6 and FB-1 luciferases were combined, incubated in either 0.05 M (●) or 0.35 M (○) phosphate buffer, pH 7, with 1 mM dithiothreitol, at 25°C, and assayed at later times for luminescence activity. Samples of AK-6 and FB-1 in 0.05 M phosphate buffer were mixed individually with equal volumes of buffer and incubated separately at 25°C. The sum of the activities of these two control samples is also plotted (×) as a function of time. Each data point is an average of three measurements. (Reproduced with permission from C. Anderson, S.-C. Tu and J. W. Hastings. 1980. Biochem. Biophys. Res. Commun. **95:** 1180.)

To determine and compare the specific activities of the three luciferases, different amounts of luciferases were added to a constant amount of antiluciferase antiserum (FIG. 5). With small amounts, the added luciferase was completely inactivated by the antiserum. With greater amounts, the excess luciferase was active; by extrapolation, the amount of luciferase which was complexed and inactivated by the antibody was calculated. Assuming that the antigen-antibody reaction is the same irrespective of the exact type of luciferase (wild type, AK or FB) and its state of purity, then any differences in the total luciferase activity inactivated by a constant level of antibody can be attributed to differences in specific activities of these luciferases.

FIGURE 5A shows that 10^{12} q/sec activity units of wild type luciferase, both with crude and purified enzyme, were neutralized by 0.05 ml of antiserum. We can thus conclude that crude extracts of wild type cells did not contain immunologically cross-reacting but inactive luciferase, unless it failed to separate during purification. With FB luciferase a given amount of antiserum (0.05 ml) inactivated 4×10^{10} activity units in a crude preparation and 5.4×10^{10} units in purified material. The specific activities were 9.7 and 7.2×10^{12} q/sec/mg, respectively, about 20-fold less than for the wild type.

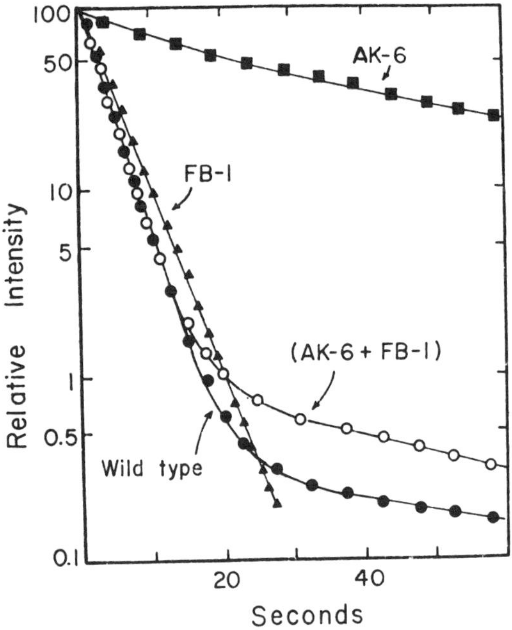

FIGURE 4. Kinetics of *in vitro* reactions at 23°C for luciferases purified from AK-6 (■), FB-1 (▲), and wild type (●) *B. harveyi*, as well as the activity formed as in FIGURE 3 in a mixture of FB-1 and AK-6 luciferases (○). The initial maximal intensities were normalized for plotting. No suitable interpretation for the second slower decay phase observed in some cases is available at the present time, but it constituted only about 1% of the reaction. (Reproduced with permission from C. Anderson, S.-C. Tu and J. W. Hastings. 1980. Biochem. Biophys. Res. Commun. **95:** 1180.)

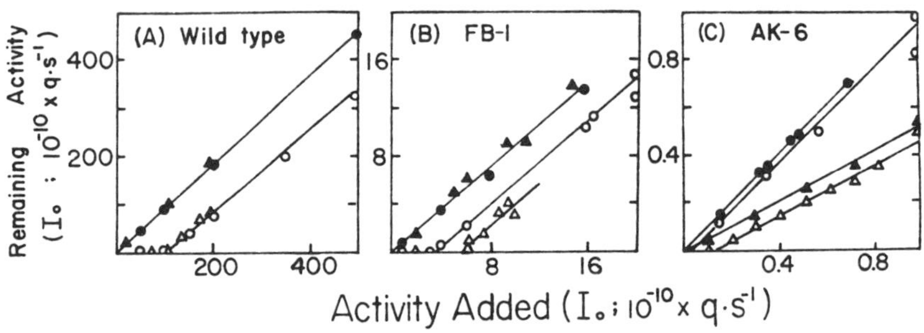

FIGURE 5. Titrations of antiluciferase antiserum with purified and crude luciferases. Different amounts of purified (○) and crude (△) luciferases from *B. harveyi* (A, wild type; B, FB-1; and C, AK-6) were admixed with 50 μl of antiserum in a final volume of 0.2 ml with 0.2 M phosphate, pH 7, 0.2% bovine serum albumin, 1 mM dithiothreitol. After 4 hours of incubation at room temperature, the activity remaining in each samples was determined and plotted against the activity added. For controls, various amounts of purified (●) and crude extract (▲) of each of the luciferases were added to the same buffer containing no antiserum, incubated and assayed as described above. Controls using normal rabbit serum gave similar results. *Ordinate*, remaining luciferase activity; *abscissa*, activity added. (Reproduced with permission from C. Anderson, S.-C. Tu and J. W. Hastings. 1980. Biochem. Biophys. Res. Commun. **95:** 1180.)

AK luciferases had even lower specific activities: in crude and purified preparations, 10 and 2×10^8 q/sec activity units, respectively, were inactivated, representing specific activities of 18 and 3.6×10^{10} q/sec/mg. It is interesting to note that although AK luciferase exhibited a much lower apparent specific activity in purified as compared to crude preparations (possibly due to inactivation during purification), the purified was more stable during the 4-hour incubation in the experiment of FIGURE 5C. A slope of 1 means that the amount of activity measured after 4 hours was equal to the amount initially added. The fact that the slope was less than 1 with AK crude enzyme, both with and without antibody, is attributable to loss of activity during the 4-hour incubation. The reason for this is not known; it may have been due to the presence of an endogenous protease which specifically attacks the α subunits.[33] Since AK is an α-subunit mutant, it may be more susceptible to this proteolysis than wild type or β mutant luciferases.

The use of mutant luciferases for the study of subunit exchange offers the possibility for determining quantitative parameters of subunit equilibria and how these are altered in various mutant luciferases. This would be an especially interesting subject for investigation with temperature-sensitive luciferases,[23,24] where one could study the possible relationship between the temperature of thermal inactivation and the equilibria for subunit dissociation.

REACTION MECHANISM: THE EMITTER

A key question concerns the reaction mechanism along with the formation and nature of the excited molecular species responsible for light emission. Since FMN occurs as a product of the reaction (FIG. 6), it had seemed at first that its excited

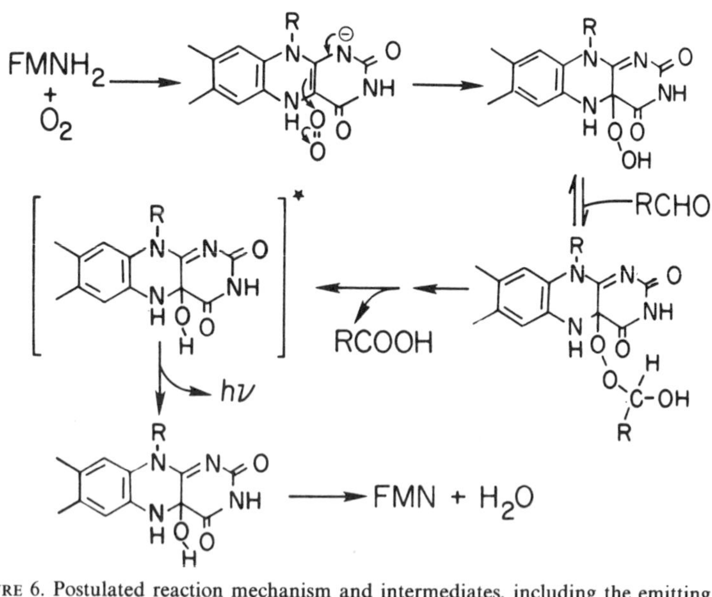

FIGURE 6. Postulated reaction mechanism and intermediates, including the emitting flavin chromophore in bacterial bioluminescence. (Reproduced with permission from J. W. Hastings and K. H. Nealson. 1977. Annu. Rev. Microbiol. **31:** 549.)

singlet might be designated as the emitting species. But the fact that its fluorescence emission is centered at about 525 nm while that of the *in vitro* bioluminescence is maximal at about 490 nm militates against this specific proposal.[9] In early work several possibilities were entertained.[34] One was that the emission derives from an enzyme-bound excited state of FMN with the blue shift attributed to the protein environment. It was also considered possible that the flavin was chemically altered in the course of the reaction to given a species with the appropriate emission properties. A third possibility considered was that the emission results from some (undetected) luciferase-bound group, even perhaps an aromatic amino acid residue of the luciferase itself.[35]

In order to evaluate these ideas, we tested a number of flavin isomers and analogues for activity with luciferase and found several to be active.[34] As shown in TABLE 3, bioluminescence emission with isoFMN and 2-thioFMN exhibited colors very different from that of the reaction with the FMN itself, effectively pinpointing the emitting species as a flavin. Such a result would be hard to explain by the third alternative; if some other molecule were serving as the emitter and the flavin functioned only as a reductant in the system, then altering the flavin should not change the emission spectrum of bioluminescence. Nevertheless, this kind of possibility continues to be seriously considered.[36]

It is well known that the color of bioluminescence *in vivo* differs in different species and strains of luminous bacteria.[37] In addition to the "blue shifted" emission of *Photobacterium phosphoreum*,[38] there is the recently discovered and quite spectacular "yellow" emitting strain of *P. fischeri.*[39] From recent studies, both of these now appear to be attributable to secondary emitters, involving prosthetic fluorescent groups attached to a protein other than luciferase.[40,41] In the case of *P. phosphoreum*, the emitter has been identified as a pteridine,[42] while in *P. fischeri* it is a flavin.[43] In the *P. fischeri* system, the purified luciferase emits in the 490-nm region in the absence of added fluorescent protein. Thus the luciferase-flavin intermediate can be designated as the primary emitter in the pure *in vitro* system.

It seems clear that the emitter is protein bound, both in the case of secondary emitters and in the luciferase-flavin intermediate. In the latter case, this was clearly shown by the fact that in cells possessing luciferase altered by mutation, the color of the light was red-shifted, and that the reaction catalyzed by luciferase purified from these cells was similarly color-shifted.[44]

TABLE 3
COLOR OF BIOLUMINESCENCE AND FLUORESCENCE EMISSION WITH DIFFERENT FLAVINS

Emission	Peak Wavelength (nm)		
	FMN	isoFMN	2-thioFMN
Bioluminescence with luciferase*	490	472	534
Fluorescence of flavin in H_2O*	525	543	nonfluorescent
Fluorescence of the flavin cation†	480–493	530	510–530
Fluorescence of the luciferase-peroxy-flavin intermediate‡	490	465–470	540–545

* From reference 34
† From reference 54
‡ Uncorrected values, Perkin-Elmer MPF 4

What then is the primary flavin emitter? The postulate that the emitter is an altered form of FMN, generated as a reaction intermediate, has been supported by many recent experimental observations. In the mechanism proposed for the luciferase reaction, a first step involves the reaction with molecular oxygen to form a hydroperoxide.[45,46] In the absence of the second substrate (long-chain aldehyde), this 4a-hydroperoxide[47] is quite stable; it may have a half-life of about 50 min at 0°C.[48] An interesting property of this intermediate is that its fluorescence emission spectrum closely matches that of the bioluminescence[49] (FIG. 7). It has recently been shown that in the presence of long-chain alcohols the intermediate is even more stable, exhibiting a half-life of about 300 min or more under similar conditions.[50] The alcohol-stabilized intermediate exhibited, at 0°C in neutral phosphate buffer, absorption and fluorescence properties closely resembling those observed with this intermediate at subzero temperatures in 50% ethylene glycol,[45,49] indicating that no artifacts in the spectral measurements were introduced by the latter cryoenzymological conditions. Furthermore, the fluorescence quantum yield of 0.17 for the alcohol-luciferase intermediate species[50] corresponds well to the *in vitro* bioluminescence quantum yield.[51] Recently we have observed that long chain alcohols also markedly stabilize the oxygenated intermediate formed with reduced flavin derivatives, e.g., riboflavin, 2-thioFMN, 2′,3′-diacetyl FMN, ω-carboxypentylflavin, and 3-carboxymethyl FMN.[52] The use of long-chain alcohols appears to be a valuable method for the isolation, and subsequent characterizations, of flavin-analog luciferase intermediates in aqueous media.

Although this 4a-hydroperoxyflavin intermediate exhibits fluorescence similar to that of the excited species in bioluminescence, it is not the emitter. However, the emitter could be a 4a-substituted flavin, possibly the 4a-hydroxy species (FIG. 6).

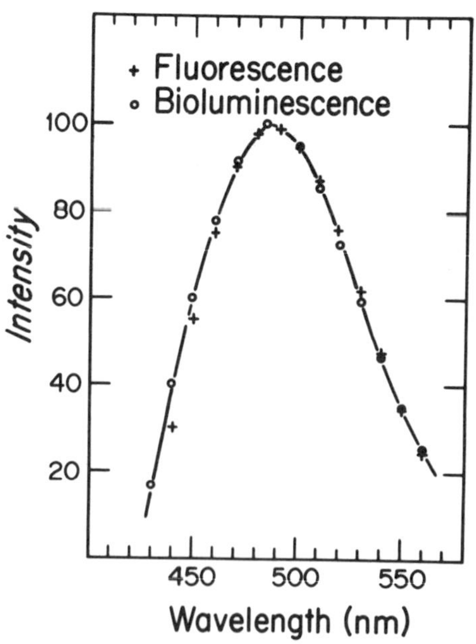

FIGURE 7. Emission spectrum of bioluminescence (O), measured directly from the cuvette during warming of the oxygenated luciferase-flavin intermediate in the presence of aldehyde, plotted together with the fluorescence emission spectrum of the luciferase intermediate in the absence of added aldehyde (×). Ordinate, intensity of bioluminescence and fluorescence, normalized at the peak. (Reproduced with permission from C. Balny and J. W. Hastings. 1975. Biochemistry **14:** 4719.)

In the experiments mentioned above,[34] it was shown that the color of the emission in the luciferase reaction differed depending upon the nature of the flavin used in the reaction, but that the spectra of the bioluminescence in these different cases were not at all correlated with the fluorescence of the oxidized forms of these flavins. We have now examined the fluorescence of the luciferase-4a-peroxy adducts of these flavins (TABLE 3); they exhibit fluorescence speaking at wavelengths corresponding to the bioluminescence.[53] We believe that this provides strong support for the postulate that the primary emitter in luminescence with pure luciferase is a 4a-substituted flavin formed as an intermediate in the reaction.

REFERENCES

1. HASTINGS, J. W. 1968. Bioluminescence. Annu. Rev. Biochem. **37:** 597–630.
2. HASTINGS,J. W. 1976. Bioluminescence. Oceanus **19**(2): 17–27.
3. HASTINGS, J. W. & T. WILSON. 1976. Bioluminescence and chemiluminescence. Photochem. Photobiol. **23:** 461–473.
4. HERRING, P. J., Ed. 1978. Bioluminescence in Action. Academic Press. New York.
5. NEALSON, K. H. & J. W. HASTINGS. 1979. Bacterial bioluminescence: Its control and ecological significance. Microbiol. Rev. **43:** 496–518.
6. NEALSON, K. H., T. PLATT & J. W. HASTINGS. 1970. Cellular control of the synthesis and activity of the bacterial luminescent systems. J. Bacteriol. **104:** 313–322.
7. EBERHARD, A., A. BURLINGAME, C. EBERHARD, G. L. KENYON, K. H. NEALSON & N. J. OPPENHEIMER. 1981. Structural identification of the autoinducer of *Photobacterium fischeri* luciferase. Structural identification of the autoinducer of *Photobacterium fischeri* luciferase. Biochemistry. In press.
8. GIBSON, Q. H. & J. W. HASTINGS. 1962. The oxidation of reduced flavin mononucleotide by molecular oxygen. Biochem. J. **83:** 368–377.
9. HASTINGS, J. W. & Q. H. GIBSON. 1963. Intermediates in the bioluminescent oxidation of reduced flavin mononucleotide. J. Biol. Chem. **238:** 2537–3554.
10. GIBSON, Q. H., J. W. HASTINGS, G. WEBER, W. DUANE & J. MASSA. 1966. The interaction of flavin mononucleotide with an enzyme system from *Photobacterium fischeri. In* Flavins and Flavoproteins. E. C. Slater, Ed. pp. 341–366. Elsevier. Amsterdam.
11. DUANE, W. & J. W. HASTINGS. 1975. Flavin mononucleotide reductase of luminous bacteia. Mol. Cell. Biochem. **6:** 53–64.
12. NE'EMAN, Z., S. ULITZUR, D. BRANTON & J. W. HASTINGS. 1977. Membrane polypeptides co-induced with the bacterial bioluminescent system. J. Biol Chem. **252:** 5150–5154.
13. TU, S.-C. & J. W. HASTINGS. 1980. Physical interaction and activity coupling between two enzymes induced by immobilization of one. Proc. Natl. Acad. Sci. USA **77:** 249–252.
14. MOSBACH, K. & B. MATTIASSON. 1976. Multistep enzyme systems. Methods Enzymol. **44:** 453–478.
15. JABLONSKI, E. & M. DELUCA. 1976. Immobilization of bacterial luciferase and FMN reductase on glass rods. Proc. Natl. Acad. Sci. USA **73:** 3848–3851.
16. GERLO, E. & J. CHARLIER. 1975. Identification of NADH-specific and NADPH-specific FMN reductases in *Beneckea harveyi.* Eur. J. Biochem. **57:** 461–467.
17. JABLONSKI, E. & M. DELUCA. 1977. Purification and properties of the NADH-and NADPH-specific FMN oxidoreductases from *Beneckea harveyi.* Biochemistry **16:** 2932–2936.
18. WATANABE, H. & J. W. HASTINGS. 1980. The use of the bacterial luminescent system for the quantitative determination of dehydrogenases. *In* Proceedings of the Second International Congress of Chemiluminescence and Bioluminescence. M. DeLuca & W. D. McElory, Eds. Academic Press, New York. In press.
19. FRIEDLAND, J. & J. W. HASTINGS. 1967. Nonidentical subunits of bacterial luciferase: Their isolation and recombination to form active enzyme. Proc. Natl. Acad. Sci. USA **58:** 2336–2342.
20. GUNSALUS-MIGUEL, A., E. A. MEIGHEN, M. Z. NICOLI, K. H. NEALSON & J. W. HASTINGS. 1972. Purification and properties of bacterial luciferase. J. Biol. Chem. **247:** 398–404.

21. Meighen, E. A., M. Z. Nicoli & J. W. Hastings. 1971. Hybridization of bacterial luciferase with a variant produced by chemical modification. Biochemistry **10:** 4062–4068.
22. Meighen, E. A., M. Z. Nicoli & J. W. Hastings. 1971. Functional differences of the nonidentical subunits of bacterial luciferase. Properties of hybrids of native and chemically modified bacterial luciferase. Biochemistry **10:** 4069–4073.
23. Cline, T. & J. W. Hastings. 1971. Temperature-sensitive mutants of bioluminescent bacteria. Proc. Natl. Acad. Sci. USA **68:** 500–504.
24. Cline, T. W. & J. W. Hastings. 1972. Mutationally altered bacterial luciferase. Implications for subunit functions. Biochemistry **11:** 3359–3370.
25. Cline, T. W. 1973. Mutational alteration of the bacterial bioluminescence system. Ph.D. Thesis, Harvard University. Cambridge, MA.
26. Meighen, E. A. & I. Bartlett. Complementation of subunits from different bacterial luciferases. Evidence for the role of the β subunit in the bioluminescent mechanism. 1980. J. Biol. Chem. **255:** 11181–11187.
27. Friedland, J. & J. W. Hastings. 1967. The reversibility of the denaturation of bacterial luciferase. Biochemistry **6:** 2893–2900.
28. Tu, S.-C., J. C. Makemson, J. E. Becvar & J. W. Hastings. 1977. Bacterial luciferase subunits are synthesized in equal quantities. J. Biol. Chem. **252:** 803–805.
29. Hastings, J. W., W. H. Riley & J. Massa. 1965. The purification, properties, and chemiluminescent quantum yield of bacterial luciferase. J. Biol. Chem. **240:** 1473–1481.
30. Hastings, J. W., Q. H. Gibson, J. Friedland & J. Spudich. 1966. Molecular mechanisms in bacterial bioluminescence: On energy storage intermediates and the role of aldehyde in the reaction. *In* Bioluminescence in Progress. F. H. Johnson & Y. Hanada, Eds. pp. 151–186. Princeton University Press. Princeton, NJ.
31. Anderson, C., S-C. Tu & J. W. Hastings. 1980. Subunit exchange between and specific activities of mutant bacterial luciferases. Biochem. Biophys. Res. Commun. **95:** 1180–1186.
32. Nicoli, M. Z., T. O. Baldwin, J. E. Becvar & J. W. Hastings. 1976. The interaction of oxidized flavin mononucleotide with bacterial luciferase. *In* Flavins and Flavoproteins. T. P. Singer, Ed. pp. 87–93. Elsevier. Amsterdam.
33. Baldwin, T. O., J. W. Hastings & P. L. Riley. 1978. Proteolytic inactivation of the luciferase from the luminous marine bacterium *Beneckea harveyi.* J. Biol. Chem. **253:** 5551–5554.
34. Mitchell, G. & J. W. Hastings. 1969. The effective of flavin isomers and analogues upon the color of bacterial bioluminescence. J. Biol. Chem. **244:** 2572–2576.
35. Cormier, M. J. & J. R. Totter. 1964. Bioluminescence. Annu. Rev. Biochem. **33:** 431–458.
36. Hart, R. C. & M. J. Cormier. 1979. Recent advances in the mechanisms of bio- and chemi-luminescent reactions. Photochem. Photobiol. **29:** 209–215.
37. Harvey, E. N. 1952. Bioluminescence. Academic Press. New York.
38. Spruit-Van Der Burg, A. 1950. Emission spectra of luminous bacteria. Biochim. Biophys. Acta **5:** 175–178.
39. Ruby, E. G. & K. H. Nealson. 1977. A luminous bacterium that emits yellow light. Science **196:** 432–434.
40. Gast, R., I. R. Neering & J. Lee. 1978. Separation of a blue fluorescence protein from bacterial luciferase. Biochem. Biophys. Res. Commun. **80:** 14–21.
41. Leisman, G. & K. H. Nealson. Unpublished results.
42. Koba, P. & J. Lee. 1979. Separation and structure of the prosthetic group of the blue fluorescence protein from the bioluminescent bacterium *Photobacterium phosphoreum.* Proc. Natl. Acad. Sci. USA **76:** 3068–3072.
43. Leisman, K. H. Nealson & J. W. Hastings. Unpublished results.
44. Cline, T. W. & J. W. Hastings. 1974. Mutated luciferases with altered bioluminescence emission spectra. J. Biol. Chem. **249:** 4668–4669.
45. Hastings, J. W., C. Balny, C. Lepeuch & P. Douzou. 1973. Spectral properties of an oxygenated luciferase-flavin intermediate isolated by low-temperature chromatography. Proc. Natl. Acad. Sci. USA **70:** 3468–3472.
46. Hastings, J. W. & K. H. Nealson. 1977. Bacterial bioluminescence. Annu. Rev. Microbiol. **31:** 549–595.

47. GHISLA, S., J. W. HASTINGS, V. FAVAUDON & J.-M. LHOSTE. 1978. Structure of the oxygen adduct intermediate in the bacterial luciferase reaction: ^{13}C nuclear magnetic resonance determination. Proc. Natl. Acad. Sci. USA **75:** 5860–5863.
48. BECVAR, J. E., S.-C. TU & J. W. HASTINGS. 1978. Activity and stability of the luciferase-flavin intermediate. Biochemistry **17:** 1807–1812.
49. BALNY, C. & J. W. HASTINGS. 1975. Fluorescence and bioluminescence of bacterial luciferase intermediates. Biochemistry **14:** 4719–4723.
50. TU, S.-C. 1979. Isolation and properties of bacterial luciferase-oxygenated flavin intermediate complexed with long-chain alcohols. Biochemistry **18:** 5940–5945.
51. BECVAR, J. E. & J. W. HASTINGS. 1975. Bacterial luciferase requires one reduced flavin for light emission. Proc. Natl. Acad. Sci. USA **72:** 3374–3376.
52. TU, S.-C. 1980. Isolation and reaction product characterization of an oxygenated bacterial luciferase intermediate formed with a flavin analog. *In* Proceedings of the Second International Congress of Chemiluminescence and Bioluminescence. M. DeLuca & W. D.McElroy, Eds. Academic Press. New York. In press.
53. HASTINGS, J. W., S. GHISLA, M. KURFURST & P. HEMERICH. 1980. Fluorescence properties of luciferase peroxyflavins prepared with isoFMN and 2-thioFMN. *In* Proceedings of the Second International Congress of Chemiluminescence and Bioluminescence. M. DeLuca & W. D. McElroy, Eds. Academic Press. New York. In press.
54. ELEY, M., J. LEE, J.-M. LHOSTE, C. Y. LEE, M. J. CORMIER, & P. HEMMERICH. 1970. Bacterial Bioluminescence. Comparisons of bioluminescence emission spectra, the fluorescence of luciferase reaction mixture, and the fluorescence of flavin cations. Biochemistry **9:** 2902–2908.

PART IV. LUMINESCENCE FROM SYNTHETIC POLYMERS IN BULK

SOLID-STATE POLYMERS AS STUDIED BY FLUORESCENCE POLARIZATION: MOBILITY AND ORIENTATION MEASUREMENTS

Lucien Monnerie and Jean-Pierre Jarry

Laboratoire Physico-Chimie Structurale et Macromoléculaire
Ecole Supérieure de Physique et de Chimie Industrielles de Paris
75231 Paris Cedex 05, France

Fluorescence depolarization is a means of probing molecular rotational processes occurring in the range of 10^{-7}—10^{-10} sec. A number of experiments have been made on polymer solutions, but very few with bulk polymers.

When dealing with oriented polymers in such a temperature range that no motions occur in the range of 10^{-7}—10^{-10} sec, the second and fourth moments of the segmental orientation distribution can be derived from fluorescence polarization (F.P.) measurements. In particular, Nishijima has applied this technique to many polymer films.[1]

For oriented polymers in the glass–rubber transition region or in the rubbery state, short time motions take part in the fluorescence polarization measurements, both with orientation effects. In such cases, the usual F.P. methods cannot lead to any information.

In the present paper, we first show that F.P. applied to isotropic polymers is a tool for studying glass–rubber relaxation phenomena. Secondly, for uniaxially oriented polymers undergoing fast motions, we present an extension of the theory of fluorescence polarization and original equipment suitable for determining the second moment of orientation and three mobility parameters.

Results obtained for polyisoprene networks are shown to demonstrate the potentialities of these two new applications of F.P.

MATERIALS

The commercially available polymer investigated is an anionic polyisoprene (Shell IR 307) of high molecular weight (Mn ≃ 500,000) with high *cis*-1,4 configuration (92%). It was carefully purified by extraction with ethanol as a solvent and mixed with small amounts of dicumyl peroxide as a cross-linking agent. Samples were molded and cross-linked at 147°C. Fluorescent label and probes were added as follows:

(a) The synthesis of labeled anionic polyisoprene chains is analogous to that reported previously for polystyrene.[2] Monofunctional "living" chains of molecular weight 300,000 are prepared and deactivated by 9,10-bis-(bromomethyl)anthracene.

0077-8923/81/0366-0328 $01.75/0 ©1981, NYAS

The resultant chains of molecular weight 600,000 contain dimethylanthracene (DMA) fluorescent groups in their middle, as shown in FIGURE 1. The advantage of this method of labeling is that the fluorescent transition moment, which is relevant to orientation and mobility measurements, lies along the chain axis. Consequently, fluorescence polarization will be insensitive to the rotation of the label around the 9, 10 axis of the anthracene moiety and reflect only the motions of the backbone. Besides, the second moment of orientation of the fluorescent transition moment is directly proportional to the second moment of the chain segments; the labeled polymer (1%) and purified IR 307 polyisoprene (99%) were mixed in solution before adding dicumyl peroxide. The final concentration of DMA in bulk samples was 10^{-5} M.

(b) Four probes of varying size were employed: 9-10-dimethylanthracene (DMA) and three trans-diphenyl-polyenes (DP). Their formulas and transition moments are shown in TABLE 1. Cross-linked IR 307 samples were swollen with solutions of the probes in benzene and then dried *in vacuo*. The final probe concentration in bulk samples was approximately 2×10^5 M.

GLASS–RUBBER RELAXATION

For an isotropic distribution of fluorescent molecules at sufficiently low concentration to avoid energy transfer, fluorescent polarization measurements under continuous excitation[3] yield the quantity:

$$\bar{M}(\tau) = \int_0^\infty M(t) \exp(-t/\tau)\, d(t/\tau) \tag{1}$$

where τ is the mean fluorescent lifetime ($\tau \simeq 10^{-8}$ *s*) and $M(t)$ is the orientational autocorrelation function

$$M(t) = \frac{3 \cos^2 \theta(t) - 1}{2} \tag{2}$$

where $\theta(t)$ is the angle through which the emission transition moment of the fluorescent molecule rotates between time 0 and time t. The brackets mean an average of all fluorescent molecules.

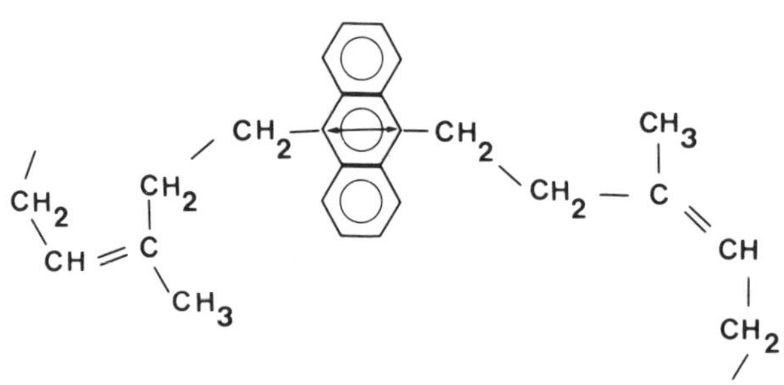

FIGURE 1. Position of the 9,10-dimethylanthracene (DMA) label in the middle of a chain skeleton. The double arrow indicates the direction of the transition moment.

TABLE 1

FORMULAS, APPROXIMATE LENGTHS, AND TRANSITION MOMENTS (DOUBLE ARROWS) OF THE PROBES*

DMA 6.5 Å	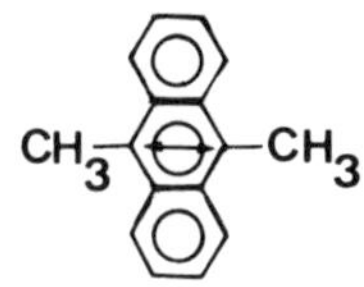
DPBD 11.5 Å	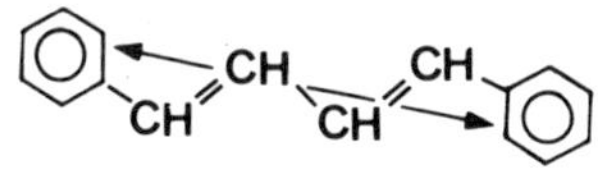
DPHT 14 Å	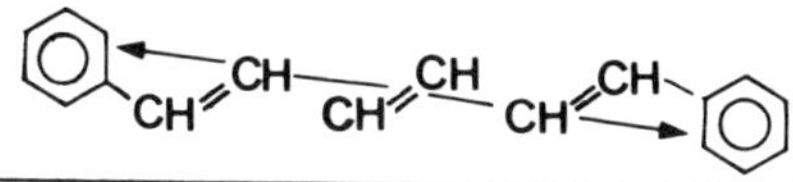
DPOT 16.5 Å	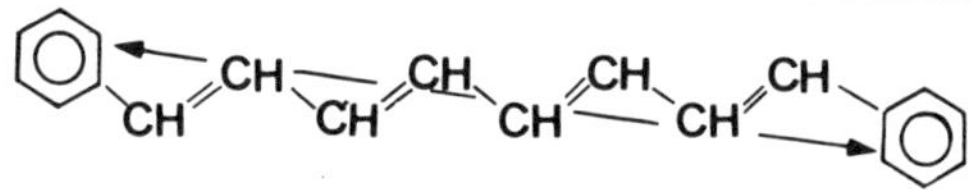

* DMA, 9,10-dimethylanthracene; DPBD, 1,4-diphenylbutadiene, DPHT 1,6-diphenylhexatriene; DPOT, 1,8-diphenyloctatetraene.

Experimentally $\bar{M}(\tau)$ is obtained from the mean fluorescent anisotropy

$$\bar{r}(T) = (I_{\parallel} - I_{\perp})/(I_{\parallel} + 2I_{\perp}) \quad (3)$$

measured at each temperature, by:

$$\bar{M}(\tau(T), T) = r(T)/r_0 \quad (4)$$

where r_0 is the limit anisotropy. Also we can get the mobility parameter m:

$$m = \bar{M}^{-1} - 1 \quad (5)$$

which increases from zero (molecular relaxation times much larger than the lifetime) to infinity (relaxation times much smaller than the lifetime).

More quantitative information can be obtained from m by using a model of Brownian rotational motion to derive the relaxation times of the molecular movements. Two models have proved convenient. The well-known Einstein-Stokes model for a spherical body yields a monoexponential autocorrelation function:

$$M(t) = \exp(-t/\tau_R) \quad (6)$$

where τ_R is a unique relaxation time which is simply obtained from:

$$\log \tau_R = \log \tau - \log m \quad (7)$$

A more specific model of chain dynamics[4] assumes segmental motions resulting in diffuse propagation of the segmental orientations along the chain backbones. Accordingly, the autocorrelation function is expressed as:

$$M(t) = \exp(t/\rho)\ \mathrm{erfc}(t/\rho)^{1/2} \quad (8)$$

where the characteristic time ρ is inversely proportional to the frequency of the motions and is given by:

$$\log \rho = \log \tau - 2 \log m \quad \textbf{(9)}$$

RESULTS

By using fluorescence polarization measurements, we have studied[5] the temperature dependence of the mobility parameter m and of the relaxation times of a bulk polyisoprene containing either fluorescent labeled chains or probes.

In the case of the probes given in TABLE 1 the plots of m *vs.* temperature are shown in FIGURE 2. For the DP probes, the onset of mobility occurs at about −10°C, that is, 50°C above the usual glass–rubber transition temperature Tg measured at 1 Hz. Such a shift agrees with what can be expected from the time-temperature superposition principle[6]: the higher the frequency of the experimental technique, the higher

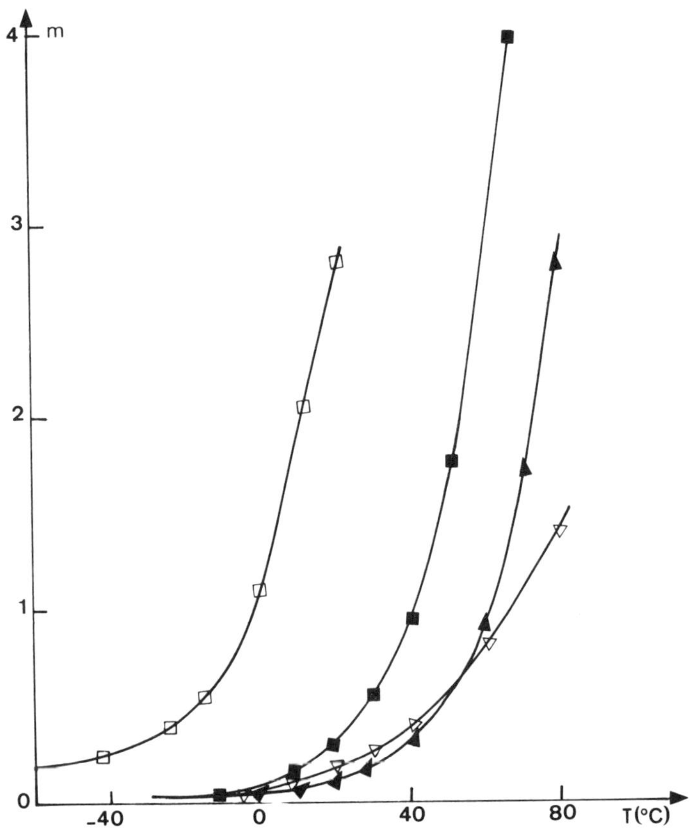

FIGURE 2. Mobility parameter m *versus* temperature for the four probes: DPBD (▽), DPHT (■), DPOT (▲), DMA (□).

the temperature at which the transition is observed. Typical frequencies for fluorescence polarization, related to $1/\tau$, are in the range of 10^8 to 10^9 Hz.

For the four probes, fluorescence anisotropy decay experiments at room temperature show that $M(t)$ is remarkably monoexponential and that the simple Einstein-Stokes model is valid. The relaxation times τ_R can be calculated from Equation 7 as a function of temperature. From the plots of log τ_R vs $1/T$, shown in FIGURE 3, it is possible to check more quantitatively if the molecular motions which are responsible for fluorescence depolarization belong to the glass–rubber relaxation of polyisoprene. Indeed, the glass–rubber relaxation theory, based on the free-volume concept,[6] predicts a temperature dependence of relaxation times which is given by the equation:

$$\log \tau_R \text{ (or } \log \rho) = -\frac{A(T - Tg)}{B + T - Tg} + \text{constant} \qquad \textbf{(10)}$$

where $Tg = -60°C$ is the glass-transition temperature of IR 307 polyisoprene as measured by differential thermal analysis. Mechanical data[7] in the low frequency range for the same polymer lead to the values $A = 16$, $B = 57$. The corresponding curve is plotted in dotted line in FIGURE 3. The slopes of log τ_R for the three DP

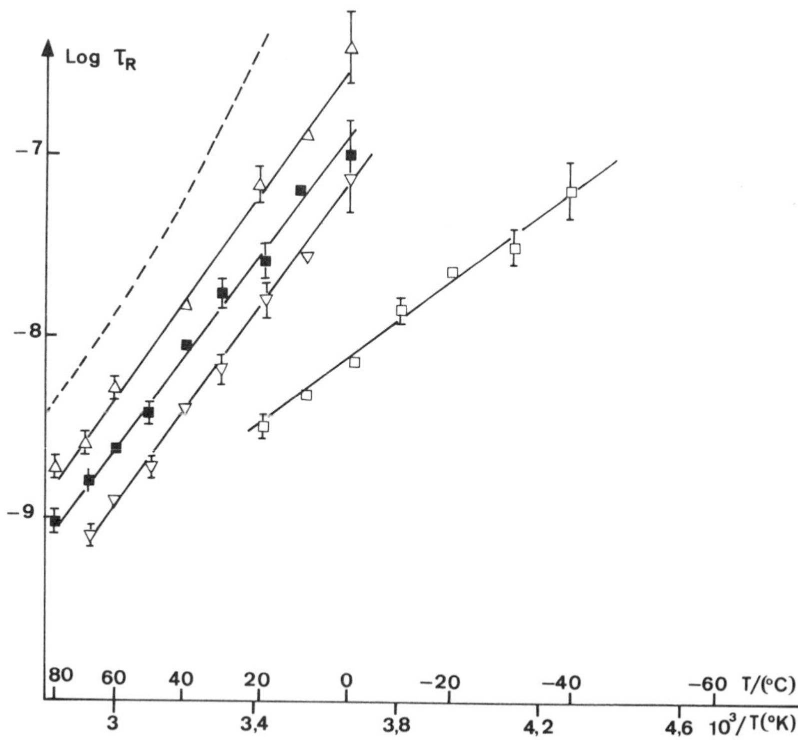

FIGURE 3. Logarithm plots of τ_R *versus* $1/T$ for the four probes in uncross-linked polymer: DPBD (▽), DPHT (■), DPOT (△), DMA (□). The dotted line is calculated from Equation 10 with a constant equal to 3.

probes are in good agreement with that derived from Equation 10. Such an agreement means that these large probes reflect molecular motions of the polymer which are involved in the glass–rubber relaxation.

From FIGURES 2 and 3, it is apparent that the DMA behavior is quite different from that of the DP probes. It seems that, because of its smaller size, the mobility of DMA is related to the secondary transition of polyisoprene occurring around 160°C at 1 Hz. A quite similar probe size effect has been reported in ESR studies[8] of the rotational mobility of nitroxide probes.

Concerning the DMA label behavior, the temperature dependence of the mobility parameter m is very similar to that of the DP probes. On the other hand, $M(t)$ is not a single exponential for the DMA label, in such a way that we derived the relaxation times by using Equation 9. The plot of log ρ versus $1/T$ is shown in FIGURE 4 with the dotted line corresponding to Equation 10. A rather good agreement in the slopes is obtained between +20 and +80°C. However, in the low mobility range lying between -10 and +20°C, the slope of log ρ is smaller than that predicted. This disagreement could be due to the occurrence of a secondary transition which could prevail over the primary glass transition.

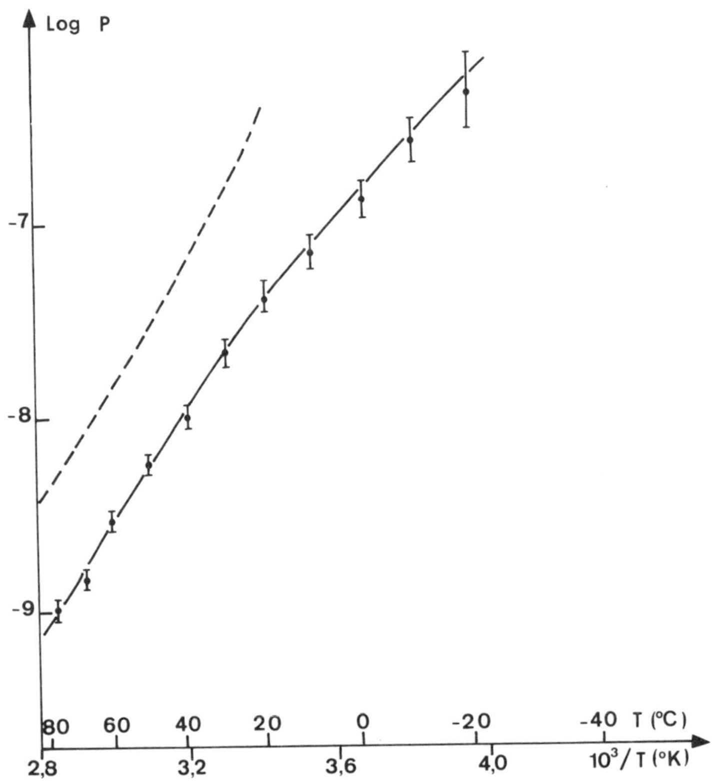

FIGURE 4. Log ρ *versus* temperature for DMA label in uncross-linked polymer. The dotted line is calculated from Equation 10 with a constant equal to 3.

Further information can be obtained by comparing label and probe mobilities. Indeed, since the auto correlation $M(t)$ is most sensitive to the smallest relaxation times, the characteristic time ρ of the DMA label principally describes the smallest segment motions. Besides, on the basis of the free-volume theory, the length a_s of the smallest segments involved in the glass–rubber relaxation should be comparable to that of a probe, the relaxation time of which equals ρ. By comparing with the DP probes, we obtained $a_s \simeq 14$ Å, which represents about five monomers in average conformation.[9]

Orientation and Molecular Dynamics in Uniaxial Polymers

When dealing with oriented polymeric samples (films, fibers, sheets) in a temperature range where the relaxation times of the molecular motions are much larger than the lifetime of the fluorescent label or probe, classical F.P. technique yields the second and fourth moments of the segmental orientation distribution[10]

$$< \cos^2 \alpha > = \frac{i(z,z) + 2\, i(x,z)}{(8/3)\, i(x,x) + 4\, i(x,z) + i(z,z)}$$
$$< \cos^4 \alpha > = \frac{i(z,z)}{(8/3) i(x,x) + 4\, i(x,z) + i(z,z)} \tag{11}$$

For oriented polymers in the glass–rubber transition region and in the rubbery state, molecular motions occur during the lifetime of the fluorescent molecules and can be influenced by stress, leading to depolarization effects which superimpose on the polarizing effects resulting from the orientation distribution.

Principle

We have extended the F.P. theory[11] to the case where both orientation and mobility effects are present, with the restricting assumption of uniaxial symmetry, but with no restriction as to the nature of molecular mobility.

We first assume that the transition moments in both absorption and emission coincide with a molecular axis M of the fluorophore and that the polarization of the exciting radiation and fluorescence is not affected by the birefringence of the anisotropic sample. The effects arising from delocalization of the transition moments and from birefringence will be dealt with subsequently.

Let us consider a uniaxial distribution of rotating molecular axes M whose direction is specified by the spherical polar angles $\Omega_0 = (\alpha_0, \beta_0)$ at time t_0 and $\Omega = (\alpha, \beta)$ at time $(t_0 + u)$ with respect to a reference frame x, y, z. Oz is the symmetry axis (FIGURE 5). The time scale t_0 is introduced to describe the macroscopic evolution of the sample, for example, during a rheological experiment, whereas u denotes the time scale of the microscopic reorientational motions occurring during the lifetime of the molecules in the excited state. In most cases, we can assume that the orientation distribution does not change appreciably during the lifetime and is independent of u.

The four mean values available from a fluorescence experiment under continuous illumination are:

$$G^0_{20} = \frac{1}{2} < 3\cos^2\alpha_0 - 1 >$$

$$G^0_{22} = \frac{1}{4}\int_0^\infty < (3\cos^2\alpha_0 - 1)(3\cos^2\alpha - 1) > e^{-u/\tau}\, du/\tau \tag{12}$$

$$G^1_{22} = \frac{9}{16}\int_0^\infty < \sin\alpha_0 \cos\alpha_0 \sin\alpha \cos\alpha \cos(\beta - \beta_0) > e^{-u/\tau}\, du/\tau$$

$$G^2_{22} = \frac{9}{64}\int_0^\infty < \sin^2\alpha_0 \sin^2\alpha \cos^2(\beta - \beta_0) > e^{-u/\tau}\, du/\tau$$

The quantity G^0_{20} represents the second moment of the orientation distribution at time t_0. When orientation and mobility are simultaneously present, the autocorrelation functions G^m_{22} depend on both. Nevertheless, the two effects can be distinguished and information on mobility can be gained. In particular, the mean mobility amplitude is given by:

$$\bar{M}(\tau) = \int_0^\infty M(u)\, e^{-u/\tau}\, du/\tau \tag{13}$$

with $M(u) = < (3\cos^2\theta(u) - 1)/2 > = G^0_{22} + \frac{16}{3} G^1_{22} + \frac{16}{3} G^2_{22}$

where $\theta(u)$ is the angle through which the M axis rotates between t_0 and $t_0 + u$ (see FIGURE 5). Three other quantities, which describe the anisotropy of the motions can be defined.[11]

In many dye molecules, it is well known that the transition moments in absorption and emission do not coincide with a molecular axis. In isotropic systems, this delocalization effect is usually described in terms of the limiting emission anisotropy r_0. When delocalization occurs in oriented systems, we have shown that it is possible to correct for this effect with an additional measurement of r_0 in the isotropic state.

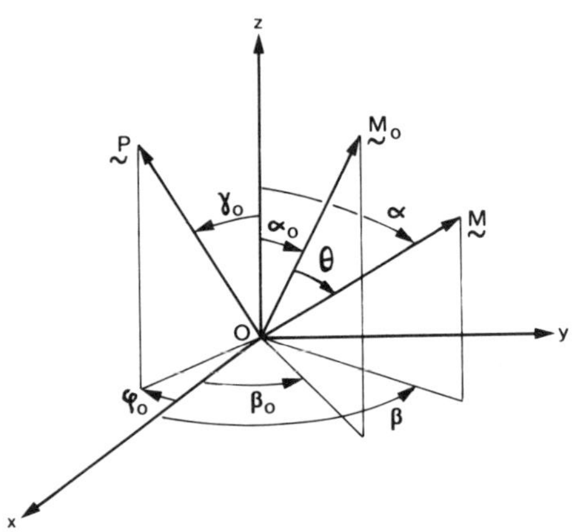

FIGURE 5. Illustration of the angles which define the position P of the polarizer and the initial and final positions of a moving molecular axis.

The birefringence effect occurs when the direction of the polarizer P and of the analyzer A does not coincide with a principal direction of the refractive index tensor. Only the G_{22}^{1} autocorrelation function is affected by birefringence and a correction factor b must be determined[11,12] from independent birefringence measurements performed on the polymer under study.

A fluorescence intensity $i(P,A)$ corresponding to given directions of the polarizer and the analyser (P and A) can be written in the form:

$$i(P, A) = K \sum C_{kl}^{m} \gamma_{kl}^{m} \tag{14}$$

where K is an instrumental constant. The coefficients C_{kl}^{m} depend on the spherical polar angles (γ_0, ϕ_0) and (γ, ϕ) of P and A, respectively (see FIGURE 5).

The quantities γ_{kl}^{m} are related to the G_{kl}^{m} by the equations:

$$\gamma_{20}^{0} = (5\ r_0/2)^{1/2}\ G_{20}^{0}$$
$$\gamma_{22}^{0} = (5\ r_0/2)\ G_{22}^{0}$$
$$\gamma_{22}^{1} = (5\ r_0/2) b\ G_{22}^{1}$$
$$\gamma_{22}^{2} = (5\ r_0/2)\ G_{22}^{2}$$

The γ_{kl}^{m} are derived from intensity measurements performed with different polarizer and analyzer directions, as indicated below.

APPARATUS

In order to get all the intensity measurements required, we have designed[12] specific equipment. The purpose of the apparatus is to measure simultaneously a convenient set of intensities $i(P, A)$ yielding at least five independent equations.[14] This requires the use of two orientations of P and three orientations of A that are not contained in the same plane.

The optical system is illustrated in FIGURE 6. In order to characterize simply the involved directions of polarizers and analyzers, we introduce the u, v and t' axes defined by the angles γ and ϕ such that $\gamma = \pi/4$ and $\phi = 0$ for u, $\gamma = \pi/4$ and $\phi = \pi$ for v, and $\gamma = \pi/2$ and $\phi = -\pi/4$ for t'.

The sample is excited by a double-beam device. The beams are modulated by choppers at different frequencies F_1 and F_2 and polarized by rotatable polarizers P_1 and P_2. P_1 can be oriented along z or u and P_2 along x or u.

The state of polarization of the fluorescence light is analyzed by two identical Wollaston prisms W_1 and W_2. The axes of the rotatable prism W_1 can be directed along x and z or u and v. The axes of W_2 are fixed along z and t'. The two signals from each photomultiplier are separated by active filters, each modulation frequency corresponding to one polarizer direction. The various measured intensities are related to the quantities γ_{kl}^{m} through expressions like:

$$i(z, x) = (K/9)\ (1 + \gamma_{20}^{0} - 2\ \gamma_{22}^{0})$$

$$i(z, u) = (K/9)\ (1 + \frac{5}{2}\ \gamma_{20}^{0} + \gamma_{22}^{0})$$

$$i(x, u) = (K/9)\ (1 - \frac{1}{2}\ \gamma_{20}^{0} - \frac{1}{2}\ \gamma_{22}^{0} + 4\ \gamma_{22}^{2})$$

$$i(u, u) = (K/9)\ (1 + \gamma_{20}^{0} + \frac{1}{4}\ \gamma_{22}^{0} + 2\ \gamma_{22}^{2} + 8\ \gamma_{22}^{1})$$

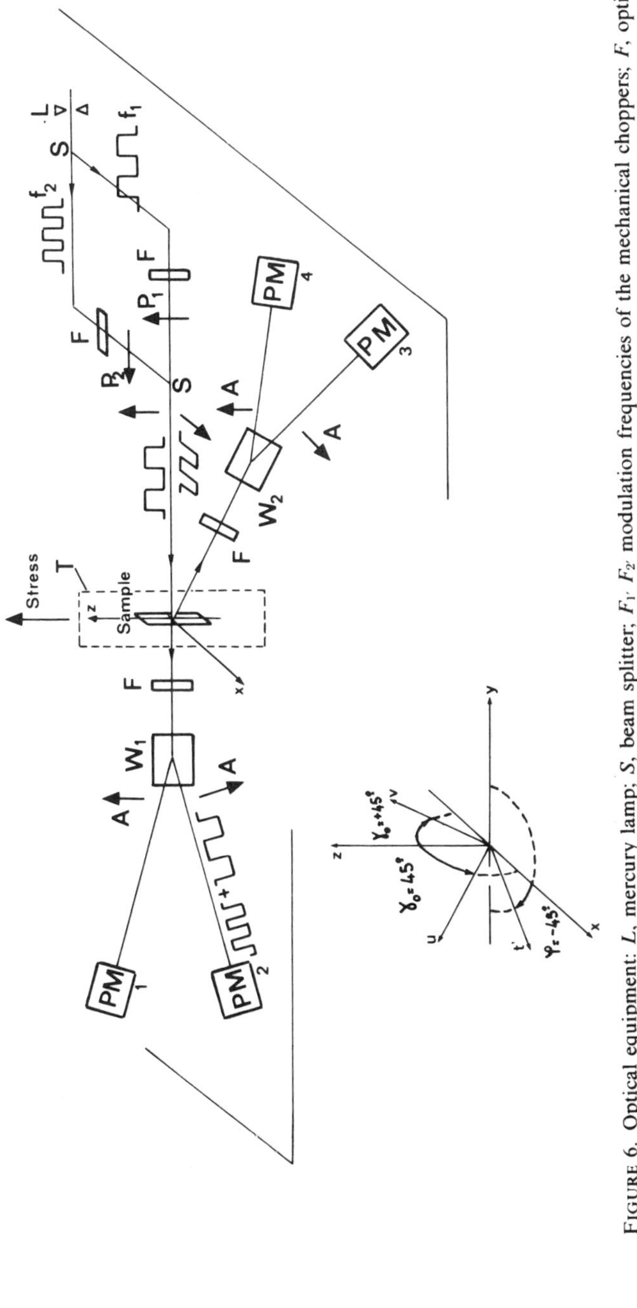

FIGURE 6. Optical equipment: L, mercury lamp; S, beam splitter; F_1, F_2, modulation frequencies of the mechanical choppers; F, optical filter; P_1, P_2, polarizers; W_1, W_2, Wollaston prisms; A, analyzing directions; PM, photomultiplier; T, temperature chamber.

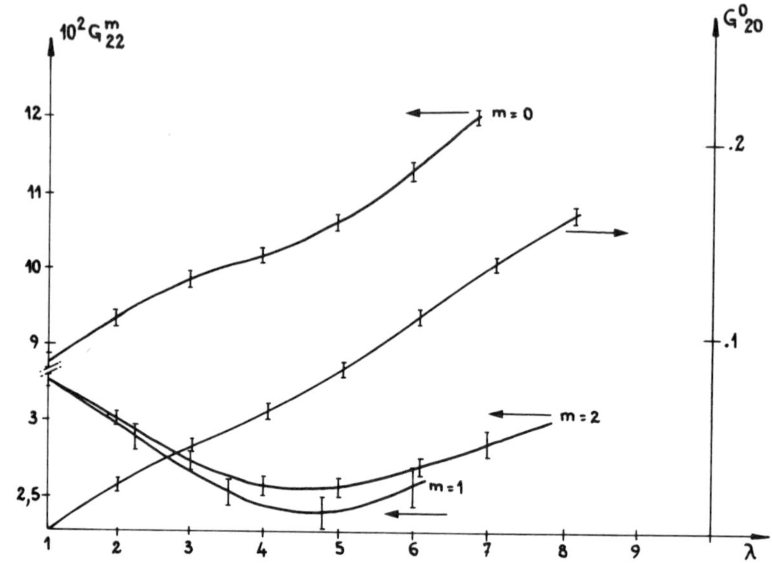

FIGURE 7. Strain dependence of the G_{kl}^m for a DMA-labeled polyisoprene network at 25°C. λ and M_c are defined in text.

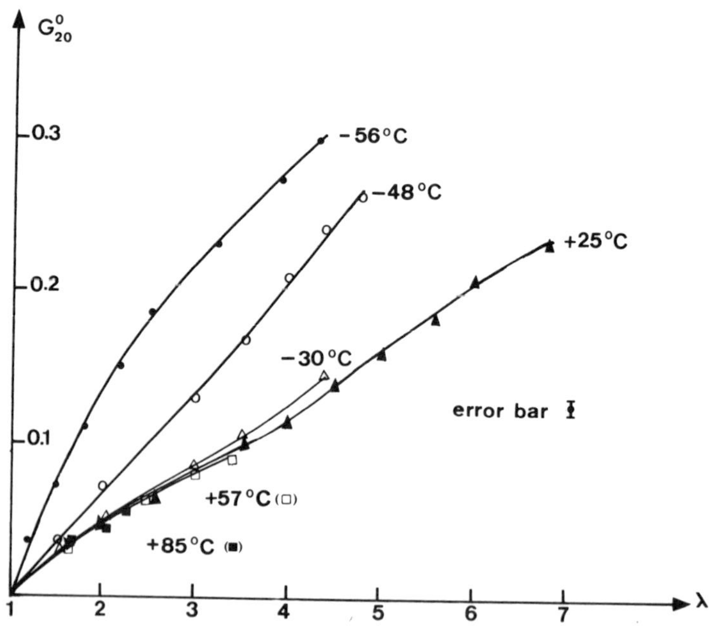

FIGURE 8. Temperature dependence of G_{20}^0 vs the extension ratio for a DMA-labeled polyisoprene network. λ and M_c are defined in text.

From this set of values and by using Equations 15 and 13, the second moment of the orientation distribution G^0_{20} and the mean mobility amplitude $\bar{M}(\tau)$ can be derived and three quantities describing the anisotropy of the motions can be characterized.

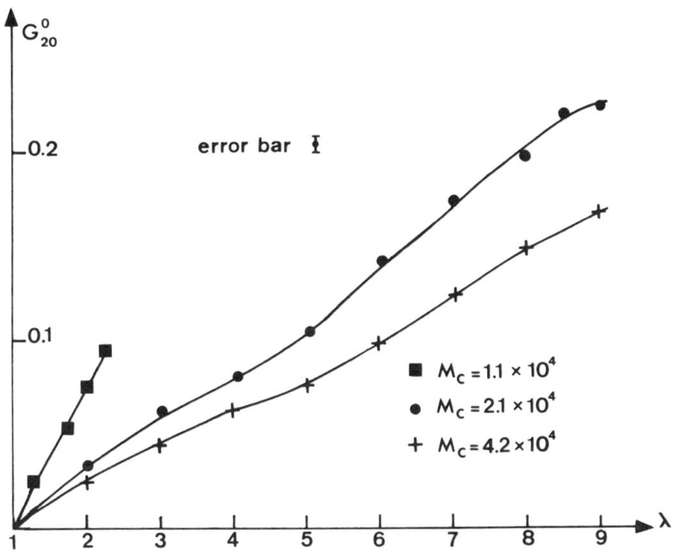

FIGURE 9. Effect of cross-linking density on the strain dependence of G^0_{20} for DMA-labeled polyisoprene networks at 25°C. λ and M_c are defined in text.

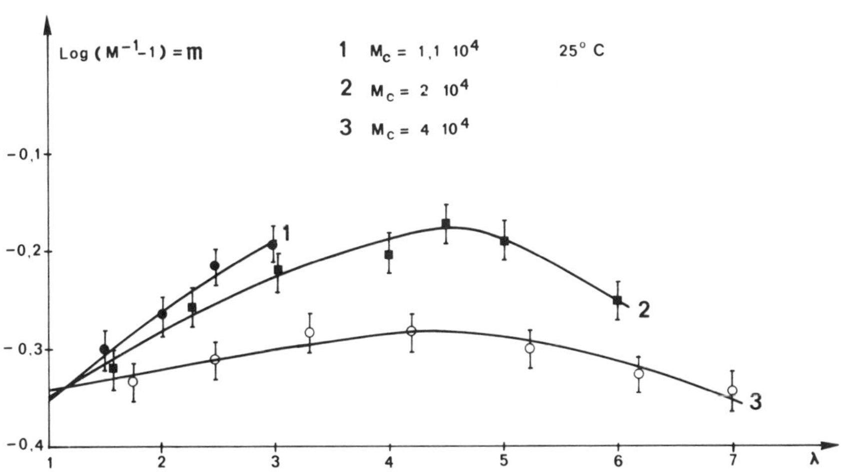

FIGURE 10. Mean mobility amplitude at 25°C for DMA-labeled polyisoprene networks *versus* the extension ratio. λ and M_c are defined in text.

Results

We have used this apparatus for studying[13] polyisoprene IR 307, during stretching at a constant elongation rate (100% min^{-1}). Labeled polyisoprene chains were incorporated in polyisoprene IR 307 as described in the material section. As an illustration, FIGURE 7 shows the strain dependence of the G_{kl}^{m} at 298 K, $\lambda = l/l_0$ is the extension ratio. Looking at the second moment of the orientation distribution G_{20}^{0}, the influence of temperature and cross-linking density are shown respectively in FIGURES 8 and 9 where M_c denotes the average molecular weight of chain strands between adjacent cross-links. As for the mean mobility amplitude $\bar{M}(\tau)$, its strain dependence is reported in FIGURE 10.

Conclusion

In the present paper, we have shown that fluorescence polarization measurements, performed on solid state polymers either in frozen state or in the glass–rubber transition region or in the rubbery state, can be used for studying molecular mobility and segmental orientation.

By using fluorescent labels or large probes, the glass–rubber relaxation can be investigated and useful information about the nature and size of the involved segmental motions have been obtained.

On the basis of the extended theory of fluorescence polarization presented above, the second moment of the orientation distribution and the mean amplitude mobility can be determined for stretched polymers in which orientation and molecular motions occur at the same time. The apparatus specifically designed for this purpose can be applied to quantitative investigation of both orientation and mobility in a time-dependent uniaxial distribution of moving molecules. Satisfactory results have been obtained with polyisoprene networks stretching and similar experiments on polystyrene in the rubbery state are now under study. The extended fluorescence polarization technique and the corresponding optical equipment can be used, not only for polymers, but also for other uniaxial media, the anisotropy of which is either spontaneous (liquid crystals) or induced by an external field.

References

1. NISHIJIMA, Y. 1970. J. Polym. Sci., **C31**: 353.
2. VALEUR, B., P. REMPP & L. MONNERIE. 1974. C. R. Acad. Sci. Ser. C. **179**:1009.
3. PERRIN, F. 1929. Ann. Phys. (Paris). **12**: 769.
4. VALEUR, B., J. P. JARRY, F. GENY & L. MONNERIE. 1975. J. Polym. Sci., Polym. Phys. Ed. **13**: 667.
5. JARRY, J. P. & L. MONNERIE. 1979. Macromolecules **12**: 927.
6. FERRY, J. D. 1970. Viscoelastic Properties of Polymers. 2nd edit. John Wiley & Sons. New York.
7. SHEN, S. P. & J. D. FERRY. 1968. Macromolecules **1**: 270.
8. TÖRMÄLÄ, P. 1979. J. Macromol. Sci.—Rev. Macromol. Chem. **C17**: 297–357.
9. ABE, Y. & P. J. FLORY. 1972. J. Rubber Chem. Technol. **45**: 982.
10. KIMURA, I. & C. R. DESPER. 1967. J. Appl. Phys. **38**: 4225.
11. JARRY, J. P. & L. MONNERIE. 1978. J. Polym. Sci., Polym. Phys. Ed. **16**: 443.
12. JARRY, J. P., P. SERGOT, C. PAMBRUN & L. MONNERIE. 1978. J. Phys. E. **11**: 702.
13. JARRY, J. P. & L. MONNERIE. 1980. J. Polym. Sci., Polym. Phys. Ed. In press.

PHOTO-PHYSICAL PROCESSES IN ORGANIZED SYSTEMS

C. David and D. Baeyens-Volant
University of Brussels
Brussels, Belgium.

Introduction

The effect of organization on energy migration and transfer has not been systematically investigated up to now except for crystals. Order is, however, expected to modify the interaction parameters between neighboring excited and nonexcited chromophores.

Different types of organization are considered in the present work:

1. Organization at the molecular level: tacticity in polymers.
2. Long and short range order in polymers: crystallinity and orientation.
3. Long and short range order in thermotropic and lyotropic liquid crystals.

The present study is concerned with:

1. The analysis of the fluorescence of polystyrene of different tacticity, crystallinity and orientation;

2. The study of the efficiency of transfer to dicumyl peroxide (CP) of polyoxyethyleneglycol (9) nonylphenol CH_3—$(CH_2)_8$- 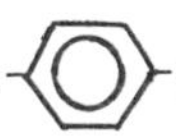 -$(OCH_2$—$CH_2)_9$—OH(NPh-EO-9) in bulk phase, organized in micelles, or in rods or lamellae; and

3. The study of the efficiency of transfer to 1,8-diphenyl-l, 3,5,7-octatetraene (DPOT) or perylene of 4-cyano-4′-octyloxybiphenyl (8COB) which exists according to the temperature in the crystalline, smectic, nematic or isotropic phase.

Owing to the exploratory nature of the present study, the results will be interpreted according to very simple kinetic models for the purpose of comparison.

Experimental

Polymer Samples

Atactic polystyrene was a commercial sample purified by successive dissolutions in benzene and precipitations in methanol. Films 20–30 μm thick were cast under nitrogen on Pyrex from CH_2Cl_2 solution and dried under vacuum for 48 hr at 110°C. Identical fluorescence spectra were obtained with various samples prepared in the laboratory by free radical polymerization.

Oriented polystyrene was obtained from 200 μm thick films of the commercial sample described above and dried under the same conditions. Strips 5 cm long were then stretched in an Instron tensile machine at 120°C. The rate of stretching was 20–50 cm/min. The ratio of final to initial length of the sample was 4.75. The elongated sample was cooled under tension, as rapidly as possible with a stream of dry nitrogen.

0077-8923/81/0366-0341 $01.75/0 © 1981, NYAS

Orientation of the sample was revealed by X-ray diffraction. Quantitative evaluation of the degree of orientation was possible by birefringence measurement.

According to Stein,[1] Δn, the birefringence of the sample is given by:

$$\Delta n = (0.194 - 0.51\,\overline{\cos^2\omega})F$$

$$F = \frac{3\,\overline{\cos^2\theta} - 1}{2}$$

where F is the orientation factor; θ is the angle between the axis of the segment and the direction of stretching; ω is the angle between the axis of the segment and the normal to the plane of the benzene ring. When all the segments are oriented $\theta = 0$ and $F = 1$. The relation between Δn and the orientation factor measured by ir dichroism was experimentally determined by Milagin *et al.*[2] It corresponds to $\Delta n = 0.16$ F. This means that ω is approximately equal to 34° for an elongation ratio ranging from 1 to more than 5. This relation was used to calculate the orientation factor of the samples. The measured birefringence and the corresponding orientation factor were, respectively, −0.011 and 0.07 for the lowest elongation rate and −0.017 and 0.1 for the highest one.

Isotactic polystyrene was prepared with Ziegler-Natta catalyst and kindly supplied by Drs. Berghmans and Overbergh from the University of Leuven. Films 20–30 μm thick were cast from CH_2Cl_2 solution and dried under vacuum. These films, heated *in vacuo* (10^{-5} torr) at 180°C for 1-2 hr, gave crystalline samples. The crystallinity, measured by X-ray diffraction according to Challa *et al.*,[3] ranges from 20 to 35 per cent. Similar degrees of crystallinity were obtained by maintaining the amorphous sample in CH_2Cl_2 vapor at room temperature for 3 hr. The crystalline samples obtained by these two different methods gave similar emission spectra.

Lyotropic Liquid Crystals

Lyotropic liquid crystal (NPh-EO-9) was research grade Arkopal 90 kindly supplied by Hoechst Belgium. Its water content was less then 0.2%. It was shown to be free from fluorescent impurity and was used without further purification. Dicumyl peroxide was purified through recrystallization in ethanol and dissolved in NPh-EO-9 under vigorous stirring. Tri-distilled water was used to prepare the micellar and lyotropic systems. Organization was detected by observation of the birefringence between crossed polarizers.

Thermotropic Liquid Crystals

Thermotropic liquid crystal (8COB) purchased from B.D.H. was shown to be free from impurities by HPLC and by spectrofluophosphorimetry. DPOT and perylene were purissimum-grade products of Fluka and Aldrich, respectively. The phase transition temperatures were controlled by DSC and by microscopic examination in a micro-furnace Mettler FP52 between crossed polarizers.

Emission Spectra

The emission spectra of polymer films were recorded with a spectrofluorimeter of our own design consisting of: an Osram HBO 200W super-pressure mercury lamp, a Bausch & Lomb high-intensity monochromator equipped with a 200–400 nm grating (2700 groves/mm) for selecting the excitation wavelength, a grating monochromator coming from a Hitachi Perkin-Elmer model 139 UV-Vis spectrophotometer for analysis of the emission spectrum, a Hamamatsu R 106 photomultiplier from the same spectrophotometer, and a Perkin-Elmer 165 recorder.

Polymer films 20–30 μm thick (absorbing all the incident energy) were cast on Pyrex discs or fixed on a frame. They were inserted into a copper block, the bottom of which was dipped in liquid nitrogen. The same holder was contained in a quartz Dewar flask and a stream of pre-cooled dry gaseous nitrogen was blown on the surface of the film. The same sample holder was used at room temperature with no Dewar flask. The exciting wavelength was 253.7 nm. The ratio of normal to excimer fluorescence was independent of the exciting wavelength.

The fluorescence spectra of the lyotropic mesophases and micelles were measured by reflection with a Perkin-Elmer MPF2A spectrofluorimeter. The exciting wavelength was 265 nm. Decay times have been measured at the University of Leuven according to the method of Van Bockstaele *et al.*[4]

The thermotropic liquid crystals were studied by reflection with the spectrofluorimeter used for the polymer films. A thin layer of liquid crystal was spread between two quartz plates which were then inserted in a thermostat copper block. The exciting wavelength was 315 nm.

Results and Discussion

Polystyrene

The fluorescence of the polystyrene samples was studied at room temperature and −196°C.[5] At room temperature, only excimer emission with λ_{max} at 324 nm is observed. All the samples give qualitatively the same emission spectra. The fluorescence spectrum of polystyrene films is shown to depend on the organization at −196°C. At this temperature (Figures 1 and 2), the emission spectrum of polystyrene consists of excimer fluorescence with λ_{max} at 324 nm and another emission at shorter wavelength (± 280 nm) that can be assigned to normal fluorescence by comparison with the emission spectrum of polystyrene in 2-MTHF glass at −196°C. The emissions are poorly resolved. The ratio of the intensity of both emissions depend on the tacticity, crystallinity and orientation of the samples. Table 1 gives the ratio of these intensities measured at λ_{max} (324 and 280 nm). The overlapping of both emissions was taken into account. The measured intensities were corrected for the contribution of excimer fluorescence I_D at 280 nm (12 per cent of its maximum value at 324 nm) and for the contribution of normal fluorescence I_M at 324 nm (4 per cent of its maximum value at 280 nm). It appears that the contribution of normal fluorescence (I_M) is the most important in atactic amorphous unoriented samples. It decreases significantly in oriented samples. For isotactic films, the ratio I_M/I_D is smaller than for atactic samples and crystallization reduced this ratio further (Table 1).

These results show that the contribution of normal fluorescence at −196°C decreases in the series atactic, atactic oriented, isotactic amorphous, isotactic crystallized.

The results can be discussed assuming a very simple model for photo-physical processes in polymers. Energy is absorbed by one aromatic chromophore. It migrates

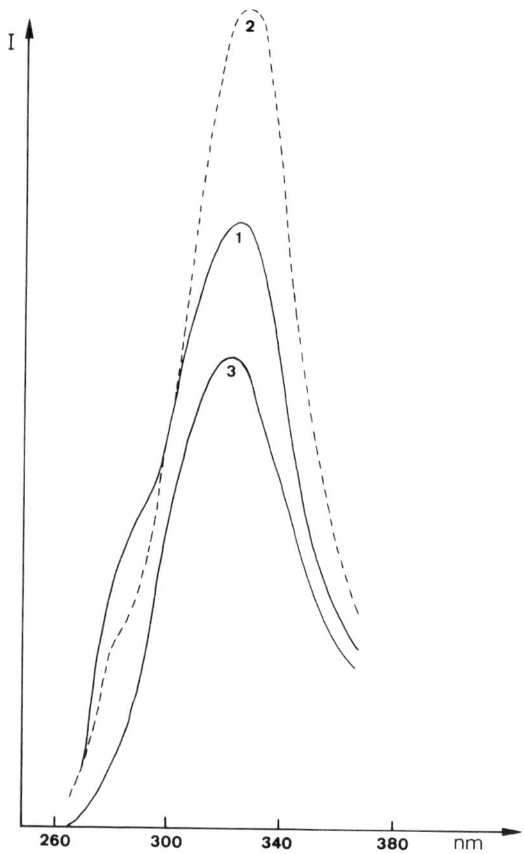

FIGURE 1. Fluorescence spectrum of polystyrene films at −196°C (I in arbitrary units). (1) Amorphous isotactic; (2) crystallized isotactic (25 per cent cryst.); (3) crystallized isotactic (35 per cent cryst.).

to excimer sites and is trapped there. Excimer deactivates by dissociation or by radiative and non-radiative processes. The ratio I_M/I_D is then given by:

$$\frac{I_M}{I_D} = \frac{k_{FM}}{k_{FD}}\left(\frac{k_D + k_{MD}}{k_{DM}(M - M)}\right) \quad (1)$$

where k_{FM} and k_{FD} are the rate constants for fluorescence emission, k_D is the sum of the rate constants of all excimer decay processes except for k_{MD} that is the rate constant for excimer dissociation, k_{DM} is the rate constant for excimer formation including excitation migration and (M−M) is the concentration of preformed excimer sites.

The rate constant k_{MD} is lower than k_D between room and liquid nitrogen temperatures.[6] According to this scheme, a decrease in I_M/I_D can be assigned to an increase of k_{DM} (migration efficiency) or an increase of the number of preformed excimer sites.

The change in the number of excimer sites as a function of orientation is difficult to predict in atactic polystyrene. The structure of atactic polystyrene has been studied by X-ray diffraction. The interchain Bragg distance ranges from 0.845 nm to 1.0

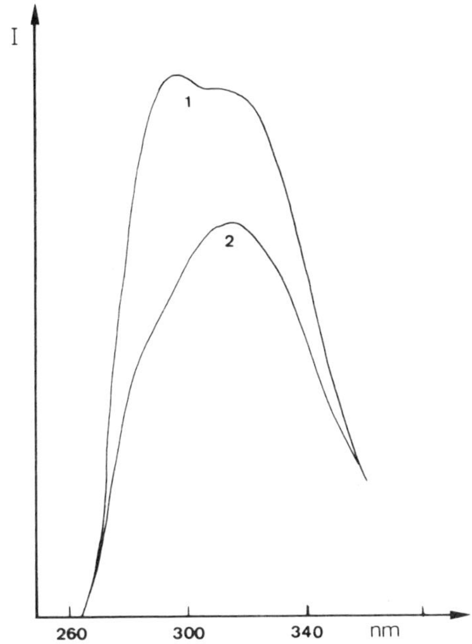

FIGURE 2. Fluorescence spectrum of polystyrene films at −196°C (I in arbitrary units). (1) Non-oriented atactic; (2) oriented atactic (F = 0.1).

TABLE 1
RATIO OF NORMAL TO EXCIMER FLUORESCENCE INTENSITY IN POLYSTYRENE AT −196°C

Polystyrene	I_M/I_D
Atactic	0.70
Oriented (F = 0.07)	0.60
Oriented (F = 0.1)	0.48
Isotactic amorphous	0.28
Isotactic crystalline (25%)	0.10
Isotactic crystalline (35%)	0.01

nm.[7] Excimers are thus formed intramolecularly. The intrachain distance, mainly associated with the benzene rings, has a mean value of 0.463 nm. In a detailed analysis of the diffraction patterns of amorphous polystyrene, Kilian and Boueke[8] have suggested that helical structures of the 11/3 type similar to those observed in crystalline poly-*m*-methylstyrene are formed over very short distances. Orientation reduces the intermolecular spacing (from 0.887 to 0.835 nm) for 300 per cent stretching according to Tobolsky *et al.*[9] and decreases the spread over the average distance but this does not give valuable information on the number of excimer sites. In isotactic polystyrene, the presence of isotactic pairs was shown to increase the probability of excimer formation in solution.[10] Crystalline isotactic polystyrene has been studied by Natta *et al.*[11] Threefold helices are formed. The distance between nearest parallel aromatic rings in the same chain is 0.650 nm and thus too large to allow excimer formation. Aromatic rings on neighboring helices in the elementary cell do not overlap and excimer cannot form intermolecularly. Excimer sites are thus probably trapped at crystalline or structural defects or in the amorphous phase. It seems thus unlikely that these should be more numerous in crystalline or oriented samples than in an amorphous polymer.

Migration could however be much more efficient in organized polymers. Indeed, in order to occur, excitation migration requires interaction between excited and unexcited neighboring molecules. This interaction energy can be represented by a sum of multipole interactions between the transition charge distributions of both molecules and by exchange interactions between the electronic clouds. It depends on the mutual orientation of the molecules and its law of decay is a function of the distance between chromophores which depends on the nature of the interactions.[12]

Lyotropic Mesophases

The physico-chemical parameters of the organized phases obtained at room temperature for mixtures of NPh-EO-9 and dicumyl peroxide are given in FIGURE 3.[13,14] The solubility of the peroxide was shown to be negligible in water in all the cases studied.

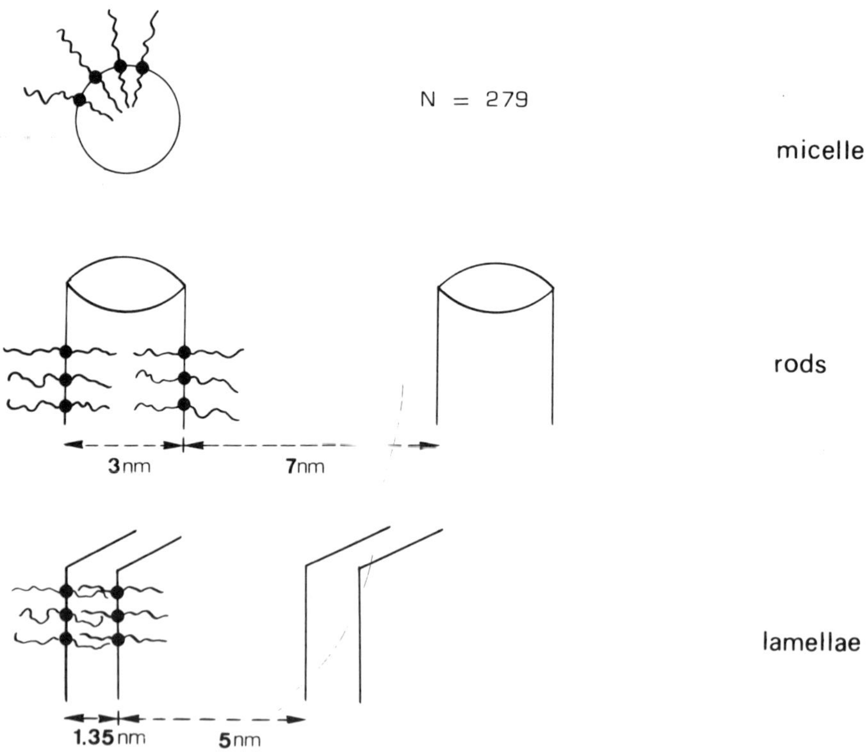

FIGURE 3. Physico-chemical parameters of water + NPh-EO-9 systems. Micelles: CMC = 6.5×10^{-5} mol. λ^{-1}-Aggregation number N = 279. Rods: from 45 to 59 weight percent NPh-EO-9. Lamellae: from 65 to 75 weight percent NPh-EO-9.

Two types of measurements were performed:

1. Fluorescence intensity of the donor as a function of acceptor concentration under steady-state illumination; and

2. Fluorescence decay of the donor as a function of acceptor concentration under flash excitation.

Energy transfer from the donor to the acceptor in solution occurs by a diffusion controlled collisional process which results in the decomposition of the peroxide.[6] The value of $(I_o/I) - 1$, where I_o and I are the intensity of the donor in the absence and presence of acceptor, respectively, is given in FIGURE 4 as a function of acceptor concentration in the surfactant phase. A linear relation with a slope related to the organization is obtained in all cases. The slope increases in the order: bulk liquid < micelles < rods < lamellae. The corresponding decay curves are exponential in all cases. These results can thus be interpreted according to simple Stern-Volmer kinetics. Indeed, owing to the size of the aggregates, the number of donors and acceptors in the present organized phase* is sufficiently high to allow statistical mixing of donors and acceptors during the lifetime of the donor. Adding the step:

$$M^* + Q \xrightarrow{k_Q} M + Q^*$$

to the usual sequence of photo-physical processes gives, if excimer formation and dissociation do not have to be taken into account:

$$\frac{I_o}{I} = 1 + k_Q\tau_M\,[Q] \qquad (2)$$

$$I_M(t) = I_{oM}\,\exp.-(k_M + k_Q\,[Q])\,t \qquad (3)$$

The obtained values for τ_M $(1/k_M)$ and k_Q are given in TABLE 2; k_Q is a complex parameter that includes diffusion, migration and quenching efficiency. Values of diffusion coefficients in lyotropic phases have, to our knowledge, never been reported. Furthermore, energy migration and material diffusion are three- dimensional processes in bulk phase whereas migration, and most probably diffusion occurs at the surface and is close to a two-dimensional process in micelles, rods and lamellae. In lamellae, the distance between opposite aromatic groups situated at the two sides of the lamella is smaller than in rods or in micelles, but energy transfer between these opposite groups is, however, less probable than between neighboring aromatic chromophores which are closer to each other.

In any case, order favors quenching efficiency in lyotropic mesophases, most probably by increasing the migration efficiency as discussed in the previous section.

Thermotropic Mesophases

Energy transfer has been studied from a donor (8COB) that exists in the crystalline, smectic, nematic, or isotropic phase according to the temperature.[15]

$$C \overset{54.5°C}{\rightarrow} S \overset{67°C}{\rightarrow} N \overset{80°C}{\rightarrow} I$$

* For example: A concentration of 5.10^{-2}M in dicumyl peroxide corresponds to a number of quenchers per micelle equal to an average value of 8, if we suppose an aggregation number N of ~279.[14]

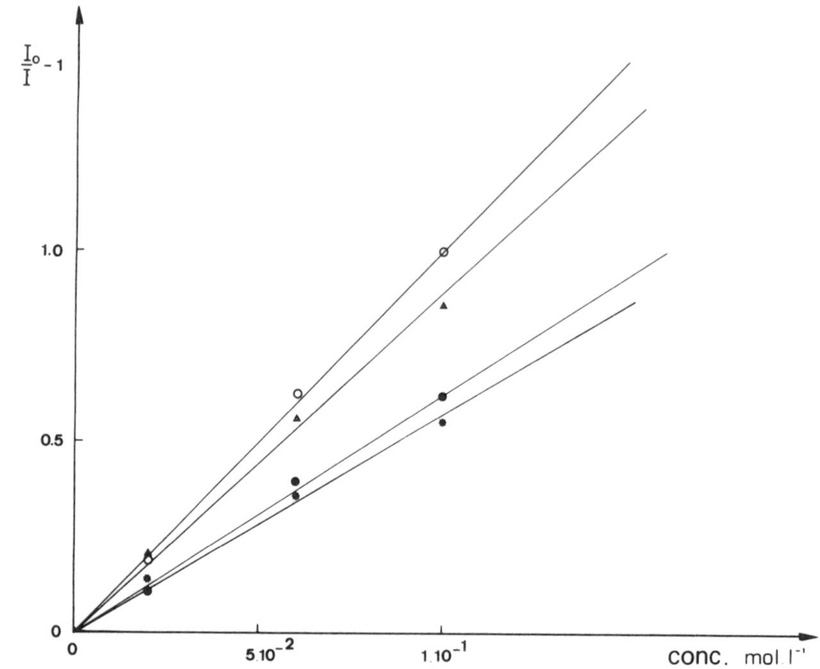

FIGURE 4. Water + NPh-EO-9 systems. ○: 70% NPh-EO-9 (lamellae); △: 50% NPh-EO-9 (rods); ●: 0.2% NPh-EO-9 (micelles); *: 100% NPh-EO-9 (bulk phase).

TABLE 2
RATE CONSTANT k_Q OF TRANSFER TO CP

System		10^{-9} k_Q (1 mol^{-1} sec^{-1}) a	b
NPh-EO-9	100%	1.0	
	0.2%micelles	1.1	1.0
	50%rods	1.5	
	70%lamellae	1.8	2.0
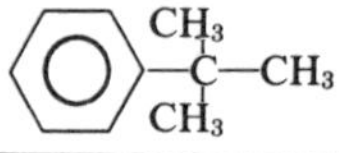	100%	10	

from
a: steady illumination (I_o/I) $\tau_M = 5.7 \times 10^{-9}$ sec^{-1}
b: decay curves

1,8-diphenyl-1,3,5,7-octatetraene and perylene have been used as acceptors. The fluorescence spectrum of 8COB in solution is shifted toward longer wavelength when compared with biphenyl.[16] The λ_{max} are respectively 340 and 315 nm in cyclohexane. Furthermore, the λ_{max} of 8COB is strongly dependent on the polarity of the solvent in fluid solution at RT (FIGURE 5) while it is independent of it in the glassy state at

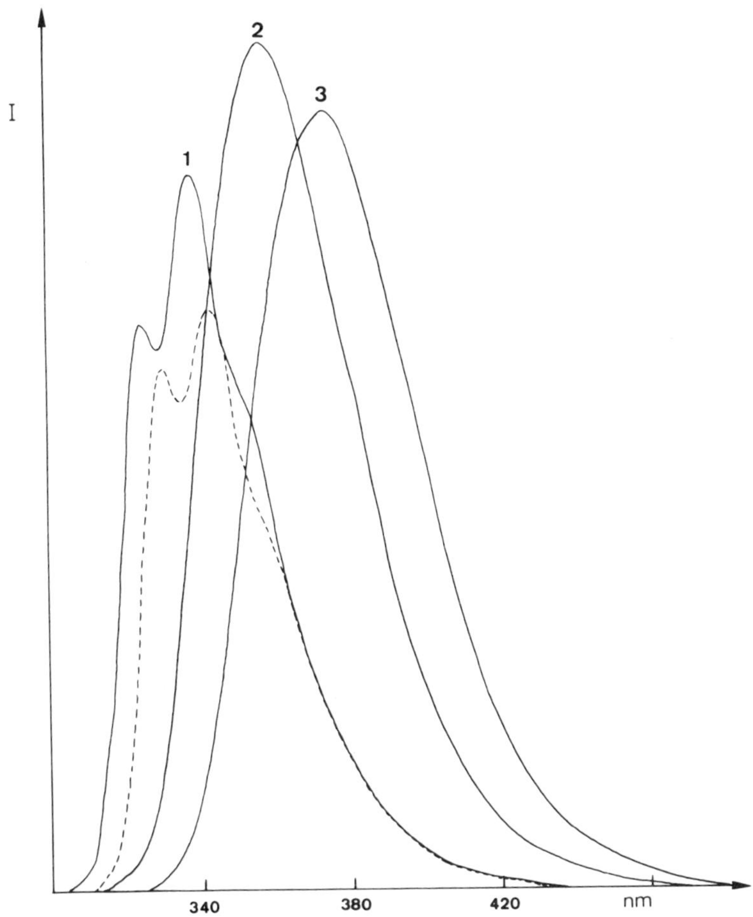

FIGURE 5. Solvent effect on the emission spectrum of 8COB at RT. (1) heptane; (2) ethylacetate; (3) dimethylformamide; and (– – – – –) ethanol at −196°C. (Intensity in arbitrary units.)

−196°C. The progression in the vibrational structure (1300 cm^{-1}) is also different from that of biphenyl (1000cm^{-1}). The lifetime is very short ($\tau \leq 10^{-9}$ sec) and the quantum yield is high (0.54 in cyclohexane). Therefore, the fluorescence of 8COB has been assigned to emission from the planar ^1L_a state to the non-planar ground state, the transition moment being parallel to the long axis of the molecule. The red shift of the emission in polar solvents has been assigned to an orientation relaxation process of the solvent cage in the electric field of the excited solute that has a higher dipole moment than the ground state.[16] The fluorescence of biphenyl, which is only weakly solvent dependent and is characterized by longer life time (15×10^{-9} sec in cyclohexane) and lower quantum yield (0.18), is generally assigned to a transition from the lowest ^1L_b state characterized by a weak dipole moment.

As the concentration increases (from 5×10^{-2} to 10^{-1} mol. l^{-1}), excimer fluorescence

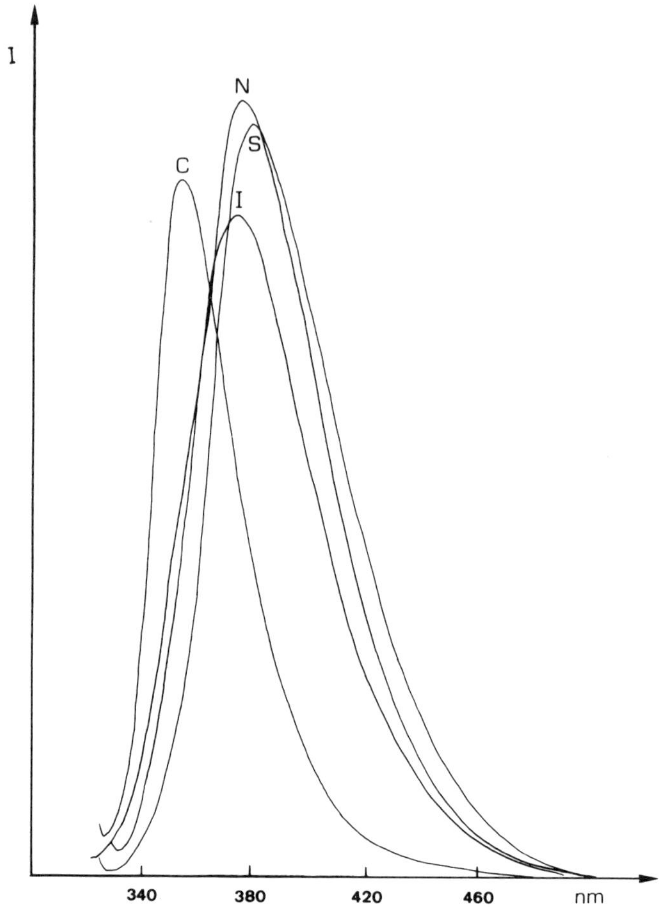

FIGURE 6. Fluorescence spectrum of 8COB in the crystalline (C), smectic (S), nematic (N), and isotropic (I) phases. (Intensity in arbitrary units.)

is observed from 8COB[16] while it has never been reported for biphenyl. The λ_{max} of the excimer (380–385 nm) is independent of solvent polarity.

The fluorescence of 8COB in crystalline, mesomorphic and isotropic liquid phases is given in FIGURE 6. The λ_{max} in the smectic, nematic, and isotropic phase corresponds to the λ_{max} of excimer fluorescence observed in concentrated solution.†

The fluorescence intensity of 8COB has been measured as a function of temperature between 20 and 90°C for different concentrations of acceptor. Discontinuities (less than 20%) corresponding to small differences in fluorescence intensity appear at the phase transitions S → N → I (FIGURE 7) but the ratio I_o/I of the fluorescence intensities in the absence and presence of acceptor is nearly identical in smectic,

†Recently obtained results indicate the formation of preformed dimers in mesomerphic 8COB (to be published).

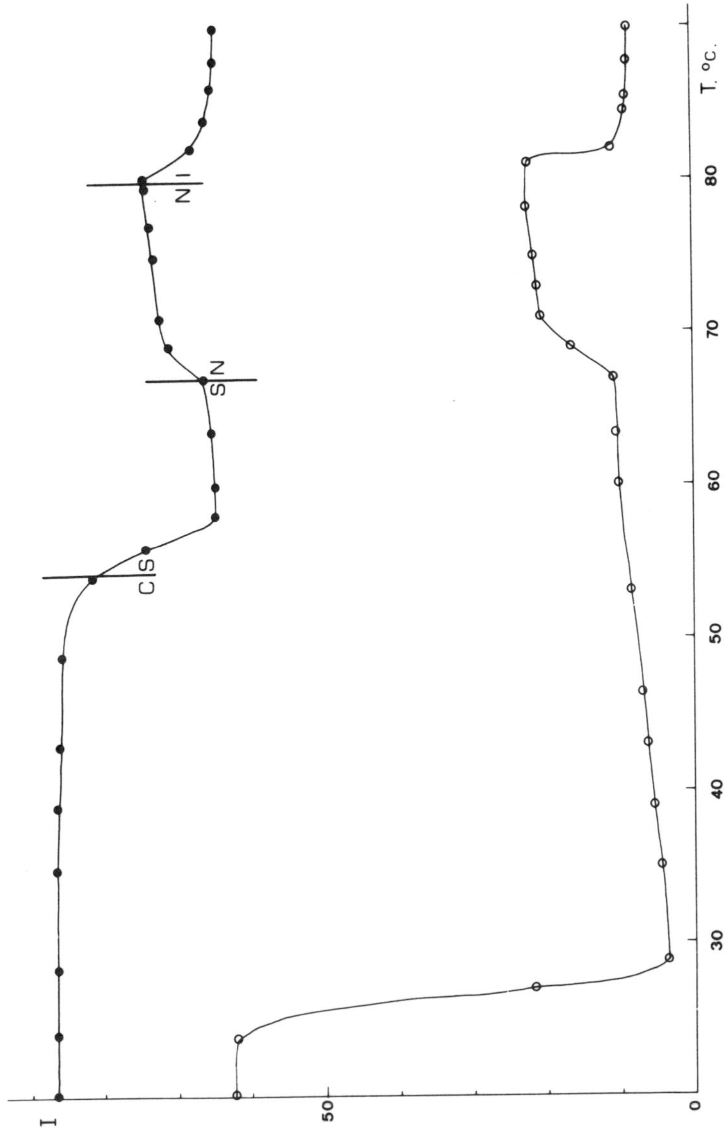

FIGURE 7. Fluorescence intensity of 8COB as function of temperature. (Intensity in arbitrary units.) ●: increasing temperature; ○: decreasing temperature. $C \xrightarrow{54.5°C} S \xrightarrow{67°C} N \xrightarrow{80°C} I$.

nematic, and isotropic phases at a given acceptor concentration. The results obey Stern-Volmer kinetics. Straight lines with identical slopes are obtained with DPOT as acceptor in the different phases. A somewhat smaller slope is obtained for 8COB when the acceptor is perylene (FIGURE 8). The values of k_Q calculated assuming a decay time of 10^{-9} sec for the donor are, respectively, 8.5×10^{11} and 2.8×10^{11} l mol^{-1} sec^{-1} for DPOT and perylene.

This system is thus characterized by:

1. Excimer formation in mesomorphic and isotropic liquid phase;
2. Stern-Volmer kinetics and very high value of k_Q in S, N and I phases; and
3. Identical values of I_o/I in these phases.

The structure of the smectic phase of 8COB has been studied by different authors.[15,17,18] According to recent results obtained by X-ray and neutron diffraction experiments,[15] the smectic phase is formed of pairs of cyanobiphenyl units arranged according to FIGURE 9. Excimer interaction between pairs of neighboring chromophores is thus highly probable. The possibility of rotation of the molecule about its long axis has been discussed but no definitive conclusion seems to have been obtained concerning the smectic phase. Since Stern-Volmer kinetics is obeyed, statistical mixing of donors and acceptors has to occur during the life-time of the excited donor. This statistical mixing can be due either to material diffusion of donors and acceptors or to energy migration. Diffusion-limited quenching is very improbable since the life time is very short and the coefficient for self-diffusion is small[19] in mesomorphic phases. Furthermore, k_Q is independent of the phase whereas the coefficient for self-diffusion increases from 10^{-7} to 10^{-5} cm^2 sec^{-1} on passing from smectic to isotropic phase.[20] The high value of k_Q can thus be assigned to efficient energy migration through the rather strong $^1L_a \rightarrow {}^1A$ transition. The efficiency of migration can be quantitatively characterized and compared with values obtained for crystals if the hopping model is used.[21]

Equation 2 can be written:

$$\frac{I_o}{I} = 1 + p\, k'_{mig}\, \tau \frac{[Q]}{[M]} \tag{4}$$

where p is the probability of an exciton being captured by Q at any encounter and k'_{mig} is the exciton jump frequency. Assuming $p = 1$, values of k'_{mig} of 12.3×10^{12} and 4.1×10^{12} sec^{-1} are respectively found for 8COB containing DPOT or perylene assuming that $\tau \sim 10^{-9}$ sec. A value of k'_{mig} of 6.7×10^{12} sec^{-1} is reported for anthracene crystals.[21]

It remains now to explain why migration is independent of the nature of the phase since it can reasonably be supposed that the interaction between neighboring chromophores which is responsible for migration will decrease with the degree of order. An interpretation of this can be found in recent experiments which are sensitive to long and short range order in the systems. X-ray diffraction[22] in the nematic and isotropic phases has shown that ordered structures of 100 to 150 molecules similar to those observed in the smectic phase still exist in the nematic and isotropic liquid phases even for temperatures that far exceed the N-I transition. The same conclusion was obtained by studies of dielectric relaxation in the nematic and isotropic phases.[23] By measuring the density as a function of temperature, Leadbetter *et al.*[24] showed that the changes in local molecular organization at the transitions are small while changes in long range order occur. Also, the order parameter determined

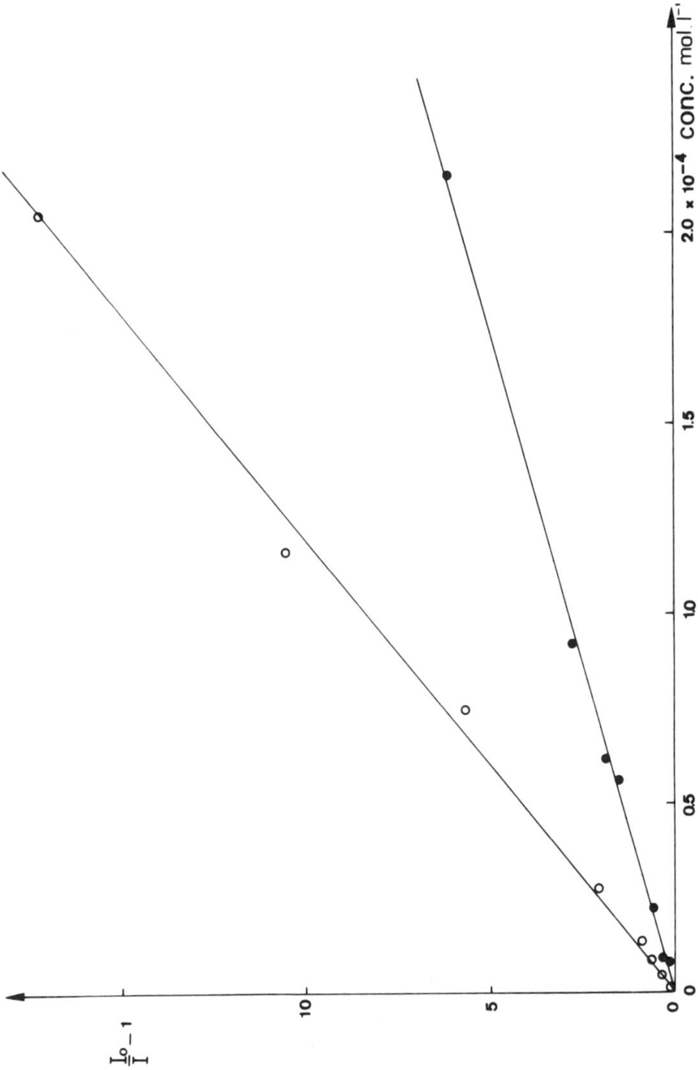

FIGURE 8. $\frac{I_o}{I} - 1$ as a function of acceptor concentration. ○: DPOT; ●: perylene.

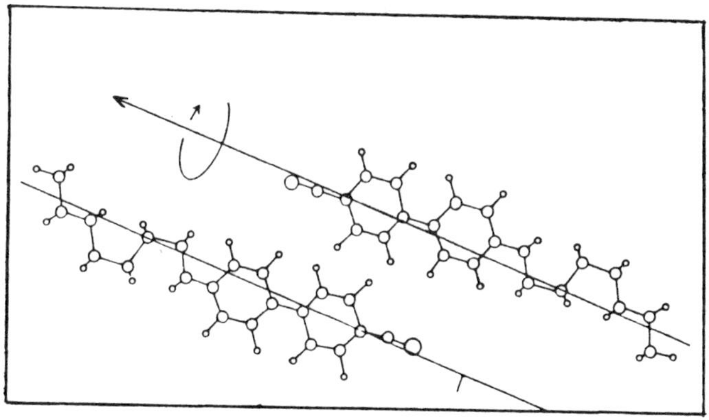

FIGURE 9. Arrangement of the cyanobiphenyl units in the smectic phase.

by ESR in another family of nematogens continues to decrease up to 40°C above the N → I transition.[25]

CONCLUSION

These results show that energy migration and transfer in organized systems do not obey any general rule. Various parameters are involved and have been clearly identified. Tacticity increases the probability of excimer formation in polymer solution. Excimer formation in the crystalline part of solid polymers is generally difficult since the formed helices are unfavorable to the formation of overlapping neighboring chromophores. Excimer could, however, form in the amorphous phase or at structural defects of the crystalline parts of polymers. The increase of excimer fluorescence with the degree of order of polystyrene could thus be most probably assigned to increased efficiency of energy migration. Indeed, the interaction energy between neighboring units which is responsible for energy migration depends on intermolecular distance, on the multipole character of the interaction and on the overlapping of the electronic clouds. This is also valid for the lyotropic and thermotropic systems. However, in systems like 8COB, where the coupling between neighboring units is very efficient, no difference could be found in the rate constants of energy transfer to additives in the S, N and I phases since ordered structures similar to those existing in the smectic phase still exist in the nematic and isotropic domains.

ACKNOWLEDGMENTS

We are grateful to the "Fonds National de la Recherche Scientifique" for financial support to the laboratory. We also thank Professor de Schrijver from the University of Leuven for decay-time measurements and helpful discussion.

REFERENCES

1. STEIN, R. S. 1961. J. Appl. Phys. **32:** 1280.
2. MILAGIN, M. F., A. D. GABARAYERA & I. I. SHISHKIN. 1970. J. Polym. Sci. U.R.S.S., *12,* 577 and 1970. Vysokomolek. Socdin. **A12:** 513.

3. CHALLA, G. P. H. HERANS & A. WEIDINGER. 1962. Makromolek. Chem. **56:** 169.
4. VAN BOCKSTAELE, M., J. GELAN, H. MARTENS, J. PUT, J. C. DEDEREN, N. BOENS F. C. DE SCHRIJVER. 1978. Chem. Phys. Letts. **58:** 211, and references cited.
5. DAVID, C., N. PUTMAN-DE LAVAREILLE & G. GEUSKENS. 1974. Eur. Pol. J. **10:** 617.
6. DAVID, C. & D. BAEYENS-VOLANT. To be published.
7. BOYER R. F. & H. KESKKULA. 1970. *In* Encyclopedia of Polymer Science and Technology. Interscience. New York. p. 224.
8. KILIAN H. G. & K. BOUEKE. 1962. J. Polym. Sci. **58:** 311.
9. KRIMM, S. & A. V. TOBOLSKY. 1951. Text Res. J. **21:** 805.
10. BOKOBZA, L., B. JASSE & L. MONNERIE. 1977. Eur. Pol. J. **13:** 921.
11. NATTA, G., P. CORRADINI & I. W. BASSI. 1960. Nuovo Cim. Suppl. **15:** 68.
12. FÖRSTER, TH. 1960. Excitation transfer. *In* Comparative Effects of Radiation. M. Burton, J. S. Kirby-Smith & J. L. Magee, Eds. Wiley. New York.
13. HUSSON, F., H. MUSTACCHI & V. LUZZATI. 1960. Acta Cryst. **13:** 668.
14. SCHICK, S. M. ATLAS & F. R. EIRICH. 1962 J. Am. Chem. Soc. **66:** 1326
15. LEADBETTER, A. J., J. C. FROST, J. P. GAUGHAM, G. W. GRAY & A. MOSLEY. 1979. J. Phys. **40:** 375.
16. DAVID, C. & D. BAEYENS-VOLANT 1980. Mol. Cryst. Liq. Cryst. **59:** 181.
17. LYDON, J. E. & C. J. COAKLEY. 1975. J. Phys. **36:** C1–45.
18. CLADIS, P. E., R. K. BOGARDUS, W. B., DANIELS & G. N. TAYLOR. 1977 Phys. Rev. Lett. **39:** 720.
19. CHU, K. S. & D. S. MOROI. 1975. J. Phys. **36:** C1–99.
20. LEADBETTER, A. J., F. P. TEMME, A. HEIDEMANN & W. S. HOWELLS. 1975. Chem. Phys. Letts., **34:** 363.
21. BIRKS, J. B. Photophysics of Aromatic Molecules, 1970. Wiley, New York. p. 5.
22. LEADBETTER, A. J., R. M. RICHARDSON & C. N. COLLING. 1975. J. Phys. **36:** C1–37.
23. BUKA, A., P. G. OWEN & A. H. PRICE. 1979. Mol. Cryst. Liq. Cryst. 273.
24. LEADBETTER, A. J., J. L. A. DURRANT & M. RUGMAN. 1977. Mol. Cryst. Liq. Cryst. **36:** 231.
25. SHIMOYAMA, Y., M. SHIOTANI & J. SOHMA. 1977. Jap. J. Appl. Phys. **16:** 1437.

LANTHANIDE ION FLUORESCENCE AS A PROBE OF IONOMER STRUCTURE*

E. Banks, Y. Ueba, and Y. Okamoto

Department of Chemistry and the Polymer Research Institute
Polytechnic Institute of New York
Brooklyn, New York 11201

INTRODUCTION

Recently, the large family of ion-containing polymers has attracted considerable attention.[1,2] The primary reason for the upswing of interest, which is reflected in the appearance of several thousand articles and patents within the past 10 years, is undoubtedly the enormous range of application of these materials. Of these ion-containing polymers, intensive investigation on ionomers has been carried out. The word ionomer was originally used to describe polymers containing a pendant ionic group (usually carboxylate) on a polyolefin backbone. However, today it is generally used to describe many different polymers which contain metal ions and form intermolecular ionic bonds.

Rare earth metal containing polymers, however, have received little attention in the past. Low molecular weight organic complexes of rare earth metals have unique and interesting properties.[3] Interest in rare earth metal complexes has been stimulated by their fluorescence and laser properties. Rare earth ions have also successfully been utilized as fluorescent probes to elucidate conformations and metal binding sites of certain biopolymers.[4] The structures of ionomers have recently been extensively studied by several investigators.[2] It is proposed that there are submicroscopic regions of the polymer in which lamellar domains are interspersed with ionic aggregates. In these ionic regions, there may be several intermolecular associations involving carboxyl, carboxylate and the metal cations. If such structures exist in the polymer systems, metal ions such as Eu^{3+} and Tb^{3+} in the ionic aggregates would be expected to be strongly coupled and as a result, concentration quenching of the fluorescence emission would be observed. Thus, we prepared several lanthanide metal salts of ionomers such as copolymers of styrene-acrylic acid (PSAA) and methyl methacrylate-methacrylic acid (PMMA/MA)[5]

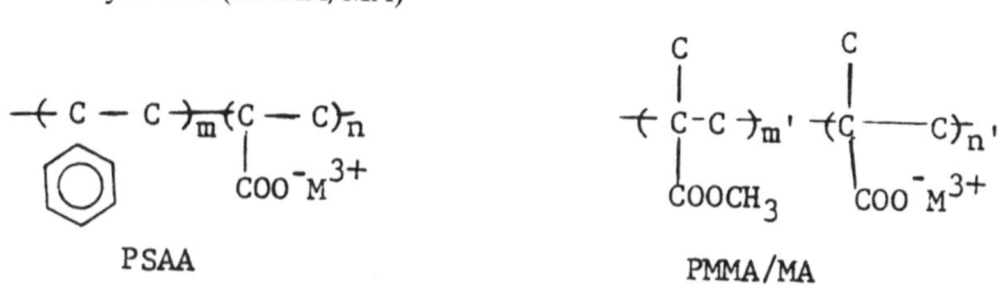

where $m > n$, $m' > n'$. The structure is as shown for M^{3+}, Eu^{3+}, Tb^{3+}, etc.

* This research was supported by a generous grant from the National Science Foundation, Polymer Program, Grant No. DMR 78-09764.

0077-8123/81/0366-0356$01.75/0 © 1981, NYAS

POLYMER-LANTHANIDE METAL SALT SYSTEMS

The fluorescence intensity was measured on dried and finely powdered samples. The scatter of data was found to be 5%. Typical excitation and emission spectra of the Eu^{3+}-polymer complex are shown in FIGURE 1. The bright transitions from the $^5D_0$ levels to all the multiplet levels of ^7F can be seen. This emission spectrum was obtained under excitation at 395 nm (25340 cm^{-1}), corresponding to the Eu^{3+}-$^5L_7$ level.[4]

We also prepared samples in which Eu^{3+} acetate is uniformly dispersed in polymer matrices such as polystyrene and polymethyl methacrylate. In both Eu^{3+}-polymer salts and Eu^{3+} acetate, Eu^{3+} was coordinated with carboxylates. As shown in FIGURES 2 and 3, for the europium acetate system, the fluorescence intensity was found to increase linearly with Eu content. However, in the polymer-Eu^{3+} complex system, the fluorescence intensity reached a maximum at Eu content as small as 4–5 wt % and subsequently decreased with increasing Eu content. The tendency of fluorescence intensity to decrease with increasing Eu^{3+} content is a typical fluorescence quenching pattern[5] and the result provides supporting evidence that ionomers have ionic aggregates in which metal ions are close together. In other words, the local concentration of Eu^{3+} ions in these regions is much higher than the overall average concentration. A similar trend in Tb^{3+}-polymer salts was also observed. These results suggest that the acrylic copolymers contain ionic aggregates, in agreement with the proposed model.

Considerable effort has been devoted to the study of nonradiative energy transfer between fluorescent lanthanide donor-acceptor pairs in the solid state.[6–11] When the emission spectrum of the donor overlaps the absorption spectrum of the acceptor, excitation energy absorbed by the donor can be transferred to the acceptor over considerable distances.[12] Thus, we attempted to elucidate the size of the ion aggregates using Tb-Eu and Tb-Co metal probes. The energy transfers from Tb^{3+} to Co^{2+} and Eu^{3+} were measured by the Tb^{3+} fluorescence quenching behavior.

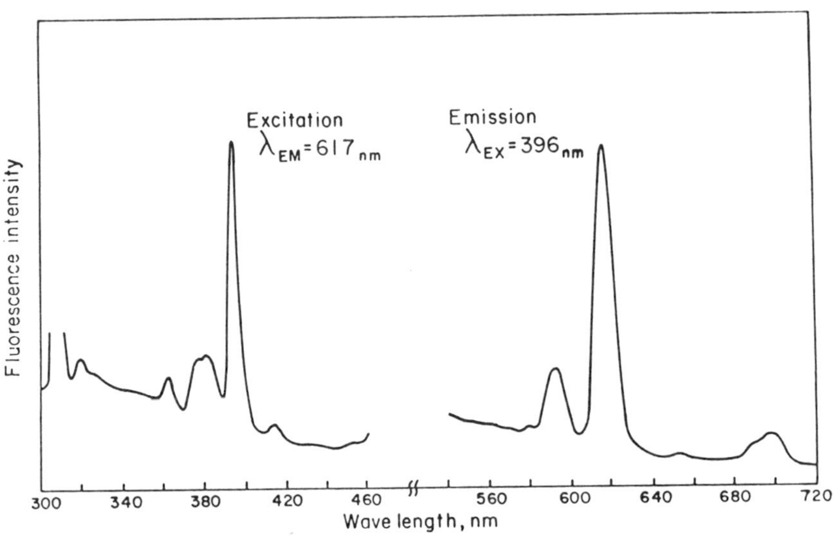

FIGURE 1. Fluorescence spectra of poly(acrylic acid) Eu^{3+} salt.

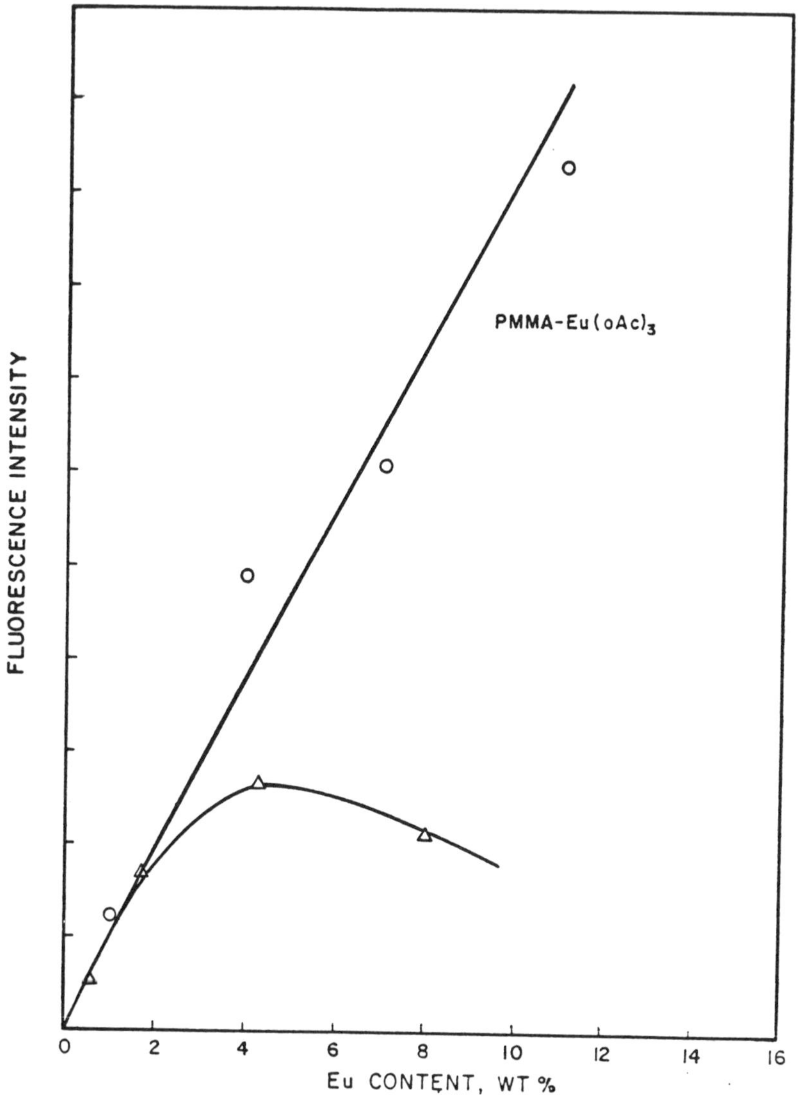

FIGURE 2. Relationship between fluorescence intensity and Eu content in PMMA/MA-Eu (△) and PMMA-Eu(OAc)₃ systems (○).

In the poly(styrene-acrylic acid) system, effective Tb^{3+} fluorescence quenching by Co^{2+} and also by Eu^{3+} was observed (FIGS. 4 and 5). The results show that Tb^{3+} fluorescence is completely quenched by the presence of less than equimolar amounts of Co^{2+}. The Tb^{3+} fluorescence was found to be effectively quenched by Eu^{3+}. We expect that the distance between the two metals must be smaller than the critical transfer distance which is of the order of 10Å.[10] In polymer systems such as partially sulfonated (**1**) or carboxylated (**2**) styrene, neither similar energy transfer between these metal ions nor the fluorescence quenching pattern (see FIG. 6) were detected.

These results provided further evidence that the sulfonated polystyrene ionomers do not contain ion aggregates.[13]

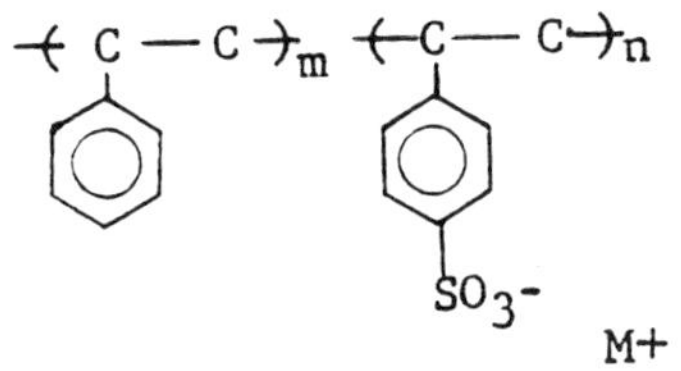

(1)

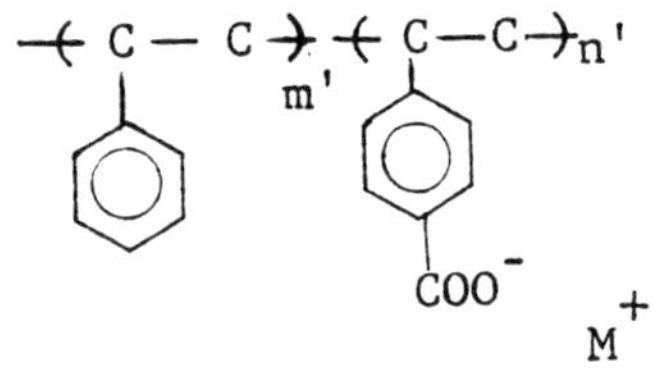

(2)

where m ≫ n, m′ ≫ n′. The structure is as shown for M+, Eu^{3+}, Tb^{3+}, Co^{2+}.

The emitting level of the UO_2^{2+} ion is about 480 nm and the fluorescence extends

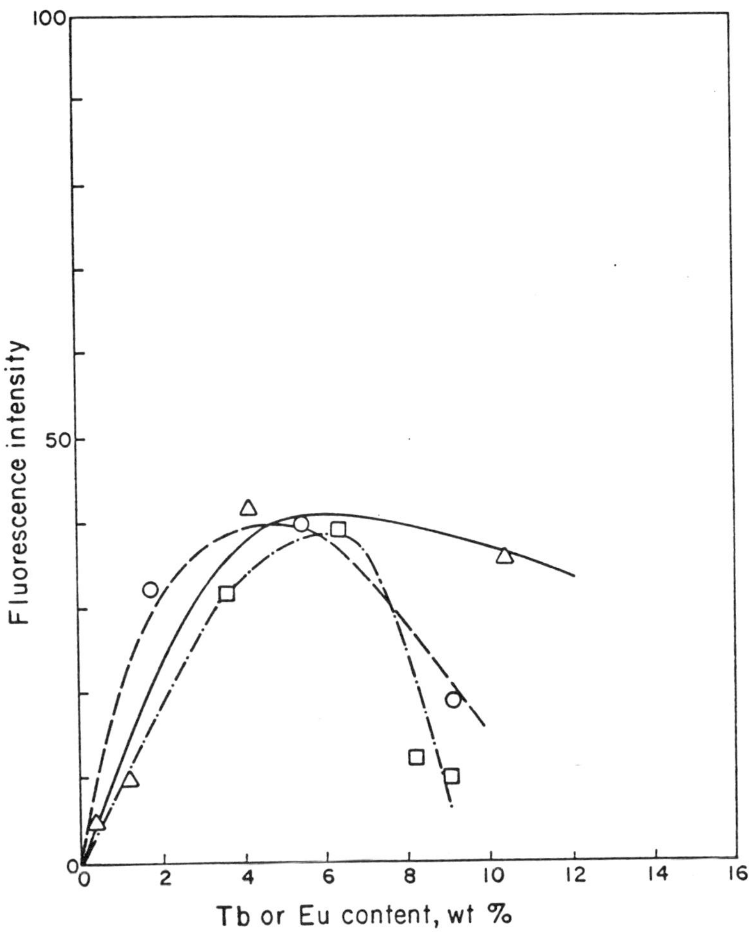

FIGURE 3. Relationship between fluorescence intensity and metal content. -○- PSSA- Eu salt (COOH content 13 mole %); -△- PSAA- Eu salt (COOH content 9 mole %); -□- PSAA- Tb salt (COOH content 13 mole %).

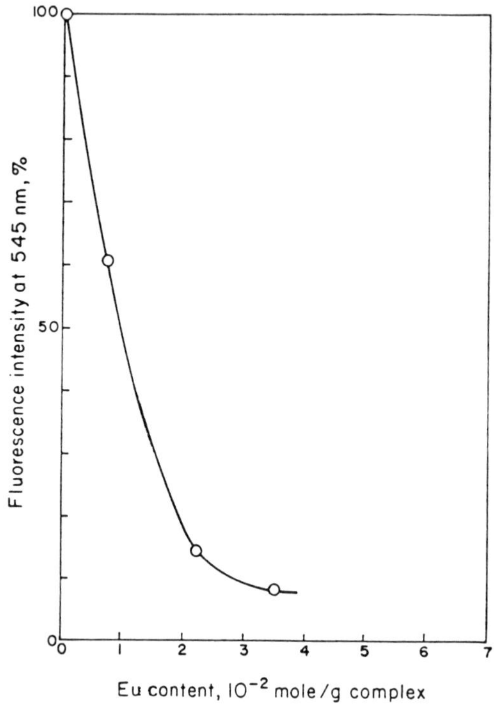

FIGURE 4. Energy transfer from Tb^{3+} to Eu^{3+} in PSAA.

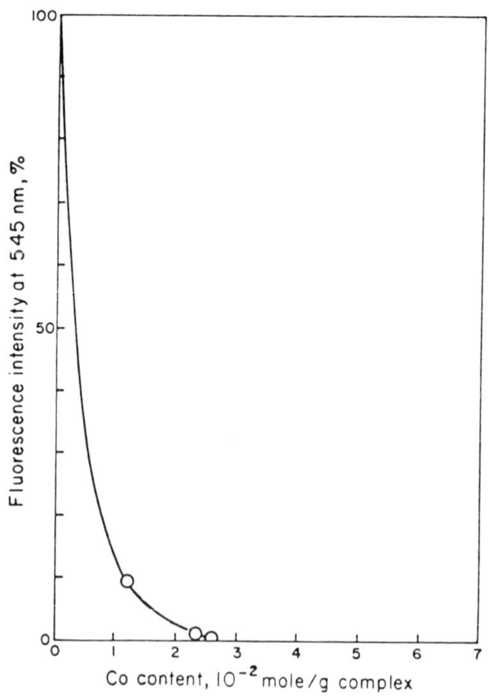

FIGURE 5. Energy transfer from Tb^{3+} to Co^{2+} in PSAA.

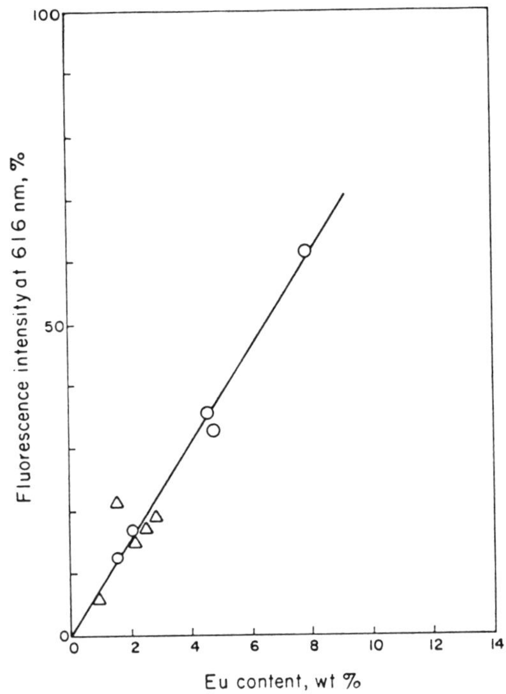

FIGURE 6. Fluorescence intensity *vs* Eu content in sulfonated polystyrene-Eu salt. -△- sulfonic acid content 2.6 mole %; -○- 6.1 mole %.

from 480 to about 570 nm. Three transitions of Eu^{3+} ion observed at room temperature lie within this energy range. Thus it has been shown that UO_2^{2+} transfers energy to Eu^{3+} in borosilicate glass, resulting in a fivefold increase in the Eu fluorescence.[14] If the ion-containing polymer forms ion aggregates, it is very likely that Eu^{3+} and UO_2^{2+} pairs are located close together within the aggregates and that efficient energy transfer would be observed. Thus, we prepared Eu^{3+} and UO_2^{2+} mixed salts of PSAA and PMMA/MA. The fluorescence intensity was found to be greatly increased (~30 fold) when equal molar concentrations of UO_2^{2+} and Eu^{3+} were present in these polymers systems when irradiated at 480nm.

ENERGY TRANSFER FROM ORGANIC LIGANDS TO Eu^{3+} IN POLYMERS CONTAINING β-DIKETONE LIGANDS

The fluorescent characteristics of europium chelates are of considerable interest in connection with electronic energy transfer processes and with their use in laser systems.[2] Chelating compounds containing the β-diketone moiety were widely used for this purpose.[3] An attempt to utilize a polymer as a matrix was reported by Wolff and Pressley.[4] They found that a solid solution of europium-tris-4,4,4-trifluoro-1-(2-thienyl)-1,3-butenedione in polymethylmethacrylate showed optical maser action. However, polymeric systems in which rare earth metals are directly bonded to the polymer have not yet been investigated.

The rare earth complexes of ionomers described above were homogeneous glassy phases with no grain boundaries to scatter light. This suggested that polymers with the β-diketone moiety built in should have interesting fluorescent properties. The polymers investigated contain the β-diketone moiety, and are linear and branched types described by formulas (**4**) and (**5**). Their fluorescent properties were compared with those of the corresponding complex of Eu^{3+} with the monomeric chelating compound; dibenzoylmethane (**3**).

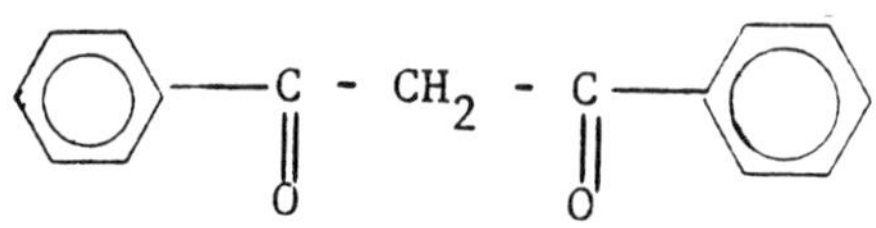

(3)
DIBENZOYLMETHANE

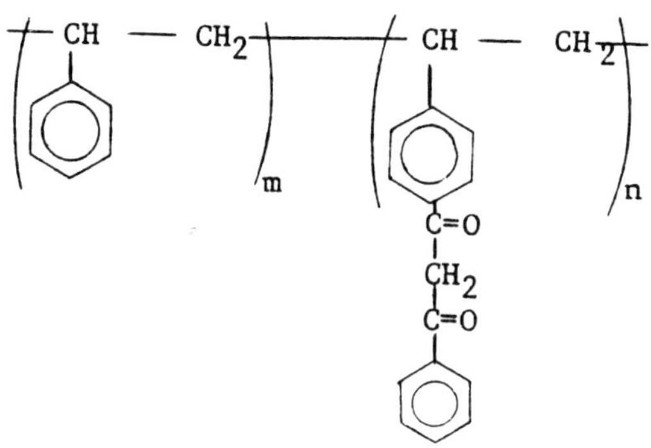

m > n

(4)
POLY(p-BENZOYLACETYL STYRENE)

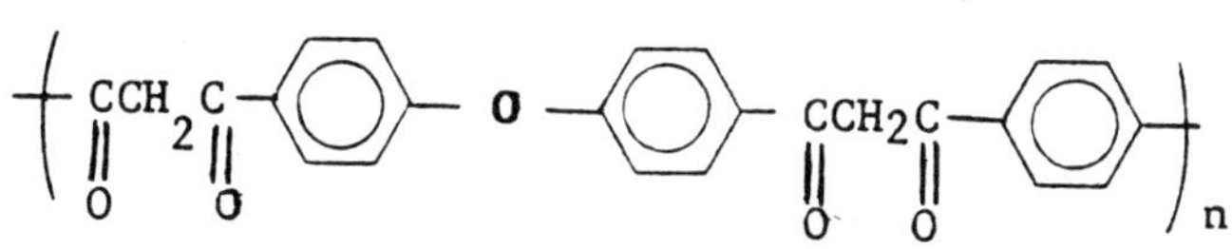

(5)
POLY(ARYL-β-DIKETONE)

The fluorescent properties of europium dibenzoyl methides have been intensively investigated.[12] The sharp emission spectrum is detected at 613 nm even at room temperature (FIG. 7). The peak is due to the $^5D_o \rightarrow {}^7F_2$ transition. The same emission peak was also detected in these Eu-polymer systems at room temperature. A typical Eu-polymer spectrum is shown in FIGURE 8.

In addition to the lines of the emission spectrum in the $Eu(DBM)_4$ (FIG. 7), as shown in FIGURE 8, a line at 580 nm was detected for the Eu-polymer complex, corresponding to the forbidden $_5D_o \rightarrow {}_7F_o$ transition.[13] The appearance of the $0 \rightarrow 0$ transition has been explained in terms of low symmetry around Eu^{3+}.[14] Thus, the spectrum indicated that the Eu-polymer complex may have a more asymmetrical structure around Eu^{3+} than that of the $Eu(DBM)_4$.

For the composite sample of Eu-$(DBM)_4$ in polystyrene, the fluorescence emission

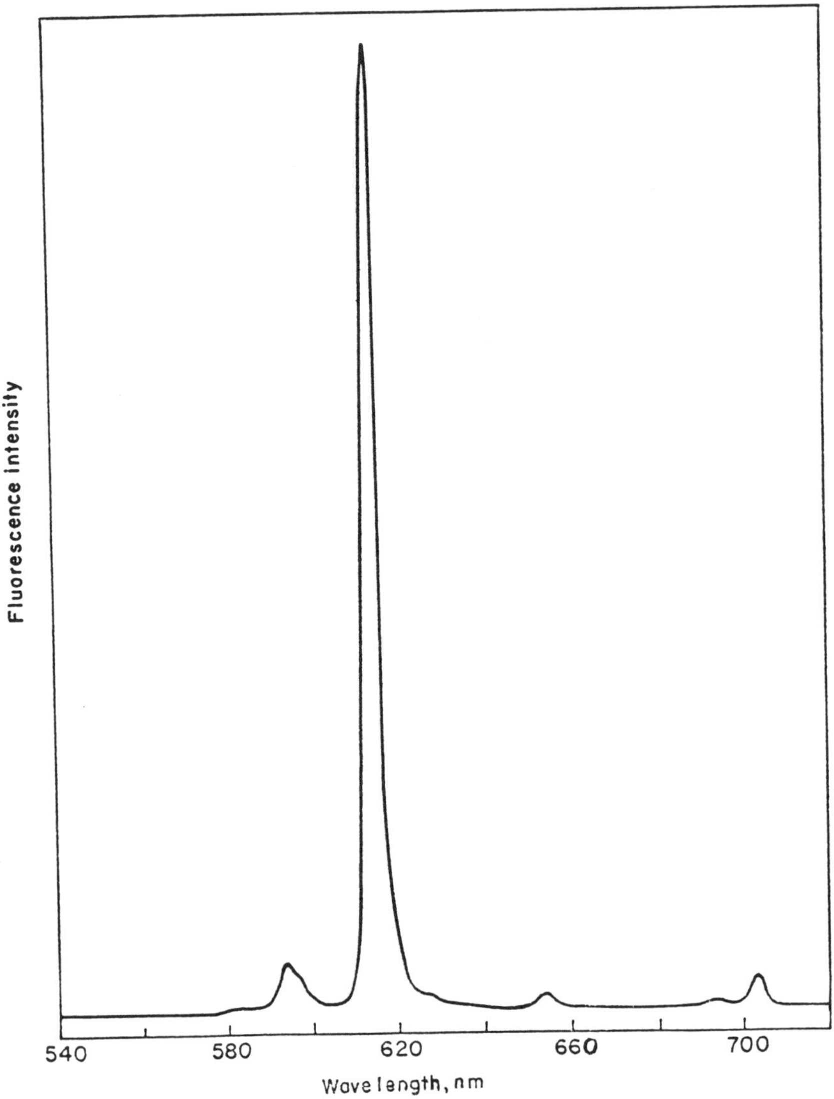

FIGURE 7. Emission spectrum of $Eu(DBM)_4$, λ_{EX} = 396 nm.

intensity was found to increase linearly with increasing Eu content (FIG. 9). However, for the Eu-coordination polymer complexes, the intensity reached a maximum at Eu^{3+} content as small as 1 wt % and remained constant on further increasing the Eu^{3+} content (See FIGS. 10 and 11).

When the fluorescence intensities of these Eu complexes were compared under the same conditions, 1 wt % Eu, at which Eu-(**4**) and Eu-(**5**) show nearly maximum fluorescence intensity, the order was found to be: $Eu(DBM)_4$ > Eu-(**4**) > Eu-(**5**). The samples' intensities differed by a factor of about 20–30. The fluorescence emission

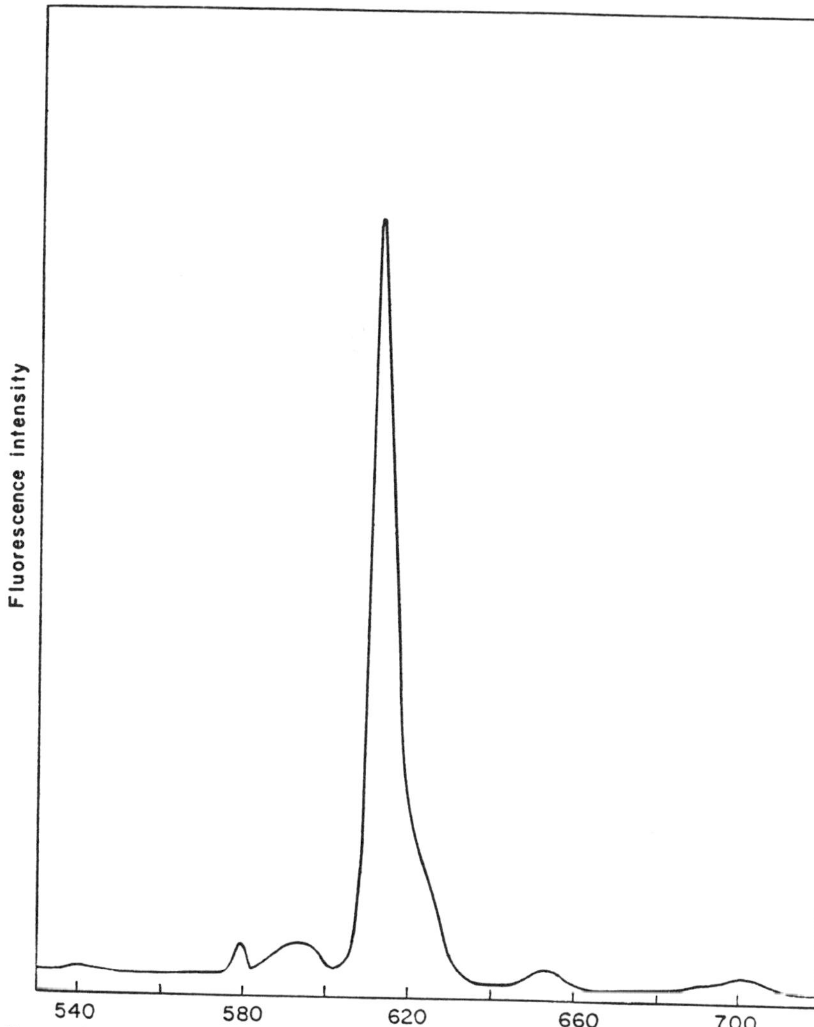

FIGURE 8. Emission spectrum of poly(*p*-benzoyl acetyl styrene), Eu complex $\lambda_{EX} = 398$ nm.

intensity is known to vary with environmental factors such as the nature of coordination bond, solvent and temperature.

In order to obtain data on the effect of the coordination number on the fluorescence intensity in the Eu-dibenzoylmethide system, we have prepared Eu-tetrakis, tris- and bis-dibenzoylmethides and found that the fluorescence intensities were in the order: Eu-$(DBM)_4$ > Eu-$(DBM)_3$ > Eu$(DBM)_2$.[16] The intensity ratios were 40–50-fold among these complexes.

When Eu ion reacts with excess dibenzoylmethane in basic solution, generally the tetrakis coordination complex is obtained. However, when β-diketone groups are incorporated in the polymer chain, the formation of multiple coordination bonding between Eu and the β-diketone may be restricted by the increasing steric hindrance and the decrease of the freedom of bond rotation.

The fluorescence intensity data discussed above suggest that the polymer (**4**) forms tris-coordination complexes with a small amount of Eu^{3+}, (~1 wt % Eu) and subsequently forms bis-and mono-coordination complexes with additional Eu^{3+}. Since the fluorescence intensity of the tris-coordination compound Eu-$(DBM)_3$ was

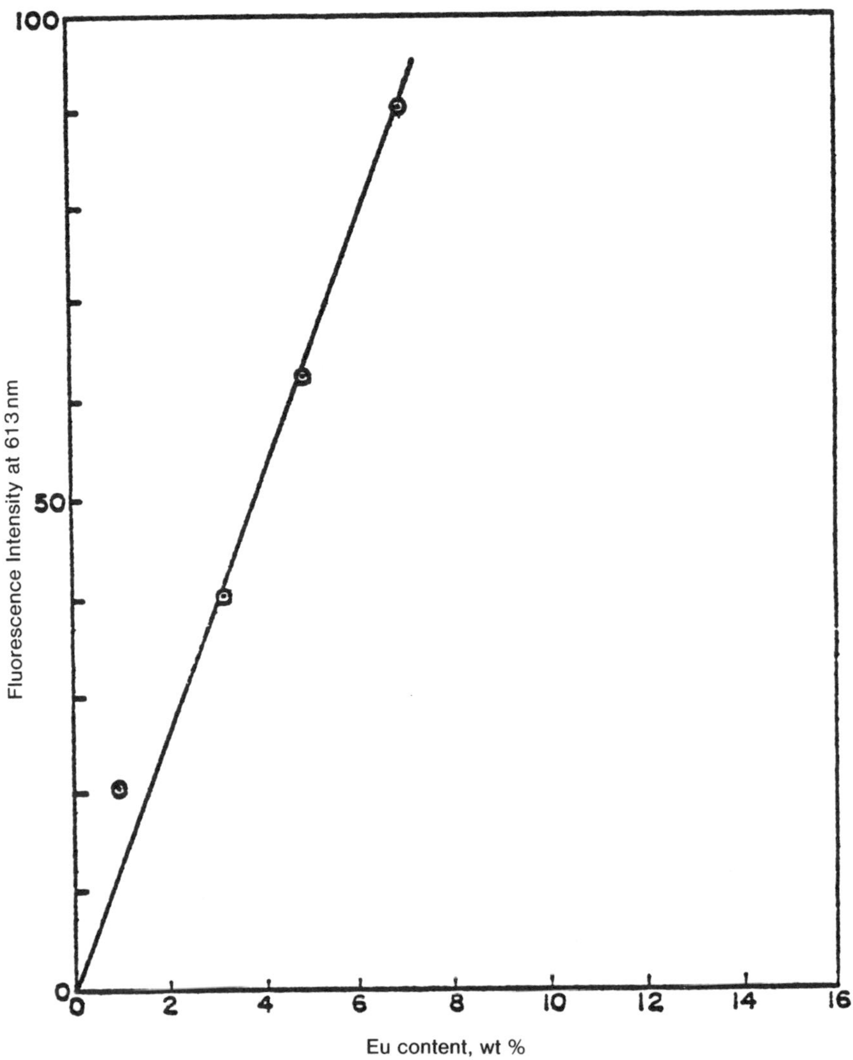

FIGURE 9. Relationship between fluorescence intensity and Eu content of $EU(DBM)_4$ in polystyrene. Slit EX/EM = 1.5/1.5 λ_{EX} = 398 nm.

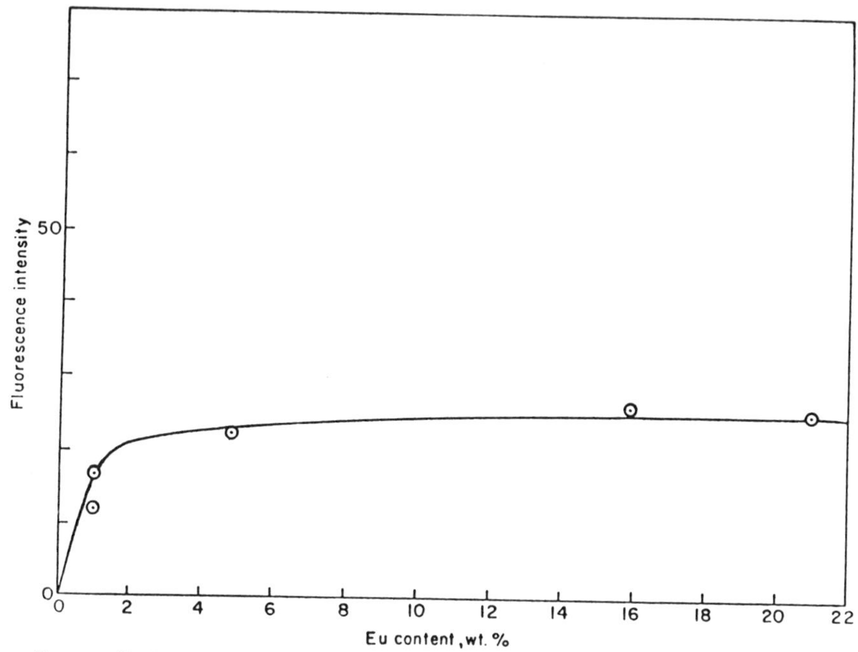

FIGURE 10. Relationship between fluorescence intensity and Eu content of poly(aryl β-diketone) Eu complex.

far greater than those of bis- and mono-coordination compounds, the intensity remained essentially constant after reaching a maximum at low Eu^{3+} concentration as shown in FIGURE 11. In other words, the maximum intensity obtained is due mostly to the tris-coordination complex between Eu^{3+} and the polymer (**4**). In the case of the polymer (**5**), the β-diketone groups are incorporated in the linear chain; consequently the formation of multiple coordination bonds between Eu ion and β-diketone moiety is expected to be more restricted as compared with the polymer (**5**) in which β-diketone is attached on the side chain. Therefore, the polymer (**5**) may form mainly the bis-coordination complex with limited amounts of Eu^{3+} followed by the formation of mono-coordination complex with excess Eu^{3+}.

Thus, the maximum fluorescence intensity obtained in the Eu-polymer (**5**) system may be due to that of the Eu-bis diketone complex and the intensity remained nearly constant even with increasing Eu content (FIG. 10).

POLYMERS CONTAINING CARBOXYLBENZOYL AND CARBOXYLNAPTHOYL LIGANDS

Tanner and Thomas prepared Eu^{3+}-o-benzoylbenzoate and demonstrated energy transfer in ethanol-water solution,[16] since carboxylic acid easily forms a salt with Eu ion. We prepared partially 2-carboxylbenzoyl (**6**) and 3-carboxyl 2-naphthoyl (**7**) substituted polystyrene and their Eu^{3+} salts.

where m > n, m′ > n′

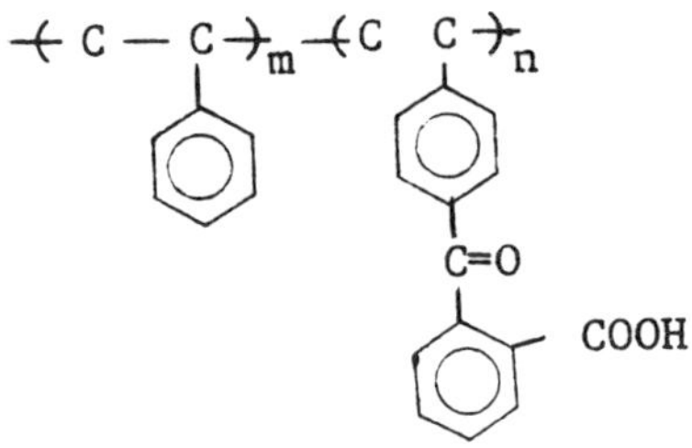

(6)

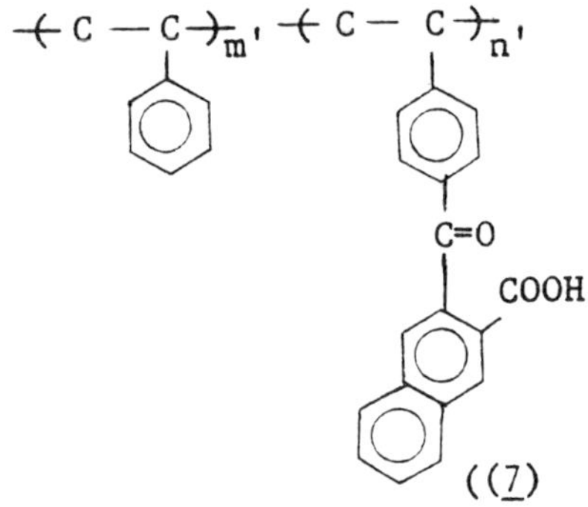

((7)

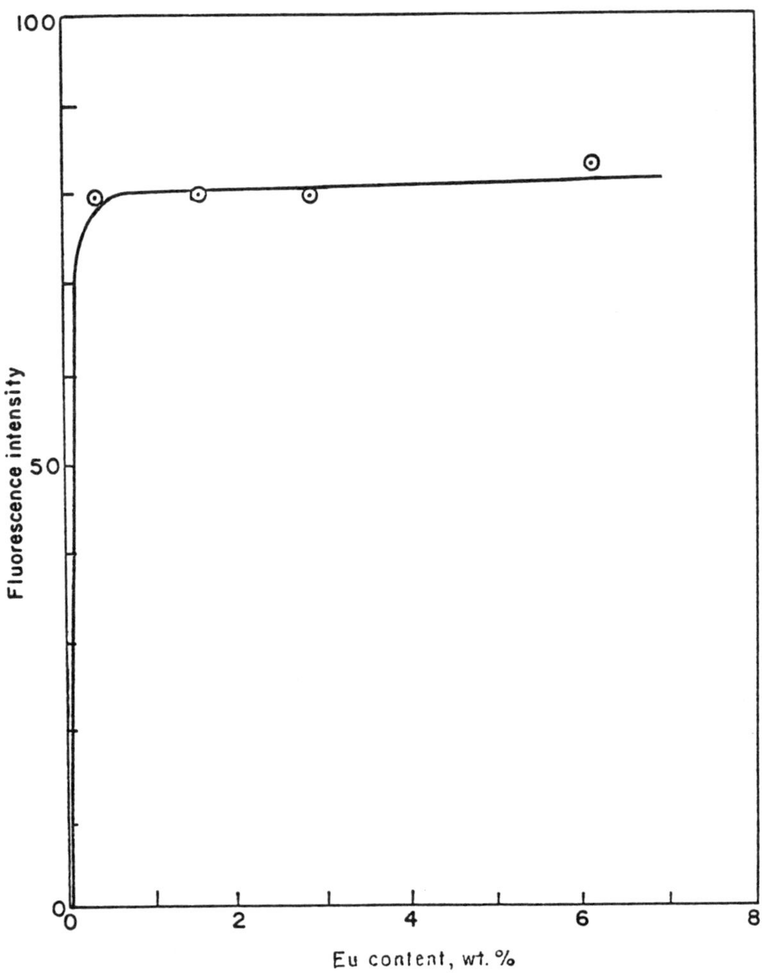

FIGURE 11. Relationship between fluorescence intensity of Eu content of poly(*p*-benzoylacetyl styrene)-Eu complex.

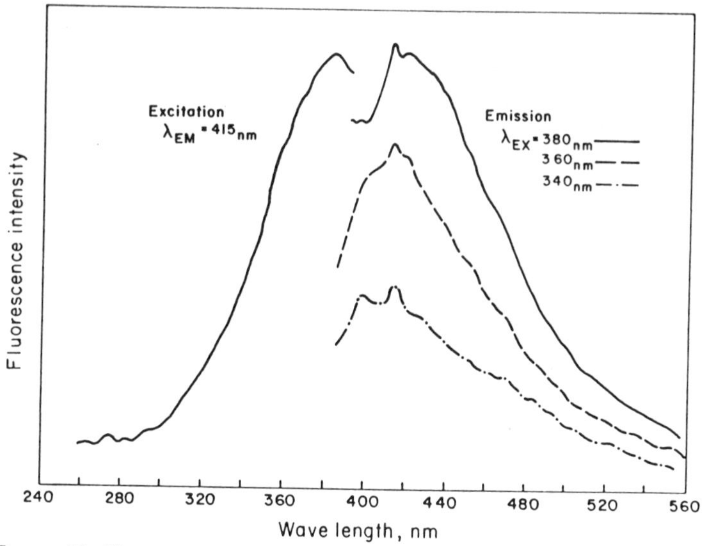

FIGURE 12. Fluorescence spectra of *o*-carboxybenzoyl substituted polystyrene.

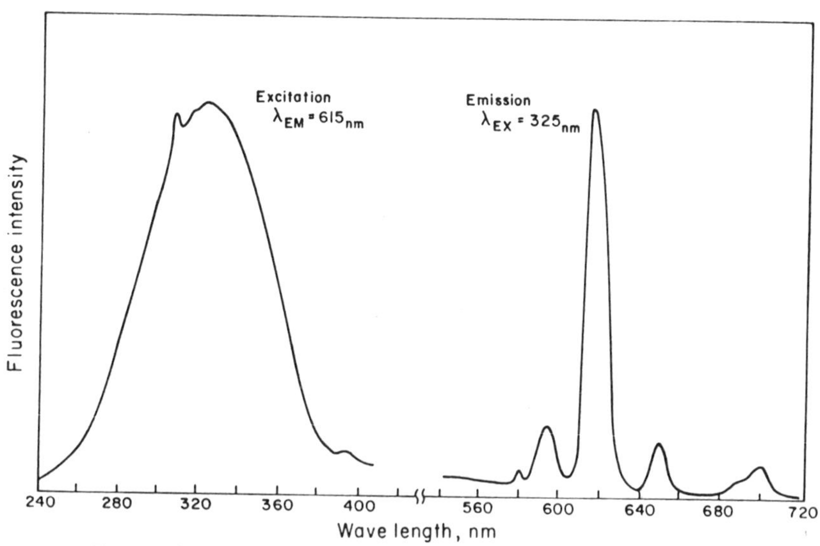

FIGURE 13. Fluorescence spectra of *o*-benzoylbenzoic acid Eu complex.

The fluorescence spectra of (**6**) and (**6**)-Eu^{3+} salt are shown in FIGURES 12 and 13. (**6**) has its excitation maximum at 380 nm and the emission maximum at 400 to 410 nm, while the (**6**)-Eu^{3+} salt has the excitation maximum at 328 nm and shows characteristic line emission from Eu^{3+} ion upon excitation at 328 nm. The Eu^{3+} ion does not show such absorption at 328 nm. This result suggests that energy transfer is taking place between the absorbing (**6**) and the emitting Eu^{3+}. The fluorescence intensity of (**6**)-Eu^{3+} salt was compared with that of $Eu(OAc)_3$ in polystyrene at the

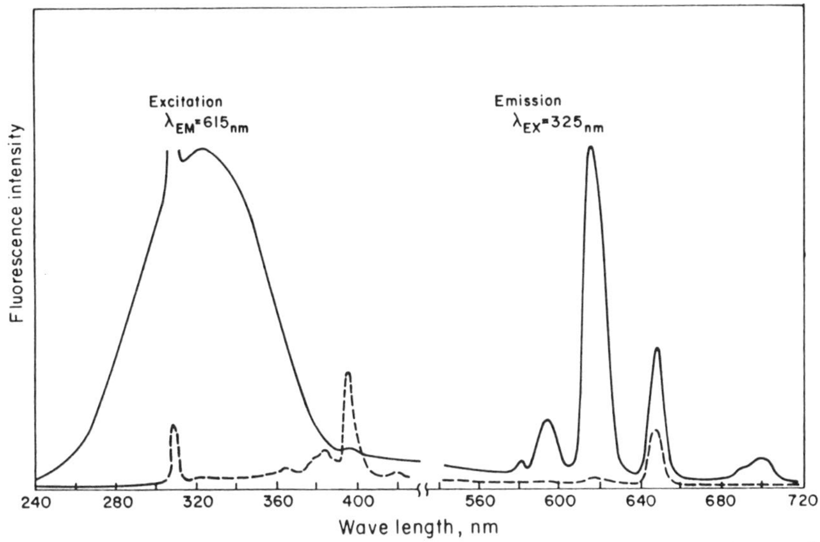

FIGURE 14. Fluorescence spectra of *o*-carboxybenzoyl substituted polystyrene-Eu complex (——) and polystyrene-Eu acetate mixture (- - - - -).

same Eu content. The intensity at 615 nm of (**6**)-Eu^{3+} salt was found to be 60–70 times greater than that of $Eu(OAc)_3$ (FIG. 14). This indicates that the benzoylbenzoate is an efficient energy donating ligand to Eu^{3+} ion in the intermolecular energy transfer process.

The relationship between the fluorescence intensity at 616 nm and Eu^{3+} content in (**6**)-Eu^{3+} salt is shown in FIGURE 15. The increase of the fluorescence intensity with Eu^{3+} content levels off at around 0.6–0.7 wt % of Eu^{3+} and the intensity remains constant on further increasing the Eu^{3+} content. The calculated maximum Eu^{3+} content in (**6**)-Eu^{3+} salt; 1:3 coordination Eu^{3+} to benozylbenzoate ligand is 0.94 wt % and for the 1:1 coordination compound is about 3 wt %. The fluorescence intensity of (**6**)-Eu^{3+} salt at the maximum was 4–5 times lower than that of the corresponding monomeric compound; Eu^{3+}-2-benzoylbenzoate, $Eu(BBA)_3$. Therefore, the leveling off of the fluorescence intensity at 0.6–0.7 wt % range indicates that the benzoylbenzoate moiety incorporated in the polymer system is restricted from forming multiple coordination bonds between Eu^{3+} ion and the ligand beyond a certain Eu^{3+} concentration. The restriction may also be due to the increasing steric hindrance and decreasing of the freedom of bond rotation in the polymer system. Thus, on further addition of Eu^{3+} ion beyond 0.6–0.7 wt %, the Eu^{3+} ion forms the 1:2 and 1:1 coordination complex with the ligand. The fluorescence intensities of lower coordination complexes were found to be much weaker than that of the maximum coordinating complex.[15] The (**7**)-Eu^{3+} salt exhibits a similar relationship between the fluorescence intensity and the Eu^{3+} content to that shown by (**6**)-Eu^{3+} salt. However, the intensity leveled off at a lower Eu concentration as compared with (**6**)-Eu^{3+} salt. This suggests that, due to the bulkiness of the naphthoyl group, the 1:3 coordinate bond formation between Eu^{3+} and the ligand in the polymer system was restricted at even lower concentrations of Eu^{3+}.

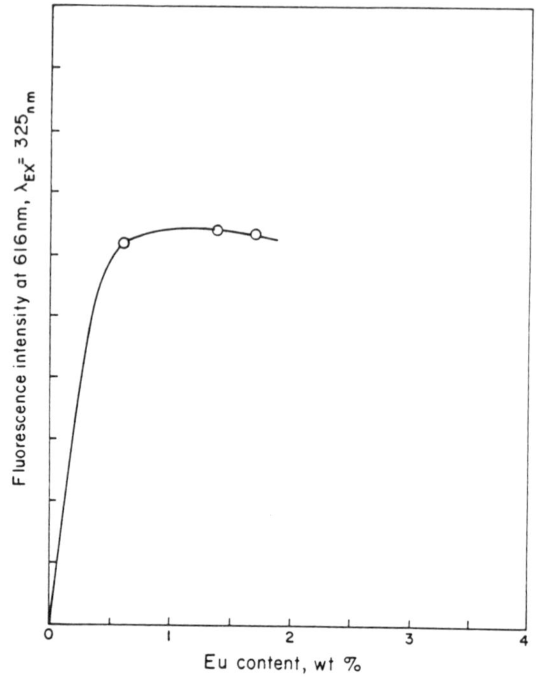

FIGURE 15. Relationship between fluorescence intensity of Eu content of polymer (**6**)-Eu complex.

These results together with previous findings in the β-diketone containing polymeric system showed that when a ligand group is incorporated in a polymer system, unless the ligand and polymer chain are bonded with a long and flexible chain, the formation of the maximum coordination bond between a metal ion and ligand is restricted. As a result, the fluorescence intensity of the polymer system always shows lower intensity as compared to that of the corresponding monomeric compound.

These results suggested that the lanthanide metal probes are also useful in the elucidation of the nature and number of coordinate bonds between metal ions and ligands in polymeric systems.

EXPERIMENTAL SECTION

Materials

Polymers

Copolymers of styrene-acrylic acid (PSAA) and methyl methacrylate-methacrylic acid PMMA/MA were prepared and purified by standard methods. Sulfonated polystyrene (**1**) (sodium sulfonate contents 2.6 and 6.1 mole %) were supplied by Dr. H. S. Makowski, Exxon Research and Engineering Co. Two carboxylated polystyrenes (**2**) were used. One of the polymers (carboxylic acid content 6.4 mole %) was obtained from Dr. R. D. Lundberg, Exxon. Another polymer (carboxylic acid content, 4.6 mole %) was prepared in our laboratory by the method described in the literature.[17] Polymer(**5**) was prepared by a method similar to that described in the literature.[18] Poly (*p*-benzoylacetyl styrene (**4**) was prepared from a partially aceylated polystyrene by condensation with methyl benzoate.[15] Partially substituted 2-carboxylbenzoyl polystyrene (**6**) was prepared by condensation of polystyrene with phthalic anhydride in the presence of $AlCl_3$. Polymer (**7**) was also prepared by a similar method with 2,3-naphthalene dicarboxylic anhydride.

Lanthanides

Europium chloride and terbium chloride were purchased from Ventron Company and were used without further purification.

Lanthanide - Polymer Complexes

The lanthanide salts and chelating complexes of these polymers were prepared by a similar method to that described in our reports.[5,15] The metal contents were determined by atomic absorption measurements.

Fluorescence Measurements

The measurements were carried out at room temperature on dried and ground fine powder samples.[5] The sample was well compressed against the quartz window of the powder sample holder. The experimental error in fluorescence intensity was usually less than 5%.

Acknowledgment

The authors thank Professor H. Morawetz for many helpful discussions.

References

1. Eisenberg, A. 1979. Polym. Prepr. Am. Chem. Soc. Div. Polym. Chem. **20:** 1.
2. Eisenberg, A. & M. King. 1977. Ion-containing Polymers. Academic Press. New York. p. 23.
3. (a) Weissman, S. I. 1942. J. Chem. Phys. **10:** 214.
   (b) Heller, A. & E. Wasserman. 1965. J. Chem. Phys., **42:** 949.
   (c) Tanner, S. P. & D. L. Thomas. 1974. J. Am. Chem. Soc. **96:** 706.
4. Horrocks, Jr., W. DeW, B. Holmquist & B. L. Vallee. 1975. Proc. Natl. Acad. Sci. USA **72:** 4764.
5. The data described in these polymer systems has been or will be published as follows:
   (a) Banks, E., Y. Okamoto & Y. Ueba. 1980. J. Appl. Polym. Sci. **25:** 359.
   (b) Okamoto, Y., E. Banks & Y. Ueba. Macromolecules. To be published.
6. Van Uitert, L. G. 1960. J. Electrochem. Soc. **107:** 803.
7. Reisfeld, R., E. Greenberg, R. Velapoldi & B. Barnett. 1972. J. Chem. Phys. **56:** 1698.
8. Dawson, R. W., L. J. Kropp & M. W. Windsor. 1966. J. Chem. Phys. **45:** 2410.
9. Nakazawa, E. & S. Shionoya. 1967. J. Chem. Phys. **47:** 3211.
10. Reisfeld, R. & L. Boehm. 1972. J. Solid State Chem. **4:** 417.
11. Reisfeld, R., E. Greenberg & E. Biron. 1974. J. Solid State Chem. **9:** 224.
12. Förster, T. 1959. Discuss. Faraday Soc. **27:** 7.

13. Rigdahl, M. & A. Eisenberg. 1979. Polym. Prepr. Am. Chem. Soc. Div. Polym. Chem. **20:** 269.
14. DeShazer, D. & A. Y. Cabezas. 1964. Proc. IEEE **52:** 1355.
15. Ueba, Y., E. Banks & Y. Okamoto. J. Appl. Polym. Sci. In press.
16. Tanner, S. P. & D. L. Thomas. 1974. J. Am. Chem. Soc. **96:** 906.
17. Lundberg, R. D. & H. S. Makowski. 1978. Polym. Prepr. Am. Chem. Soc. Div. Polym. Chem. **19:** 287.
18. Nose, Y., M. Hatano & S. Kambara. 1966. Makromol. Chem. **98:** 136.

ENERGY TRANSFER IN FILMS OF POLYMERS WITH AROMATIC SIDE-GROUPS

Walter Klöpffer

Battelle-Institut e.V.
D-6 Frankfurt am Main, Federal Republic of Germany

First Observations

Energy transfer in solid aromatic polymers was first observed in plastic scintillators, most of which have been prepared by bulk polymerization of styrene or vinyltoluene containing a small amount of sensitizer.[1-6] The sensitizers or guest molecules, mostly aromatic hydrocarbons or heterocyclic compounds known from liquid scintillation counting, have to provide efficient luminescence in the region of maximum sensitivity of the photomultiplier used as detector. The polymer, on the other hand, absorbs the energy provided by the high-energy particles and photons, the lower excited electronic states being intermediates in the deactivation cascade following the primary excitation processes. It soon turned out that nonradiative energy transfer plays an important role in passing the energy transiently stored in the excited electronic states of the polymer to the guest molecules which ultimately emit this energy in the form of near-uv of visible radiation.[3,5,6] Although in some cases, emission-reabsorption (the "trivial" or radiative process of energy transfer) cannot be neglected,[2,-4] the focus of interest has always been nonradiative transfer which apart from its interesting physical features also offers the opportunity of increasing the overall quantum efficiency of the process in addition to shifting the emission to the optimal spectral region of the detector. The possible types of nonradiative transfer of electronic excitation energy have been identified:[5]

- energy migration—a random exciton hopping or multi-step process;[7] and
- single-step transfer which in case of singlet—state transfer can be described by Förster's theory.[8,9]

Diffusion of excited molecules rather than of excitation energy is an unlikely process in solid polymers and will not be considered further although small molecules can transfer electronic excitation energy in polymers, e.g., singlet molecular oxygen.[10]

Fluorescence of Pure Polystyrene (PS)

In order to exclude possible side effects due to high-energy radiation, near-uv excitation has become the favorite mode of stimulating polymer and sensitized-guest luminescences. In this way, fluorescence spectroscopy has been introduced in energy transfer research for polymers, as has been done much earlier for dilute molecular systems[18,19] and aromatic molecular crystals.[7,20]

Identifying the prevailing mechanism of energy transfer requires, first of all, knowledge of the electronic states involved as well as the concentration of the

0077-8923/81/0366-0373$01.75/0 © 1981, NYAS

molecules or parts of macromolecules carrying them. It was to the merit of Basile[11] to have shown us for the first time the fluorescence spectrum of pure PS (FIG. 1) thus opening the way for a better understanding of energy transfer in this important polymer. The true fluorescence of PS—a broad and structureless emission which is red-shifted with regard to uv-absorption—has been obtained by purifying the thermally prepared crude polymer containing monomeric styrene. The older spectra doubtless show the fluorescence of styrene, thus complicating any attempt to interpret the energy transfer behavior in the systems studied in early experimental work. The red—shifted fluorescence of pure PS was interpreted as excimer fluorescence a few years later by Hirayama[12] and Vala, Haebig & Rice[13] on the basis of fluorescence studies performed with model compounds and PS in solution.

This interpretation of the PS solid-state fluorescence immediately suggests exciton diffusion as the most likely energy transfer mechanism in this polymer, since excimers can only be formed at suitable sites (sandwich structure)[14] whereas optical absorption occurs randomly. Thus, the excitation energy has to migrate from the side groups which originally absorb photons to suitably oriented pairs of phenyl groups[15] (later on named "excimer-forming sites"[16]).

Although this mechanism has been basically recognized, it seems that PS is a difficult polymer for quantitative studies on energy transfer.[15] This may be due to a relatively high concentration of excimer-forming sites which favors the alternative single-step transfer.[17]

ENERGY TRANSFER IN POLY (*N*-VINYLCARBAZOLE) (PVCA) AT ROOM TEMPERATURE

It turned out that for experimental reasons PVCA is a very convenient polymer for studying quantitatively the energy transfer.[16] It should be mentioned that the

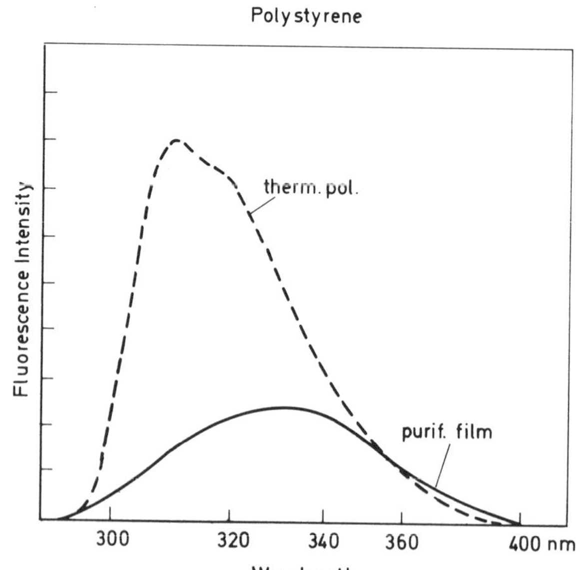

FIGURE 1. Fluorescence of solid polystyrene, after L. J. Basile.[11]

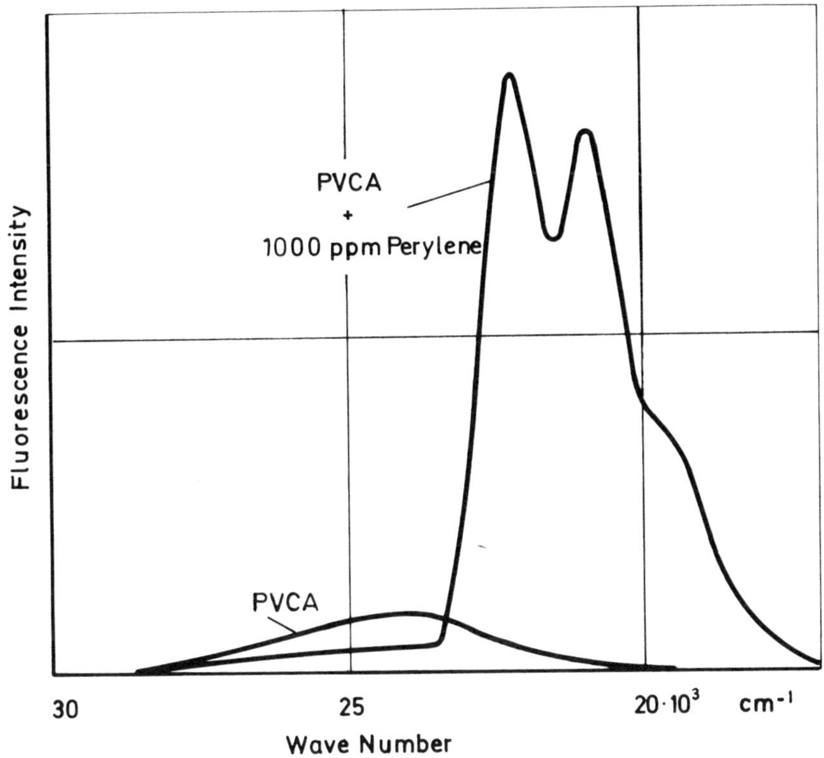

FIGURE 2. Energy transfer in solid PVCA/perylene[16] visualized by comparing the fluorescence spectra of pure polymer and polymer/guest.

great interest attributed to energy transfer processes in this polymer is rooted in Hoegl's discovery[21] of photoconductivity in PVCA and its sensitization by electron acceptors. Solid PVCA at room temperature shows excimer fluorescence[14,16] (FIG. 2) if fluorescent impurities have been removed by repeated dissolution/precipitation cycles (the bright blue fluorescence of some technical grade samples is due to anthracene impurities). No carbazole-like fluorescence can be detected in solid PVCA, indicating efficient energy transfer from carbazole side groups to excimer-forming sites. At this point, we should look for analogous systems in order to learn more about possible mechanisms of energy transfer which could be applied to aromatic polymers. The nearest possible analogue to the polymers considered here are amorphous layers of aromatic molecules which in general tend to crystallize at room temperature. More information has been available on single crystals and polycrystalline layers of aromatic molecules.[7,22-26] Most studies have been done on anthracene-type crystals,[26] showing weak overlap between neighboring molecules in the lattice and consequently monomer fluorescence. This behavior is shown, e.g., by *N*-isopropylcarbazole, the monomeric model of PVCA.[25] Crystals yielding excimer fluorescence, on the other hand, are characterized either by structures favoring a pair-like arrangement of molecules,[26] e.g. pyrene, or by structures resembling a stack of coins.[26] Energy transfer in anthracene-like crystals at room temperature is effected by the exciton-hopping mechanism[7] which at low temperature and unperturbed crystal structure may change to a wavelike exciton[20] picture, implying coherent energy spreading over many lattice sites.* This latter model is clearly inappropriate for the mostly amorphous polymers considered here. In excimer-forming, pyrene-like crystals, energy transfer is due to a thermally activated exciton-hopping process.[27,30] This mechanism is also not relevant to our case since a sterical arrangement involving all, or at least the majority, of the polymer side-groups in sandwich-pairs (excimer-forming sites) is highly unlikely for energetical reasons. A close analogue to polymer behavior is offered, however, by experimental work on externally disturbed crystals of the anthracene type: high pressure, for example, introduces defect sites[28] of suitably oriented pairs of molecules which can act as traps for the excitons and gradually changes the fluorescence spectrum from monomer to excimer characteristics depending on the concentration of traps created.

The following model has finally been adopted in order to explain energy transfer experiments in PVCA using different guest molecules (fluorescent and nonfluorescent ones)[16]:

- Absorption of uv radiation takes place randomly (but observing Lambert-Beer's Law), forming excited singlet states of side groups (carbazole groups in case of PVCA);
- The lowest excited singlet state forms an exciton (see asterisked footnote (*)) which can be trapped or quenched either at excimer-forming sites or, depending on their concentration, by guest molecules;
- The concentration of excimer-forming sites (c_E) is very small, so that direct excitation is negligible;
- Quantitatively, the excimer fluorescence intensity in doped films $I_F{}^E$ relative to that in the pure polymer $I^E_{F_0}$ can be expressed in the following equation:

$$Q_E = (I^E_{F_0} - I_F{}^E)/I_F{}^E \approx c_G/c_E \qquad \textbf{(1)}$$

where Q_E is the quenching factor[23] of excimer fluorescence, and c_G, c_E is the concentration of guest molecules and excimer-forming sites, respectively, in mol/mol basic unit.

Equation 1 is valid if there is complete energy transfer to excimer-forming sites in the absence of competing guest molecules and if there is no efficient single step transfer excimer → guest. This latter process is to be expected if the absorption spectrum of the guest strongly overlaps with the excimer fluorescence spectrum of the polymer, thus giving optimal conditions for Förster's mechanism[19] to apply. It has been found experimentally that the energy transfer PVCA to perylene (FIG. 2; Förster-allowed) is somewhat more efficient than from PVCA to hexachloro-*p*-xylene which is not suited for single-step transfer.[16] The linear increase of Q_E as a function of c_G has been confirmed experimentally.[16,29] The best estimate of the concentration

* Due to this double aspect of the exciton, there is some confusion about the correct use of the term which is used by theoretical physisists predominantly in its wave aspect. In experimental papers, however, the term is used for both aspects, since these can be strictly separated only in theory. We therefore propose to use the term "exciton" in polymer research synonymously with "energy migration" or "exciton diffusion" keeping in mind that in highly ordered structures and at very low temperatures coherent excitons may be encountered.

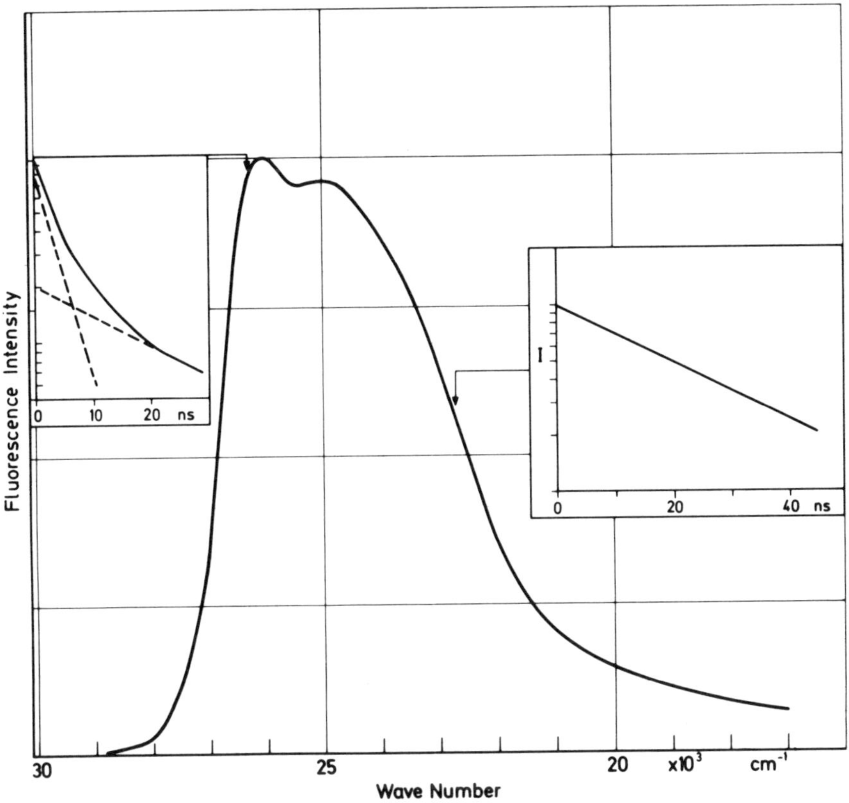

FIGURE 3. Low Temperature fluorescence spectrum of PVCA and fluorescence decay at two wavelengths.[33]

of excimer-forming sites in PVCA available is

$$c_E \approx 2 \times 10^{-3} \text{ mol/mol basic unit}$$

and seems to be independent to the mode of polymerization.

LOW TEMPERATURE LUMINESCENCE OF SOLID PVCA

The binding energy of carbazole excimers in PVCA, in other words, the singlet exciton trap depth is not known exactly, although a rough value of $E \approx 1000 \text{ cm}^{-1} \approx 5 \text{ kT}$ (room temperature) can be estimated from the spectra. At low temperature, shallow traps can become effective which at room temperature are hardly noticed by the excitons. P.C. Johnson and Offen[31] discovered a second fluorescence band in PVCA at 77°K (FIG. 3) which is due to a second intrinsic singlet exciton trap,[32,33] most probably a second excimer-forming site. It should be noted that a weak shoulder at the short wavelength side of the excimer fluorescence[34] and a second component in the fluorescence decay[33] have occasionally been observed even at room temperature. As a consequence of the appearance of the new "Trap II", singlet energy transfer should be less efficient at low temperature.

At low temperature, triplet-state processes which are quenched at ordinary temperatures often become observable. In PVCA, phosphorescence as well as delayed fluorescence can be observed at 77°K. As in the case of fluorescence, polymer phosphorescence is red-shifted with regard to monomer (carbazole) phosphorescence[32,33,35] and has been interpreted as excimer phosphorescence.[32] Triplet excitons have been evidenced by sensitized phosphorescence studies[32,37] and delayed fluorescence.[33,36–38] As will be discussed below in more detail, the range of the triplet excitons is restricted by phosphorescent and non-phosphorescent traps. There are indications that at least part of the triplet trap is shallow[39] so that triplet energy transfer in the absence of other quenching processes may be more efficient at higher temperatures.[40]

Early Energy Transfer Studies in Polymers Other Than PS and PVCA

Triplet-triplet energy transfer in glassy solutions was discovered by Terenin and Ermolaev in 1952.[41] Extending this work to polymer films, Lashkov and Ermolaev[42] studied poly-(*N*-vinylphthalimide). This polymer shows very efficient inter-system crossing to the triplet state, giving rise to phosphorescence with a maximum near 500 nm. Efficient energy transfer by triplet excitons has been observed[43] which was further enhanced by purification of the polymer using the dissolution/precipitation technique.[44]

Polyvinylbenzophenone is another polymer inclining predominantly to triplet energy transfer, as has been found by David, Demarteau and Geuskens.[44] Efficient energy transfer has been found using naphthalene as guests at 77°K and an exciton hopping rate $k_h{}^T = 7 \times 10^4 \text{ sec}^{-1}$ has been estimated using a somewhat modified theory by Voltz *et al.*[45]

Poly-(1-vinylnaphthalene) has been studied by the same authors[46] and exclusive excimer fluorescence has been taken as evidence for singlet energy migration. Triplet excitons have been detected in this polymer by Fox and coworkers using TT-annihilation as a probe.[47]

Experimental work reviewed up to this point provides plenty of evidence that energy transfer by singlet as well as triplet excitons is quite a common phenomenon in polymers with aromatic side groups; it is not restricted to solid polymers but also operates in dilute, rigid solution,[48] not discussed here. The same is true for liquid solutions, the most complicated case due to the superposition of electronic and molecular dynamics.

In the following, some recent results on low temperature energy transfer in PVCA will be reported, demonstrating a fairly complex behavior which may be typical for this polymer (e.g., more open-structured carbazole polymers do not show "Trap II").[49] On the other hand, more traps could be involved in low temperature energy transfer of other polymers as well, thus invalidating the simple models applicable for the room temperature behavior.

Energy Transfer in PVCA at Low Temperature

In 1976, Itaya, Okamoto and Kusabayashi[37,50] observed that the fluorescence of PVCA depends on the mode of polymerization which has been used: cationic polymerization favors "true" excimer fluorescence, whereas in radically polymerized samples "Trap II" fluorescence dominates. There is obviously a relationship between polymerization and the ability of the resulting polymer for developing the different excimer-forming sites in solution[50] or, frozen in during film formation, in the solid state.[37] It has been claimed by the same group[51] that the differences observed are due to predominant syndiotactic configuration of radicalic PVCA whereas the cationic polymers are said to contain up to 50% isotactic sequences. From molecular models, it can be deduced that the 1, 3-sandwich conformation responsible for the long wavelength excimer can easily be formed at isotactic sequences. Similarly for Trap II, a weakly overlapping 1, 3-structure has been proposed[50] which can be formed without difficulty in syndiotactic regions. Whatever the exact nature of this trap and the tacticity of PVCA samples may be, the relationship between fluorescence spectrum and means of polymerization has been confirmed.[52] In a quantitative study on energy transfer in PVCA at 77°K,[52,53] therefore, both types of polymers have been used. Films of carefully purified radicalic and cationic PVCA have been investigated with regard to their fluorescence, phosphorescence and delayed fluorescence as a function of the intensity of exciting radiation (313 nm). Furthermore, the decay of phosphorescence and delayed fluorescence has been studied.

The result of this work can be summarized as follows (see also TABLE 1):

- The 77°K fluorescence spectrum is dominated by "Trap II" emission. In addition, a strong "true excimer" fluorescence is observed in cationic PVCA samples, but only a weak one (often hardly observable) in radicalic polymers.
- Delayed emission consists of phosphorescence and delayed fluorescence in both types of polymers. The relative importance of the two emissions is a function of excitation intensity:
  —the phosphorescence is proportional to excitation intensity up to about 10^{14} photons sec^{-1} cm^{-2} and tends to saturate above this level; and

TABLE 1
SPECTROSCOPIC PROPERTIES OF PVCA AT 77°K

	PVCA Cat.	PVCA Rad.
Fluorescence	410 nm + 380 nm $\tau_F = 28 \pm 2$ nsec*	380 nm $\tau_F = 4 \pm 1$ nsec*
Phosphorescence	500 nm $\tau_P = 9.6 \pm 0.9$ and 2.1 ± 0.3 sec	
Delayed fluorescence	435 nm (corr.)† + 380 nm (sh, uncorr.) $\tau_{DF} = 40$ to 50 msec	380 nm (uncorr.) + 420 nm (uncorr.)
Approximate Concentration of EFS		
I	2×10^{-3}	
II	2×10^{-4}	6×10^{-3}
III	1.2×10^{-5}‡	2.6×10^{-4}‡
(mol/mol basic unit)		
Triplet exciton hopping rate	$k_h^T \approx 10^4$ sec^{-1}	

* Both components can be observed in the two types of polymers, although in different strengths.

† Corr. means that delayed Trap II fluorescence has been subtracted.

‡ This concentration can only be obtained as a product with k_e (excimer phosph., rad. rate) which has been taken as 0.02 sec^{-1} in order to obtain the values given.

—delayed fluorescence depends quadratically on radiation intensity below 10^{14} photons sec^{-1} cm^{-2} and only linearly above. Fluorescence, on the other hand, is linear over the whole range investigated (10^{10} to 10^{17} photons sec^{-1} cm^{-2}).

- The phosphorescence spectrum is broad and structureless with a maximum near 500 nm; the quantum efficiency is much smaller than that of (prompt) fluorescence.
- There are two exponential phosphorescence decay times of about 10 and 2 sec (see TABLE 1). The decay of delayed fluorescence is much faster (40 to 50 msec) and seems to be independent on polymer preparation.
- The delayed fluorescence spectrum is red-shifted with regard to prompt fluorescence.

In order to describe these facts quantitatively, a kinetic model has been developed which will be published elsewhere.[54] The basic assumptions of this model are (see also FIG. 4):

- Energy transfer can be described with a simple hopping model,[7,16] the probability of trapping (or annihilation) being given by k_h (sec^{-1}) × c_E (mol per mol basic unit) where k_h is the triplet or singlet exciton hopping rate and c_E the concentration of one of at least three different excimer-forming sites.
- The differences between radicalic and cationic polymers consist only in different concentrations of the excimer-forming sites, not in their nature. All hopping rates are assumed to be equal in both types of polymers.
- Annihilation (mostly TT) occurs between excitons and filled traps (excimers), the trap depth being much larger than kT at 77°K.

A set of rate equations describing the formation and the decay of singlet and triplet excitons and excimers has been developed and solved for stationary conditions. The most important result of this kinetic analysis of the experiments is the existence of separate phosphorescent excimers which are not related to one of the main singlet excimers, as proposed earlier by us.[14,27] Otherwise, the linearity of fluorescence *versus* excitation intensity could not be explained: fluorescence should saturate at high intensities due to singlet-triplet annihilation, a Förster-type, spin-allowed process which is bound to occur if saturated triplet traps are present in the polymer in a high concentration. Direct experimental proof for this fact besides the linear dependence of I_F *vs* I_A (see FIG. 5) is the red-shifted delayed fluorescence, indicating that those singlets created in TT-annihilation and staying at the site of their formation (some of them are migrating away) have a lower energy, i.e., higher excimer binding energy, and are therefore different from the states causing the major portion of prompt fluorescence at 77°K.

Unless an unreasonably long radiative phosphorescence lifetime (III) on the order of hours is accepted (caused by small quantum efficiency and slow decay), an efficient nonradiative triplet trap has to be present in the PVCA samples. This has tentatively been ascribed to the triplet state of excimer I. Trap II (EFS II), on the other hand, is unlikely to provide a binding triplet excimer state, since neither additional quenching of triplet exciton diffusion nor saturation of fluorescence is observed in radicalic PVCA.

In this model, delayed fluorescence is due to TT-annihilation between migrating triplet excitons and immobile triplet excimers (III). In the low intensity regime, increasing the excitation intensity increases both the triplet exciton density (by intersystem crossing at EFS II or at the monomer level) and the probability of triplet

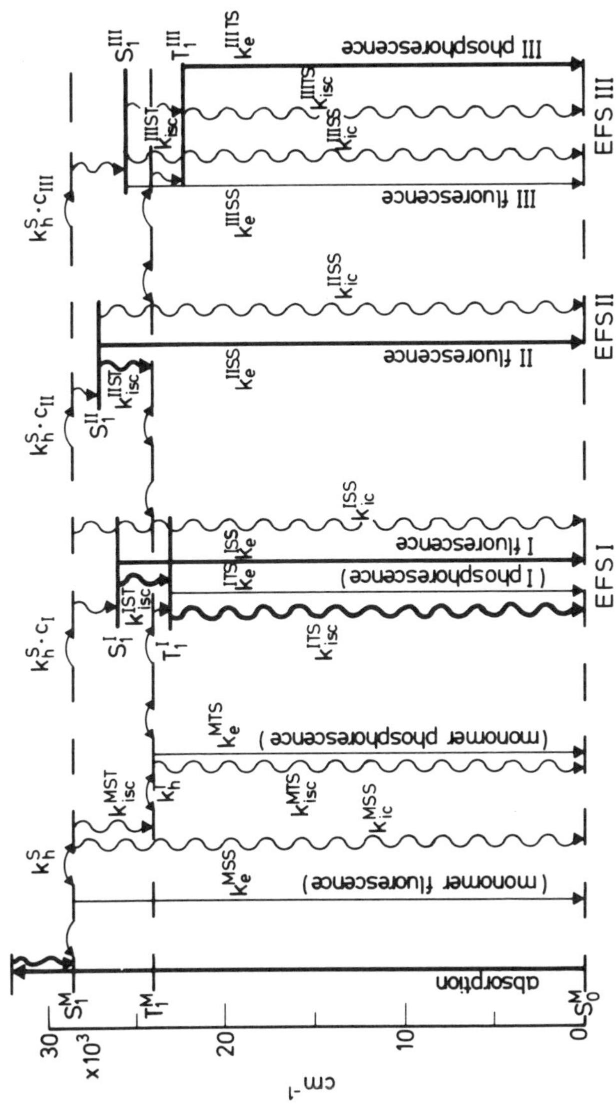

FIGURE 4. Schematic view of exciton-diffusion and trapping in PVCA. Only one triplet excimer-forming site is shown (EFS III). EFS I corresponds to the "true" excimer, EFS II to trap II. Processes not observed experimentally are shown in brackets.

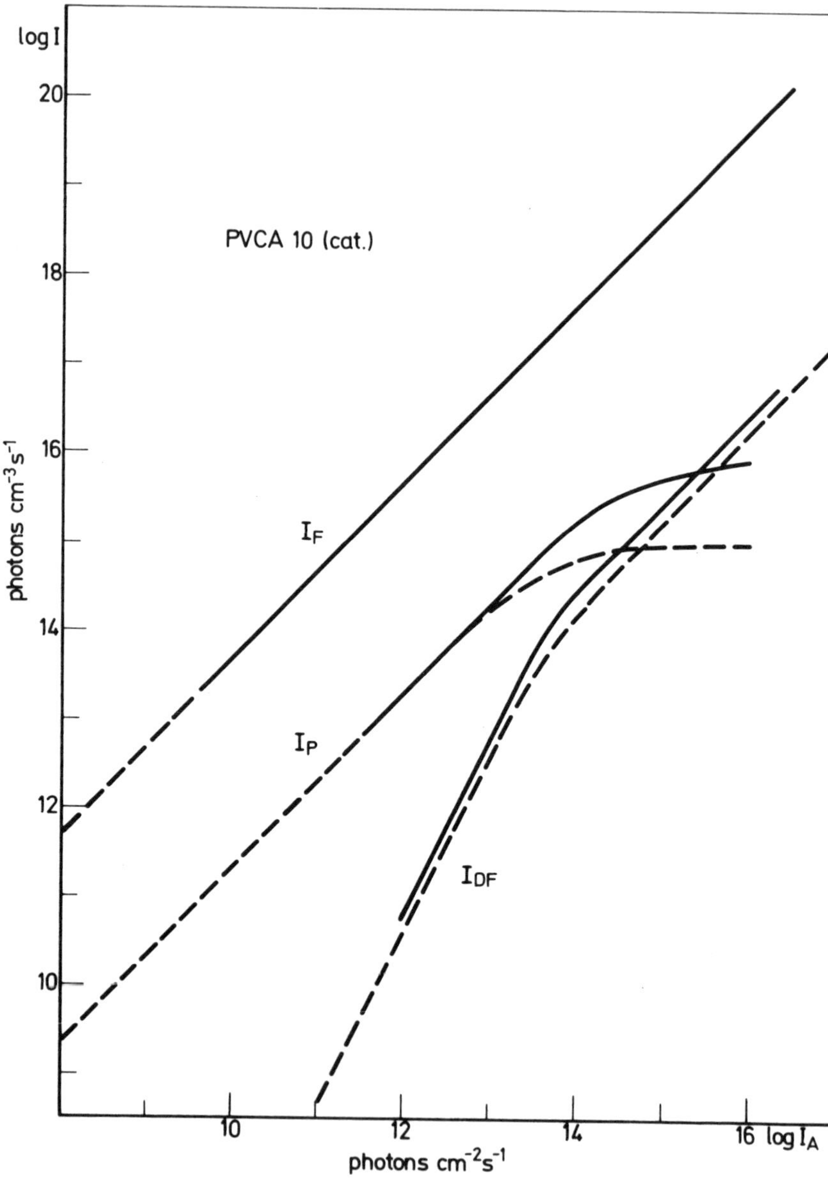

FIGURE 5. Luminescence intensities of PVCA at 77° K as a function of the excitation intensity, experimental (*solid lines*) and fitted according to the kinetic model (*dashed lines*).

excimer formation (both via singlet and triplet excitons diffusing to EFS III). Thus, delayed fluorescence depends quadratically on excitation intenisty in this range. Under high excitation intensity, the long-lived triplet excimer saturates and so does phosphorescence intensity, whereas the delayed fluorescence switches to linear behavior since only the triplet exciton production increases further with intensity. Quantitatively, this behavior can be satisfactorily reproduced by the kinetic model using some realistic estimates for those parameters which have not yet been measured.

FIGURE 5 shows experimental and — fitted using fluorescence quantum efficiency — calculated curves for the luminescence intensity of a cationic PVCA as a function of excitation intensity. It is important to note that using the same parameters except the concentrations of EFS II and III (the density of EFS I seems to be independent on polymer preparation[50,51]) a very reasonable fit of experimental and theoretical data has been found for the PVCA (rad.) investigated in detail.[54] The kinetic model therefore seems to be essentially correct, although some assumptions certainly will turn out to be oversimplified. It can be seen in FIGURE 5 that the calculated curve for I_P as a function of I_A levels off at lower intensity than found experimentally; this is due to the neglect of the second, shorter-lived phosphorescence which, of course, saturates at higher excitation intensity. Considering the limited amount of experimental data available and the quite complex pattern of traps in PVCA, a further refinement of theory did not seem promising.

The experiments presented in this section have been performed at 77°K and the EFS have been have been assumed to be deep traps for both singlet and triplet excitons. It has been mentioned that EFS II is a shallow trap at room temperature which can be identified by means of decay measurements[33] and only in a few cases in the fluorescence spectrum.[34,52,59] There is no reason to believe that the phosphorescent traps are not shallow ones with regard to higher temperatures, the binding energy of triplet excimers being generally believed to be lower than those of the corresponding singlet excimers. Actually, Burkhart and Avilés[55–57] found evidence for detrapping of the phosphorescent states in PVCA in the temperature range between 77°K and room temperature. They also observed for the first time delayed luminescence in pure PVCA at room temperature,[55] whereas energy transfer to phosphorescent guest molecules had been detected earlier using the chemiluminescence technique;[40] room temperature phosphorescence has recently been confirmed by Chow.[58] The pressure dependence of PVCA fluorescence has recently been studied up to 40 kbar[59] confirming the existence of two independent excimers.

Finally, it should be mentioned that partly similar effects as described here for PVCA at 77°K have also been observed in solid poly(2-vinylnaphthalene) by Kim and Webber.[60]

OUTLOOK

This presentation has been restricted to the behavior of pure solid polymers, investigated mostly in the form of thin films, monomeric guest molecules being introduced as probes whenever necessary. As has been shown, these probes are not required in studying triplet excitons by TT-annihilation. I did not mention the important case of polymer blends where the excimer-forming polymer (guest) is used for sensitive probing of the interaction between host and guest polymer.[61,62] Similar applications have recently been reviewed by Morawetz.[63] It is a great pleasure to see that some work on energy transfer and excimer formation, aiming at the elucidation of the mechanisms of scintillation, radiation protection and photoconductivity (by the way, another type of energy transfer)[64] turn out to be a useful tool in solving problems of polymer chemistry. At this point, it must also be remembered how important the contributions from studies of biopolymers[65] have been in the early stage of work on synthetic polymers. The relationship between energy transfer processes in synthetic polymers, molecules in solution, and biological systems was recognized more than 20 years ago and was the topic of a famous discussion of the Faraday Society.[66]

Going back to energy transfer in solid, synthetic polymers, it seems appropriate to ask what should be done in the near future in order to improve our rudimentary knowledge about the basic processes involved. Clearly, the polymers we are using should be much better characterized with regard to tacticity, molecular mass (M) and endgroups. Ideally, purely iso- and syndio-tactic polymers with narrow M-distribution should be available. This constitutes a challenge for synthetic polymer chemistry. Furthermore, nearly nothing is known about the sterical arrangement in polymers with large pendant groups in solution as well as in the solid state. Molecular models as well as diffraction studies on partly crystalline PVCA suggested some kind of helical order in this polymer, albeit a perturbed one.[67]

Using well-defined polymers, it would be worthwhile to extend time-dependent studies down to the picosecond range in order to probe the detailed mechanism of exciton diffusion which has recently been claimed to be dispersive, i.e., faster in the first stage of the (singlet-) exciton's lifetime.[68]

Finally, the nature of the traps controlling the solid-state energy transfer behavior of aromatic polymers must be elucidated. This goal can hardy be accomplished without appropriate dimeric or even trimeric model compounds of known configuration and conformation. The work of the teams headed by Monnerie[69] and Staab[70] may be considered exemplary in this regard. Questions related to molecular and electronic structure of excimers (singlet and triplet) clearly have priority in exploring solid-state traps. It seems, however, that in rigid solutions of aromatic polymers additional, shallow traps can be identified[48] not showing the typical emission spectra of excimers but rather weakly red-shifted luminescences showing vibrational structure similar to that of monomers although in most cases a less resolved one. There is no reason to believe that these traps, though not detected so far, are absent in solid polymers.

References

1. Schorr, M. G. & F. L. Torney. 1950. Phys. Rev. **80:** 474.
2. Koski, W. S. 1951. Phys. Rev. **82:** 230.
3. Swank, R. K. & W. L. Buck. 1953. Phys. Rev. **91:** 927.
4. Kreuz, F. H. 1955 Trans. Faraday Soc. **51:** 172.
5. Brown, F. H., M. Furst & H. Kallmann. 1959. Disc. Faraday Soc. **27:** 43.
6. Birks, J. B. & K. N. Kuchela. 1959. Disc. Faraday Soc. **27:** 57.
7. Wolf, H. C. (1965). *In* Festkörperprobleme. F. Sauter, Ed. Vol. **4:** 57. Vieweg & Sohn Braunschweig, West Germany.
8. Förster, Th. 1946. Naturwissenschaften **33:** 166.
9. Förster, Th. 1948. Ann. Phys. **2:** 55.
10. Kaplan, M. L., P. G. Kelleher. 1970. J. Pol. Sci. **A** 1–8: 3163.
11. Basile, L. J. 1962. J. Chem. Phys. **36:** 2204.
12. Hirayama, F. 1965. J. Chem. Phys. **42:** 3163.
13. Vala, Jr., M. T. J. Haebig & S. A. Rice. 1965. J. Chem. Phys. **43:** 886.
14. Klöpffer, W. 1973. *In* Organic Molecular Photophysics. J. B. Birks, Ed. Vol. **1:** 357. J. Wiley & Sons. London.
15. Hirayama, F., L. J. Basile & C. Kikuchi. 1968. Molecular Cryst. **4:** 83.

16. KLÖPFFER, W. 1969. J. Chem. Phys. **50:** 2337.
17. JONES, P. F., S. SIEGEL & R. S. NESBITT. 1972. Singlet- and triplet-state electronic energy transfer in poly(styrene). Paper presented at the IUPAC Symposium on Photochemical Processes in Polymer Chemistry, Leuven.
18. PERRIN, J. 1923. Compt. rend. (Paris). **177:** 469.
19. FÖRSTER, TH. 1951. *In* Fluoreszenz Organischer Verbindungen. Vandenhoeck & Ruprecht. Göttingen.
20. A. S. DAVYDOV. 1951. Theory of the Absorption of Light in Molecular Crystals. Kiev. Translated by M. Kasha & M. Oppenheimer. 1962. *In* Theory of Molecular Excitons. Mc Graw-Hill Comp. New York
21. HOEGL, H. 1965. J. Phys. Chem. **69:** 755.
22. WINDSOR, M. W. 1965. *In* Physics and Chemistry of the Organic Solid State. D. Fox, M. M. Labes, A. Weissberger Eds. Vol. **2:** 343. J. Wiley & Sons. New York.
23. NORTHROP, D. C. & O. SIMPSON. 1956. Proc. Roy. Soc. (London) **A 234:** 136.
24. HOCHSTRASSER, R. M. 1962. Rev. Mod. Phys. **34:** 531.
25. KLÖPFFER, W. 1969. J. Chem. Phys. **50:** 1689.
26. STEVENS, B. 1962. Spectrochim. Acta **18:** 439.
27. KLÖPFFER, W., H. BAUSER, F. DOLEZALEK & G. NAUNDORF. 1972. Mol. Cryst. Liqu. Cryst. **16:** 229.
28. JONES, P. F. & M. NICOL. 1968. J. Chem. Phys. **48:** 5457.
29. OKAMOTO, K., A. YANO, S. KUSABAYASHI & H. MIKAWA. 1974. Bull. Chem. Soc. (Japan) **47:** 749.
30. YAKHOT, V. 1979. *In* Advances in Photochemistry, J. N. Pitts, Jr., G. S. Hammond, K. Gollnick, Eds. Vol. **11:** 489. J. Wiley & Sons. New York.
31. JOHNSON, P. C. & H. W. Offen. 1971. J. Chem. Phys. **55:** 2945.
32. KLÖPFFER, W. & D. FISCHER. 1973. J. Pol. Sci. Symp. No. **40:** 43.
33. KLÖPFFER, W. & H. BAUSER. 1976. Z. Phys. Chem. NF. **101:** 25.
34. POWELL, R. C. & Q. KIM. 1973. J. Luminescence **36:** 351.
35. LASHKOV, G. I. 1970. Khim. Vysok. Energii **4:** 405.
36. BURKHART, R. D. 1976. Macromolecules **9:** 234.
37. ITAYA, A., K.-I. OKAMOTO & S. KUSABAYASHI. 1976. Bull. Chem. Soc. (Japan). **49:** 2037.
38. YOKOYAMA, M., T. TAMAMURA, T. NAKANO & H. MIKAWA. 1976. J. Chem. Phys. **65:** 272.
39. BURKHART, R. D. & R. G. Avilés. 1979. Macromolecules **12:** 1078.
40. KLÖPFFER, W., N. J. TURRO, M. F. CHOW & Y. NOGUCHI. 1978. Chem. Phys. Lett. **54:** 457.
41. TERENIN, A. N. & V. L. ERMOLAEV. 1952. Dokl. Akad. Nauk SSR. **85:** 547.
42. LASHKOV, G. I. & V. L. ERMOLAEV. 1967. Opt. Spectr. **22:** 848. English Translation **22:** 462.
43. LASHKOV, G. I. & V. L. ERMOLAEV. 1968. Theoret. i. elsperim. Khim. **4:** 918.
44. DAVID. C., W. DEMARTEAU & G. GEUSKENS. 1970. Eur. Pol. J. **6:** 537.
45. VOLTZ, R., G. LAUSTRIAT & A. COCHE. 1963. J. Chim. Phys. **63:** 1253.
46. DAVID, C., W. DEMARTEAU & G. GEUSKENS. 1970. Eur. Pol. J. **6:** 1397.
47. FOX, R. B., T. R. PRICE, R. F. COZZENS & J. R. MCDONALD. 1972. J. Chem. Phys. **57:** 2284.
48. KLÖPFFER, W. 1978. Spectroscopy Letters. **11:** 863.
49. JOHNSON, G. E. 1975. J. Chem. Phys. **62:** 4697.
50. ITAYA, A., K.-I. OKAMOTO & S. KUSABAYASHI. 1976. Bull. Chem. Soc. Japan **49:** 2082.
51. OKAMOTO, K.-I., M. YAMADA, A. ITAYA, T. KIMURA & S. KUSABAYASHI. 1976. Macromolecules **9:** 645.
52. RIPPEN, G. & W. KLÖPFFER. 1979. Ber. Bunsenges. Phys. Chem. **83:** 437.
53. RIPPEN, G., G. KAUFMANN & W. KLÖPFFER. 1980. Chem. Phys. **52:** 165.
54. KLÖPFFER, W. 1980. See,[52] Part II. Kinetic Model. Chem. Phys. In press.
55. BURKHART, R. D. & R. G. AVILÉS. 1979. J. Phys. Chem. **83:** 1897.
56. BURKHART, R. D. & R. G. AVILÉS. 1979. Macromolecules **12:** 1073.
57. BURKHART, R. D. & R. G. AVILÉS. 1979. Macromolecules **12:** 1078.
58. TURRO, N. J. 1979. Private communication.
59. CHRYSSOMALIS, G. & H. G. DRICKAMER, 1979. J. Chem. Phys. **71:** 4817.
60. S. E. WEBBER. 1980. Private communication.
61. FRANK, C. W. & M. A. GASHGARI. 1979. Macromolecules **12:** 163.
62. FRANK, C. W. & M. A. GASHGARI. 1981. Excimer fluorescence as a molecular probe of polymer-polymer interactions in the amorphous solid state. Ann. N.Y. Acad. Sci. **366:** 387–403.

63. MORAWETZ, H. 1979. Science **203:** 405.
64. TERENIN, A., E. PUTZEIKO & I. AKIMOV. 1959. Disc. Faraday Soc. **27:** 83.
65. LONGWORTH, J. W. & F. A. BOVEY. 1966. Biopolymers **4:** 1131.
66. Discussions of the Faraday Society. 1959. Vol. **27.**
67. GRIFFITHS, C. H. 1975. J. Pol. Sci. Pol. Phys. Ed. **13:** 1167.
68. SCHÖNHERR, G., R. EIERMANN, H. BÄSSLER & M. SILVER. 1980. Dispersive exciton transport in a hopping system with gaussian energy distribution. Chem. Phys. In press.
69. BOKOBZA, L., B. JASSE & L. MONNERIE. 1977. Eur. Pol. J. **13:** 921.
70. STAAB, H. A. & M. HAENEL. 1973. Chem. Ber. **106:** 2190.

EXCIMER FLUORESCENCE AS A MOLECULAR PROBE OF POLYMER-POLYMER INTERACTIONS IN THE AMORPHOUS SOLID STATE*

C.W. Frank and M.A. Gashgari

Department of Chemical Engineering
Stanford University
Stanford, California 94305

INTRODUCTION

A fundamental question for the analysis of both the thermodynamic and kinetic behavior of polymer blends is the extent to which the blend components are mixed on the molecular level.[1,2] Unfortunately, this is inextricably bound up with the choice of experimental method and depends upon the distance scale over which random mixing may be assumed. Use of one technique may lead to the conclusion that a particular blend is compatible while use of a more sensitive method may allow detection of phase separation into domain sizes below the sensitivity limits of the first. In addition, standard techniques such as measurement of glass transition temperatures by dynamic mechanical spectroscopy (DMS) and differential scanning calorimetry (DSC) are insensitive to the presence of a second polymer phase at concentrations less than about 10%, regardless of domain size. Finally, in the region where extensive interfacial mixing occurs, transitions are exceedingly difficult to detect by either DMS or DSC. In view of these difficulties, there is strong motivation to search for new techniques with greater sensitivity.

One new method which shows promise as a molecular probe of polymer-polymer interactions in the solid state is excimer fluorescence. An excimer is an electronically excited molecular complex formed between two coplanar aromatic rings.[3] Since the ground state of the transient species is repulsive, the excimer may be examined only by fluorescence spectroscopy. This is one of the major advantages of the technique because fluorescence is extremely sensitive to low concentrations. Another special feature of the excimer probe which makes it attractive for studies of molecular structure is that electronic stabilization of the complex will result only for fairly specific geometrical configurations of the rings. Nominal excimer-forming sites in the aromatic vinyl polymers may arise intramolecularly between either adjacent or nonadjacent repeat units and intermolecularly between rings on different polymer chains. The most convenient experimental measure of excimer fluorescence under photostationary-state conditions is the quantity I_D/I_M, where I_D and I_M are the emission intensities of the excimer and monomer bands, respectively.

The objective of this work is to review and extend recent studies of excimer fluorescence in blends of poly(2-vinyl naphthalene) (P2VN), an excimer-forming

*This work has been supported by the Polymers Program of the Division of Materials Research, National Science Foundation under Grants 77-09372 and 79-16477.

0077-8923/81/0366-0387$01.75/0 © 1981, NYAS

polymer, with a homologous series of poly(alkyl methacrylates) and with polystyrene. Three experiments will be described. The first two emphasize blend thermodynamics through consideration of enthalpic interactions and concentration effects on the aggregation of guest aromatic vinyl polymer chains in a host polymer matrix. Their common element is the interest in characterizing the blend morphology in the glassy state where molecular mobility is highly restricted. In the third study, attention is focused on molecular relaxation processes which result upon heating the blends to temperatures in the vicinity of the glass transition.

EXPERIMENTAL

The poly(2-vinyl naphthalene) sample used in the host matrix and concentration studies was prepared by bulk thermal polymerization of doubly sublimed 2-vinyl naphthalene monomer obtained from Polysciences. Polymerization of the dried monomer was carried out at 80°C for 72 hours in a sealed glass ampule which had been purged with nitrogen and then evacuated to a pressure of 5×10^{-6} torr before sealing. Benzoyl peroxide, 0.1% by weight, was used as a free radical initiator. The polymer was freed of residual monomer by three times precipitating from toluene solution. The purity of the P2VN was verified by spectroscopic means. The P2VN sample used in the relaxation experiments was prepared by bulk thermal polymerization for 96 hours at 120°C, using 0.1% di-tertiary butyl peroxide initiator, as described previously.[4]

The poly(alkyl methacrylate) hosts for the concentration study, poly(methyl methacrylate) (PMMA), poly(ethylmethacrylate) (PEMA), and poly (*n*-butyl methacrylate) (P*n*BMA) were freed of residual monomer by precipitating three times from acetone by the addition of deionized water. PMMA (Cat. #4553, medium molecular weight), (Cat. #4552, Lot #1889-125, high molecular weight), (Cat. #4554, Lot #1888-125, low molecular weight), and P*n*BMA (Cat. #2061, Lot #21-38-7, high molecular weight) were supplied by Polysciences and precipitated from 5% by weight acetone solutions with a water: acetone ratio of 10:1. A secondary standard sample of PEMA (Cat. #1-8335-0, Lot #01, high molecular weight) was supplied by Aldrich and used without further purification since fluorescence spectra of a blank film-showed no monomer impurities. A similar purification procedure was used for the remaining poly(alkyl methacrylate) hosts, as described previously.[6] The polystyrene host for the relaxation experiments was a broad distribution sample of Dylene 3, supplied by Sinclair Koppers Company.[5]

Molecular weights for the host polymers of the concentration study were estimated from intrinsic viscosity measurements to be: PMMA (80,400; 125,400; 163,400), PEMA (395,000) and P*n*BMA (254,300). The P2VN guest polymer molecular weight was 71,400 for the host and concentration studies and 178,000 for the relaxation experiments. Densities, measured by flotation, were: PMMA (1.106 g/cm^3), PEMA (1.059 g/cm^3), P*n*BMA (0.992 g/cm^3) and P2VN (1.01 g/cm^3). Glass transition temperatures, determined by probe penetration using a Perkin-Elmer TMS 1 were: PMMA (105°C), PEMA (68°C), P*n*BMA (20°C) and P2VN (132°C).

Solid films of the blends were prepared by solvent casting from spectrophotometric grade toluene on glass plates which were covered and left to air dry for approximately 24 hours. Each PMMA host film was cast by spreading two milliliters of 0.0833 gram/ml solution onto a 7 cm × 10 cm rectangle. PEMA and P*n*BMA host films were cast in the same way from 0.0667 gram/ml solutions. All films were then removed from the casting plates and drying was continued under vacuum until a constant weight was obtained. The final film thickness was 15–25 microns.

The spectrofluorimeters for the thermodynamics[6] and relaxation measurements[7] have been described previously.

RESULTS

Effect of Host Matrix

Host matrices for the poly(2-vinyl naphthalene) guest polymer were selected on the basis of both thermodynamic and spectroscopic criteria. The classical work in thermodynamics of polymer solutions is due to Flory[8,9] and Huggins.[10,11] Scott[12] and Tompa[13] were the first to apply Flory-Huggins theory to polymer blends, obtaining the following expression for the free energy of mixing:

$$\Delta G_{mix} = \frac{RTV}{V_r}\left[\frac{\varphi_1}{x_1}\ln \varphi_1 + \frac{\varphi_2}{x_2}\ln \varphi_2 + \chi_{12}\varphi_1\varphi_2\right] \quad (1)$$

where V_r is a reference volume, taken to be the smaller of the repeat unit molar volumes for the two blend components, φ_i is the volume fraction of component *i*, x_i is the degree of polymerization of component *i*, and χ_{12} is the binary interaction parameter. The first two terms in Equation 1 arise from the entropic contribution which is negative and generally small for blends of high molecular weight polymers. Under such conditions, the enthalpic contribution from the third term governs whether ΔG_{mix} will be negative, a necessary but insufficient condition for a blend to be compatible. If regular solution theory is applicable to the polymer blends, the binary interaction parameter is given by

$$\chi_{12} = \frac{V_r}{RT}(\delta_1 - \delta_2)^2 \quad (2)$$

where δ_i is the Hildebrand solubility parameter of component *i*.[14,15]

Krause[16,17] indicates that the solubility parameter scheme works best for polymers with similar polarity and hydrogen bonding characteristics as would be found in a homologous series. The requirement that such a series be used is the first criterion for host matrix selection. From Equation 2, it follows that a broad spectrum of enthalpic interactions between guest and host will be obtained if the host matrices cover a wide range of solubility parameters. When the solubility parameters of the blend components are equal, the interaction parameter vanishes and ΔG_{mix} is determined by the entropic term. Thus, the second criterion is to select members of the series which have solubility parameters bracketing that of the guest aromatic vinyl polymer. Finally, the host matrix must be optically transparent both to the incident radiation at 290 nanometers and to the resulting fluorescence between 320 and 550 nanometers. All of these criteria are met by the poly(alkylmethacrylates) whose structures are shown in TABLE 1.

Fluorescence results for blends of P2VN (0.2% by weight, solvent cast at 22°C) with each of the poly(alkyl methacrylates) and with polyvinyl acetate (PVAc) and

TABLE 1
HOST POLYMERS

Poly(alkyl methacrylates)

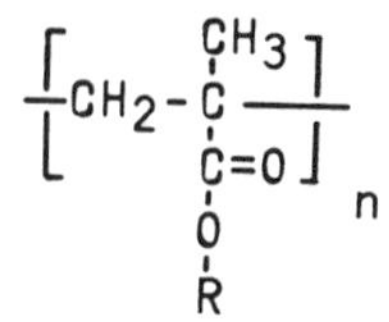

R	CODE	R	CODE
$-CH_3$	PMMA		
$-CH_2CH_3$	PEMA	$-C(CH_3)_3$	PtBMA
$-CH_2CH_2CH_3$	PnPMA		
$-CH(CH_3)_2$	PiPMA	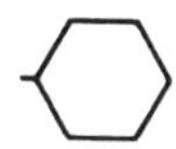	PCMA
$-CH_2CH_2CH_2CH_3$	PnBMA	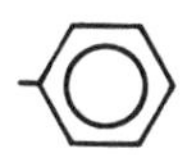	PPhMA
$-CH_2CH(CH_3)_2$	PiBMA	$-CH_2$ 	PBzMA
$-CH(CH_3)CH_2CH_3$	PsBMA	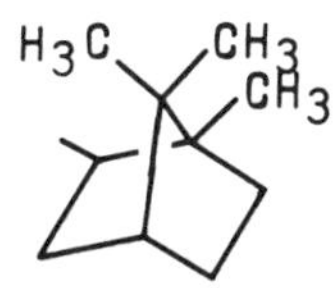	PiBoMA

polystyrene (PS) are shown in FIGURE 1. Here the excimer to monomer fluorescence intensity ratio, I_D/I_M, has been plotted against the difference in guest and host solubility parameters. The major observation about these results is that the smooth curve drawn through the data is perfectly symmetric in the solubility parameter difference, indicating that it is the absolute magnitude and not the sign that is important. Furthermore, it is significant that the extremum in $I_D I_M$ is a *minimum*. Finally, even though there is moderate variation in host molecular weights over the homologous series, there is no appreciable deviation from the smooth curve behavior. These observations lead to the conclusion that enthalpic interactions predominate. Moreover, since these interactions are expected to be mainly dispersive in nature, the use of regular solution theory and Hildebrand solubility parameters should be quite justified for these blends. Similar fluorescence behavior has been observed for poly(acenaphthalene) and poly(4-vinyl biphenyl) in the same host series.[18]

EFFECT OF CONCENTRATION

Three of the poly(alkyl methacrylate) hosts in TABLE 1 were selected for further study: PMMA, PEMA and PnBMA. These particular polymers were selected primarily because they exhibit a modest range of solubility parameter differences with values of $\delta_{host} - \delta_{guest}$ equal to -0.15 cal$^{1/2}$/cm$^{3/2}$ for PnBMA, $+0.26$ cal$^{1/2}$/cm$^{3/2}$ for PEMA and $+0.52$ cal$^{1/2}$/cm$^{3/2}$ for PMMA, and they were relatively easy to prepare in solid film form by solvent casting. Typical data illustrating the effect of concentration on I_D/I_M for films cast at 22°C are presented in FIGURE 2. Each data point represents an average of several fluorescence determinations with the vertical bars indicating the standard deviation. The numbers in parentheses beside the identifying host codes for each smoothed line through the data are the absolute values of the solubility parameter differences.

There are several significant qualitative features of these data. First, each blend exhibits a linear dependence of I_D/I_M on guest concentration over the range of 0.1 to 8.0% by weight. Second, the slopes of the lines increase as the absolute value of the solubility parameter difference and, hence, the binary interaction parameter, increase. Third, if the solid lines were extrapolated to infinite dilution, different intercepts would be obtained for each blend.

The final observation is of a different character and is related to the optical clarity of the blends. It is not possible to state unequivocally that an optically clear film of a polymer blend is thermodynamically compatible. This is because two separate transparent layers may form, or the two polymers may have equal refractive indices

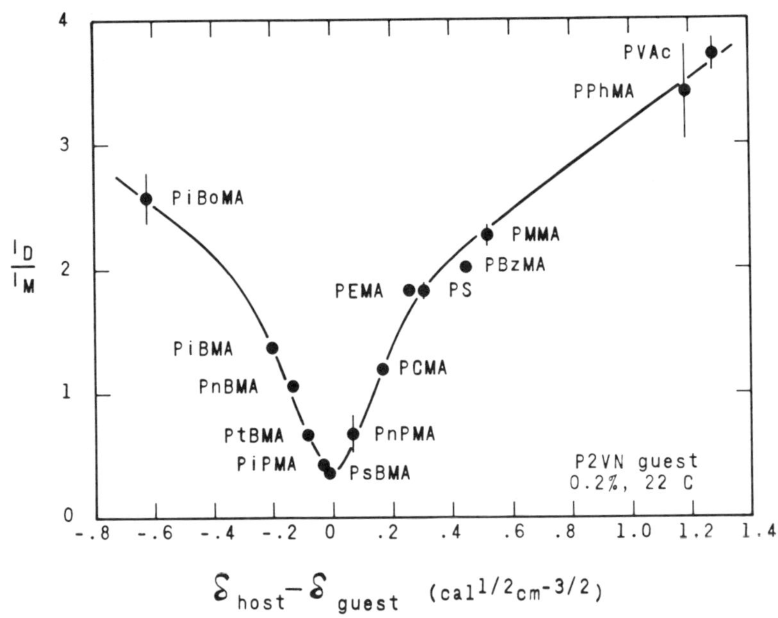

FIGURE 1. Dependence of observed excimer fluorescence in the guest polymer on enthalpic interactions with the host polymer.

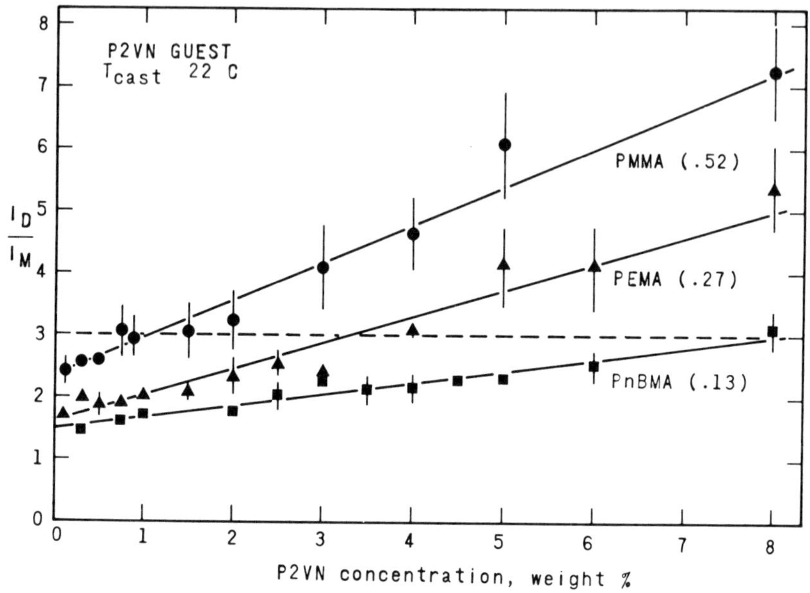

FIGURE 2. Dependence of observed excimer fluorescence on guest concentration.

or the dispersed phase may have dimensions smaller than the wavelength of visible light. However, optically cloudy films almost certainly represent thermodynamically incompatible blends. It is interesting to note that the demarcation between clear and cloudy films occurs approximately at $I_D/I_M = 3.0 \pm 0.5$ for each of the three host matrices. This region is indicated by the dashed line in FIGURE 2. Blends with I_D/I_M ratios above this line are cloudy and assuredly phase-separated whereas clear films of, as yet, unknown morphology on the molecular level are observed below the line. An important aspect of these results is that the subjective determination of optical quality is essentially bimodal, i.e., the film is either cloudy or it is not, whereas the quantitative fluorescence measurement allows continuous changes in I_D/I_M to be followed. It is particularly significant that there is no break in the fluorescence results at $I_D/I_M = 3.0$. What is causing the increase in I_D/I_M at high concentrations in phase-separated blends must also be occurring even when there is no visible indication. Thus, the excimer fluorescence method may provide information about the early stages of phase separation.

One of the more interesting questions regarding the results of FIGURE 2 is whether the very low concentration data, less than 0.1% by weight, agree with the extrapolation of the higher concentration points. To determine this, the fluorescence of P2VN blends with two of the poly(alkyl methacrylate) hosts, PMMA and P*n*BMA, and with polystyrene, were examined over the concentration range of 0.003% to 0.1%. These results are shown in FIGURE 3. The linear portions of the solid lines above concentration of 0.2% are the extrapolations of the high concentration data. Clearly, the fluorescence behavior at very low concentrations is quite different than at high values. The most significant feature is that the value of I_D/I_M at infinite dilution now appears to be independent of the host matrix for each of the three systems.

Taking the results of FIGURES 2 and 3 together, there are three features that must be analyzed: the common value of I_D/I_M at infinite dilution, the abrupt rise in excimer fluorescence between 0 and 0.1% and the linear increase in I_D/I_M above 0.1%. To simplify the later analysis of the high concentration results, the observed I_D/I_M values for each of the five poly(alkyl methacrylate) blends have been corrected by subtraction of the intercept based upon extrapolation of the high concentration data. The smoothed results are shown in FIGURE 4. Linear increases in I_D/I_M are observed for all blends. Moreover, it is clear that as the molecular weight of the host matrix increases, so does the slope.

Relaxation Near the Glass Transition

The underlying assumption for both of the previous studies on the effect of host matrix and concentration on excimer fluorescence in glassy blends is that the solvent casting process permits the blend morphology to be "frozen in" upon solvent evaporation. Although solvent casting is a common method for preparation of polymer films, some measure of caution must be used in its application. First, the choice of casting solvent can affect the morphology and compatibility behavior of the resulting blend. This has been found to be true for blends of polystyrene with poly(vinyl methyl ether).[19] More important, however, are the kinetic factors associated with the glass transition region. By way of illustration we consider a hypothetical ternary polymer 1-polymer 2-solvent system which is compatible at high solvent concentration but thermodynamically incompatible at very low solvent concentra-

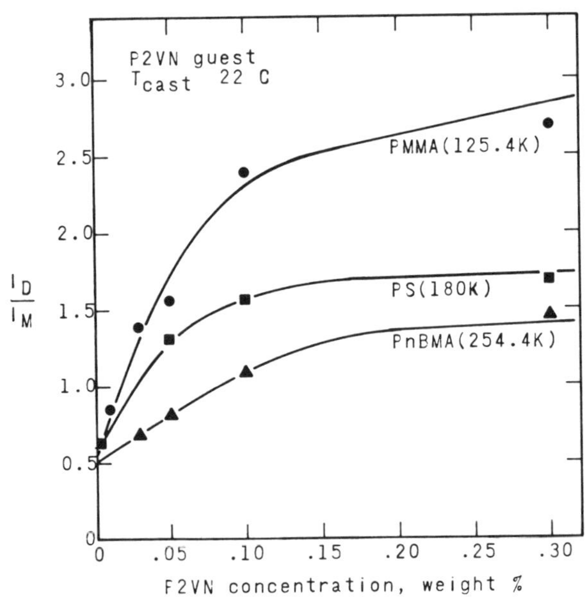

FIGURE 3. Dependence of observed excimer fluorescence on the host polymer at very low values of concentration.

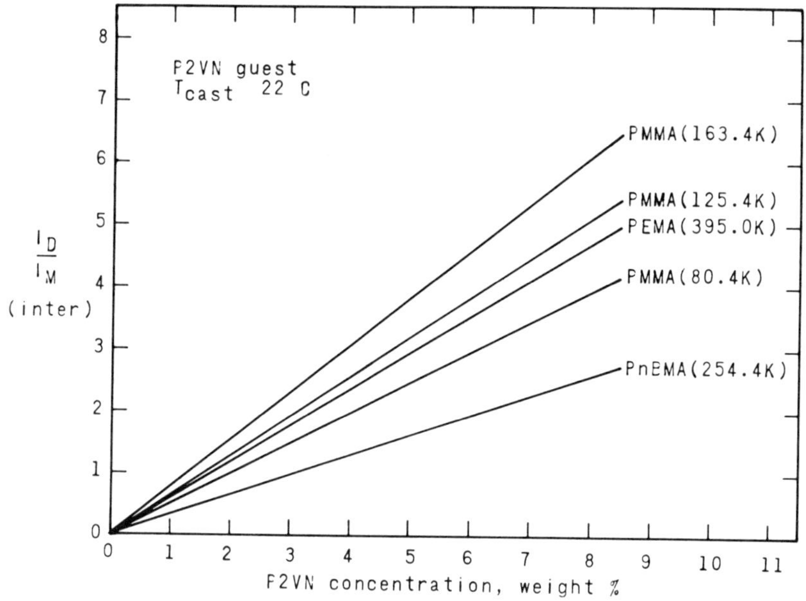

FIGURE 4. Dependence of intermolecular portion of observed excimer fluorescence on guest concentration.

tion. At the beginning of the casting process, the compatible system will have a low glass transition temperature due to the presence of the solvent. However, as the solvent evaporates the T_g of the system will increase. This poses no problems with regard to maintenance of thermodynamic equilibrium in the blend as long as the casting temperature, T_c, remains sufficiently greater than the glass transition temperature, T_g. However, in the region where T_g exceeds T_c, segmental mobility will be drastically reduced and a nonequilibrium state may be obtained. This is likely to be a common phenomenon for the blends under consideration since all of the host matrices except P*n*BMA have glass transition temperatures greater than 22°C, the casting temperature in the host and concentration studies. Nevertheless, the essential point is that, regardless of whether or not the film is in a state of thermodynamic equilibrium, excimer fluorescence may be used to characterize the polymer-polymer interactions existing at the time that the film passes into the glassy state.

The corollary to the use of excimer fluorescence to examine static glassy state morphology is to apply it to the dynamic situation where motion of the polymer chains is no longer restricted. Before considering the relaxation behavior, however, it is useful to consider the phenomenological effect of casting temperature on fluorescence. In earlier work,[5] blends of 0.2% P2VN in polystyrene were prepared at different casting temperatures from 22 to 100°C, the glass transition temperature of the polystyrene host matrix. The fluorescence results are shown in FIGURE 5 where I_D/I_M is plotted against the reciprocal of the casting temperature on semilogarithmic coordinates. A detailed consideration of this Arrhenius-like behavior is pursued elsewhere.[5] The significant observation for this work is that it is possible to prepare blends with remarkably different fluorescence behavior and also, possibly, different molecular morphologies simply by selection of the appropriate casting temperature.

The effect of an increase in sample temperature on 0.2% P2VN/PS blends previously cast at different temperatures is shown in FIGURE 6.[7] The solid curves represent smoothed fits of the fluorescence data for films prepared by solvent casting at 51, 77 and 100°C. The arrows denote the course of the experiment. The blends are

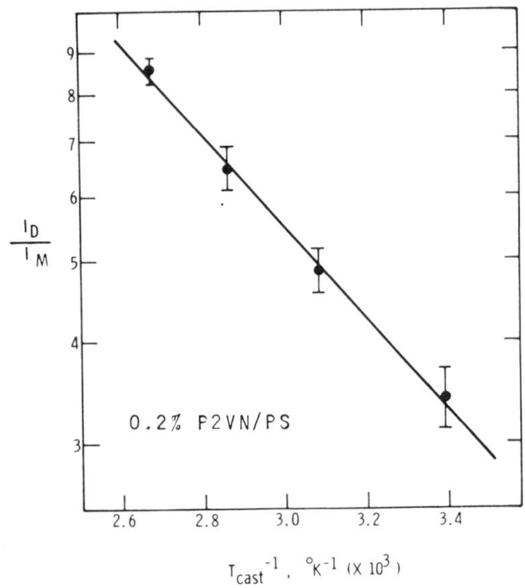

FIGURE 5. Dependence of observed excimer fluorescence on casting temperature.

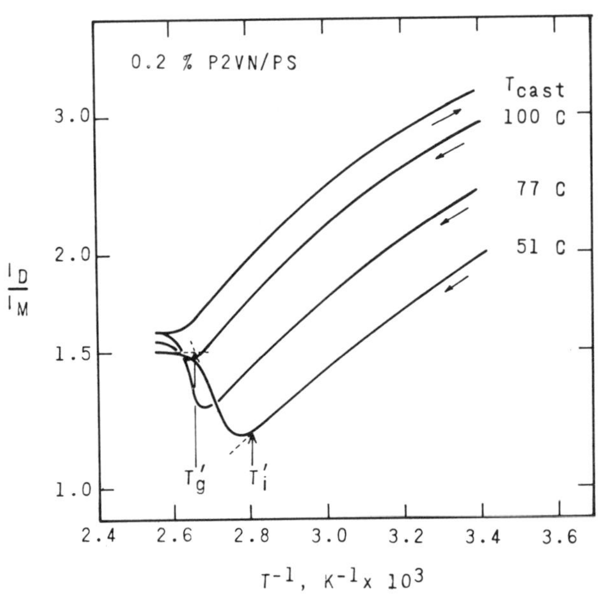

FIGURE 6. Dependence of observed excimer fluorescence on sample temperature near the glass transition region.

first heated at a constant rate of 5°C/min to 120°C, allowed to anneal at that temperature for about thirty minutes and then cooled at 5°C/min to room temperature. The parameter T_i' is the point at which I_D/I_M first begins to deviate from a smooth monotonic decrease with increasing sample temperature. T_g' is defined by a knee formed from the intersection of the line extrapolated from the level region at high temperatures and the line of negative slope beyond T_i'.

A number of observations may be made for these results. The first, rather trivial, point is that the shift in the I_D/I_M value for films cast at different temperatures, but measured at the same temperature, 22°C, is due to an alternative method of presenting the data of FIGURE 5. A more important feature is that each blend exhibits the same general temperature dependence, regardless of casting temperature. In fact, in the region between 22 and 84°C, the three "heating" curves are exactly parallel. The third point is that the T_g' fluorescence parameters are the same as the glass transition temperatures of the blend. Finally, all blends, if allowed to anneal at 120°C for at least thirty minutes, yield the same I_D/I_M value which, upon cooling, is parallel to the three heating curves.

It seems clear that whatever the morphology of the initial blend, annealing above the glass transition temperature has allowed rearrangement to a common final state. In order to follow the kinetics of this rearrangement, experiments were performed in which the blend was heated to a temperature between T_i' and T_g' and annealed for periods up to two hours, during which time I_D/I_M was monitored. Typical results are shown in FIGURE 7 for a 0.2% P2VN/PS blend cast at 22°C and annealed at 73°C. The solid line ia a single exponental fit to the data; thus, the data are consistent with a first-order process. The activation energy for the relaxation process, determined by repetition of this procedure for other annealing temperatures, is 18 ± 5 kcal.

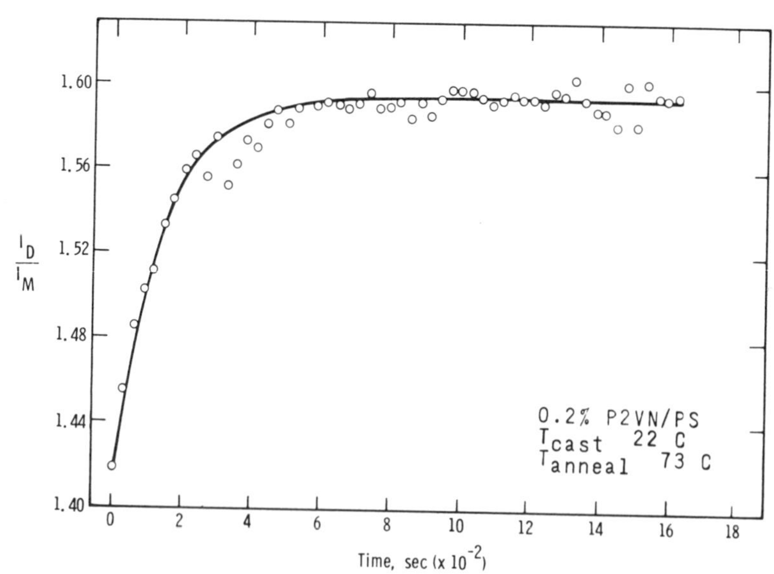

FIGURE 7. Time dependence of observed excimer fluorescence during annealing near the glass transition region.

Discussion

In order to interpret the excimer fluorescence observed in the three studies, it is necessary first to consider a brief outline of the photophysics of aromatic vinyl polymers. There are three major aspects to examine: the number of each type of excimer-forming site, the process by which these sites are "sampled" by exciton migration in the solid state and the stability of the excited complex once it is formed. The interrelation of these features may best be seen with reference to an earlier analysis of photostationary-state kinetics and the migrational sampling process leading to intramolecular excimer fluorescence.[5] The result may be expressed in the form

$$\frac{I_D}{I_M} = \left[\frac{k_{FD}}{k_{FM}}\right]\left[\frac{1}{k_D+k_{MD}}\right]k_E f(C) \qquad (3)$$

The first bracket in Equation 3 includes the fluorescence decay constants of the excimer, k_{FD}, and monomer, k_{FM}. Their ratio has been reported to be independent of solvent for intermolecular excimer formation in solution; thus, it is expected to be independent of the host matrix in the solid blends. In the second bracket, k_D is the overall excimer decay constant given by $k_D = k_{FD} + k_{ID}$ where k_{ID} represents all internal quenching processes. Since k_D for intermolecular excimer formation of naphthalene and several of its derivatives is solvent dependent, it may vary from host to host in the solid blends. However, this effect should be minimized in the poly(alkyl methacrylate) host matrices since, as members of a homologous series, they have similar chemical structures. k_{MD} represents dissociation of the excimer complex to an excited chromophore plus a ground-state chromophore. In an earlier study on P2VN, k_{MD} was determined to be much less than k_D so that it may be ignored.[5] The major factors which may change due to alteration of the physical characteristics of the blend are the energy-transfer rate constant, k_E, and the number of excimer sites, which is represented in general by the functional form $f(C)$. In earlier work, which only considered intramolecular excimer formation, $f(C)$ was given by $f_R/1\text{-}f_R$, where f_R is the fraction of aromatic rings in excimer-forming sites. Note that this quantity is independent of the bulk concentration C. The bulk concentration must be included, however, when intermolecular interactions are possible. Since k_E and $f(C)$ appear as multiplicative factors in Equation 3, it is difficult to separate their effects, a fact which has led to some polarization of views on their relative roles. To define more clearly the nature of the problem, we first consider the types of excimer-forming sites and their dependence on environmental factors such as matrix or concentration effects.

As noted in the introduction, there are three types of excimer-forming sites. The first arises from association between rings on adjacent repeat units in the same chain backbone. Since the interaction is intramolecular, excimer formation is independent of chain concentration. Stabilization of the excimer complex requires that the aromatic rings be in a fairly precise coplanar sandwich geometry. Such a configuration is possible in the polymer only for racemic g^-t and tg^- dyads in syndiotactic sequences and meso *tt* and *g-g+* dyads in isotactic sequences. However, from the conformational energy maps of polystyrene determined by Yoon *et al.*,[20] it is clear that only the *tt* meso dyad will be populated appreciably in the aromatic vinyl polymers. Thus, intramolecular excimer formation between adjacent chain segments is restricted to a single dyadic type. Although this may lead to a very small number of intramolecular (adjacent) excimer-forming sites, a more important consideration is whether the host matrix has any effect on their population. The concept of chain expansion in a good solvent is well established. However, Flory has shown that modest changes in the radius of gyration will not be felt at the level of the local dyadic population.[21] Thus, intramolecular (adjacent) excimer formation should be independent of the host matrix.

The second type of excimer site is also intramolecular, but in this case the complex is formed between aromatic rings from remote segments of the chain which are contiguous due to the chain bending back upon itself. Since the aromatic vinyl polymers are fairly stiff, the likelihood of this occurrence is rather low. Nevertheless, if such sites do exist, they would increase in population as the radius of gyration of the random coil decreases in a thermodynamically poorer host matrix. They also are independent of guest chain concentration.

The third site is intermolecular between segments on different chains. In contrast to the other two, it does depend upon the guest polymer concentration. Two situations may be considered. If the guest and host polymers are molecularly compatible, placement of guest chain segments will be random. On the other hand, if the blend is incompatible, clustering of guest molecules will occur so that the distribution of aromatic rings will no longer be statistically random. The extent of this aggregation must depend upon the thermodynamic nature of the guest/host interaction.

To summarize, the intramolecular (adjacent) excimer-forming site is independent of both concentration and host; the intramolecular (remote) site is independent of concentration but perhaps not of the host and the intermolecular site always depends upon concentration and probably also depends upon the host for incompatible blends.

Energy migration is the other essential element of the photophysics of excimer formation in polymers. It is generally accepted that facile exciton migration between identical chromophores may occur following absorption of radiation by an aromatic ring which is pendent to the chain.[22] However, development of a quantitative model for the exciton hopping is a difficult task. A major problem involves accounting for the multidimensional energy migration between different chains or across loops of the same chain as well as the one-dimensional random walk along a single chain. To our knowledge, no comprehensive model exists at this time. In the absence of such an analysis, the observed behavior may be interpreted in terms of limiting cases. At one extreme, the number of excimer sites is presumed fixed and the observed I_D/I_M is attributed mainly to variations in efficiency of energy migration. At the other extreme, the energy migration process is presumed efficient enough to sample all sites equally well; however, the number of such sites is changing due to environmental conditions. To help clarify the relative contributions of excimer-site population and energy migration efficiency, it is useful to focus on one of the guest polymer chains in which a particular aromatic ring has just been promoted to an excited singlet state. A reference volume V_0 surrounding the excited ring may be defined such that it encloses all possible aromatic rings which can be sampled by the energy migration process during the lifetime of the excited state. Two physical situations are then possible; the guest polymer chain may be molecularly dispersed with no other guest chain segments within the coil volume or coil interpenetration may have occurred as a result of local increases in concentration. The consequences of these two morphological conditions on interpretation of the fluorescence behavior will be considered in turn.

In the case of high dilution and molecular dispersion of the guest and host polymer chains, energy migration will only occur intramolecularly over the guest random coil. Such a hopping process may be treated as a one-dimensional random walk. If energy migration is slow or if the molecular weight and, hence, radius of gyration of the coil is too large, portions of the random coil will remain unsampled. For highly efficient exciton migration, or short polymer chains, the entire coil should be accessible and the size of V_0 will be limited by the radius of gyration of the coil. If the host matrix is changed to one which is thermodynamically poorer, the radius of gyration of the guest polymer chain should be reduced. If the guest concentration, casting temperature, and host molecular weight are selected such that a compatible blend is maintained, the guest chain should still be isolated and excimer formation should remain intramolecular. The consequences of this for the fluorescence behavior may be interpreted in two ways. If all photophysical behavior is attributed to excimer site population, there should be no change in the number of intramolecular (adjacent) excimer-forming sites but the number of intramolecular (remote) sites could increase due to an increase in the number of times the chain coils back upon itself. This would result in an increased I_D/I_M if the energy migration were efficient enough to sample the entire coil. If, however, energy migration were not very effective, the new remote sites may never be sampled and I_D/I_M would be relatively unchanged. Alternatively, it could be argued that no new excimer sites of any kind are formed but that such "backbending" leads to sufficiently close approach of remote aromatic rings such that exciton hopping across the coil junctions occurs.[23] Such short-circuits would introduce alternative paths which could allow the randomly hopping singlet exciton access to a greater portion of the chain and, thus, more of the fixed number of excimer-forming sites. This would also cause I_D/I_M to increase but the effect would probably be larger for the lower efficiency migration process.

In the case of moderately concentrated solid solutions, e.g., in the semidilute range and higher, the reference volume will include segments from more than one guest polymer chain simply due to coil overlap. A similar effect results if phase separation takes place such that the local concentration of guest polymer chains within a separate domain is higher than the nominal bulk concentration. From one viewpoint, the observed fluorescence behavior due to interaction between different chains could result solely from an increase in the number of intermolecular excimer sites, regardless of whether the distribution of guest polymer chains is uniform or clustered. In this case, I_D/I_M should increase in direct proportion to the concentration of aromatic rings within the reference volume. In contrast to the intramolecular case, it is much more difficult to support the argument that the resulting photophysical behavior at high guest concentration is due solely to energy migration. The most important situation is probably in the very early stages of phase separation. As intermolecular contact occurs, the dimensionality of the energy migration will increase from one-dimensional to three-dimensional, with a consequent increase in efficiency of sampling. At high compatible concentrations, or at a sufficiently advanced stage of phase separation, all regions within V_0 are equally accessible so that the limiting factor must be the number of excimer-forming sites.

In light of the preceding discussion, it is now possible to propose a model for molecular interaction which is consistent with the host and concentration dependence.

At the lowest guest concentrations of 0.003% or less, the P2VN guest polymer chains are molecularly dispersed in the host matrix and are noninteracting. This is supported by I_D/I_M being independent of the type of host polymer. Energy migration in this case should be one-dimensional along the chain. The rapid rise in I_D/I_M over the concentration range of 0.003% to 0.1% is consistent with incipient phase separation leading to aggregation of guest polymer chains. Note that this is at a concentration which is far below the sensitivity of any classical method for detecting multiple phases. The abruptness of the transition may indicate that the major contribution to the photophysical behavior is a change in the efficiency of the energy migration process, presumably due to a change in the dimensionality of exciton hopping. At the upper concentration limit of the transition zone, all regions of V_0 should be equally accessible to the exciton. At that point, the number of excimer-forming sites becomes the limiting factor.

As a first step toward the development of the excimer probe as a tool to analyze intermolecular interaction quantitatively, a phenomenological clustering model has been developed for the concentration range where energy migration is facile and the number of intermolecular excimer-forming sites is the limiting factor. The objective of the model, which is an extension of earlier work,[18] is to describe empirically the two possible forms of intermolecular interaction giving rise to excimer fluorescence. On the one hand, if guest and host polymers are completely miscible, the distribution of guest chain segments within the reference volume V_0 will be random. Thus, the local concentration should be the same as the bulk and the contribution to I_D/I_M should be of the form

$$\frac{I_D}{I_M}\text{(uniform mixing)} = AC \tag{4}$$

where A is a constant related to the photophysical rate constants of Equation 3 and C is the bulk concentration of the guest polymer. On the other hand, if the blend components are thermodynamically incompatible, phase separation will occur at equilibrium. If V_0 is located in a domain that is rich in guest polymer, the local concentration of aromatic rings will be higher than the bulk. This preferential clustering is expected to depend both upon the binary interaction parameter, χ_{12}, and upon the host molecular weight, M_{host}. Since a decrease in either χ_{12} or M_{host} could lead to a compatible blend, it is reasonable to include their contributions to I_D/I_M in the form of a product

$$\frac{I_D}{I_M}\text{(cluster)} = BM_{\text{host}}\chi_{12}C \tag{5}$$

where B is another collection of photophysical rate constants. Thus, if χ_{12} vanishes, so should the cluster contribution. Although the molecular weight dependence probably applies only above some critical molecular weight, that modification cannot be verified by the available data. The combined expression becomes

$$\frac{I_D}{I_M}\text{(intermolecular)} = (A + BM_{\text{host}}\chi_{12})C \tag{6}$$

If Equation 6 is valid, a plot of $d(I_D/I_M)/dC$ against $M_{\text{host}}\chi_{12}$ should yield a straight line of slope B and intercept A. This is indeed the case, as seen in FIGURE 8. A least squares fit of the data yields values of $A = 3.2 \times 10^3$ and $B = 1.16$ for C in cm^3 guest segments/cm^3 blend.

The physical meaning of Equation 6 is that the local concentration of aromatic rings within V_0 will be the same as in the bulk for compatible blends but will be enhanced if V_0 is located within a phase separated domain, i.e.,

$$C_{\text{local}} = \left(1 + \frac{B}{A} M_{\text{host}}\chi_{12}\right) C_{\text{bulk}} \tag{7}$$

Using Equation 7, it is possible to calculate the bulk concentration at which the local concentration is pure guest polymer and thereupon calculate the expected I_D/I_M value. These results are shown in FIGURE 9 as the solid points which terminate the

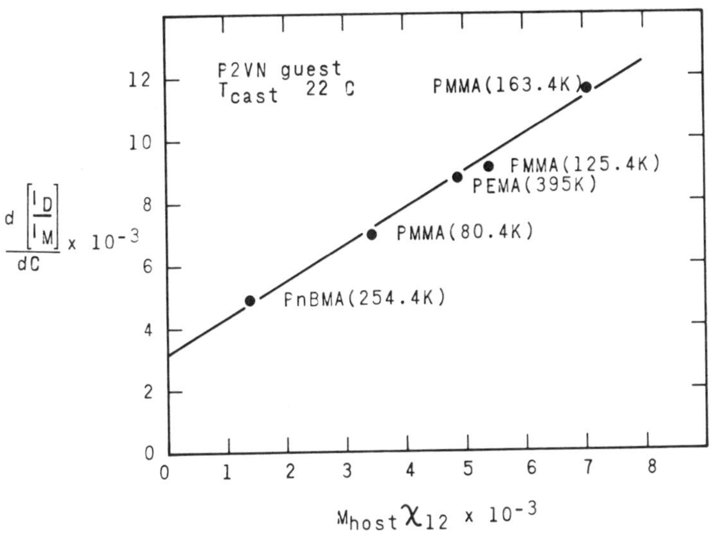

FIGURE 8. Dependence of slope of curves in FIGURE 4 on host molecular weight and binary interaction parameter.

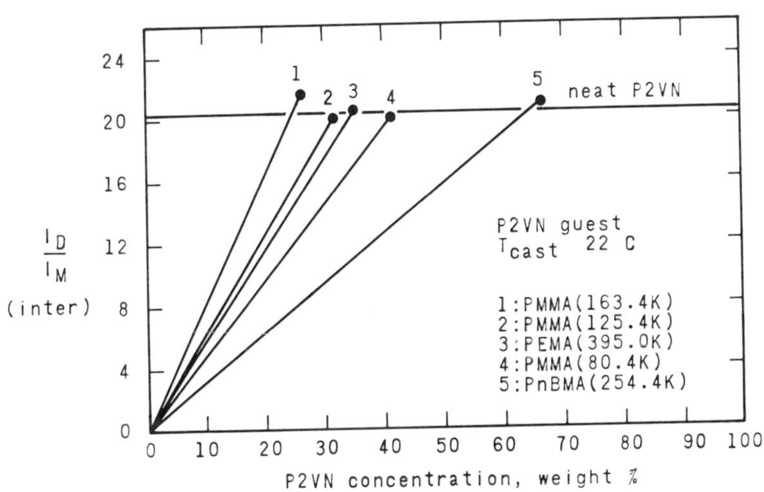

FIGURE 9. Verification of cluster model.

extrapolated concentration data for each blend. The close agreement of the predicted I_D/I_M values among blends signifies only that the fitting model is a good empirical representation of the data. What is noteworthy, however, is that the predicted I_D/I_M for pure guest within V_0 is essentially identical to what is experimentally observed for neat P2VN films, as indicated by the horizontal line. This excellent agreement lends strong support to the rather simple physical concepts embodied in the model.

The final point for consideration is the physical significance of the relaxation behavior in the region of the glass transition. In the original work,[7] the P2VN concentration was thought to be low enough that only intramolecular excimer formation was taking place. The emphasis was then placed on the effect of temperature on the conformational statistics. The concentration results described in this study make it seem likely that initial guest clusters exist in 0.2% P2VN/PS films. Moreover, since there is evidence that these blends exhibit lower critical solution temperatures,[24] it is possible that the observed time-dependent increase in I_D/I_M reflects the dynamic aspects of phase separation. This does not alter the central conclusion of the original work that the observed fluorescence behavior is a manifestation of a redistribution of excimer-forming sites due to onset of increased segmental mobility. What may change is the relative contributions of the intramolecular (adjacent) and intermolecular excimer-forming sites. More detailed studies are in progress.

In conclusion, excimer fluorescence has been shown qualitatively to be a sensitive molecular probe of polymer-polymer interactions in the amorphous solid state. The most significant experimental aspect of the excimer probe is that measurements of intermolecular association may be performed at extremely low concentrations, thus providing an advantage over classical methods such as DSC and DMS. The long term objective of these studies is to determine quantitatively the domain size and its dynamic behavior during the early stages of phase separation. Before this is possible, however, it will be necessary to eliminate the ambiguity associated with distinguishing between the contributions of energy migration efficiency and the number of excimer-forming sites. In addition to extending the photostationary-state measurements described in this work, transient experiments must be performed to determine the radiative rate constants.

Acknowledgment

The authors wish to thank S. N. Semerak for the use of the polystyrene data of Figure 3 and for much fruitful discussion.

References

1. Paul, D. R. & S. Newman, Eds. 1978. *In* Polymer Blends, Vol. 1 and 2. Academic Press. New York.
2. Olabisi, O. J., L. M. Robeson & M. T. Shaw. 1979. *In* Polymer—Polymer Miscibility. Academic Press. New York.
3. Birks, J. B. 1970. *In* Photophysics of Aromatic Molecules. Wiley-Interscience. New York.
4. Harrah, L. A. 1972. J. Chem. Phys. **56**: 385.
5. Frank, C. W. & L. A. Harrah. 1974. J. Chem. Phys. **61**: 1526.
6. Frank, C. W. & M. A. Gashagari. 1979. Macromolecules **12**: 163.

7. FRANK, C. W. 1975. Macromolecules **8**: 305.
8. FLORY, P. J. 1941. J. Chem. Phys. **9**: 660.
9. FLORY, P. J. 1942. J. Chem. Phys. **10**: 51.
10. HUGGINS, M. L. 1941. J. Chem. Phys. **9**: 440.
11. HUGGINS, M. L. 1942. Ann. N.Y. Acad. Sci. **43**: 1.
12. SCOTT, R. L. 1949. J. Chem. Phys. **17**: 279.
13. TOMPA, H. 1949. Trans. Faraday Soc. **45**: 1142.
14. HILDEBRAND, J. H. & R. L. SCOTT. 1950. *In* The Solubility of Nonelectrolytes, 3rd Ed. Van Nostrand-Reinhold. Princeton, NJ.
15. HILDEBRAND, J. H. & R. L. SCOTT. 1962 *In* Regular Solutions. Prentice-Hall. Englewood Cliffs, NJ.
16. KRAUSE, S. 1972. J. Macromol. Sci. Rev. Macromol. Chem. **C7**: 251.
17. KRAUSE, S. 1978. *In* Polymer Blends. D. R. Paul & S. Newman, Eds., Vol. **1**: 15–113. Academic Press. New York.
18. FRANK, C. W., M. A. GASHGARI, P. CHUTIKAMONTHAM & V. J. HAVERLY. 1980. Second Cleveland Symposium on Macromolecules: Structure and Properties of Amorphous Polymers. Elsevier.
19. YOON, D. Y., P. R. SUNDARARAJAN & P. J. FLORY. 1975, Macromolecules **8**: 776.
20. BANK, M., J. LEFFINGWELL & C. THIES. 1971. Macromolecules **4**: 43.
21. FLORY, P. J. 1975. Science **188**: 1268.
22. KLOPFFER, W. 1978. Spectrosc. Lett. **11**: 863.
23. GUILLET, J. 1979. Polym. Prepr. Am. Chem. Soc. Div. Polym. Chem. **20**: 395.
24. FRANK, C. W. & M. A. GASHGARI. 1980. Polymer. Prepr. Am. Chem. Soc. Div. Polym. Chem. **21:** 42.

STUDIES OF POLYMER COMPATIBILITY BY NONRADIATIVE ENERGY TRANSFER*

H. Morawetz

Polymer Research Institute
Polytechnic Institute of New York
Brooklyn, New York 11201

The phenomenon of nonradiative energy transfer has been used extensively to characterize intramolecular distances in biological macromolecules.[1,2] A number of studies have also been reported dealing with intramolecular energy transfer in synthetic oligomers or polymers[3-9] but no previous work has come to our attention which explores energy transfer between two polymeric species labeled with donor and acceptor chromophores, respectively. Such a study is of interest since the mutual approach of polymer-bound chromophores depends on the extent to which chains of different polymers are able to interpenetrate. Thus, the emission spectrum of blends of two chromophore-labeled polymers should depend on their compatibility, while emission spectra of dilute solutions containing donor- and acceptor-labeled polymers should be sensitive to any aggregation of the macromolecules.

Materials and Methods

In most of our experiments, the fluorescent label was introduced into the polymer by copolymerization using 1-(2-naphthyl)ethyl methacrylate (NEMA) or 2-(N-carbazole)ethyl methacrylate (CEMA) as the donor and 1-(2-anthryl)ethyl methacrylate (AEMA) or 9-anthrylmethyl methacrylate (AMMA) as the acceptor. The preparation of these monomers has been described previously.[10] In the case of anionically prepared polystyrene or poly (α-methylstyrene), the lightly chloromethylated polymer was condensed with potassium carbazole or 9-anthrylmethanol. Poly (2,6-dimethyl-1,4-phenylene ether) was slightly chlorinated before condensation with potassium carbazole or 9-anthrylmethanol. The content of the fluorescent labels in the polymer was determined by UV spectroscopy.

Films were cast from tetrahydrofuran or methylene chloride solutions containing equal concentrations of two polymers carrying donor or acceptor fluorescent labels. Emission spectra from both films and solutions were recorded on a Hitachi-Elmer MPF-2A spectrophotometer with the exciting beam at 60° and the observation of the emission at 30° to the sample surface. Energy transfer was characterized by the ratio of the emission intensity of the naphthyl and anthryl labels, I_N/I_A, or the carbazolyl and anthryl label, I_C/I_A, using emissions at 336 nm for NEMA, 347 nm for CEMA, 408 nm for AEMA and 413 nm for AMMA residues. Excitation wavelengths were 276 nm for blends containing NEMA and AEMA, 285 nm for

* This study was supported by Grant DMR 77-05210, Polymer Program, of the National Science Foundation.

0077-8923/81/0366-0404$01.75/0 © 1981, NYAS

blends with NEMA and AMMA and 296 nm for blends containing the carbazolyl donor.

RESULTS

The characteristic distance R_o for energy transfer was calculated from the overlap integral of the donor emission spectrum and the acceptor absorption spectrum according to Förster's theory.[1] It was assumed that the donor and acceptor chromophores were oriented at random towards each other, so that $\kappa^2 = 2/3$ could be used for fluid media, while $\kappa^2 = 0.476$ was assumed[11] for systems in which the labels were incorporated in glassy polymer films. This led to $R_o = 1.95 \pm 0.05$ nm for the NEMA-AEMA pair and $R_o = 2.8 \pm 0.1$ nm for the CEMA-AMMA pair in rigid films, while $R_o = 2.2$ nm was estimated for the NEMA-AMMA pair in aqueous solution.

The validity of the expectation that energy transfer between chromophore-labeled polymers could be used to characterize polymer compatibility was first demonstrated on blends of poly(methyl methacrylate) (PMMA) with methyl methacrylate-ethyl methacrylate or methyl methacrylate-butyl methacrylate copolymers.[10] As seen in FIGURE 1, the I_N/I_A ratio is very low in a blend of donor- and acceptor-labeled PMMA, but rises gradually as the PMMA is blended with copolymers containing an increasing fraction of ethyl or butyl methacrylate residues, tending to a plateau for copolymers with more than 40 mole % of the comonomer. This change in the emission spectrum may be taken as reflecting an increasing incompatibility of the PMMA and the copolymers which leads to the segregation of the donor and acceptor in separate phases.

Although PMMA is incompatible with polystyrene and with polyacrylonitrile, it

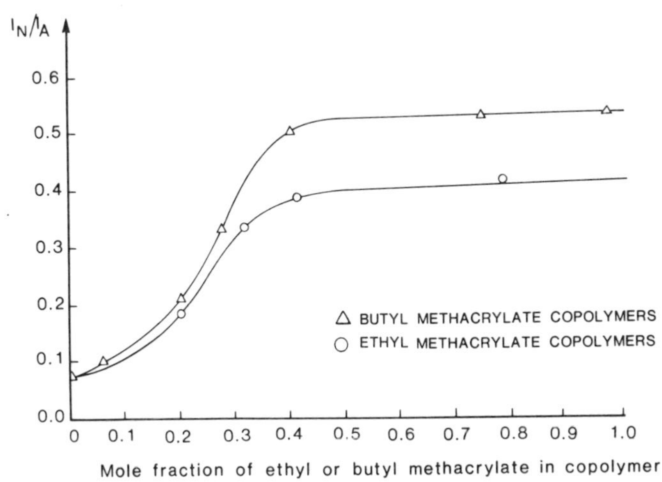

FIGURE 1. Energy transfer in blends of poly(methyl methacrylate) containing 1.4 weight % AEMA and methyl methacrylate copolymers with ethyl methacrylate or butyl methacrylate containing 1.2 weight % NEMA.

has been known for some time that it is compatible with styrene-acrylonitrile copolymers within a certain range of copolymer composition.[12–14] FIGURE 2 shows that the range of composition within which the copolymer is miscible with PMMA can be determined conveniently by the fluorescence technique; the results are quite similar to those obtained in the earlier studies using a variety of methods. It may be seen also that samples of the films which had been heated to 145°C and rapidly quenched in dry ice-methanol exhibited higher I_C/I_A ratios than films which had not been subjected to this treatment.[15] This is consistent with the general expectation that compatible polymer blends will exhibit phase separation on heating.[16]

Blends of polystyrene (PS) with poly(2,6-dimethyl-1,4-phenylene ether) (POP) have been characterized by a variety of methods as exhibiting perfect miscibility[17–20] and it was, therefore, of special interest to investigate this system by the fluorescence method.[21] It can be seen in FIGURE 3 that I_C/I_A is, in this case, almost independent of the composition of PS-PPO blends; yet, the experimental points lie consistently on a curve concave to the abscissa and we believe that this indicates a slight fluctuation in the local composition of the blend. Blends of POP with poly(α-methylstyrene) (PMS) exhibit a decreased energy transfer in systems rich in the PMS component, indicating phase separation as also observed by Robeson *et al.*[22] who detected two glass transitions by mechanical loss measurements.

It would be of special interest to establish principles which would allow us to predict polymer compatibility. Blends of a methacrylate homopolymer with methacrylate copolymers were believed to be very likely to lend themselves to such predictions. Here, a matching of their cohesive energy densities, obtained when the mean length of the alkyl groups in the copolymer equals their length in the

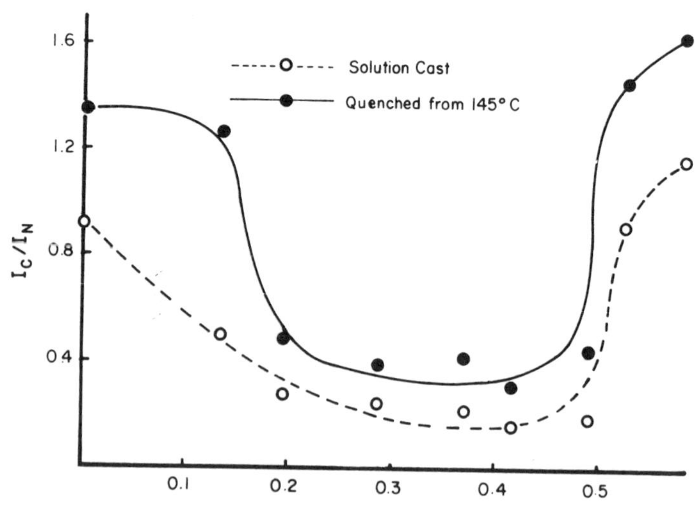

FIGURE 2. Energy transfer in blends of poly(methyl methacrylate) labeled with 1.0 wt % AMMA with styrene-acrylonitrile copolymers labeled with 1.45 wt % CEMA as cast from solution and after quenching from 145°C.

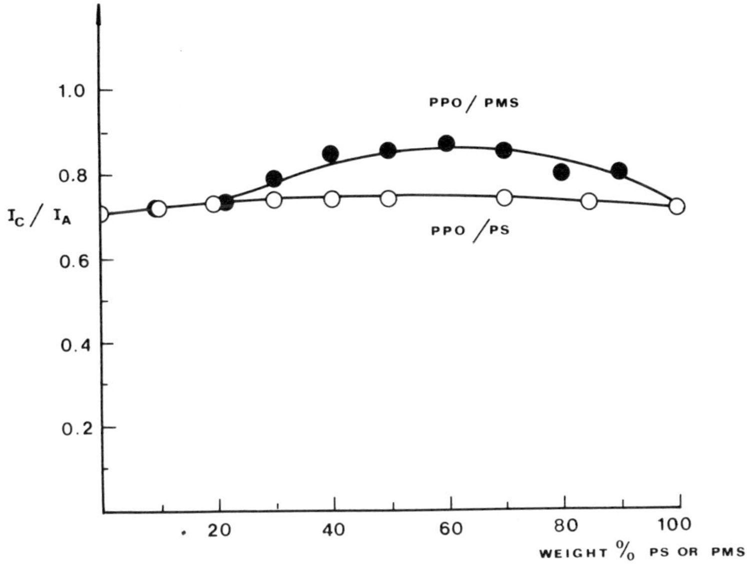

FIGURE 3. Energy transfer in blends of PPO with 1.83% of monomer units labeled with anthryl residues and PS with 2.63% of monomers carrying carbazole residues of PMS with 1.62% of monomers carrying carbazole residues.

homopolymer, should be sufficient to ensure miscibility. FIGURE 4 shows that this conjecture is valid; when poly(ethyl methacrylate) or poly(propyl methacrylate) were blended with methyl methacrylate-butyl methacrylate copolymers, energy transfer was maximized when the copolymer contained one third and two thirds of butyl methacrylate residues, respectively. However, in many other cases the matching of cohesive energy densities does not lead to polymer miscibility.[15] This is illustrated in FIGURE 5 in which I_C/I_A is plotted for carbazole-labeled styrene-acrylonitrile copolymers blended with anthracene labeled homopolymers of methyl methacrylate, ethyl methacrylate and *p*-chlorostyrene. It can be seen that the I_C/I_A minimum for the PMMA blend is substantially lower than for the blends containing the other two polymers. This demonstrates that no copolymer of the nonpolar styrene and the highly polar acrylonitrile will produce homogeneous blends with poly(ethyl-methacrylate) or poly(*p*-chlorostyrene).

Energy transfer between chromophore-labeled polymers in solution will be enhanced if the polymers form association complexes. This is illustrated in FIGURE 6[23] on aqueous solutions containing acidic copolymers of dimethylacrylamide and basic copolymers of N-vinylpyrrolidone. Energy transfer decreases on addition of simple salts, which shield the polyion charges and reduce the Coulombic attractions between the polymer chains. It is interesting that in the two polyanion-polycation pairs with different densities of ionized groups, energy transfer is similar when the cloud point is reached. This suggests that phase separation occurs for a similar size of the macromolecular aggregate.

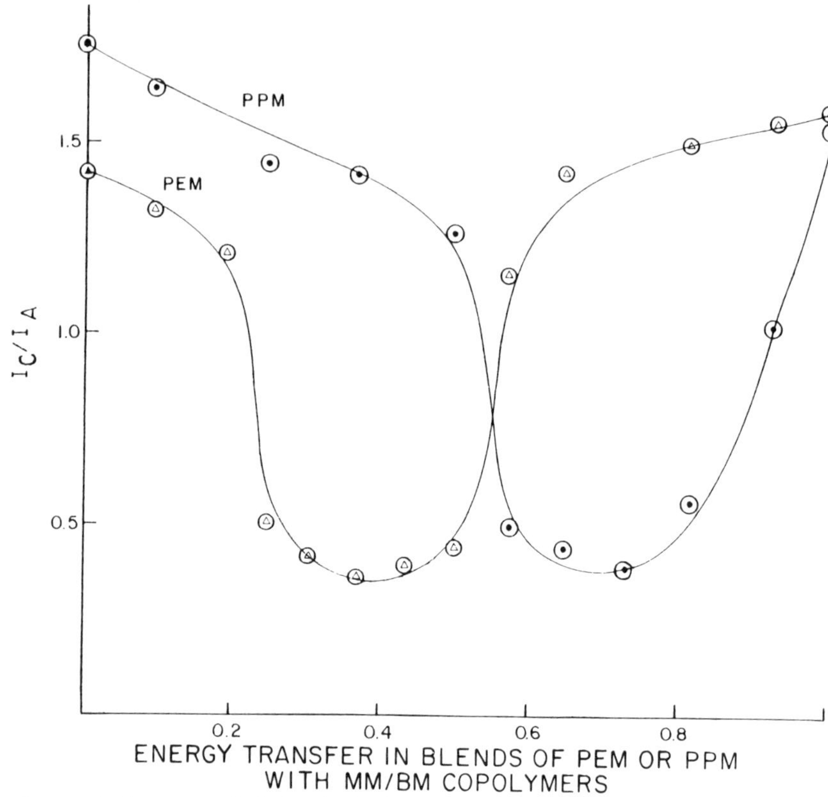

FIGURE 4. Energy transfer in blends of poly(ethyl methacrylate) or poly(propyl methacrylate) labeled with 1.1 wt % of AMMA and a methyl methacrylate-butyl methacrylate copolymer labeled with 0.8 wt % CEMA.

Discussion

One of the most interesting aspects of our technique for the study of the interpenetration of flexible polymer chains is the gradual transition from the behavior of a compatible to that of an incompatible blend. This is in sharp contrast to the claim[24] that the transition from compatibility to incompatibility occurs sharply over a very narrow range of copolymer composition and is an indication of the higher sensitivity of the energy transfer method as compared to characterizations based on glass transition phenomena. We believe that this gradual transition reflects the "fuzziness" of phase boundaries in polymer blends which has previously been observed by electron microscopy[25] and phase-contrast microscopy.[26] Thus, we may assume that even when two polymers are segregated, the chains of each may penetrate to some extent into the region of the other, with this depth of mutual interpenetration shrinking gradually as the enthalpy of mixing increases.†

†Helfand and Tagami (*J. Polym. Sci. B 9:* 741, 1971) reached this conclusion on theoretical grounds. Their analysis predicts that the characteristic depth of interpenetration is inversely proportional to the square root of the polymer-polymer interaction parameter.

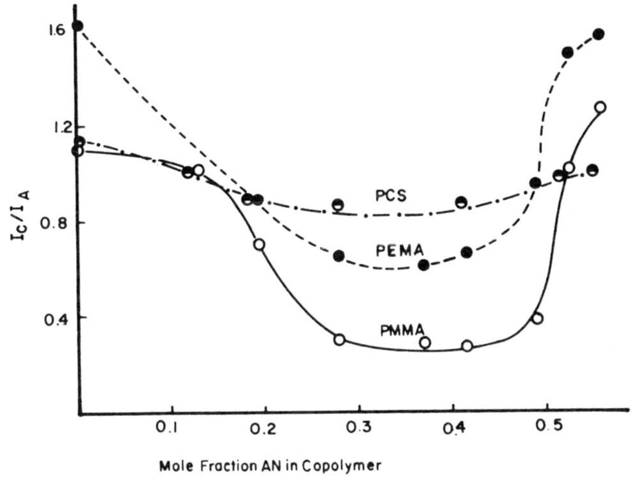

FIGURE 5. Energy transfer in blends of styrene-acrylonitrile copolymers labeled with 1.45% CEMA and poly(methyl methacrylate), poly(ethyl methacrylate) or poly(*p*-chlorostyrene labeled with 1.0 wt %, 1.1 wt % and 1.05 wt % AMMA, respectively.

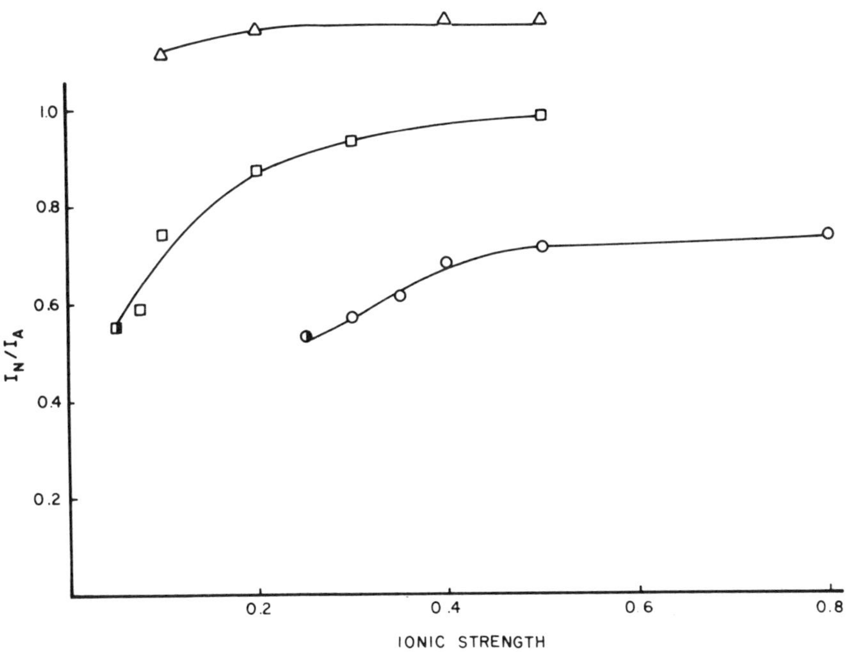

FIGURE 6. Energy transfer in solutions containing: N,N-dimethylacrylamide copolymers with 0.54 mole % NEMA and 0.42 mole % AMMA, respectively (△); N,N-dimethylacrylamide copolymer with 10.7 mole % acrylic acid and 0.48 mole % NEMA + N-vinylpyrrolidone copolymer with 9.7 mole % N,N-dimethylaminoethyl methacrylate and 0.60 mole % AMMA (□); N,N-dimethylacrylamide copolymers with 19.3 mole % acrylic acid and 0.47 mole % NEMA + N-vinylpyrrolidone copolymer with 16.6 mole % N,N-dimethylaminoethyl methacrylate and 0.56 mole % AMMA (○). The half-filled points indicate incipient phase separation. Concentration of naphthyl 3.08×10^{-4} M, anthryl 3.28×10^{-4} M; pH 25°C.

Efforts to predict polymer compatibility have met so far with only limited success. Olabisi[27] suggested that in a search for compatible polymer pairs four parameters should be determined, characterizing dispersion forces, dipole interactions, hydrogen bond donor and hydrogen bond acceptor capability. He suggested that the miscibility of poly(vinyl chloride) with polycaprolactone is due to hydrogen bonding and the same principle seems to account for the compatibility of a number of other polymer pairs.[28] Where these specific interactions cannot be invoked, predictions are most uncertain; Sanchez finds that neither cohesive energy density nor an equation of state approach is a reliable guide.[29] In the case of the much studied PS-PPO blends the situation appears to be quite complex; a spectroscopic study indicated that mixing with PS forces the PPO chains to assume a higher energy conformation, but that this unfavorable factor is overbalanced by the attraction between the phenylene ether and the phenyl groups.[30] An investigation of PMMA blends with styrene-acrylonitrile copolymers by ir spectroscopy indicated that an interaction of the carbonyl groups with phenyl residues is significant[14] and compatibility could then be due to a combination of a close matching of dispersive forces supplemented by this specific interaction. It is, however, surprising that it is impossible to obtain such a favorable balance when PMMA is substituted by poly(ethyl methacrylate) (PEMA). Yet, the nonradiative energy transfer method allowed us to locate the composition of the styrene-acrylonitrile copolymer at which miscibility with PEMA is optimized, although true compatibility cannot be achieved.

We have found that nonradiative energy transfer provided a more sensitive criterion of polymer compatibility than the characterization of polymer blends by their glass transitions. Still, it should be conceded that energy transfer data cannot be interpreted in detail since they depend both on the mutual interpenetration of the polymer chains and on the size and geometry of phase domains. This limitation exists also for other methods for the characterization of polymer compatibility and only electron microscopy is free of these ambiguities. However, electron microscopy is applicable only to systems which lend themselves to suitable staining and the fluorescence technique recommends itself by its experimental simplicity.

Finally, it should be pointed out that the extent of mixing obtained with a polymer pair may vary with the method of sample preparation. Two polymers that are incompatible in bulk in a state of thermodynamic equilibrium may form a single phase up to a critical solution concentration, so that the extent of phase separation will depend on the rate at which the solvent is evaporated from the film casting solution. The converse situation may also arise, i.e., two polymers which are compatible in bulk may form two immiscible solution phases. This phenomenon is to be expected if the solvent interacts much more powerfully with one of the polymeric components.[31,32] The system polystyrene-poly(vinyl methyl ether)-chloroform behaves in this manner[31] and this is undoubtedly the reason why PS-PVME blends cast from chloroform exhibit two-phase behavior, while films of the same composition appear homogeneous when toluene is used as the casting solvent.[33] It is improbable that such conditions were encountered in any of the systems used in our investigation. When a polymer blend is to be characterized in the glassy state (as was the case in all our systems) sample preparation by film casting should yield a closer approach to equilibrium than polymer mixing in the melt, since phase equilibria are frozen as the melt is cooled through the glass transition temperature.

ACKNOWLEDGMENTS

I am indebted to Drs. F. Amrani, J. M. Hung, F. Mikeš and I. Nagata for their collaboration in this investigation.

REFERENCES

1. STEINBERG, I. Z. 1971. Ann. Rev. Biochem. **40:** 83.
2. SCHILLER, P. W. 1975. *In* Biochemical Luminescence R. F. Chen, & E. Edolhoch, Eds. Marcel Dekker. New York. Vol. 2, Chapter 5.
3. STRYER, L. 1961. Science. **162:** 526.
4. EDELHOCH, H., L. BRAND & M. WILCZEK. 1967. Biochemistry. **6:** 547.
5. LONGWORTH, J. W., & R. O. RAHN. 1967. Biochim. Biophys. Acta **147:** 526.
6. FOX, R. B., T. R. PRICE & R. F. COZZENS. 1971. J. Chem. Phys. **54:** 79.
7. FOX, R. B., T. R. PRICE, R. F. COZZENS & J. R. MCDONALD. 1972. J. Chem. Phys. **57:** 534.
8. HAAS, E., E. KATCHALSKI-KATZIR & I. Z. STEINBERG. 1978. Biopolymers **17:** 11.
9. ASPLER, J. S., J. E. GUILLET & C. E. HOYLE. 1978. Macromolecules **11:** 925.
10. AMRANI, F., J. M. HUNG & H. MORAWETZ. 1980. Macromolecules **13:** 649.
11. ROZMAN, I. M. 1962. Opt. Spectrosc. **12:** 337.
12. STEIN, D. J., R. H. JUNG, K.-H. ILLER & H. HENDUS. 1974. Angew. Makromol. Chem. **36:** 89.
13. KRUSE, W. A., R. G. KIRSTE, J. HAAS, B. J. SCHMIDT & D. J. STEIN. 1976. Makromol. Chem. **177:** 1145.
14. NAITO, K., G. E. JOHNSON, D. L. ALLERA & T. K. KWEI. 1978. Macromolecules **11:** 1260.
15. HUNG, J.-M. 1981. Ph. D. Thesis, Polytechnic Institute of New York.
16. MCMASTER, L. P. 1973. Macromolecules **6:** 760.
17. MACKNIGHT, W. J., J. STOELTING, & F. E. KARASZ. 1971. *In* Multicomponent Polymer Systems. Adv. Chem. Ser. No. **99:** 29.
18. SHULTZ, A. R. & B. M. GENDRON. 1972. J. Appl. Polym. Sci. **16:** 461.
19. PREST, W. M. JR., & R. S. PORTER. 1972. J. Polym. Sci. Polym. Phys. Ed. **10:** 1639.
20. SHULTZ, A. R. & B. M. BEACH. 1978. Macromolecules **11:** 158.
21. MIKEŠ, F., H. MORAWETZ & K. S. DENNIS. 1980. Macromolecules **13:** 969.
22. ROBESON, L. M., M. MATZNER, L. J. FETTERS & J. E. MCGRATH. 1973. Polym. Prepr., Am. Chem. Soc. Div. Polym. Chem. **14:** 1063.
23. NAGATA, I., & H. MORAWETZ. Macromolecules. In press.
24. FRIED, J. R., F. E. KARASZ & W. J. MACKNIGHT. 1978. Macromolecules **11:** 150.
25. VOYUTSKII, A. N. KAMENSKII & N. M. FODIMAN. 1966. Kolloid-Z. **215:** 36.
26. LETZ, J. 1969. J. Polym. Sci. Polym. Phys. Ed. **7:** 1987.
27. OLABISI, O. 1975. Macromolecules **8:** 316.
28. AUBIN, M. & R. E. PRUD'HOMME. 1980. Macromolecules **13:** 365.
29. SANCHEZ, I. C. 1978. *In* Polymer Blends. D. R. Paul & S. Newman, Eds. Academic Press. Vol. **1**, p. 115–139.
30. WELLINGHOFF, S. P., J. L. KOENIG & E. BAER. 1977. J. Polym. Sci. Polym. Phys. Ed. **15:** 1913.
31. ROBARD, A., D. A. PATTERSON & G. DELMAS. 1977. Macromolecules **10:** 706.
32. ROBARD, A. & D. A. PATTERSON. 1977 Macromolecules **10:** 1021.
33. BANK, M., J. LEFFINGWELL & C. THIES. 1972. J. Polym. Sci. **A-2: 10:** 1097.

Index of Contributors

Anglister, Jacob, 125–139
Austin, Robert H., 176–196

Baeyens-Volant, D., 341–355
Banks, E., 356–366
Bartholdi, Marty, 176–196
Betteridge, David, 285–299
Blumberg, W. E., 155–175
Brand, Ludwig, 197–207

Cantor, Charles R., 246–252
Cheng, Sheve-Yann, 253–264
Corda, D., 265–273
Cuniberti, Carla, 35–43

Dale, R. E., 155–175
David, C., 341–355
Demeyer, K., 93–108
DeSchryver, F. C., 93–108
Dieterich, Ann E., 246–252

Edelhoch, Harold, 253–264
Eisinger, J., 155–175

Frank, C. W., 387–403

Gashgari, M. A., 387–403
Graceffa, Philip, 285–299
Guillet, J. E., 11–23

Haas, E., 44–61
Hastings, J. Woodland, 315–327
Henis, Yoav I., 217–236
Holden, D. A., 11–23

Irace, Gaetano, 253–264

Jarry, Jean-Pierre, 328–340
Jovin, Thomas, M., 176–196

Katchalski-Katzir, E., 44–61
Klöpffer, Walter, 373–386

Laws, William R., 197–207
Lehrer, Sherwin S., 285–299
Levitzki, Alexander, 217–236
Longworth, J. W., 237–245

Macgregor, Robert B., 140–154
Mita, Itaru, 62–74
Monnerie, Lucien, 328–340
Morawetz, H., 404–411

O'Connor, D. V., 109–124
Okamoto, Y., 356–366

Pasternak, C., 265–273
Pecht, Israel, 208–216
Perico, Angelo, 35–43
Phillips, D., 109–124
Prior, Christopher P., 246–252

Quanten, E., 93–108

Redpath, A. E. C., 75–92
Rendall, W. A., 11–23
Roberts, A. J., 109–124
Ross, J. B. Alexander, 197–207

Schlessinger, Joseph, 274–284
Shinitzky, M., 265–273
Soutar, Ian, 24–34
Steinberg, I. Z., 44–61, 125–139
Stieve, H., 265–273

Tu, Shiao-Chun, 315–327

Ueba, Y., 356–366

Van der Auweraer, M., 93–108
Vaz, Winchil L. C., 176–196

Webb, W. W., 300–314
Weber, Gregorio, 140–154
Weill, G., 1–10
Winnik, Mitchell A., 75–92